Nanomaterials, Nanotechnologies and Design

Nanomaterials, Nanotechnologies and Design

An Introduction for Engineers and Architects

Michael F. Ashby

Paulo J. Ferreira

Daniel L. Schodek

AMSTERDAM • BOSTON • HEIDELBERG • LONDON
NEW YORK • OXFORD • PARIS • SAN DIEGO
SAN FRANCISCO • SINGAPORE • SYDNEY • TOKYO

Butterworth-Heinemann is an imprint of Elsevier

Butterworth-Heinemann is an imprint of Elsevier
30 Corporate Drive, Suite 400, Burlington, MA 01803, USA
Linacre House, Jordan Hill, Oxford OX2 8DP, UK

Library of Congress Cataloging-in-Publication Data
Application submitted

British Library Cataloguing-in-Publication Data
A catalogue record for this book is available from the British Library.

ISBN: 978-0-7506-8149-0

For information on all Butterworth–Heinemann publications visit our
Web site at www.elsevierdirect.com

Contents

The first three chapters in the book are introductory in nature. Chapter 1 provides an overview of the general characteristics and field of nanomaterials.

Chapter 2 presents a retrospective look at materials, nanomaterials in nature, and nanomaterials as they have long been found in works of artisans.

Chapter 3 discusses the general use of materials in architecture and product design, general product and building types, and design processes—all with a special emphasis on the emerging role of nanomaterials.

v

Chapter 4 presents the basic properties of materials from a qualitative and quantitative perspective. Topics should be familiar to engineers and material scientists and can be of interest to technically oriented designers.

Chapter 5 presents a structured design process based on material property charts that can be used for material selection and resolving conflicting design objectives. The approach is of great interest to design-oriented engineers and technically oriented architects.

Chapter 6, 7, and 8 provide an introduction to the classes and fundamental characteristics of nanomaterials, how they are made and characterized, and their properties. The treatment generally assumes some level of prior technical background in materials. These chapters are of interest to engineers and material scientists who are entering the field of nanomaterials. Technically oriented architects and designers who want a deeper understanding of nanomaterials will find these sections useful.

Chapter 9 looks at characteristics of primary design environments that are important in architecture and product design. The role of nanomaterials and nanotechnologies within these environments are explored in a qualitative way. For those not in the design fields, the chapter provides an insight into how nanomaterials impact real applications and how an applications perspective can stimulate ideas about developing novel materials.

Chapter 10 explores the many different
and exciting functional and form
characteristics of nanomaterials that are
related to architecture, engineering, and
product design as well to the several
industries that produce them. Chapter
10 is suitable for readers with any type
of background.

Chapter 11 surveys the many positive applications of nanomaterials and nanotechnologies in the medical and pharmaceutical world as well as in improving the environment. Health questions are also addressed. Chapter 11 is suitable for readers with any type of background.

Chapter 12, the final chapter of the book, succinctly reviews drivers (economic, societal, ethical) that are leading to increased uses of nanomaterials and nanotechnologies in various industries.

Preface

There is currently an extraordinary amount of interest in *nanomaterials* and *nanotechnologies*, terms now familiar not only to scientists, engineers, architects, and product designers but also to the general public. Nanomaterials and nanotechnologies have been developed as a consequence of truly significant recent advances in the material science community. Their use, in turn, is expected to have enormous consequences on the design and engineering of everything from common consumer products and buildings all the way through sophisticated systems that support a wealth of applications in the automotive, aerospace, and other industries. Hopes exist for being able to make things smaller, lighter, or work better than is possible with conventional materials. Serious problems facing society might also be positively addressed via the use of nanomaterials and nanotechnologies. In the energy domain, for example, nano-based fuel cells or photovoltaics can potentially offer greater efficiencies than are possible with conventional materials. Developments in nanomaterials and nanotechnologies have consequently aroused the interest of many individuals involved in engineering, architecture, and product design, whether in the automotive, building, or even the fashion industries.

In the excitement surrounding these new materials and technologies, however, their potential can, and has been, frequently over-hyped. A mystique surrounds these words that can cloud understanding of what nanomaterials and nanotechnologies really are and what they can deliver. One of the purposes of this book is to demystify the subject and distinguish what is real from what is not. Though there is a need to better understand what benefits and costs might be associated with using nanomaterials, in the design fields little true understanding exists about what these new materials and technologies actually are and how they might be used effectively. In the science and engineering domain the situation is often the converse. The fundamental science-based knowledge is there, but not necessarily an understanding of how these materials and technologies might be used to address real societal needs or, more simply, to provide a basis for useful applications—a situation often best described by the classic phrase "a technology looking for an application." Relatively few applications in architecture and product design exist at this point, so there are few role models or case studies to explore. Still, the core belief reflected in this book is that nanomaterials and nanotechnologies *do* offer new ways of addressing many current needs as well as providing the basis for a positive vision for future developments. The book explores these kinds of forward-looking potential applications.

The field is clearly in an emerging state, but its potential is evident. The question is one of how to move forward. The position reflected in this book is that this can only be accomplished by designers, engineers, and material scientists working together. Yet simply suggesting that these groups "collaborate" is usually ineffective. There needs to be a common basis for communication and idea exchange. The language of understanding spoken by one group can be foreign to the other. Within this context, a primary goal of this book is to provide both a common ground and a common language for better understanding how to use nanomaterials and nanotechnologies effectively. It seeks to convey the necessary technical understanding of the essential ideas underlying these new materials and technologies to designers so that they might better be able to understand how to exploit their characteristics effectively within a design context. At the same time it seeks to introduce material scientists and engineers knowledgeable about nanomaterials to design issues and design processes. What design problems need to be solved, and how do designers think about solving them? Only with this mutual understanding can they effectively become engaged in developing meaningful applications.

To meet these aspirations, the scope of the book is inherently large, seeking to explore what nanomaterials and nanotechnologies are and how they might be applied in diverse industries. However, though the scope is large, the approach followed in this book is based on the well-explored premise that most effective design applications occur when there is a true in-depth understanding of the materials or technologies involved and not when an overly simplified approach is taken wherein discussions of technological issues are reduced to an almost meaningless state.

The book has many unique features. It summarizes discussions of many different ideas and technologies normally found distributed all across the literature. A case in point occurs in Chapter 8 of this book, which presents coverage of the many different synthesis processes used for making nanomaterials. There are many in-depth discussions throughout the literature of specific synthesis methods, such as soft lithography or sol-gel deposition methods, but these same discussions do not comprehensively cover other techniques. The approach here is to explore all primary methods of nanosynthesis and the mechanical, thermal, optical, and electrical properties they create—a comprehensive view that will be welcomed by teachers of engineering and material scientists. A broad knowledge of resulting properties is important, too, for any designer or engineer trying to use nanomaterials. It can be difficult to locate real values for properties of nanomaterials; a mechanical property may be reported in one source, a thermal property in another. In writing this book, considerable effort was put into sifting through the literature to identify credible values for primary properties.

Even once some of these properties are known, experience in the design world suggests that these values have to be placed in comparison to more traditional materials before they will be used, particularly given the higher costs of nano-based materials. A good way to explore materials and to select them to meet specific design objectives is to present their properties as "material property charts." These charts give a graphical overview of material attributes and allow comparisons to be made between them, as well as serving as a basis for more advanced material selection and related design techniques. Critical properties of various nanomaterials have now been incorporated into these kinds of charts, facilitating an understanding of where and how to effectively use nanomaterials within a design context.

In a broader sense, another unique feature of the book is the overall orientation to applications. Discussions are focused around thinking about products and buildings via the nature of the involved physical *environments, systems,* and *assemblies.* Environments may alternatively be considered either as an integral part of a design, such spaces as within a building, or as defining the context within which a product operates or is reflective of its salient mode of operation. Systems are generally defined as physical components or parts that work interactively to provide some particular type of *function.* In both products and buildings, broad considerations of *thermal, sound, lighting,* and *mechanical* environments and their related functional systems are invariably important and invariably influence the definition of needed material properties; consequently they influence the role that nanomaterials and nanotechnologies might or might not effectively play. This kind of perspective reflects the way designers think about how and where to introduce new materials or technologies—a perspective that would be well worth scientists and engineers interested in applications of nanomaterials knowing something about.

The book is organized in a direct way and makes extensive use of illustrations. The first two chapters are self-contained nontechnical overviews of the general characteristics of nanomaterials and provide examples of ways in which nanomaterials can be found in both nature and in an historical context. Chapter 3 sets the broad design context and explores fundamental design processes. Chapter 4 then reviews basic technical properties of materials as a prelude for focusing on nanomaterials. This is followed in Chapter 5 by an introduction to material property charts, material indices, and their use for optimized material selection. Formal material screening and selection procedures are also presented, providing a methodology for comparing the benefits of using nanomaterials to other kinds of more traditional high-performance materials.

This introduction leads into an in-depth treatment of nanomaterials that begins in Chapter 6. A major contribution here is that of clarifying ways that various nanomaterial types can be formally classified. Size effects and other defining characteristics of nanomaterials are explored. Specific mechanical, thermal, and electrical properties of common nanomaterial types are detailed in Chapter 7. Processes for synthesizing and characterizing nanomaterials are next described and illustrated in Chapter 8. Several material property charts illustrating various properties of nanomaterials are included.

The book then turns to design or application-oriented considerations. A discussion of specific applications in common design environments—thermal, mechanical, electrical, and optical—follows in Chapter 9. These sections broadly explore a range of nanomaterial and nanotechnology applications as they relate to these different design environments. Chapter 10 looks at applications from specific product or industry perspectives—nanotextiles, nanopaints, nanosealants, nanotechnologies for self-cleaning, antimicrobial or self-healing, and others. Chapter 11 briefly examines nanomaterials in relation to health and the environment. Chapter 12 concludes with a whirlwind tour of the driving forces underlying nanomaterial and nanotechnology developments in several industries.

The target audience for this book includes students in engineering and architecture, professional engineers, architects, and product designers who are interested in exploring how these new materials and technologies might result in better products or buildings. Firms and industry groups who are planning to be intimately involved in the nanotechnology field will find the broad perspective interesting. Obviously, individuals in these fields have quite different backgrounds, interests, and understanding of materials and their properties as well as technologies. Having a single book address these multiple audiences is inherently difficult. Designers have primary goals of making products and buildings interesting, efficient, and useful, with materials being an important enabling design component. Their interests revolve more around final design qualities and applications than on achieving a quantitative understanding of materials and their properties. Many design-oriented engineers have similar interests but are more focused on the way material properties affect design, and they employ quantitative methods to achieve these ends. Material engineers and scientists are clearly interested in fundamental properties and how to create new materials for broad usage. Therefore, depending on the reader's background, the following two paragraphs suggest general guidelines on how to read this book.

In the field of engineering, second- or third-year undergraduate students and first-year graduate students will find the book useful as supplementary reading. These students should find familiar ground in the more technically oriented discussions of basic material properties in Chapter 4, as will engineers in firms. Chapter 5 on material selection processes based on material attribute charts will be of great interest to any who have never been exposed to this approach before and will provide a concise review for those who have already used them. Chapters 6, 7, and 8 provide a good introduction to classification systems, basic fundamentals, synthesis, characterization methods, and properties of nanomaterials—information that is currently scattered among many sources and difficult to find in one place. Obviously, whole in-depth courses could be built around many of the topics succinctly addressed in these chapters. The intent, however, is to provide a solid conceptual understanding of these topics, not to go into depth. Individuals interested in further depth should refer to the entries described in "Further Readings." The application-oriented sections in Chapters 9, 10, 11, and 12 provide overviews of ways that nanomaterials and nanotechnologies are either used or poised to be used in practice. Engineers and material scientists not actively engaged in design practices will benefit from reviewing these chapters as a way of understanding what material characteristics are needed and considered useful by the design community. Some of the basic ways designers conceptualize the use of materials in products and buildings are also introduced in Chapter 3, which is useful reading for anyone who does not have a design background.

Architects and product designers range widely in their interests and need to understand material properties in depth. For typical architects in firms or students who have had basic courses in materials and who have a general curiosity about nanomaterials and how they might ultimately impact design, quantitative treatments are rarely needed. Chapters 1, 2, and 3 provide an insight into what nanomaterials are all about and are useful here. Chapter 5 on material selection processes is highly relevant to designers. Chapters 9, 10, 11, and 12 contain discussions on nanomaterials that are particularly appropriate for the design community. These chapters contain only qualitative technical descriptions and should be generally accessible to all interested readers. Many designers, however, are indeed quite technically inclined and have focused interests on materials and on technologically based innovations and are willing to put in the time necessary to understand them in some depth. This group would indeed find it useful to review the chapters on basic nanomaterial types and processing methods mentioned previously. Indeed, it is to this group that the design focus in this book is largely oriented, not to the casual reader.

We hope that you enjoy the book!

Mike Ashby
Engineering Design Centre
Engineering Department
Cambridge University

Paulo Ferreira
Materials Science and Engineering Program
University of Texas at Austin

Daniel Schodek
Graduate School of Design
Harvard University

Acknowledgments

The authors would like to thank students Mr. Joshua Sahoo for his dedicated work in collecting and analyzing the data required to build the chart diagrams containing nanomaterials; and Alda Black, Elizabeth Bacon, Nathan Fash, Ben Schodek, Xue Zhou and others who helped draw many of the figures.

For some of the transmission electron microscopy images, the authors would like to acknowledge the use of the Microscopy Facilities at the High Temperature Materials Laboratory (HTML) at Oak Ridge National Laboratory. The research was sponsored by the Assistant Secretary for Energy Efficiency and Renewable Energy, Office of FreedomCAR and Vehicle Technologies, as part of the High Temperature Materials Laboratory User Program, Oak Ridge National Laboratory, managed by UT-Batelle, LLC, for the U.S Department of Energy under contract number DE-AC05OO0R22725.

The author would like to thank his students, Mr. Joshua Schoo for his dedicated work in collecting and analyzing the data required to finalize the experiment containing nanomaterials, and ... ... win-behind data many of the frames.

For some of the transmission electron microscope images, the authors would like to acknowledge the use of the Microscopic facilities at the High Temperature Materials Laboratory (HTML) at Oak Ridge National Laboratory. The research was sponsored by the Assistant Secretary for Energy Efficiency and Renewable Energy Office of FreedomCAR and Vehicle Technologies, as part of the High Temperature Materials Laboratory User Program, Oak Ridge National Laboratory, managed by UT-Battelle, LLC, for the U.S. Department of Energy under contract number DE-AC05-00OR22725.

CHAPTER 1

Nanomaterials and Nanotechnologies: An Overview

1.1 WHY NANOMATERIALS?

Imagine dissociating a human body into its most fundamental building blocks. We would collect a considerable portion of gases, namely hydrogen, oxygen, and nitrogen; sizable amounts of carbon and calcium; small fractions of several metals such as iron, magnesium, and zinc; and tiny levels of many other chemical elements (see Table 1.1). The total cost of these materials would be less than the cost of a good pair of shoes. Are we humans worth so little? Obviously not, mainly because it is the arrangement of these elements and the way they are assembled that allow human beings to eat, talk, think, and reproduce. In this context, we could ask ourselves: What if we could follow nature and build whatever we want, atom by atom and/or molecule by molecule?

In 1959, Physics Nobel Laureate Richard Feynman gave a talk at Caltech on the occasion of the American Physical Society meeting. The talk was entitled, "There's Plenty of Room at the Bottom." In this lecture, Feynman said:

> What I want to talk about is the problem of manipulating and controlling things on a small scale … What are the limitations as to how small a thing has to be before you can no longer mold it? How many times when you are working on something frustratingly tiny like your wife's wrist watch have you said to yourself, "If I could only train an ant to do this!" What I would like to suggest is the possibility of training an ant to train a mite to do this … A friend of mine (Albert R. Hibbs) suggests a very interesting possibility for relatively small machines. He says that, although it is a very wild idea, it would be interesting in surgery if you

Table 1.1 Approximate Chemical Composition of a 70 Kg Human Body

Element	Symbol	Number of Atoms	Element	Symbol	Number of Atoms
Hydrogen	H	4.22×10^{27}	Cadmium	Cd	3×10^{20}
Oxygen	O	1.61×10^{27}	Boron	B	2×10^{20}
Carbon	C	8.03×10^{26}	Manganese	Mn	1×10^{20}
Nitrogen	N	3.9×10^{25}	Nickel	Ni	1×10^{20}
Calcium	Ca	1.6×10^{25}	Lithium	Li	1×10^{20}
Phosphorous	P	9.6×10^{24}	Barium	Ba	8×10^{19}
Sulfur	S	2.6×10^{24}	Iodine	I	5×10^{19}
Sodium	Na	2.5×10^{24}	Tin	Sn	4×10^{19}
Potassium	K	2.2×10^{24}	Gold	Au	2×10^{19}
Chlorine	Cl	1.6×10^{24}	Zirconium	Zr	2×10^{19}
Magnesium	Mg	4.7×10^{23}	Cobalt	Co	2×10^{19}
Silicon	Si	3.9×10^{23}	Cesium	Ce	7×10^{18}
Fluorine	F	8.3×10^{22}	Mercury	Hg	6×10^{18}
Iron	Fe	4.5×10^{22}	Arsenic	As	6×10^{18}
Zinc	Zn	2.1×10^{22}	Chromium	Cr	6×10^{18}
Rubidium	Rb	2.2×10^{21}	Molybdenum	Mo	3×10^{18}
Strontium	Sr	2.2×10^{21}	Selenium	Se	3×10^{18}
Bromine	Br	2×10^{21}	Beryllium	Be	3×10^{18}
Aluminum	Al	1×10^{21}	Vanadium	V	8×10^{17}
Copper	Cu	7×10^{20}	Uranium	U	2×10^{17}
Lead	Pb	3×10^{20}	Radium	Ra	8×10^{10}
					Total = 6.71×10^{27}

Adapted from R. Freitas Jr., in Nanomedicine, Volume I: Basic Capabilities, *Landes Bioscience, 1999, http://www.nanomedicine.com/NMI. htm.*

could swallow the surgeon. You put the mechanical surgeon inside the blood vessel and it goes into the heart and "looks" around. It finds out which valve is the faulty one and takes a little knife and slices it out.

Although Feynman could not predict it, this lecture was to become a central point in the field of *nanotechnology*, long before anything related with the word *nano* had emerged. Feynman's vision was amazing, particularly if we consider that a broad spectrum of

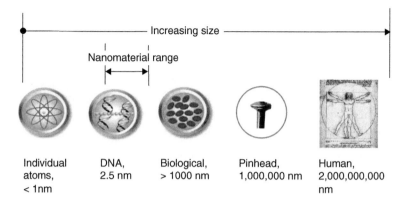

Individual atoms, < 1nm | DNA, 2.5 nm | Biological, > 1000 nm | Pinhead, 1,000,000 nm | Human, 2,000,000,000 nm

FIGURE 1.1

Sequence of images showing the various levels of scale. (Adapted from Interagency Working Group on Nanoscience, Engineering and Technology, National Science and Technology Council Committee on Technology, "Nanotechnology: Shaping the World Atom by Atom." Sept.1999.)

currently mature scientific and engineering areas are based on concepts described in Feynman's lecture 50 years ago.

The term *nanotechnology* was first used in 1974 by Norio Taniguchi to refer to the precise and accurate tolerances required for machining and finishing materials. In 1981, K. E. Drexler, now at the Foresight Nanotech Institute for Molecular Manufacturing, described a new "bottom-up" approach, instead of the top-down approach discussed earlier by Feynman and Taniguchi. The bottom-up approach involved molecular manipulation and molecular engineering to build molecular machines and molecular devices with atomic precision. In 1986, Drexler published a book, *Engines of Creation*, which finally popularized the term *nanotechnology*.

The term *nano* derives from the Greek word for *dwarf*. It is used as a prefix for any unit such as a second or a meter, and it means a billionth of that unit. Hence, a nanometer (nm) is a billionth of a meter, or 10^{-9} meters. To get a perspective of the scale of a nanometer, observe the sequence of images shown in Figure 1.1. Despite the wide use of the word *nanotechnology*, the term has been misleading in many instances. This is because some of the technology deals with systems on the micrometer range and not on the nanometer range (1–100 nm). Furthermore, the research frequently involves basic and applied science of nanostructures and not basic or applied technology. Nanomaterials are also not undiscovered materials, but nanoscale forms of well-known materials such as gold, silver, platinum, iron and others. Finally, it is important to keep in mind that some past technology such as, for example, nanoparticles of carbon used to reinforce tires as well as nature's photosynthesis would currently be considered a form of nanotechnology. Photosynthesis is in fact an excellent example of the role of nanostructures in the world's daily life. Leaves contain millions of chloroplasts, which

FIGURE 1.2
A green leaf is composed of chloroplasts inside which photosynthesis occurs.

make a green plant green (see Figure 1.2). Inside each chloroplast, hundreds of thylakoids contain light-sensitive pigments. These pigments are molecules with nanoscale dimensions that capture light (photons) and direct them to the photo reaction centers. At every reaction center, there are light-sensitive pigments that execute the actual photon absorption. When this happens, electrons become excited, which triggers a chain reaction in which water and carbon dioxide are turned into oxygen and sugar.

What is so special about nanotechnology? First, it is an incredibly broad, interdisciplinary field. It requires expertise in physics, chemistry, materials science, biology, mechanical and electrical engineering, medicine, and their collective knowledge. Second, it is the boundary between atoms and molecules and the macro world, where ultimately the properties are dictated by the fundamental behavior of atoms. Third, it is one of the final great challenges for humans, in which the control of materials at the atomic level is possible. So, are nanoscience and nanotechnology real, are they still fiction? During the 1990s Drexler's book *Engines of Creation* inspired many science fiction writers to think about the prospects of nanotechnology. For example, in the episode "Evolution" of *Star Trek*, a boy releases "nanites," which are robots at the nanoscale fabricated to work in living cells. These machines end up evolving into intelligent beings that gain control the starship *Enterprise*. In the book *Queen of Angels*, humans with psychological problems can be treated by an injection of nanodevices. These fictional ideas could become reality if in the future we become able to control matter at the atomic or molecular level. Yet, some skeptical questions may arise, such as:

- Are molecular entities stable?

- Are quantum effects an obstacle to atomic manipulation?

- Is Brownian motion a significant effect in nanocomponents?

- Are friction and wear relevant for nanocomponents?

The answer to the first question is yes. The human population is the best living example. Each human being contains approximately 10^{27} atoms that are reasonably stable, and due to cellular multiplication, the human body is able to build itself using molecular mechanisms. With respect to the quantum effects, the uncertain atomic position (Δx) can be estimated from classical vibrational frequency calculations. As discussed by R. Freitas in his book *Nanomedicine*, for a carbon atom in a single C-C bond, Δx is approximately 5% of the electron cloud diameter. Hence, the manipulation of nanostructures should not be affected by quantum effects. In terms of Brownian motion, we should consider the nanocomponents as harmonic oscillators. In this context, molecular collisions are as likely to absorb energy as to release it, and thus no net effect is expected. Finally, in regard to friction and wear effects at the nanoscale, the repulsive fields between the molecules should provide the necessary lubrication. Products such as oil should be avoided because they cause contamination. For example, a molecular bearing of 2 nm rotating at 1 MHz dissipates about 10^{-6} picowatts due to friction, whereas a micron-size mechanical component dissipates 1–1000 picowatts (R. Freitas, *Nanomedicine*). In general, the skeptical questions we've posed can be disqualified because so far, nanotechnology does not seem to violate any physical law.

What are the possible approaches to making nanomaterials and nanotechnologies? There are basically two routes: a top-down approach and a bottom-up approach. The idea behind the top-down approach is the following: An operator first designs and controls a macroscale machine shop to produce an exact copy of itself, but smaller in size. Subsequently, this downscaled machine shop will make a replica of itself, but also a few times smaller in size. This process of reducing the scale of the machine shop continues until a nanosize machine shop is produced and is capable of manipulating nanostructures. One of the emerging fields based on this top-down approach is the field of nano- and micro-electromechanical systems (NEMS and MEMS, respectively). MEMS research has already produced various micromechanical devices, smaller than 1 mm^2, which are able to incorporate microsensors, cantilevers, microvalves, and micropumps. An interesting example is the microcar fabricated by Nippondenso Co. (see Figure 1.3). The car is 4800 μm long, 1800 μm wide, and 1800 μm high. Each tire is 690 μm diameter and 170 μm wide, whereas the license plate is 10 μm thick. The car has 24 components, including tires, wheels, axles, bumpers, all parts assembled by a micromechanical manipulator.

Which technologies can be used to produce nanostructures using a top-down approach? At the moment, the most used is photolithography. It has been used for a while to manufacture computer chips and produce structures smaller than 100 nm. This process is discussed extensively in Chapter 8. Typically, an oxidized silicon (Si) wafer is coated with a 1μm thick photoresist layer. After exposure to ultraviolet (UV) light, the photoresist undergoes a photochemical reaction, which breaks down the polymer by rupturing the polymer chains. Subsequently, when the wafer is rinsed in a developing solution, the exposed areas are removed. In this fashion, a pattern

FIGURE 1.3

Microcar produced by Nippondenso Co. The latest model has a micromotor 1 mm in diameter. With power supplied by a 25 micron copper wire, the car runs smoothly at a speed of about 1 cm/s with 3 V voltage and 20 mA current. (Courtesy of Nippondenso.)

is produced on the wafer surface. The system is then placed in an acidic solution, which attacks the silica but not the photoresist and the silicon. Once the silica has been removed, the photoresist can be etched away in a different acidic solution. Though the concept of photolithography is simple, the actual implementation is very complex and expensive. This is because (1) nanostructures significantly smaller than 100 nm are difficult to produce due to diffraction effects, (2) masks need to be perfectly aligned with the pattern on the wafer, (3) the density of defects needs to be carefully controlled, and (4) photolithographic tools are very costly, ranging in price from tens to hundreds of millions of dollars.

As a response to these difficulties, electron-beam lithography and X-ray lithography techniques have been developed as alternatives to photolithography. In the case of electron beam lithography, the pattern is written in a polymer film with a beam of electrons. Since diffraction effects are largely reduced due to the wavelength of electrons, there is no blurring of features, and thus the resolution is greatly improved. However, the electron beam technique is very expensive and very slow. In the case of X-ray lithography, diffraction effects are also minimized due to the short wavelength of X-rays, but conventional lenses are not capable of focusing X-rays and the radiation damages most of the materials used for masks and lenses.

Due to these limitations, the most recent lithography methods, such as printing, stamping, and molding, use mechanical processes instead of photons or electrons. These methods are normally called *soft lithography* methods because they involve the use of polymers. They are discussed in Chapter 8. The process starts with the fabrication of a mold by photolithograpy or e-beam lithography. Subsequently, a chemical precursor to polydimethylfiloxane (PDMS) is applied over the mold and cured. As a result, a solid PDMS stamp that matches the pattern of the mold is produced. The stamp can then be used in several ways to make structures at the nanoscale. For example, in the case of microcontact printing the stamp is coated with a solution consisting of organic molecules (thiol ink) and compressed into a thin film of gold (Au) deposited on a silicon (Si) wafer. In this fashion the thiol ink is transferred from the stamp to the gold surface, reproducing the original pattern of the stamp. In the case of micromolding, the PDMS stamp sits on a flat, hard surface and a liquid polymer is injected into any available cavity between the stamp and the surface. Subsequently, the polymer cures into the patterns provided by the cavities. The advantages of using soft lithography methods are (1) no special equipment is required, and (2) versatility in the range of materials and geometries produced. As for disadvantages, due to the inherent presence of polymers, soft lithography processes often exhibit significant strains that are not ideal for certain applications, such as nanoelectronics.

From the discussion so far, it is obvious that the top-down approach is not a friendly, inexpensive, and rapid way of producing nanostructures. Therefore, a bottom-up approach needs to be considered. The concept of the bottom-up approach is that one starts with atoms or molecules, which build up to form larger structures. In this context, there are three important enabling bottom-up technologies, namely (1) supramolecular and molecular chemistry, (2) scanning probes, and (3) biotechnology. The supramolecular and molecular chemistry route is based on the concept of self-assembly. This is a strategy for nanofabrication that involves designing molecules so that they aggregate into desired structures. The advantages of self-assembly are that (1) it solves the most difficult steps in nanofabrication, which involve creating small structures; (2) it can directly incorporate and bond biological structures with inorganic structures to act as components in a system, and (3) it produces structures that are relatively defect-free.

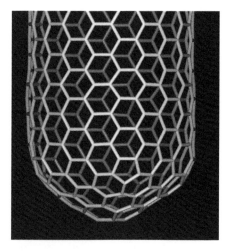

FIGURE 1.4
Computer simulation of a single-wall carbon nanotube with a diameter of 1.4 nm. Carbon nanotubes can be thought of as wrapped sheets of graphene.

One of the best examples of self-assembly is the fabrication of carbon nanotubes. These nanostructures are composed of C atoms that assemble into cylinders of approximately 1.4 nm in diameter (see Figure 1.4). In the last decade, the idea of using carbon nanotubes to fabricate simple gears evolved by bonding ligands onto the external surfaces of carbon nanotubes to produce "gear teeth" (see Figure 1.5). The efficiency of these gears depends on placing the gear teeth just right in atomically precise positions. Researchers at the National Aeronautics and Space Administration (NASA) performed a molecular dynamics simulation to investigate the properties of molecular gears made from carbon nanotubes. Each gear is made of a 1.1 nm diameter nanotube with seven benzene teeth. The distance between two nanotubes is 1.8 nm. The simulations show that the gears can operate up to 70 GHz without overheating. As speed increases above 150 GHz, the gears overheat and begin to stall.

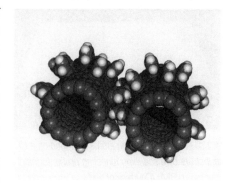

FIGURE 1.5 *Computer simulation of nanogears made of carbon nanotubes with teeth added via a benzine reaction. (Courtesy of NASA.)*

Another very important bottom-up approach using molecular chemistry is employed in the fabrication of quantum dots. Quantum dots are crystals composed of a few hundred atoms. Because electrons in a quantum dot are confined to widely separated energy levels, the dots emit only one type of wavelength of light when they are excited (see Figure 1.6). A typical procedure used to make quantum dots involves a chemical reaction between a metal

FIGURE 1.6
Cadmium selenide (CdSe) quantum dots of different sizes and shapes in solution emit light at different wavelengths. (Courtesy of M. J. Bawendi, MIT.)

ion—cadium (Cd), for example—and a molecule that is able to donate a selenium ion. This reaction generates crystals of cadmium selenide (CdSe). The shape and size of quantum dots can be controlled by tuning the ratio of the molecules and by adding surfactants. Their size and shape are important because they determine their electronic, magnetic, and optical properties.

The second important bottom-up method is the use of scanning probes. One of the biggest steps toward nanoscale control occurred in 1981 when researchers at the IBM research center in Switzerland, led by Gerd Binning and Heinrich Rohre, developed the *scanning tunneling microscope* (STM). The STM is, essentially, a fine tip that scans over a material's surface. Since the tip is only a few atoms wide, electrons "tunnel" across the gap between the surface and the tip as the tip sweeps over the surface (see Chapter 8). In this way scientists developed several related instruments, now known as *scanning probe microscopes* (SPMs), to collect images and analyze the identities of atoms on a material's surface. At the moment, SPMs can do more than simply look at atoms. They can also be used to move atoms. At IBM, Donald Eigler and Erhard Schweizer were able to displace 35 xenon atoms onto a crystal of Ni and write the word *IBM* (see Figure 1.7). These tools have opened up many new doors for further scientific discovery.

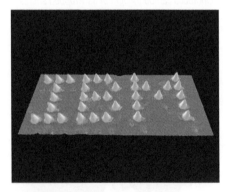

FIGURE 1.7 *Xenon atoms arranged on a Ni substrate by a scanning tunneling probe, forming the word* IBM. *(Courtesy of IBM.)*

The third relevant route for producing nanostructures using a bottom-up approach is the use of biotechnology. By 1998, biotechnologists were able to make DNA sequences and assemble artificial viruses, which are examples of molecular engineering. In addition, several devices, which looked like biomolecular motors, sensors, and actuators, could be integrated into systems made of molecular machines. One of the best biological examples of a molecular machine is the *ribosome*, which is a remarkable nanoscale assembler. The role of the ribosome is to act as a factory of proteins by combining amino acids in a very specific order. Section 2.2 discusses ribosome self-assembly methods in more detail.

Another good example of self-assembly is the abalone (see Figure 1.8), an expensive appetizer in fine restaurants. This mollusk builds a strong shell by secreting proteins, which self-assemble into a configuration that resembles walls surrounding a room. Inside each room there is seawater saturated with calcium (Ca) ions and carbonate ions, which will eventually form crystals of $CaCO_3$ (calcium carbonate). Therefore the abalone shell is a composite material composed of alternating layers of calcium carbonate and proteins.

As we have seen so far, both the top-down and the bottom-up approaches are capable of producing nanostructures, which will soon be applied in many of the engineering fields. One area of great interest for the application of nanostructures is the area of *nanoelectronics*. In 1985, Konstantin K. Likharev, a professor at Moscow State University, postulated that it would be possible to control the flow of single electrons in and out of a coulomb island. This would form the groundwork for the single-electron transistor, which was first built successfully in 1987 by researchers at Bell Labs. This phenomenon can be very useful in situations where electrical currents are passed through molecules that can act like coulomb islands, for example, if they are weakly coupled to electrodes, which will transfer the signal to the macro world. It has also been proposed that organic molecules could make good transistors. It has already been demonstrated that clusters of molecules can carry electrons from one metal electrode to another. Each molecule is about 0.5 nm wide and 1 to 2 nanometers long. The switching mechanism that allows the molecules to act as transistors seems to involve a chemical reaction, where electrons move among the various atoms within the molecule. Under certain conditions, the chemical reaction twists the molecule, preventing the flow of the electrons. On the other hand, in the absence of the reaction, the molecule will conduct electrons.

One other area of increasing interest for the application of nanomaterials is the field of nanomedicine. All of biology, even the most complicated creature, is made up of small cells that are constructed of building blocks at the nanoscale, such as proteins, lipids, and nucleic acids. There is, for example, the story of some bacteria that live in swampy waters. These bacteria can only survive in specific depths. Very close to the surface, oxygen is too abundant; too deep, oxygen is limited. So, how can the bacteria know their position? The answer lies on a string of about 20 ferromagnetic crystals, each 35–120 nm in diameter (see Figure 1.9), which act together as a compass needle influenced by the Earth's magnetic field. Why are

FIGURE 1.8
The abalone, a marine mollusk whose shell is composed of calcium carbonate layers separated by a protein.

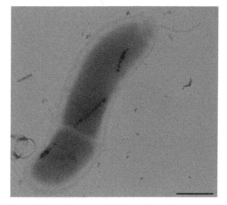

scale bar = 500 nm

FIGURE 1.9
A magnetotactic bacteria containing a chain of magnetic crystals 35–120 nm in diameter is capable of orienting itself along Earth's magnetic field lines. (Courtesy of Prozorov et al., Physical Review B, *76, 054406, 2007.*)

the particles so small? Because below a critical dimension, typically at the nanoscale, ferromagnetic crystals become single domain (see section 7.4) which helps increase the effectiveness of the bacteria as a compass.

Following a similar approach, magnetic nanoparticles are now being employed to detect particular biological species, such as microorganisms that cause disease. Magnetic nanoparticles are coupled with antibodies that bind to their target. In this case the magnetization vector of all nanoparticles becomes parallel, which results in a strong magnetic signal. On the other hand, if the illness is not present, the antibodies do not recognize the target and do not bind. Thus, all magnetization vectors will remain randomly oriented, leading to a weak magnetic signal. Another interesting application in nanomedicine is related to the use of gold particles for bio-identification. One set of gold particles carries DNA that binds to one half of the target sequence. A second set carries DNA that binds to the other half. The DNA with the complete target sequence will couple to both sets of particles, binding them together. As gold nanoparticles aggregate, there is a shift in the wavelength of light from red to blue. Thus, the red and blue lights can be used to probe whether a certain DNA corresponds to a particular individual.

These examples are only a small fraction of what will become possible in the future. Though still in their infancy, nanoscience and nanotechnology have already demonstrated that they will have a tremendous impact on various aspects of human life, such as health, environment, energy, transportation, information technology, and space exploration. This fact has been recognized by most of the industrial world, as confirmed by government funding allocated for the study of this area (see Figure 1.10). Nanoscience and nanotechnology hold great promise for future innovation and transformation. If the field's expectations are fulfilled, it will create sought-after sensors with improved detection sensitivity and selectivity; strong, lightweight bullet-stopping armor; energetic nanoparticles for fast-release explosives; miniaturization of aircraft to reduce payload; materials that perform under extreme temperatures and pressures; radiation-tolerant materials; nanostructures that lower waste disposal costs; self-healing systems for extended space missions, self-assembly, and processing in space; implants to replace worn or damaged body parts; nanomotors; and cellular implants. In the wake of these innovations, nanoscience and nanotechnology will transform and revolutionize life as we now know it.

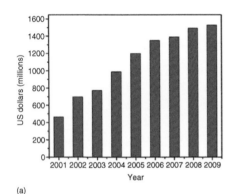

(a)

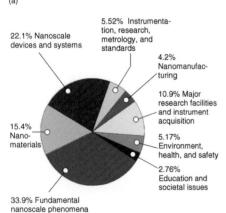

(b)

FIGURE 1.10

U.S. federal funding trends in nanotechnology, 2001–2009. (National Nanotechnology Initiative: Second Assessment and Recommendations of National Advisory Panel, April 2008.)

1.2 SCALE, STRUCTURE, AND BEHAVIOR

As discussed in the previous section, *nano* stands for *nanometer*, one billionth (10^{-9}) of a meter. This defines a length scale that is, for these materials, important in some way, influencing their mechanical, thermal, electrical, magnetic, optical, and aesthetic properties. Before embarking on a discussion of these, it is useful to have an idea of where *nano* lies in the broader scale of things.

If you ask: What is the smallest particle in matter? Most people would reply: an atom. Atoms have a diameter of about one tenth of a nanometer. Atoms are themselves made up of units that can sometimes be thought of as particles: electrons, neutrons, and protons. They are much smaller, about 10^{-15} meters (a femtometer) in diameter, but when things get this small quantum-mechanical effects make them fuzzy, their size ill defined, so we shall stop there. What is the other extreme? It used to be of global scale ("the ends of the Earth," "poles apart"), but space exploration has changed all that, enlarging the concept of scale to intergalactic dimensions: light-years, just under 10,000,000 million km (10^{16} meters, 10 pentameters).

Figure 1.11 shows this wide range, from a femtometer at the bottom to 100 light-years on the top. Its utility is much like that of the time scale used by geologists in classifying rocks (5 billion years ago to the present) or of archeologists in charting human history (500,000 BC to 2000 AD): It is a ladder on which things can be hung. Start at the human scale. It is the one from which perception expands, one that is built into our language. At the lower end, the diameter of a hair ("it missed by a hair's breadth"). The inch, originally the ell, is the length of the lower joint of the thumb. The foot is, well, the foot. A yard is a pace. A furlong is the length of a furrow, a league (about 5 km) is the distance a man or a horse can walk in an hour; thus 7-league boots take you a day's journey at one step. Perhaps that is why these are called Imperial units. You would not design anything using units like these (although NASA apparently still does). It was France, a nation that has contributed so much to rational thinking, that gave us a system with some sense in it. The National Assembly of the First Republic, by decree in 1790, created the standards of length, volume, and weight that we know today as the metric system.

The original proposal was that the meter should be some fraction of the circumference of the Earth at the equator, but this was not seen as sufficiently Gallic. The decisive vote went for a fraction of the length of the meridian that passes through Paris and both poles. The meter, then, is one ten millionth (10^{-7}) of the length of the

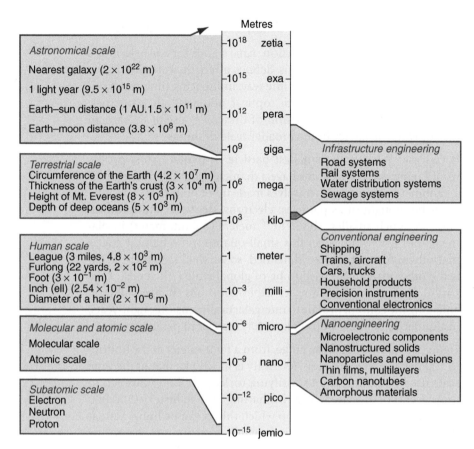

FIGURE 1.11

Length scales ranging from the subatomic through molecular, human, and terrestrial to astronomic. The nano range is shown on the lower right.

quadrant (a quarter) of the circumference of the Earth along this particular trajectory. A liter is the volume of a cube with an edge of one tenth of a meter. A kilogram is the weight of one liter of water at 4°C. Once the reformers of the First Republic got started, there was no way of stopping them. Imperial units may have charm, but metric units are logical, comprehensible, and (as Figure 1.11 shows) extendable.

The tools of science allow the study of materials over this immense range. Of the many ways in which they can be probed, the most revealing has been the interaction with radiation. Electromagnetic (e-m) radiation permeates the entire universe. Observe the sky with your eye and you see the visible spectrum, the range of wavelengths we call "light" (400–770 mm). Observe it with a detector of X-rays or γ-rays and you see radiation with far shorter wavelengths (as short as 10^{-4} nm, one thousandth the size of an atom). Observe it instead with a radio telescope and you pick up radiation with wavelengths measured in millimeters, meters, or even kilometers, known as *radio waves* and *microwaves*. The range of wavelengths of radiation is vast, spanning 18 orders of magnitude (Figure 1.12). The visible part of this spectrum is only a tiny part of it—but even that has entrancing variety, giving us colors ranging from deep purple through blue, green, and yellow to deep red.

Objects interact strongly with radiation that has a wavelength comparable to their size or, if they have internal structure, with the scale of that structure. They interact in much less interesting ways with radiation that has wavelengths that differ much from their structural scale. The strong interaction gives diffraction and scattering. The analysis of these phenomena has proved to be a tool of extraordinary power, revealing the structure of the atom, the organization of atoms and crystals and glasses, and complex organic molecules such as DNA. At the other end of the spectrum, microwaves (radar) and waves of radio frequencies give us information about the thickness of the Arctic ice sheet and even insight into the structure of deep space.

E-m waves are not the only way of probing materials. Acoustic waves, too, give information about structure that is comparable to the acoustic wavelength. Here the range is narrower (see Figure 1.13). At the lower end is extreme ultrasonic, used to probe surface structures and to detect and image cracks and defects in castings. Medical CT scanners use the range 1–20 MHz, giving a resolution of a few millimeters. Bats, dolphins, and whales communicate and locate prey using ultrasonics, 20–100 kHz. Sonar spans much of the audible range, extending down to frequencies as low as 1 Hz, typical of seismic waves. Nano, on the scale of Figure 1.11, lies near the bottom. The wavelengths of the visible spectrum (Figure 1.12) lie at the upper end of this range, so we anticipate the potential for strong interaction of light with nanomaterials; indeed, it is this interaction that gives opals their elusive and shifting colors, imparts hues to butterfly wings, and allows paints that change color as you walk past them. The ultraviolet spectrum overlaps the central part of the range, X-rays the lower part, so these, too, interact strongly with nano-materials, are used to study their structure, and can be exploited in devices to manipulate them. Acoustic waves, with wavelengths that are large compared with the nanoscale, are less useful.

This strong interaction when length scales are comparable extends to other properties. Electric currents are carried by electrons; heat is carried by both electrons and phonons. The resistance to both depends on the electron or phonon mean-free path, meaning the distance between scattering events. It is the structural length

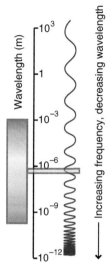

FIGURE 1.12

The spectrum of electromagnetic radiation, from gamma to radio waves, a range of 10^{15}. The visible portion is a tiny sliver of this range. The nanoscale overlaps the visible, the ultraviolet, and the long X-ray regimes.

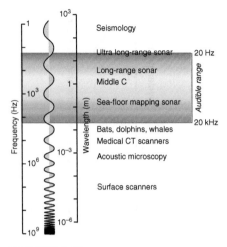

FIGURE 1.13

The acoustic spectrum from the extreme ultrasonic to the seismic, a range of 10^8. The audible part is a thick slice of this. The figure is based on acoustic waves in air, velocity 350 m/s. For water, shift the wavelength scale up relative to the frequency scale by a factor of 3; for rock, shift it up by a factor of 10.

of the material that determines this: the finer the structure, the shorter is the mean-free path and the higher the resistance. Magnetism involves domains of locally aligned magnetic moments; the domain boundaries have a characteristic thickness, again of atomic or nanodimensions. Optical properties, particularly, involve scattering or diffraction, both with specific length scales. Strength, too, is influenced by structural scale. When materials deform—bend, twist, or stretch permanently—it is because sheets of the atoms that make them up ratchet over each other. The ratchet steps are fixed by the atomic size. The lower end of the nanoscale approaches this in scale, giving, as the structural length approaches it, increasingly strong interaction, meaning greater strength. Brittleness (and its opposite, toughness) has a characteristic length—that of the flaw that will just propagate at the yield strength. The flaw size is itself related to the scale of the structure. This intrinsic dimensionality, so to speak, of material properties is one of the reasons for the great interest in nanomaterials. If we can engineer the structural scale of a material (as particles, thin layers, or the crystal size in bulk materials) in the right way, we can influence or interact with the property length scale and thus manipulate the property itself.

We live in a world in which everyday objects have a size—we shall call it *scale* and give it the symbol *L*, for length—in the range of millimeters to meters. Scale has a profound effect on the behavior of structures made from materials. A steel tuning fork 80 mm long with arms that are 3 mm thick oscillates with a frequency of middle C, 256 Hz. Vibration frequencies scale as 1/L. So change the millimeters of the tuning fork to microns and the frequency rises to 256 kHz. Change it to nanometers and it becomes 256 MHz. This is comparable to the clock speed of computers, stimulating the novel idea of mechanical computers (not such a new idea: think of the abacus). This is an example of scale-dependent mechanical response. Response time—the time it takes a mechanical latch to close or a switch to switch—scale in the same way, as 1/L ; it is this that allows automobile airbag triggers to react fast enough to save lives. Thermal response scales even faster. The time for an object to reach thermal equilibrium when the temperature is changed scales as $1/L^2$ so the thermal response time that, for mm-scale objects, is seconds or minutes becomes microseconds at the micron scale and picoseconds at the nanoscale. (There is, in reality, a limit set by the acoustic velocity that cuts in, preventing the response becoming quite that short.)

That's the big picture. Focus now for a moment on the small one: the nano. What is so different and special about it? There is more than one answer to that question of which the most obvious is that of surface area. Almost all properties of the thin layer of atoms that lie in the surface or interface of things behave in a different way than those in the interior. Surfaces, you could say, have their own properties. We experience this in their propensity to corrode, to stick (or not stick) to things, or to release electrons in plasma displays. But most of the useful properties—stiffness, strength, thermal and electrical conductivities, magnetism—come from the inside of atoms. At the familiar macroscale, the fraction of atoms that lie in the surface is minuscule. At the micron scale it is still tiny. But approach the nanoscale and it takes off (see Figure 1.14). The combination of small scale and a large fraction of "surface" atoms gives a property set that can be very different from that of the bulk. It is only now being unveiled. Nanomaterials have different mechanical, thermal, electrical, magnetic, optical, and, above all, chemical properties than those of the bulk. They can offer strong, wear-resistant coatings; they can change the ways in which heat and electricity are conducted; they have the ability, through their electrical and magnetic behavior, to store information and via their chemical behavior to catalyze chemical reactions, distribute drugs in the human body, and much more.

Some of this "science fiction" is already real. Much more is a promise, but one in which considerable confidence can be placed. As with most major innovations, there are two principle obstacles to be overcome. The first: to develop a sufficiently deep understanding of behavior to establish both the good and the bad, the benefits and the hazards, of the nanoscale. The second is that of economics. Nanomaterials are expensive and will remain so, at least for some time. Finding ways to cushion the transition to economic viability needs thought.

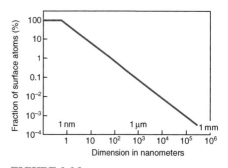

FIGURE 1.14

The fraction of atoms that lie in the surface or internal interfaces of a nanoscale or nanostructured material, expressed as a percentage (%).

FURTHER READING

R. Freitas Jr., in Nanomedicine, Vol. I: Basic capabilities, Landes Bioscience, 1999.

R. P. Feynman, There's plenty of room at the bottom, Engineering and Science (California Institute of Technology), p. 22, Feb. 1960.

N. Taniguchi, Proc. International Conf., Prod. Eng. Tokyo, Part II, Japan Society of Precision Engineering, 1974.

K. Eric Drexler, Molecular engineering: An approach to the development of general capabilities for molecular manipulation, Proc. National Academy of Sciences, p. 5275, Sept. 1981.

K. Eric Drexler, Engines of creation: The coming era of nanotechnology, Anchor Press/Doubleday, 1986.

Interagency Working Group on Nanoscience, Engineering and Technology, National Science and Technology Council Committee on Technology, Nanotechnology: shaping the world atom by atom, Sept. 1999.

Greg Bear, Queen of angels, Warner Books, 1990.

K. Eric Drexler, Nanosystems: Molecular machinery, manufacturing and computation, John Wiley & Sons, 1992.

A. Tesigahara et al., Fabrication of a shell body microcar, Proc. Third International Symposium on Micro Machine and Human Science, Nagoya, Japan, Oct. 14–16, 1992.

Jie Han, Al Globus, R. Jaffe, and G. Deardorff, Nanotechnology 8, p. 95, 1997.

B. O. Dabbousi, et al., Journal of Physical Chemistry B, 101, p. 9463, 1997.

G. Binnig and H. Rohrer, Helv. Phys. Acta, 55, p. 726, 1982.

D. M. Eigler and E. K. Schweizer, Nature, 344, p. 524, 1990.

P. Ball, Designing at the molecular world: Chemistry at the frontier, Princeton University Press, 1994.

T. A. Fulton and G. J. Dolan, Phys. Rev. Lett., 59 (1), p. 109, 1987.

M. A. Reed and J. M. Tour, Scientific American, 2000.

M. Sarikaya, Proc. National Academy of Science, 96 (25), p. 14183, 1999.

Janine Benyus, Biomimicry: Innovation inspired by nature, Perennial (HarperCollins) Publishers, 1998.

CHAPTER 2

An Evolutionary Perspective

2.1   A BRIEF HISTORY OF MATERIALS

Our species, *Homo sapiens*, differs from others most significantly, perhaps, in its ability to design and make things. *Making* implies the use of materials. Materials have enabled the advance of mankind from its earliest beginnings; indeed, the ages of man are named after the dominant material of the day: the Stone Age, the Copper Age, the Bronze Age, the Iron Age (Figure 2.1). If you have seen the movie *2001: A Space Odyssey*, you might remember one of the first scenes, in which prehistoric humans fight using animal bones. It is known that the tools and weapons of prehistory, 300,000 or more years ago, were bone and stone. Stones, particularly flint and quartz, could be shaped for tools; these materials could be flaked to produce a cutting edge that was harder, sharper, and more durable than any other material that could be found in nature.

Gold, silver, and copper, the only metals that occur in native form, must have been known from the earliest times, but the realization that they were ductile, could be beaten into complex shapes, and, once beaten, could be hardened seems to have occurred around 5500 BC—almost 300,000 years after the time stone and bone were first used. By 4000 BC there is evidence that technology to melt and cast these metals had developed, allowing more intricate shapes. However, metals in native form, such as copper, are not abundant. Copper occurs in far greater quantities as the minerals azurite and malachite. By 3500 BC, kiln furnaces, developed for pottery, could reach the temperature required for the reduction of these minerals, making copper sufficiently plentiful to usher in the Copper Age. But even in the deformed state, copper is not all that hard. Poor hardness means poor wear resistance; therefore copper weapons and tools were easily blunted.

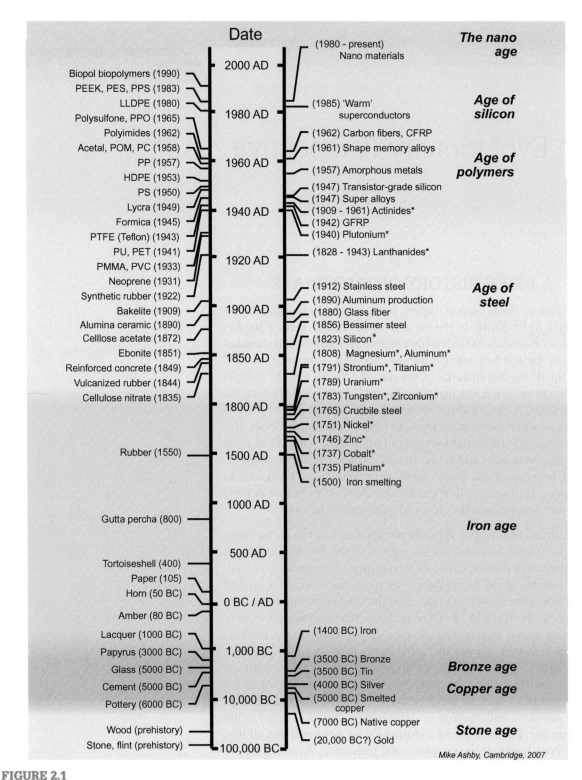

FIGURE 2.1

The materials timeline. The scale is nonlinear, with big steps at the bottom, small ones at the top. A star (*) indicates the date at which an element was first identified. Labels with no stars give the date at which the material became of practical importance.

Sometime around 3000 BC the probably accidental inclusion of a tin-based mineral, cassiterite, in the copper ores provided the next step in technology: the production of the alloy bronze, a mixture of tin and copper. Tin gives bronze an additional hardness that pure copper cannot match, allowing the production of superior tools and weapons. This discovery of alloying and solid solution strengthening—the hardening of one metal by adding another—was of such importance that it, too, became the name of a new era: the Bronze Age.

"Old-fashioned" sounds like 20th century vocabulary, but the phenomenon is as old as technology itself. The discovery, around 1450 BC, of ways to reduce ferrous oxides to make iron, a material with greater stiffness, strength, and hardness than any other then available, rendered bronze old-fashioned. Iron was not entirely new; tiny quantities existed as the cores of meteors that had impacted the Earth. The oxides of iron, by contrast, are widely available, particularly hematite, Fe_2O_3. Hematite is easily reduced by carbon, although it takes high temperatures, close to 1100°C, to do it. This temperature is insufficient to melt iron, so the material produced was a spongy mass of solid iron intermixed with slag; this was reheated and hammered to expel the slag and then forged into the desired shape.

Iron revolutionized warfare and agriculture; indeed, it was so desirable that at one time it was worth more than gold. The casting of iron, however, was a more difficult challenge, requiring temperatures around 1600°C. Two millennia passed before, in 1500 AD, the blast furnace was developed, enabling the widespread use of cast iron. Cast iron allowed structures of a new type; the great bridges, railway terminals, and civic buildings of the early 19th century are testimony to it.

But it was steel, made possible in industrial quantities by the Bessemer process of 1856, which gave iron its dominant role in structural design that it still holds today. The demands of the expanding aircraft industry in the 1950s, with the development of the jet engine, shifted emphasis to light alloys (those of aluminum, magnesium, and titanium) and to materials that could withstand the extreme temperatures of the jet combustion chamber. The development of superalloys—heavily alloyed iron- and nickel-based materials—became the focus of research, delivering an extraordinary range of alloys able to carry loads at temperatures above 1200°C. The range of their applications expanded into other fields, particularly those of chemical and petroleum engineering.

Though the evolution of metals is interesting, the history of polymers is rather different. Wood, of course, is a polymeric composite, one used for construction from the earliest times. The beauty of amber (petrified resin) and of horn and tortoise shell (the polymer keratin) attracted designers as early as 80 BC and remained attractive into the 19th century. (In London there is still a Horners' Guild, the trade association of artisans who worked horn and shell.) Rubber, brought to Europe in 1550, was already used in Mexico for the Mayan ball game Pok-a-Tok. Rubber grew in importance in the 19th century, partly because of the wide spectrum of properties made possible by vulcanization—cross-linking with sulfur that produced materials as elastic as latex and as rigid as ebonite.

The real polymer revolution, however, had its beginnings in the early 20th century with the development of Bakelite, a phenolic, in 1909 and synthetic butyl rubber in 1922. This was followed in mid-century by a period of rapid development of polymer science. Almost all the polymers we use so widely today were developed in a 20-year span from 1940 to 1960, among them the bulk commodity polymers polypropylene (PP), used for food containers; polyethylene (PE), used for children's toys; polyvinyl chloride (PVC), used for pipes and fittings; and popyurethane (PU), used for foams and car dashboards, the combined annual tonnage of which now approaches that of steel. Designers seized on these attributes—cheap and easily molded to complex shapes—to produce a spectrum of brightly colored, cheerfully ephemeral products. Design with polymers has since matured; they are now as important as metals in household products, automobiles, and, most recently, in aerospace. The polymers of transport and aerospace are, however, more complex than "pure" polymers because they do not have the stiffness and strength these applications demand. They are polymer-matrix composites, reinforced with fillers and fibers.

Composite technology is not new. Straw-reinforced mud brick is one of the earliest of the materials of architecture and remains one of the traditional materials for building in parts of Africa and Asia even today. Steel-reinforced concrete—the material of shopping centers, road bridges, and apartment blocks—appeared just before 1850. Reinforcing concrete with steel to enhance the tensile strength where previously it had none was easy. Reinforcing metals, already strong, took much longer, and even today metal matrix composites are few. But polymers, with many attractive properties but lamentable stiffness, got scientists thinking. The technology for making glass fibers had existed since 1880, when it was used to make "glass wool" insulation. Glass wool, or better, woven or layered glass fibers, could be incorporated into thermosetting polyesters or epoxies, giving them the stiffness and strength of aluminum alloys. By the mid-1940s glass fiber reinforced polymer (GFRP) components were in use in the aircraft industry. The real transformation of polymers into high-performance structural materials came with the development of aramid and carbon fibers in the 1960s. Incorporated as reinforcements, they give materials with performance (meaning stiffness and strength per unit weight) that exceeded that of all other bulk materials, earning their now-dominant position in the creation of sports equipment and in aerospace.

The period in which we now live might have been named the Polymers and Composites Era had it not coincided with a second revolution: that of silicon technology. Silicon was first identified as an element in 1823 but found few uses until the 1947 discovery that, when doped with tiny levels of impurity, it could act as a transistor. The discovery created the fields of electronics, mechatronics, and modern computer science, revolutionizing information storage, access, and transmission; imaging; sensing and actuation; numerical modeling; and much more. So we live in what is rightly called the Information Age, enabled by the development of transistor-grade silicon.

The 20th century saw other striking developments in materials technology. Superconduction, discovered in mercury and lead when cooled to $4.2\,^{\circ}K$ $(-269\,^{\circ}C)$ in 1911, found no applications outside scientific research. But in the mid-1980s, two physicists at the IBM research center in Zurich discovered a complex oxide containing barium, lanthanum, and copper that was superconducting at $30\,^{\circ}K$. This triggered a search for superconductors with yet higher transition temperatures, leading, in 1987, to a material with a transition temperature above $98\,^{\circ}K$, the temperature of readily available liquid nitrogen, making applications practical.

In the last two decades, the area of biomaterials has developed rapidly. Implanting materials in or on the human body was not practical until an aseptic surgical technique was discovered in the late 1800s. In addition, when nontoxic metal alloys were introduced, they tended to fracture in service. It was not until the development of polymer-based systems and, subsequently, with a new wave of discoveries in cell biology, chemistry, and materials science, that synthetic biomaterials became a reality.

During the late 20th century it was realized that the behavior of materials depended on scale and that the dependence was most evident when the scale was that of nanometers. Although the term *nanoscience* is new, some of the technology is not. For example, nanoparticles of carbon have been used for the reinforcement of tires for over 100 years, nanoscale proteins have been part of vaccines since the early 20th century, and nature's own nanotechnology, such as photosynthesis, has been around for millions of years. The ruby-red color of stained glass, known and used in the Middle Ages, is due to the presence of gold nanoparticles trapped in the glass matrix (see the next section). The decorative glaze known as luster, also found on medieval pottery, is the result of special optical properties provided by the gold nanoparticles. However, modern nanotechnology gained prominence with the discovery of various forms of carbon, such as the C_{60} molecule and carbon nanotubes. Simultaneously, with the advance of analytical tools capable of resolving and manipulating matter at the atomic level, scientists started asking: What if we could build things the way nature does, atom by atom and molecule by molecule? This approach involves molecular manipulation and molecular engineering in the context of building molecular machines and molecular devices with atomic precision.

We are now entering a new era—that of the nano (see Figure 2.2). The rate of development of new metallic alloys and new polymers is slowing. Much research is now focused on making the ones we have already got more reliably, more cheaply, and particularly with less damage to the environment. However, the drive for ever better and more exciting properties remains. The nanoroute appears, to many to be the most promising way forward.

If we now step back and view the timeline of Figure 2.1 as a whole, clusters of activity are apparent; there is one in Roman times, one around the end of the 18th century, one around 1940. What was it that triggered the clusters? Scientific advances, certainly. The late 18th and early 19th centuries were a time of rapid development of inorganic chemistry, particularly electrochemistry, and it was this surge that allowed new elements to be isolated and identified. The mid-20th century saw the birth of polymer chemistry, an understanding that enabled the development of the polymers we use today.

But there might be more to it than that. Conflict stimulates science. The first of these two periods coincides with that of the Napoleonic Wars (1796–1815), a time in which technology, particularly in France, developed rapidly. And the second was that of the Second World War (1939–1945), in which technology played a greater part than in any previous conflict. One hopes that scientific progress and advances in materials are possible without conflict. The competitive

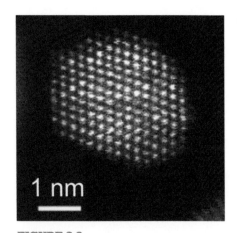

FIGURE 2.2

The coming of the Nano Age. Image of a nanoparticle composed of platinum and cobalt and obtained from an aberration-corrected scanning-transmission electron microscope. Note the scale bar and the small dimensions of the nanoparticle. Each dot on the nanoparticle corresponds to an atomic column imaged in projection. (Courtesy of P. J. Ferreira, University of Texas at Austin; L. F. Allard, Oak Ridge National Laboratory; Y. Shao-Horn, MIT.)

drive of free markets appears to be an equally strong driver. It is interesting to reflect that more than half of all the materials scientists and engineers who have ever lived are alive today, and all of them are pursuing better materials and better ways to use them. Of one thing we can be certain: There are many more advances to come.

2.2 NANOMATERIALS AND NANOSTRUCTURES IN NATURE

The timing of life arising on Earth was around 3.8 billion years ago (see Figure 2.3). Since then life has learned to reproduce, live under extreme toxic conditions, transform sunlight into oxygen, live underwater, fly, and eventually think, talk, and show emotions. As humans, we have constantly tried to supersede and control nature. However, so far we have spent little time trying to do things the way nature does. In other words, humans have not shown interest or been too proud to mimic or at least have not been motivated by nature's millions of years of evolution. This is somewhat strange, because nature is undoubtedly the most experienced and tested laboratory ever available to us and capable of making sophisticated materials, capturing energy, self-healing, and storing information with incredible efficiency. Interestingly, most of what nature does takes place at the nanoscale.

Perhaps the best-known biological example of such molecular machinery is the *ribosome*, which is a remarkable nanoscale assembler. The role of the ribosome is to act as a factory of proteins by combining amino acids together in a very specific order. A typical ribosome is located in an aqueous solution surrounded by thousands of solutes. The work is difficult. The ribosome must identify among 60 possible options a specific transfer RNA, a task that demands precision at the nanoscale (Figure 2.4). In this fashion, the ribosome machine can assemble proteins at a frequency of 20 Hz and an error rate of 10^{-3}. Thus, ribosomes can manufacture, with an accuracy greater than 99.9%, linear strings of aminoacids of great length, which then produce three-dimensional protein structures.

Another beautiful example of the role of nanostructures in the world's daily routine is photosynthesis. Every green plant and photosynthetic bacteria can take carbon dioxide, water, and sunlight (the fuel) and transform them into oxygen and sugar, a process that is accomplished with an amazing efficiency of around 95%. In the case of green plants, everything happens inside the so-called chloroplasts. These units, which are responsible for the green color of

(Billions of years)

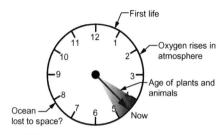

FIGURE 2.3

Earth's clock of life. (Adapted from Peter D. Ward and Donald Brownlee, The Life and Death of Planet Earth: How the New Science of Astrobiology Charts the Ultimate Fate of Our World, *Times Book Publisher, 2003.)*

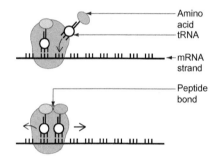

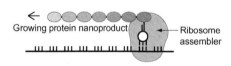

FIGURE 2.4

The ribosome assembler first selects from among 60 possible transfer RNAs (tRNA) one that allows a particular amino acid to bind into the key (Condon) of the messenger RNA (mRNA). The ribosome continues to assemble in this fashion until a particular protein is produced. (Adapted from P. Ball, Designing the Molecular World: Chemistry the Frontier, *Princeton University Press, 1994.)*

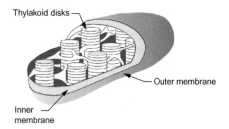

Thylakoid disks

Outer membrane

Inner membrane

FIGURE 2.5

Structure of a chloroplast. The thylakoid disks contain nanoscale pigments that convert light energy into chemical energy. (Adapted from Interagency Working Group on Nanoscience, Engineering and Technology, National Science and Technology Council Committee on Technology, "Nanotechnology: Shaping the World Atom by Atom," Sept.1999.)

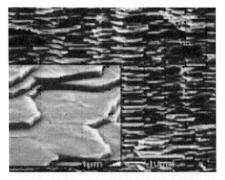

FIGURE 2.6 *An abalone shell consists of platelets of calcium carbonate forming a layered structure and separated by a protein sheet 20 nm thick. (Courtesy of Nan Yao, Princeton Materials Institute.)*

plants, exist by the millions in the leaves. Inside each chloroplast are many disks, called thylakoids, which are stacked on top of each other (see Figure 2.5). These thylakoids are fluid-filled sacs containing in their skin light-sensitive pigments. Hence, when sunlight reaches these chloroplasts, the light-sensitive pigments collect the photons and direct them to the photosynthetic reaction centers, which are also embedded in the thylakoid's skin. Each reaction center is composed of approximately 10,000 atoms and 200 pigments. Furthermore, at every reaction center, there are two highly light-sensitive pigments that execute the actual photon absorption.

Within the pigments, electrons become excited when they absorb the energy of the sun. This initiates a chain reaction whereby water is dissociated, oxygen and hydrogen ions are released, and carbon dioxide is turned into sugar. Although difficult to believe, the process of photosynthesis produces approximately 300 billion tons of sugar per year, a massive industrial operation. So why is it so difficult to reproduce the photosynthesis process? Because the 10,000 atoms in the reaction center are configured in such a way that excited electrons can jump from molecule to molecule without decaying to the old orbital. As a result, photosynthesis moves electrons to the outside of the thylakoid membrane, leaving positively charged ions inside. In other words, a battery powered by the sun is formed, generated by the membrane potential, which is then used to drive chemical reactions. Mimicking this configuration is still an immense challenge.

Another good example of natural nanostructures is the abalone, a marine mollusk, which is served in upscale restaurants. The mollusk builds the shell from traditional materials, namely calcium carbonate ($CaCO_3$) and a protein, forming a layered nanocomposite that is strong and resilient. Looking at the shell under an electron microscope at high magnifications, the shell looks like a brick wall, with calcium carbonate "bricks" separated by the protein "mortar," which acts as the "glue" (see Figure 2.6). The formation of the abalone shell starts with the secretion of proteins, which self-assemble into "room walls" with a distribution of negatively charged sites.. Inside each "room" there is seawater filled with calcium and carbonate ions, which are attracted to the walls and eventually form crystals of $CaCO_3$. The end result is a shell that exhibits twice the toughness of our best high-tech ceramics.

The reason behind these outstanding mechanical properties is the layered architecture composed of the protein material, with nanoscale thickness, and the ceramic calcium carbonate. Under applied stress, any crack in the shell can readily propagate through the ceramic material, but it will be deflected on contact with the protein material. Therefore, instead of a brittle material in which cracks can rapidly move without stopping, the shell of the abalone behaves as a ductile material.

One other quite talented mollusk is the mussel (see Figure 2.7). When these animals want to feed or reproduce, they move to tidal waves. Due to the inherent turbulence of ocean waves, mussels need to use a waterproof adhesive to anchor themselves to a solid surface, to overcome the forces created by the waves. Clearly, this is not an easy task, since we know how difficult it is to have an adhesive or glue that sticks under water. So, how do mussels do it?

A closer look at an anchored mussel reveals the presence of translucent filaments extending from the soft body toward the solid surface (see Figure 2.8). These filaments, called *byssus*, exhibit at their ends a foamy plaque, which is used to attach the mussel to the solid surface. The mussel starts the anchoring process by pressing the end of its foot on a solid surface to push the water away. Because it cannot repel all the water, the mussel attaches itself to a single point, then builds a loose structure, enclosing the water. In fact, it secretes a protein and forms a long vertical groove, which is under vacuum and acts as a mold for the following step. Next, the mussel sends out a series of proteins that self-assemble and harden, forming the plaque and the thread. Still in the vacuum space, the mussel then secretes liquid proteins that harden into a very strong adhesive, gluing the plaque to the solid surface. From a structure point of view, the filaments are quite interesting. The thread exhibits a concentration gradient of various proteins along its length, showing a rubbery-type behavior near the mussel and stiffer properties close by the plaque. The plaque itself is a solid foam that can easily expand and contract during tidal cycles without breaking.

As with mussels, geckos (a type of lizard) have an extraordinary ability to adhere to a surface. They can stick to walls or ceilings, even on a single toe (see Figure 2.9). This behavior is due to keratin hairs, 200 nm in diameter, that cover their feet. Each hair produces a very small force of 10^{-7} N (due to Van Der Waals interactions). However, half a million of these tiny hairs produce an extremely strong adhesive force, as high as 10 N/cm^2. The hairs can readily bend to conform to the topography of a surface, which allows geckos to stick to even flat surfaces such as glass. The connection breaks when the gecko shifts its foot enough to change the angle between the hairs and the surface. This discovery has launched

FIGURE 2.7

Mussel anchored to a surface by filaments called byssus. (Courtesy of Mieke C. van der Leeden, TU Delft.)

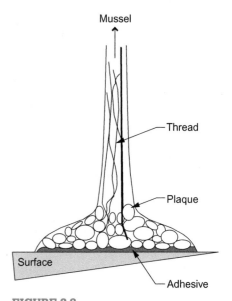

FIGURE 2.8

Schematic view of the byssus structure. The thread is rubbery-like near the mussel and stiffer close by the plaque. The plaque is a solid foam. (Adapted from Mieke C. van der Leeden, TU Delft.)

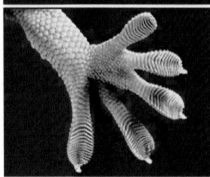

FIGURE 2.9

The base of geckos' feet is covered by half a million keratin hairs, 200 nm in diameter. (Daniel Heuclin, above and middle; Andrew Syred/Photo Researchers Inc, top.)

several research groups into developing tapes based on the gecko adhesive force. In one example, the tape is made by microfabrication of dense arrays of flexible plastic pillars, the geometry of which is optimized to ensure their collective adhesion (see Section 10.4).

Now imagine a material capable of catching an airplane traveling at cruise speed. Although at a very different scale, a spider web (see Figure 2.10) has such energy-absorbing properties, enabling the capture of flying insects at high speeds. It turns out that the secret behind the combined strength and flexibility of a spider web lies in the arrangement of nanocrystalline reinforcements embedded in a polymer matrix. A closer look at the spider web microstructure reveals the existence of strongly oriented nanocrystals, which adhere to the stretchy protein that composes their surrounding polymeric matrix. In other words, the spider web is a nanocomposite material composed of relatively stiff nanocrystals dispersed within a stretchy matrix that is now stronger and tougher.

The spider web starts as raw silk (a liquid protein) secreted by a gland. The protein is then pushed through channels at the spider's back end. What emerges is a water-insoluble, highly ordered fiber with outstanding mechanical properties. Although this process is not fully understood, it is believed that the raw silk transforms into a liquid crystal phase just before entering the channels, whereas the passage through the channels aligns the molecules, which are anisotropic in nature.

Following the properties exhibited by spider webs, several research groups have been trying to develop materials capable of emulating this nature's marvel. Recently, a group at MIT has focused on developing a commercial polyurethane elastomer (a rubbery substance) reinforced with nano-sized clay platelets. These molecular nanocomposites are likely to be good candidates for lightweight membranes and gas barriers due to the fact that the nano-sized clay platelets enhance the mechanical and thermal properties. In addition, materials based on the characteristics of spider webs are suitable for body-armor components, thin and strong packaging films, and biomedical devices.

If you've ever observed two rhinos fighting with each other, you were probably impressed by the properties exhibited by their horns. Despite the enormous forces involved when the animals clash, the horns remain intact without fracturing. The reason for this is the presence of *keratin*, a fibrous protein that is also present in hair and fingernails and that is self-assembled into a very specific structure.

The horn is a composite made of two types of keratin. One type is in the form of tubules that are densely packed (see Figures 2.11 and 2.12). These tubules, which range from 300 to 500 μm, comprise around 40 layers of cells. Surrounding these tubules is another type of keratin that is continuous and acts as a matrix. Because the matrix and the tubules are the same material, the interfacial strength is very good, which leads to a rigid material that is very tough to break. Following this concept, there is now great interest in emulating these horn structures at the nanoscale by designing materials with nanotubules embedded in a matrix. In other words, this approach is not very different from dispersing nanowires or nanotubes in a polymer matrix.

FIGURE 2.10
Spider web.

Another remarkable property of natural materials is their ability to self-repair and adapt to the environment. This is crucial for efficiency and, ultimately, survival of the fittest. By contrast synthetic materials rarely exhibit this property. Therefore, producing materials capable of repairing cracks, restoring functions, or self-producing is still a challenge that nanotechnology can help meet. Two interesting examples of natural materials that exhibit self-healing properties are bone and skin.

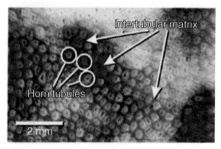

FIGURE 2.11
Microstructure of a white rhinoceros horn. The horn tubules, surrounded by the intertubular matrix, act as a composite material. (Courtesy of Tobin Hieronymus et al., Ohio University.)

Bone is essentially a composite material formed by an organic matrix (collagen) and reinforced by mineral particles (apatite). The collagen is in the form of fibrils of about 100 nm in diameter and 5–10 μm long that consist of an assembly of 300 nm long and around 1.5 nm thick. The apatite are platelike crystals with a thickness around 1.5–4.5 nm that fill the collagen fibrils. The collagen fibrils are separated by extrafibrillar material, which consists of extrafibrillar collagen and apatite that coats the fibrils (see Figure 2.13).

The remarkable result of this hierarchical configuration is that the bone can hold weight and toughness to absorb energy and not break into small fragments (see Figure 2.14). When subjected to a load, the whole bone deforms to different degrees. In fact, the tissue, fibrils, and mineral particles absorb successively lower levels of strain, in a ratio of 12:5:2. Although the mechanisms of deformation are not fully understood, it seems that the interface between the extrafibrillar matrix and the collagen fibrils breaks and reforms under load, providing a way for damage repair at the molecular scale. On the other hand, the hard nanoparticles of apatite are shielded from excessive loads, although they can achieve strains of 0.15–0.2%, which is twice the fracture strain of bulk apatite.

Bone is also capable of constant remodeling, eliminating damaged bone and replacing it with new material. Strain sensor

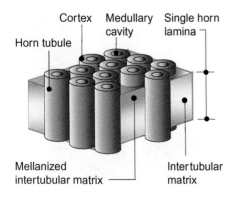

Cortex   Medullary   Single horn
cavity   lamina

Horn tubule

Mellanized
intertubular matrix

Intertubular
matrix

FIGURE 2.12

Schematic three-dimensional view of a white rhinoceros horn, where the various components of the composite material forming one horn lamina are indicated. Each horn tubule is composed of the cortex and medullary cavity, which is surrounded by melanized and non-melanized intertubular matrix. (Courtesy of Tobin Hieronymus et al. Ohio University.)

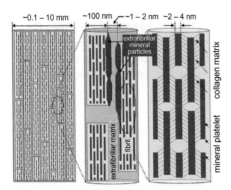

~0.1 – 10 mm   ~100 nm   ~1 – 2 nm   ~2 – 4 nm

extrafibrillar
mineral
particles

extrafibrillar matrix

fibril

collagen matrix

mineral platelet

Tissue strain   Fibril strain   Mineral strain

FIGURE 2.13

Hierarchical structure of bone. The yellow regions are the collagen fibrils and the red plates are the apatite crystals. The multidimensional configuration of the various bone components is crucial for efficient load transfer. Typically, the strain decreases from the tissue level to the mineral particle level by a factor of 6. (Courtesy of Himadri Gupta et al. Max Planck Institute of Colloids and Interfaces.)

cells (osteocytes) associated with the bone tissue first detect material that has been subjected to large deformations and damaged. Subsequently, these sensors send information to a group of cells that will remodel the old bone and form a new one. However, by far the most striking property of bone is its capacity to self-heal a fracture. Typically, a broken bone starts by experiencing a cut in the blood supply, and cells die. This process cleans up the area of dead material before reinitiating the blood supply and sending stem cells. These cells are a precursor to cells that eventually will produce cartilage, fibrous tissue, and new bone (Figure 2.14).

The skin is also an outstanding self-healing system. How many times through life have you had your wounds completely regenerated? The wound-healing process starts with the formation of a blood clot on the wound to avoid unwanted chemicals from penetrating healthy tissue. Subsequently, a network of capillaries forms, with diameters around 8– 10 μm, to supply nutrients through the blood and accomplish cell division and cellular growth.

Inspired by these ideas, researchers at the University of Illinois have recently developed self-healing materials that consist of a micro-encapsulated healing agent and a catalyst distributed throughout a composite matrix. If the material cracks, microcapsules will break and release a healing agent. The healing agent will react with the catalyst to repair the damage. This is similar to the case in which a cut in the skin triggers blood flow to promote healing.

To create these materials, the researchers begin by building a scaffold, which consists of polymeric ink in the form of continuous filaments in three dimensions. Once the scaffold has been produced, the ink filaments are embedded with a resin. After curing, the resin is heated, turning the ink into a liquid. The ink is then removed, leaving behind a substrate with a network of micro-channels. Finally, an epoxy coating is deposited on the substrate while the network of microchannels is filled with a liquid healing agent. Under these conditions, when a crack forms, it propagates until it finds one of the fluid-filled microchannels. At this point, the healing agent moves from the capillary into the crack, where it interacts with the catalyst particles and repairs the damage.

Last but not least, we ought to discuss the human brain. It is definitely the most incredible piece of machinery we have encountered so far. It allows us to think and even dream. We have tried to emulate or surpass the human brain with the development of computers. However, we now know that computers are nothing like our brains. Brains are essentially carbon; computers are mainly silicon. The brain's neurons have thousands of connections (see Figure 2.15); a computer's transistors have a few. The brain expends 10^{-15} Joules per operation, whereas a computer expends 10^{-7} Joules per operation.

But the major difference is that brains compute with molecules that "work" together and produce a result, instead of the "on" and "off" switches used by computers. And the reason for this distinct behavior is that carbon can establish a large variety of bonds with other atoms and thus can form many molecules of different shapes. On the other hand, silicon tends to be less flexible in bonding and thus cannot assume many shapes, as carbon does. The shape of molecules is crucial in identifying other species and deciding what to do, whether joining, separating, or passing messages across. In other words, each neuron is a computer controlled by nanoscale components.

2.3 NANOMATERIALS IN ART AND CULTURAL HERITAGE

Examples Found in Art

In the opening of this chapter, we noted that the ruby-red color of many stained-glass windows from the Medieval era was a consequence of embedded nanoscale metallic particles within the glass (see Figure 2.16). These rich colors in stained glass, like the metallic sheens associated with naturally embedded nanoparticles in many ceramics, were appreciated and highly valued by artisans, patrons, and laymen alike. Stained-glass artisans sometimes treasured small vials of materials that we know were metallic oxides, obtained from special mines and handed down within their families with careful instructions on how to work with them. As we will see later in this chapter and in Section 7.5, when the size of material particles is reduced to the nanoscale, optical properties—particularly color—can be dramatically affected. In such cases, the wavelength of light is very close to the size of the particles themselves, which causes the way that color is reflected, scattered, or absorbed to be dependent on the size and shape of the nanoparticles themselves.

There was no scientific understanding of these phenomena at the time, nor were there deliberate attempts to produce what we now know as nanomaterials. Early knowledge relied on craft-based trial and error to achieve effects. It was known, for example, that the introduction of certain materials from specific mines and according to empirically understood methods did indeed produce rich colors

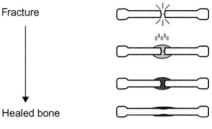

Fracture

Healed bone

FIGURE 2.14

Different phases in bone fracture healing. Upon fracture, the broken bone is initially restored by a blood clot. After removal of dead bone fragments, the healing process sets in with the formation of primitive bone tissue ordered by precursor cells. Finally, the cells that produce cartilage and fibrous tissue take over and produce new bone.

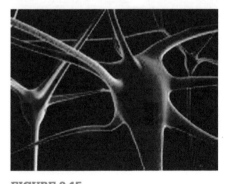

FIGURE 2.15

Network of neurons. It is estimated that a human brain contains around 100 billion neurons, each one with thousands of connections.

FIGURE 2.16
Window from Chartres Cathedral. The intense colors of many Medieval stained-glass windows resulted from nanosized metal oxide particles added to the glass during the fusion process.

in glass. Metallic compounds (which we now know were cobalt oxide) from mines in Bohemia were added during the fusion process to achieve the deep blues so widely admired in the glass at Chartres Cathedral. Likewise, the intense ruby-colored stained glass is now known to have been obtained from the dispersion of solid silica particles or of nanosized gold particles within the glass. We must keep in mind, however, that not all interesting color phenomena are a result of embedded nanomaterials.

One of the most interesting specimens is the Roman-era Lycurgus cup. Roman glassware has often been used to characterize the material cultural achievements of the late Roman Empire. Glass-producing techniques were highly developed, and workmanship was superb. In the Lycurgus cup, now housed in the British Museum, the 324 AD victory of Constantine over Licinius in Thrace was represented through the death of an enemy of Dionysius, Lycurgus, who is shown being overcome by vines. The most remarkable characteristic of this goblet is that under normal external lighting conditions the glass appears green, but when lighted from within, it assumes a strong red color (see Figure 2.17).

The Lycurgus cup has now assumed an almost iconic status in the nanomaterial field as an early example of the *surface plasmon* phenomenon, in which waves of electrons move along the surface of metal particles when light is incident onto them (see Section 7.5 for a detailed description of the surface plasmon phenomenon). Analyses have demonstrated that the glass in the Lycurgus cup contains rather small amounts of gold powder embedded within it (on the order of 40 parts per million). These tiny metallic particles suspended within the glass matrix have diameters comparable to the wavelengths of visible light. As a consequence, a form of plasmonic excitation (an oscillation of the free electrons at the surface of a metal particle at a certain frequency) can occur. Light reflections are enhanced as the waves are highly absorbed and scattered, reducing transmission. This absorption has an orientational dependence. Interestingly, other colors aside from the red and green seen in the Lycurgus cup could be achieved by altering metal particle sizes. In the cup, however, color properties depend primarily on reflection when the light is external to the cup and on absorption and transmission when the light source is internal.

Many Medieval and Renaissance ceramics have surfaces characterized by a remarkable iridescent metallic shine (see Figure 2.18). This form of ceramic decoration, a type of luster, appeared in the

Middle East in the ninth century or before and subsequently spread through Egypt, Spain, and other countries. A particularly fine period of development occurred in Spain with Hispano-Moresque ware, a glazed ceramic made by Moorish potters largely at Málaga in the 15th century and later at Manises near Valencia in the 16th century. To produce this type of lusterware, a glaze was first applied over a design and the piece fired to produce a thin, hard coating. Glazes were based on dry powdered minerals or oxides, which commonly included tin and copper. After the first firing, the luster coating, consisting of metallic pigments (normally copper or silver compounds) mixed with clays, was brushed on over the glaze. Then the piece was fired again but at a lower temperature and within a reducing atmosphere (a condition whereby a reducing agent chemically causes a change in a material with metallic compounds to a metallic state by removing nonmetallic constituents as it is itself oxidized by donating free electrons). Afterward the piece was cleaned and polished to reveal the resulting metallic sheen.

Later examples include the "tin-glazed" pottery of 15th and 16th century Italy and the "copper glazed" lusterware porcelains of Wedgwood in early 19th century England. Several studies of medieval lusterware via transmission electron microscopy (TEM) have been undertaken to understand the composition and microstructure of luster. Results have clearly indicated that various luster characteristics can be described in terms of the presence of different levels of silver or copper nanoparticles within the glassy matrix. The associated surface plasmon effects (described previously) cause the appealing metallic sheen to develop. Again, though the artisans producing lusterware lacked an understanding of the chemical processes that achieved the optical effects and were unaware that their empirical processes led to the creation of nanoparticles, the craft-based development of the requisite knowledge was remarkable.

Similarly intriguing was the development of the beautiful blue paint found in the murals and pottery of the ancient Mayan world (see Figures 2.19 and 2.20). The Mayan blue has long been admired for its marvelous color qualities as well as its inherent resistance to deterioration and wear over long periods of time. Various natural alkalis, oxidants, mineral acids, and other agents seem to have little effect on the blue paint of the Maya. Unlike the blues of Europe and Asia, largely based on ground lapis lazuli, the origin of the Mayan blue was likely a dye known as *anil*, which is found in pre-Columbian textiles and is obtained from a local plant to produce indigo. Normal indigo, however, is not acid resistant and its use

FIGURE 2.17 *The Lycurgus cup looks green when light shines on it but red when a light shines inside it. The cup contains gold nanoparticles. (Courtesy of the British Museum.)*

FIGURE 2.18
Medieval lusterware, circa 16th century, Manises, Spain. The glaze was made by firing metal oxides.

FIGURE 2.19
Detail of Feathered Serpent *from a wall painting at Cacaxtla, Mexico. (Courtesy of Barbara Fash, Peabody Museum, Harvard University.)*

would not in itself account for the many interesting properties of Mayan blue. A predominant presence of palygorskite clays in the paint that were found to have a plate- or needle-shaped internal structure have been identified.

Within these clays very small amounts of impurities (iron -Fe, manganese -Mn, chromium -Cr) were also found. The host clays appear to have originated in a mine near Merida (and were subsequently heated to high temperatures), whereas the embedded impurities came from the anil plant used to make the indigo. Most of these impurities were in metallic states. Some oxidized elements were near the surface. Detailed studies have been made of these materials (see Figures 2.21 and 2.22). Interestingly, even minor amounts of nanosized particles in the impurities can be expected to have significant effects on optical properties due to the surface plasmon excitation effects.

The particular blue color thus results from the exact type of absorption curve present, which is in turn influenced by the exact size, shape, and dispersion of the nanoparticles. These effects, along with those from the oxide metals, appear to help account for the strong blue color. Meanwhile, it is the *intercalation* (insertion of foreign atoms into the crystalline lattice of another material) of indigo molecules within the clay that may account for the corrosion resistance exhibited by the paint. Thus, the interesting properties of the final product result from a complex sequence of circumstances involving naturally occurring nanomaterials. Like the glass and ceramics examples described earlier, the process developed by Mayan artisans did not involve an understanding of chemistry or nanomaterials as we know them today but rather a persistent search in visual effects in the service of fine art.

Conservation

Conservation and restoration of works of art and other forms of cultural heritage have been a constantly evolving pursuit in which nano-based techniques play increasingly valuable roles (see Figures 2.23 and 2.24). A great number of factors can play a role in the degradation of artworks. For instance, microbial growth can have a range of detrimental effects on various media. Significant damages can be inflicted on both paintings and sculptures by the many pollutants in the atmosphere. The problem of nitric oxides in polluted atmospheres slowly degrading the surfaces of marble statues and marble buildings from ancient times is well known. In wood artifacts, acids can cause degradation of the cellulose structure present.

FIGURE 2.20
Mayan wall painting from Cacaxtla, Mexico. The intense blue results from an amorphous silicate substrate with embedded metal nanoparticles and oxide nanoparticles on the surface. (Courtesy of Barbara Fash, Peabody Museum, Harvard University.)

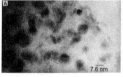

FIGURE 2.21
(Left) Metallic and oxide particles on SiO_2 support and (right) an atomic resolution image of a particle showing its cubo-octahedral shape. (Courtesy of J. Yacaman, et al.)

Paintings that are exposed to air and elements can become covered with foreign particles that change visual appearances and begin to act mechanically on the artifact. The paint itself as well as the substrate that has been painted on can begin to crack. There can be a loss of cohesion between paint layers on various media. Particles can flake off.

Among the greatest of our cultural treasures are Medieval wall paintings done in the fresco technique. They adorn many Medieval buildings, particularly in Italy. In the *buon fresco* method, pigments mixed with water were applied to freshly placed and still-wet lime mortar after initial carbonation, thus embedding the pigments well into mortar (or, more precisely, into the crystalline structure of the newly formed calcium carbonate). Paintings not only had marvelous color and visual qualities, they were quite stable as well. With time, however, salts can begin migrating through pore structures in the mortar due to dampness and other reasons, and salt crystallization can occur. The consequence is that severe degradation can occur in the form of flaking of paint layers or powdering of colors. Deeper damage can occur to the overall porous structures as well. Damaging salts or particulates can also come from wind-blown sources.

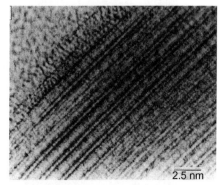

FIGURE 2.22 *Image of a palygorskite crystal showing a superlattice period of about 1.4 nm. (Courtesy of J. Yacaman, et al.)*

Conservation work was carried out in the 1970s using the Ferroni-Dini method for cleaning from sulfatation and consolidation of frescoes. Pre- and post-restoration images under raking light.

According to undocumented tradition, the face of the saint is a self-portrait of the artist, Beato Angelico. Pre- and post-restoration under glazing light.

FIGURE 2.23

Fra Angelico, Saint Dominic in Adoration Before the Crucifix, *Convento di San Marco, Florence, Italy, c. 1438. (Conservation images courtesy of Piero Baglioni, Center for Colloid and Interface Science, CGSI, University of Florence.)*

Prerestoration image.

Post-restoration image; nanostructured systems were used for conservation.

FIGURE 2.24 *Santi di Tito,* The Musician Angels, *15th century. (Courtesy of Piero Baglioni, Center for Colloid and Interface Science, CGSI, University of Florence.)*

These and other sources and types of damage have long plagued curators and owners of these delicate frescoes. Many early attempts to conserve artworks ultimately did more damage than good. At one time, for example, varnishes were used over paintings in a misguided attempt to protect them. These now-darkened varnishes are a major problem in restoration efforts today. In more recent years, less problematic acrylic polymers such as ethyl methacrylate (paraloid) as well as others have been used to consolidate wall paintings. Even the best of current techniques, however, can still result in some type of change, not be very effective, or simply be short lived.

Methods of conservation have come a long way since the use of varnishes, but the many inherent challenges have not been entirely overcome and the process remains painstaking. Current restoration and conservation standards demand that any conservation or restoration technique be reversible and should in no way alter the fundamental chemical or physical composition of the original artifact or otherwise affect its properties. New techniques were developed as a consequence of the terrible floods in Florence and Venice in 1966 that caused widespread damage to the art and architecture of these cities. One important method developed shortly thereafter, the barium or Ferroni-Dini technique, used chemical methods for consolidation and restoration of wall paintings suffering from salt crystallization. The method seeks to reconstitute the calcium carbonate layer. Surface applications of aluminum and barium hydroxides are applied so that harmful calcium sulfates are ultimately converted into inert barium sulfates. These, in turn, ultimately act to consolidate the substrate and to create a fresh binder. This technique has been widely used throughout the world and has led to the development of other sophisticated techniques as well.

Researchers at the Center for Colloid and Interface Science (CSGI) at the University of Florence have extended the Ferroni-Dini method using nanomaterial technologies. Calcium hydroxide would be especially desirable for conservation use with carbonate-based materials because of compatibility. A saturated calcium hydroxide water solution, known as *lime water*, has been used, but its low solubility has presented problems. The use of calcium hydroxide suspensions has been explored, but these are not stable enough and create surface effects. Researchers at the Center explored the use of dispersions of nanosized calcium hydroxide in nonaqueous solvents. Nanoparticles of this type can be obtained from a simple homogeneous phase reaction in hot water. Going to smaller sizes

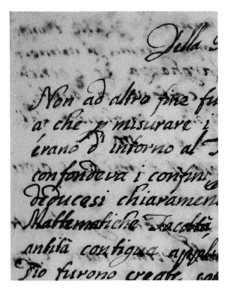

FIGURE 2.25
Acid paper. (Courtesy of Piero Baglioni.)

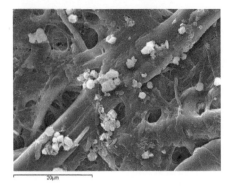

FIGURE 2.26
Nanoparticles of calcium hydroxide on paper—deacidification. (Courtesy of Piero Baglioni.)

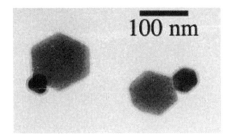

FIGURE 2.27
Calcium hydroxide nanoparticle.

changed the physical and chemical features of the calcium hydroxide particles. Nanoparticle sizes ranged from 10 nm to 200 nm and were found to be able to penetrate within pore structures of wall paintings and limestone and without leaving surface effects. The particles increased cohesion to the painted layers. After a short period of time, the calcium hydroxide particles were found to react with carbon dioxide in the air and create a greatly improved binder in the host material, thus consolidating it. The process essentially replaces calcium hydroxide lost during degradation. The same methods have been applied to many fresco paintings, including at the Maya archeological site of Calakmul. Interestingly, this consolidation method can be applied to other materials, including ancient brick mortar.

Using approaches similar to that just described, nanoparticles can also be used for conserving or restoring old textiles, paper, and wooden objects (see Figure 2.25). It is well known that paper made from wood pulp in the last few centuries is extremely prone to deterioration. With rising demand for paper, older medieval techniques of producing paper from rags (which yield quite good and long-lasting paper) were replaced with approaches that extracted cellulose from wood. Acid sizings, such as alum, were often used on the papers as well. The unfortunate combination results in acid-catalyzed degradation of the cellulose in papers that is hastened by the high reactivity of cellulose obtained from wood pulp. As a consequence, the failure of papers has become all too familiar in our archives and libraries—and even to the amateur old book collector. Deacidification processes can be used to stabilize documents and increase the life of these papers, but these methods are expensive and slow. Nanoparticle-based paper treatments have been suggested for use to achieve more efficient and long-lasting deacidification. Smaller particles allow easier penetration and more complete dispersion (see Figures 2.26 and 2.27). The approach generally suggests an improved way of dealing with one of our most delicate preservation problems.

The work on cellulose structures described previously has led to other applications. In the 17th century the royal battle galleon *Vasa* was built on the order of King Gustavus Adolphus of Sweden (see Figure 2.28). The ship was built of large oak, and its two gun decks held 64 bronze cannons. On its maiden voyage on August 10, 1628, the *Vasa* fired a farewell. A sudden squall caused it to list, and water poured through still open gun ports. The vessel capsized and sank with great loss of life. Amazingly, the ship's largely intact hull was salvaged in 1961. To prevent the hull from drying out, shrinking, and decaying, preservationists immediately treated it with polyetheylene glycol (PEG) by intermittent spraying and slow drying. It was moved to the Vasa Museum in Stockholm, where alarming rates of acidity increases in the wood were observed and again threatened the hull by acid wood hydrolysis. Sulfuric acids were proving especially harmful. The development of sulfurs was traced back to metabolic actions of bacteria in the harbor water and was subsequently oxidized by the iron released from long-corroded bolts as well as from more recent ones put in during salvaging. Preservation efforts focused on removing iron and sulfur compounds. Neutralization treatments using alkali solutions helped in only outer wood layers and can potentially cause cellulose degradation itself.

A new method of neutralizing the acids by the use of nanoparticles has recently been explored by the group from the University of Florence, mentioned previously. The immediate preservation focus was to slow the production of acids inside the wood and, if possible, remove the iron (or render it inactive) and sulfur. Prior treatments and conditions further complicated the problem. The Florence group focused on deacidification. Wood samples were obtained from the *Vasa* and then treated with an alkaline solution was used to remove the PEG, which would prevent nanoparticles from penetrating the wood and provide neutralization. Nanoparticles of calcium hydroxide were synthesized and samples soaked in an alkaline nanoparticle dispersion (see Figure 2.29). Progress of the penetration of the nanoparticles through the wood was tracked with the aid of scanning electron microscopes (SEM analysis—see Chapter 7). The nanoparticles were found to adhere to wall fibers and the dispersion medium volatized, leaving an alkaline presence. The application led to a marked decrease in the pH levels present (reduced acidity). Other benefits were noted as well.

Work with nanotechnology has only recently been emerging in the conservation area, but initial results look very promising. Later chapters will look at other nano-based techniques that may prove beneficial in conserving our cultural heritage. For instance, the airborne pollutants from traffic or smog, which are known to attack the surface of sculptures and architectural monuments made of marble and other stones that form part of the cultural experience of our finest urban environments, might be combated with the self-cleaning actions of certain kinds of nanomaterials described in Chapter 10. Indeed, the potential value of inert self-cleaning surface treatments would be literally enormous, but considerable gaps

FIGURE 2.28

The Vasa *sank in 1628 and was salvaged in 1961. Originally treated with PEG, the hull was still threatened by acidification.*

FIGURE 2.29

Experiments have been conducted with embedding nanoparticles of calcium hydroxide to stabilize wood. Ca(OH)$_2$ nanoparticles on the wall fibers of wood are shown in this scanning electron microscopy (SEM) image. (Courtesy of Piero Baglioni, Center for Colloid and Interface Science, CGSI, University of Florence.)

must still be bridged before we can safely benefit from these new nanotechnologies.

FURTHER READING

The History and Evolution of Materials

C. Singer, E. J. Holmyard, A. R. Hall, T. I. Williams, and G. Hollister-Short (eds.), A history of technology (21 volumes), Oxford University Press, 1954–2001, ISSN 0307-5451. (*A compilation of essays on aspects of technology, including materials.*)

J. Delmonte, Origins of materials and processes, Technomic Publishing Co., 1985, ISBN 87762-420-8. (*A compendium of information about materials in engineering, documenting the history.*)

R. F. Tylecoate, A history of metallurgy, 2nd ed., The Institute of Materials, 1992, ISBN 0-904357-066. (*A total-immersion course in the history of the extraction and use of metals from 6000 BC to 1976 AD, told by an author with forensic talent and love of detail.*)

www.tangram.co.uk.TL-Polymer_Plastics_Timeline.html/. (*A Website devoted to the long and full history of plastics.*)

Nanomaterials and Nanostructures in Nature

Peter D. Ward and Donald Brownlee, The life and death of planet earth: How the new science of astrobiology charts the ultimate fate of our world, Times Book Publisher, 2003.

R. Freitas Jr., in Nanomedicine, Volume I: Basic Capabilities, Landes Bioscience, 1999.

P. Ball, Designing the molecular world: Chemistry at the frontier, Princeton University Press, 1994.

Janine Benyus, Biomimicry: Innovation inspired by nature, Perennial (HarperCollins) Publishers, 1998.

Himadri S. Gupta, Jong Seto, Wolfgang Wagermaier, Paul Zaslansky, Peter Boesecke, and Peter Fratzi, Cooperative deformation of mineral and collagen in bone at the nanoscale, PNAS, 103, pp. 17741–17746, 2006.

Tobin L. Hieronymus, Lawrence M. Witmer, and Ryan C. Ridgely, Structure of white rhinoceros (*Ceratotherium simum*) horn investigated by X-ray computed tomography and histology with implications for growth and external form, Journal of Morphology, 267:1172–1176 2006.

Yao et al., Organic inorganic interfaces and spiral growth in nacre, Journal of the Royal Society Interface, 2008 (in press).

Nanomaterials in Art and Cultural Heritage

P. Baglioni and Rodorico Giorgi, Soft and hard nanomaterials for restoration and conservation of cultural heritage, first published as an advance article on the Web, Feb. 10, 2006.

DOI: 10.1039/b516442g, RSG, Soft Matter, 2, 293–303, The Royal Society of Chemistry, 2006.

Piero Baglioni, Rodorico Giorgio, and Ching-chih Chen, Nanoscience for the conservation of cultural heritage and the emergence of a powerful digital image collection, Proceedings of the 8th Russian Conference on Digital Libraries, RCDL'2006, Suzdal, Russia, 2006.

Rodorico Giorgi, David Chelazzi, and Piero Baglioni, Nanoparticles of calcium hydroxide for wood conservation: The deacidification of the *Vasa* warship, 10.1021/la0506731 CCC, American Chemical Society, 2005.

M. Jose-Yacaman, Luis Rendon, J. Arenas, and Mari Carmen Serra Puche, May Blue Paint: An Ancient Nanostructured Material, Science, 273: 223–225, 1996.

CHAPTER 3

The Design Context

3.1 MATERIALS IN DESIGN

In this chapter we begin to look more directly at the use of new nanomaterials and nanotechnologies in design. There are already many applications of nanomaterials in design, and more are expected soon (see Figure 3.1). We look at these topics from the perspective of designers and within the larger cultural-socioeconomic context in which designers operate.

Making effective design application use of new scientific findings that seemingly appear every day within the nanomaterial field is not as easy as might first appear. History is replete with examples of seemingly fabulous scientific discoveries that surely seemed to the discoverers to have huge potential applications in many areas, only to have them lie idle for many years or not be developed at all. Reams of studies in fields such as technology transfer have looked into why this is so. Reasons vary widely but are rarely of any surprise to individuals working in a real-world development context. Design ideas may often be simply naïve, or design objectives can be confused or unclear. Target markets or user audiences may be undefined or unclear, nonexistent, or simply not large enough to be of commercial interest. Other existing products may already accomplish similar desired ends better or in a more cost-effective way than the proposed product. Proposed products may never have been benchmarked against existing products. There might not be adequate test results to convince anyone of the efficacy of the product. Actual manufacturing processes for converting a science-based finding into implemental technology suitable for use in a commercial environment may either not be actually feasible or be cost-prohibitive (or not yet explored enough to know). There may be legal or institutional barriers that would prevent

FIGURE 3.1

The Kurakuen house in Nishinomya City, Japan, designed by Akira Sakamoto Architect and Associates, uses a photocatalytic self-cleaning paint—one of the many architectural products based on nanomaterials. (Courtesy of Japan Architect.)

active consideration of a new product or cause an interested developer to think twice before proceeding. There can be user resistance from sources that should have perhaps been anticipated but perhaps were not (e.g., environmental health hazards in using new materials) or come from sources that simply could not have been easily anticipated *a priori*. The list can go on and on.

In thinking about how we might use nanomaterials in design, it is useful to step back considerably and not define the issue simply as a technology-transfer problem—a self-limiting approach—but rather to think about it first in more fundamental terms. What is the role of materials in design? How do material properties influence the shape or form of objects and environments? What are we trying to do when we are using a specific material? On what basis do we compare nanomaterials to other high-performance materials? Do we really expect new products to be made of just nanomaterials, or if not, what specific role do we expect them to play? In what kind of product or building system might they be best used? What can we hope to accomplish?

The first of these questions is by no means new. Questions surrounding the way an artifact or environment has been conceived, how it was made, and the materials of which it has been made have been a particular preoccupation of designers, engineers, and builders for ages. An understanding of the potential benefits and limitations of various materials is clearly evident in early works of art, architecture, and engineering (see Figure 3.2). Examples abound. Medieval builders, for example, are often said to have clearly understood the properties of stone, and they used this knowledge to help shape the arches and vaults of history's great Romanesque and Gothic cathedrals. But what is exactly meant here by this kind of reference? Is it that the knowledge of certain properties of stone—that it is quite strong when carrying forces that cause compression within it and relatively weak when subjected to forces that cause tension to develop—somehow led directly to the creation of these complex cathedrals as we now see them? Clearly this direct line of thinking—a form of technical determinism—is highly suspect in this example, to say the very least. We *do* know that the use of arches and vaults, which we now know to naturally carry internal forces by a compression action, has been known since antiquity to be a good way of spanning large spaces with stone and would sensibly have been used by Medieval builders, and that this knowledge was clearly fundamental in the development of history's great cathedrals, but it was obviously only *one* of many contributing factors in a landscape of complex reasons that range from the symbolic to the societal and cultural. We thus need to keep in mind that the nature of our world of designed objects and environments is not dictated by a consideration of the technical properties of a material alone, no matter how fascinating they might be; but it is equally important to acknowledge their fundamental role—we know that the introduction of new materials with improved technical properties has also led to innovative new designs (see Figure 3.3).

FIGURE 3.2
The evolution of arches that act primarily in compression only was related to the inherent material properties of masonry, which can carry large stresses in compression but little in tension.

From the point of view of this book, the best approach to understanding the use of materials in design remains through an examination of the benefits and limitations associated with the specific properties of materials. For this initial discussion, material attributes can be very broadly thought of in terms of their *technical properties* that stem from the intrinsic characteristics of the material itself (its density and its mechanical, thermal, optical, and chemical properties); their *perceptual qualities* that stem from our senses (sight, touch, hearing, taste, smell); and those *culturally dependent qualities* that fundamentally stem from the way our society or culture views materials.

The intrinsic characteristics of materials are dependent primarily on the fundamental atomic structure of the materials and are discussed extensively in Chapter 4. Typical technical properties include failure strengths, elastic moduli values that relate deformations to stress levels, electrical conductivities, thermal conductivities, and a host of

FIGURE 3.3
The introduction of materials such as steel that can carry bending stresses involving both tension and compressive stresses has allowed designers to explore new shapes.

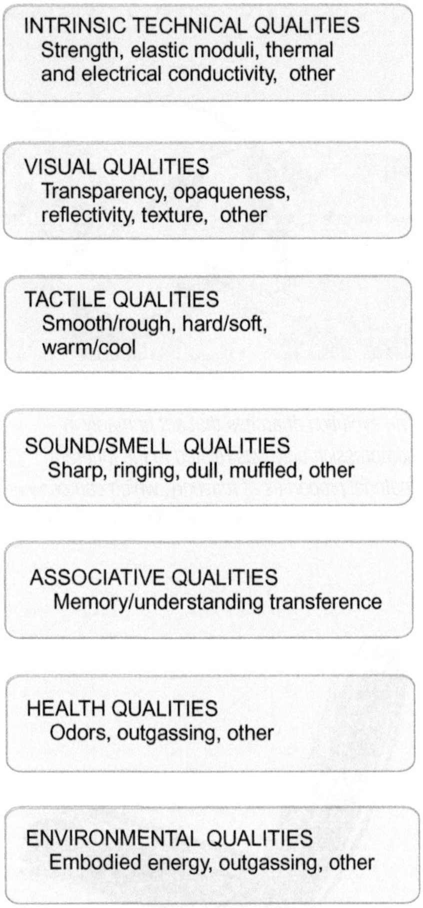

INTRINSIC TECHNICAL QUALITIES
Strength, elastic moduli, thermal
and electrical conductivity, other

VISUAL QUALITIES
Transparency, opaqueness,
reflectivity, texture, other

TACTILE QUALITIES
Smooth/rough, hard/soft,
warm/cool

SOUND/SMELL QUALITIES
Sharp, ringing, dull, muffled, other

ASSOCIATIVE QUALITIES
Memory/understanding transference

HEALTH QUALITIES
Odors, outgassing, other

ENVIRONMENTAL QUALITIES
Embodied energy, outgassing, other

FIGURE 3.4
Primary material characteristics.

other measures related to the mechanical, optical, thermal, and chemical qualities of materials (see Figure 3.4). Clearly, these kinds of properties are of fundamental importance to an engineering perspective on the use of materials in the context of designing products or buildings, and we can expect that work in the nanomaterials field can lead to dramatic improvements in these kinds of properties. At the moment, the important point here is simply that they intrinsically result from a material's internal structure and are not dependent on any kind of societal or cultural view of the material.

The perceptual qualities of a material relate to the way humans perceive them in terms of our basic senses. *Visual qualities* stem from a combination of specific characteristics such as *transparency, translucency, opaqueness, reflectivity,* and the *texture* of the surface (which in turn produces *glossy, matte,* or other appearances). *Tactile qualities* related to the sense of touch stem the texture of the surface—whether it is *rough or smooth,* its relative *hardness* or *softness,* and the feeling of *warmth* or *coldness* experienced. The qualities of materials that relate to our sense of hearing have to do with the kind of sounds—*dull, sharp, ringing, muffled, low or high pitch*—produced when the material is set in a vibrational mode, including by simply striking it. The sound of a metal object striking a sheet of lead is quite different than when it hits a piece of glass. In some design situations, the senses of smell or taste can be important as well. Certainly these qualities are directly related to the intrinsic properties and structure of a material. Polycrystalline materials are normally opaque or translucent because of the way light impinging on them is scattered. The sense of warmth or coldness depends on the way heat is conducted away at the point of touch, which in turn depends on both the thermal conductivity and specific heat of the material. We look at these kinds of relationships in more detail in Chapters 4 and 5. For the moment, what is important to note is that there is a basis for these qualities in the intrinsic properties of the material.

But here we should also note that the way we ultimately perceive these same qualities in a neurological sense is also dependent on our own receptive mechanisms. The spectrum of light that is visible to humans or the sound wavelengths we can hear, for example, are quite different than for other animals. What we actually perceive can be quite a complex topic. We can still note, however, that since these qualities that relate to the senses remain in some way linked to specific mechanical, thermal, optical, or chemical properties, enhancing or otherwise modifying these qualities through the use of nanomaterials is entirely possible. Thus, the sense of warmth on touching a surface might be enhanced using nanomaterials to modify the thermal conductivity and/or specific heat of the surface material.

The associative qualities of a material are placed on it by both its use context and its users (see Figure 3.5). As such, these characteristics are coupled with the *context* in which a material is found or used, and consequently, are ultimately dependent on the view of both individuals and society toward these materials, which are in turn culturally dependent. Specific examples are abundant and obvious here. For one, diamond is a material possessing many highly interesting intrinsic and technically oriented properties that make it extremely attractive for use in many circumstances in which extreme material hardness—an intrinsic property—is desired, as is the case in many industrial processes. At the same time, diamonds clearly possess many *associative* qualities placed on them by our culture and society that are fundamentally symbolic but nonetheless of real importance.

The role of diamond jewelry in our society, for example, is fascinating. Yet even here, intrinsic properties of diamonds remain relevant. Humans are fascinated with the stunning color characteristics that we see when we peer into the material itself. This effect is in turn directly related to the material's high index of refraction. As another example, we also see associative qualities present in the many synthetic materials that were historically widely introduced and used as substitutes for more expensive materials (for example, various synthetic materials introduced for flooring, such as linoleum, or for kitchen countertops). In some instances, these materials often became viewed as exactly that—cheap substitutes—no matter whether their physical properties are superior to those of the original or not. (Many technical properties of some new synthetics are indeed superior to common traditional materials.) In other cases, such as with the introduction and use of Bakelite in early radio housings, there were few negative perceptions of this type because

An early concrete building: Le Vesinet, 1864, by Francois Coignet.

A 19th century cast-iron facade in New York City.

A 21st century floor synthetic material: vinyl flooring designed to look like stone.

FIGURE 3.5

Many materials have strong associative qualities, such as diamonds or gold with wealth. In the 19th century, early buildings made of Portland concrete or cast iron, then new materials, were deliberately designed to look like stone buildings, since stone was the most prestigious building material of the era. Only later did architects such as Le Corbusier and others exploit the inherent technical properties of concrete and other new materials to create new and exciting forms. Some modern materials are still intended to resemble stone.

FIGURE 3.6
The introduction of Bakelite, with its many positive qualities that allowed inventive product forms, became a symbol of forward progress and an exciting future.

the synthetic material came almost immediately to be associated with exciting new technological advances and wonderful new emerging lifestyles that allowed broad connectivity to the world at large (see Figure 3.6). Countless other examples exist. Understanding and dealing with these associative qualities is an important part of a designer's role in dealing with objects used by the population at large.

Societal *health* concerns with the use of materials were not a terribly important consideration in describing or selecting a material until fairly recent years, when the impact of material selection on human health became better understood. We now know that the outgassing of materials within closed environments, for example, can produce a form of indoor air pollution that can cause occupant sickness. Nanomaterials are seen by some as posing potential health threats but by most others as a way of reducing them (see Chapter 11). Issues of sustainability and general *environmental* concerns span the spectrum from notions of how much embodied energy is used to produce a material all the way through its susceptibility to recycling are, again, responsive to societal concerns and mandates. *Economic* concerns, including the literal cost of a material, are ever-present. Costs of nanomaterials are known to be extremely high, which in turn suggests that they be selectively used. Both initial and long-term or life-cycle costs are important, but again these are values placed on the materials by their use in our society and our culture.

There are also many *forces for change* that are strong drivers for new product forms. Figure 3.7 illustrates some of these driving forces at a very broad level. Many are *opportunity based* and evolve from either changing population characteristics or the availability of new technologies that allow needs to be met that could not be met before. Indeed, in many instances nanotechnologies are expected to be disruptive technologies that radically alter product forms. In other cases, nanomaterials and nanotechnologies will afford continuous improvement opportunities. Other broad forces for change can be characterized as *concern driven*. These include a wide array of issues relating to the environment, national security, and so forth. Many new products have been spawned in response to concerns of this type. In both opportunity- and concern-driven forces, a remarkable range of new scientific understandings contribute to the evolution of new or significantly adapted products.

Given that externally based qualities are dependent on societal and cultural values, it is likewise to be understood that there is no

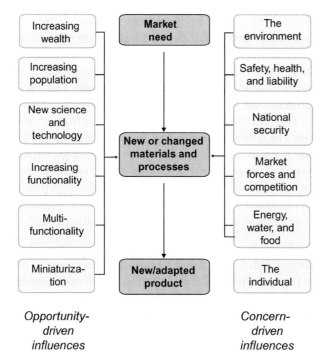

FIGURE 3.7

Forces for change. Many societal, economic, and technological drivers lead to the development of innovative new products.

uniform agreement on the relative importance of these different qualities, as is evidenced by the intense societal debates that still occur over recycling. Though reuse of materials has been accepted as a value by broad segments of our society, anyone can still go into any store and buy products that are fundamentally designed to be cast away after use, made from materials deliberately chosen with disposability rather than recycling in mind. These societal debates are ultimately played out in very specific material choices. It is the outcome of these debates over broad societal and cultural values, however, that provides impetus for some of the greatest forces for change in the development and selection of materials.

Design Focus

In thinking in more detail about how to use nanomaterials in a design context, a first consideration is simply to define *what* is being designed. In general, this book broadly addresses common products for everyday use as well as in architecture and related works. In this section, we briefly review the characteristics of each field separately. Ultimately, however, we will see that there are marked similarities in the basic design processes and issues present in each of these fields, despite the fact that there are also dramatic differences as well. In many ways, differences are more easily characterized than similarities and have to do with factors such as the markets or target audiences addressed, the product's physical size and scale, functional and assembly complexities, the number of units produced, the nature of the industry that produces the final objects, and the kind of design, engineering, and manufacturing or construction teams that bring a design from an idea to final fruition. In terms of similarities, both fields share similar issues about the need to identify product or project design objectives as well as ways to translate these needs into design requirements, generate one or more design responses (including the selection of appropriate materials) that meet requirements, resolve conflicting design objectives, and ultimately make or construct the product or project.

Both products and buildings can also be generally thought of in terms of their constituent environments, systems, and assemblies. It is here where material selection issues occur. Environments may be an integral part of a design, such as spaces within a building, or they can define the external context within which a product operates. Systems and subsystems provide specific functions (e.g., heat sources, lifting). Physical assemblies and subassemblies house the components of functional systems or otherwise provide support or enclosure functions. Products can range from very simple single-function items (where single material type responses are common) to highly complex objects involving many complex and interacting systems that provide multiple functions. Buildings are invariably complex objects when considered as a whole and do involve complex functional systems, but they also incorporate many quite elemental components.

In general, as system and assembly complexities increase, problems related to the need to resolve *conflicting design objectives* invariably increase as well, as do problems associated with optimizing design solutions. The idea that any *single* material approach will dominate a design tends to decrease. The question becomes more one of decomposing the design problem into its constituent parts that are addressable in terms of different material responses (including whether nanomaterials can be feasibly used) and orchestrating the whole to achieve design goals. We will return to these issues in Chapter 5.

Another important way of understanding how nanomaterials and nanotechnologies fit into the broader design picture is via a close look at general *design processes* and their *objectives*. Many formal design processes proceed from a determination of design needs and requirements and move to the determination of needed material properties and subsequent material type selection. Materials are seen as a factor to be chosen to implement a desired objective. Needs, objectives, and constraints are thus first established for a particular design approach, and a material is subsequently selected from among many on the basis of its material qualities best meeting specific design requirements. This is a common approach used in a great many design situations, particularly in engineering and in many product designs with a technical orientation. Any of several materials might meet requirements, and a design process objective is to find the best among them. We will look in detail at these kinds of structured design processes for materials in Chapter 5. An alternative design process approach is the inverse of this process. A designer becomes familiar, and perhaps even fascinated, with a particular property or set of properties and then quite literally imagines or identifies interesting design uses for the material.

3.2 PRODUCT DESIGN, ARCHITECTURE, AND ENGINEERING

Products

Organizations normally make products to be sold in quantity. Successful products meet market needs or demands and are associated with general user appeal and satisfaction for market targets, usability, and technical performance vis-à-vis intended functionalities and costs—all broad categories that also ultimately reflect factors related not only to direct material characteristics but to the nature of user interfaces and control systems, manufacturing process concerns, and other considerations.

All successful products are in some way responsive to a market need or problem definition. Many needs have been with us for a long time, and established product forms respond to them. Lighting fixtures that provide common room illumination serve basic functions that are not likely to go away soon. The performance, control, and forms of products such as this may change, and the use of nanomaterials or nanotechnologies might offer significant performance improvements, but many basic characteristics will not be radically altered. There are, however, many opportunity- or concern-driven forces for change that are incentives for brand-new product forms. These same drivers for change are then manifested in a series of forces that characterize different kinds of more specific product

User-driven products

Technology-driven products

Platform products

Process-intensive products

FIGURE 3.8

General types of products according to the fundamental considerations that characterize their design forms and development processes.

development activities (e.g., market push, technology driven, user driven, or platform) briefly discussed in the next section.

Characterizing Products: A Broad Perspective

There are many different ways of formally characterizing products and the design and production organizations that create them (see Figure 3.8). These distinctions become of critical importance in assessing the needed makeup of a design team, the roles played by various players at different times, the physical and use characteristics of the products, and ultimately, the selection or engineering of materials used within the final products.

A *user-driven* product is one in which the primary value of the product depends on its visual appeal or on the perceived functionality—how a user would relate to it or use it. Though underlying technologies might be either simple or sophisticated, design emphasis remains focused on factors such as appearance, ease of use, ease of maintenance, and so forth. Industrial designers are frequently engaged at an early stage in products of this type.

In a *technology-driven* product the primary value of the product lies with its technology or its functional capability to accomplish some specific task. User perceptions of the technical performance dimensions of the product influence selections more than appearance or even ease of use. Design emphasis thus initially focuses on assuring the technical performance of the product. For example, many products for making precise measurements fall into this category in that a user's paramount need for accurate measurements might outweigh an obviously clumsy interface when product selections are made. The design team for these kinds of products is usually led by an engineering group of one type or another. Industrial designers might well play a role in packaging the product, but typically they are engaged later in the overall process.

There are companies that produce technology-driven products for a clear or already established market area—for example, a company producing nano-based products (e.g., recording media) for use in the music industry. There are, however, many instances wherein a company starts with either a proprietary technology or a particular technological capability and then looks for a market in which these technologies could be effectively employed in creating a responsive product. In the current world of nanotechnology development, many individuals and organizations find themselves possessing unique and seemingly valuable technologies before they know what can actually be done with them vis-à-vis commercially viable products other than harboring a general belief in their broad applicability to an industry, such as the automotive industry. Many current startup nanotechnology firms can be characterized in this way. Many successful products have been developed via "technology-push" processes, but the dangers and pitfalls are obvious and many. Seemingly brilliant emerging technologies have died on the vine for lack of a clear application domain. For the general "technology-push" process to be successful, there must clearly be a match with identified market needs for a product based on the technology, and then development must be carefully done to assure that the resulting product does indeed meet a market need at a competitive price. Positive benefits are that many leaders of the design team needed to develop the technological aspects of the product are typically already in place and have been involved with the core technology before. Other members are added as needed.

Obviously, many products are both technology- and user-driven simultaneously or fall somewhere on the spectrum between the two extremes. Various product lines within the same family also can be targeted for positions along the spectrum. For example, many camera companies known to produce the very best of cameras with lenses designed to meet the needs of the most demanding professional photographer also make "easy-to-use" cameras designed for a consumer oriented and less technically demanding market. Interestingly, many products have been historically defined in the literature as "technology driven" on the assumption that the driving selection factor was indeed technical performance, but as these products matured and technical distinctions became less compelling, "user-driven" characteristics became more and more a determinant in users' product selections.

Many companies find themselves with existing technologies, termed *platform products*, developed for one application that they seek to apply elsewhere. In comparison with new technologies associated with "technology-push" products, these platform products generally have proven performance and customer acceptance records. Here we might have a product such as a paint employing nanoparticles that was developed for the automotive industry being advocated for use in another industry where similar functional needs are present, even though the industry sector is different.

Many products can be described as being *process intensive*. In these cases, the manufacturing and production processes involved are so intensive, and often expensive, that product characteristics are highly determined by them. (This stands in marked contrast to products for which a market is identified and a product and appropriate process determined later.) These products are often produced in bulk form and then used as parts of other products as more or less raw constituents. Many nano-based products that could be potentially produced in bulk form, such as nanocomposites using polymeric or metal matrix materials, are process intensive and can be thought of in these terms as essentially primary products to be used in other more functionally complex products.

Production Volumes

Another broad and traditional way of characterizing products has to do with long-made distinctions between high-, medium-, and low-volume production quantities—considerations that are best made in relation to the cost of the product, its technical sophistication, and its manufacturing process determinants. Clearly, many common "high-production/low-cost/low-sophistication/intensive-process" products are produced with great economy of scale in huge numbers by strongly deterministic manufacturing processes. The humble stamped pie plate provides an example. In terms of our discussion, these are "process-intensive" end products for which care must be taken in assuring that a competitive advantage would accrue before any even seemingly minor design variations are undertaken. (For example, is there really a market pull for elliptically shaped pie plates made of expensive high-strength nanocomposites as the base material?)

Value

Further product characterizations hinge directly around relative product value as perceived by the target market. Higher perceived value is often accompanied by premium final costs to the user, particularly in technologically sophisticated products. High final product costs on the market—especially if coupled with high production volumes—can in turn justify larger research and development costs (including more extensive design and analysis activities) that are oriented toward further increasing the actual or perceived value of the product. The sporting equipment field, for example, has long been a test bed for the introduction of new materials. The value perceived by the buying audience of having a better golf club made of a new material that improves its driving power can be so persuasive that final product costs to the consumer can be surprisingly high, often seemingly outrageously so. Even the mere suggestion that a new material or technology is employed can improve sales. Many nano-based products are already finding their way into sporting goods and are expected to continue to do so. Conversely, low-value products do not commonly enjoy the same advantages and rarely provide a setting for extensive research and development efforts that might employ new nano-based materials or technologies. The nano-based paper clip will probably not be available for some time.

General Product Complexity and Performance

Products can somewhat simplistically be thought of in terms of varying degrees of complexity and performance levels. At the lowest level, products are near raw materials and provide only a few basic benefits. Many are low-value, inexpensive product forms that can be used in bulk to make other, more sophisticated products. Simple product forms such as films or sheets can be dramatically increased in value if they are engineered or processed to have particularly useful specific qualities that are fundamental to the performance of other products. This would be the case with many nano-based films, paints, or sealants as well as some other bulk forms of basic metal, polymeric, or ceramic nanocomposites. Many of these simple-appearing product forms that can be used to impart particular functionalities to other products may in themselves be quite technically sophisticated and complex internally, such as dichroic films, and be of very high value. In all cases, product forms appear relatively simple but vary in levels of technological sophistication, benefits, and relative value.

To go to the other end of the value spectrum, at the highest end are finely tuned, technologically sophisticated products that provide multiple sets of controllable, specialized high-performance functions and typically involve complex electromechanical components. Extensive use of nanotechnologies are expected here.

At varying levels on the spectrum, we find also items that are low-value products and of low complexity, such as common dishware, or relatively low-value products that serve only limited and relatively low-value functions that nonetheless have electromechanical components, such as lawnmowers and many other items. There are also high-value products not involving any kind of electronic controls, which are designed to provide the very best performance available regardless of cost, such as clubs for professional golfers (or wannabe professionals).

Trends I

Where might nanomaterial technologies fit into this broad view of product development? Figure 3.9 speculatively suggests, in very general and admittedly oversimplified terms, relations among product value, complexity, and other general product descriptors. Trends suggest that the kinds of products or components that would initially benefit most from exploiting the unique characteristics of nanomaterials and nanotechnologies would be very high-value products with controllable functions accomplished via embedded electromechanical devices where extremely high performance, small sizes, and high strength-to-weight ratios are important. Nanotechnologies and nanomaterials can make real contributions here (see Chapter 9). Disruptive changes are potentially present. One would also expect nanomaterials to be used in other high-value, high-performance products, such as sporting equipment, even when functionalities

are more limited. It is also expected that many relatively simple-appearing large-scale production forms, albeit of relatively high value, will benefit strongly from nano-based technologies, such as nano-based films, optical coatings, paints, and coverings. Here there are expected high-performance improvements in products that are very widely used in many different settings and in which performance outcomes are noticeable and important. Conversely, few immediate applications are expected in large bulk-form products for low-value applications. In some low- to medium-value products, one might expect to see incremental applications of nano-based technologies in specific product parts, such as enhancements to the hardness of cutting edges of lawnmower blades made possible via special nanocoatings. Many of these expected improvements are better described as continuous improvements rather than disruptive changes.

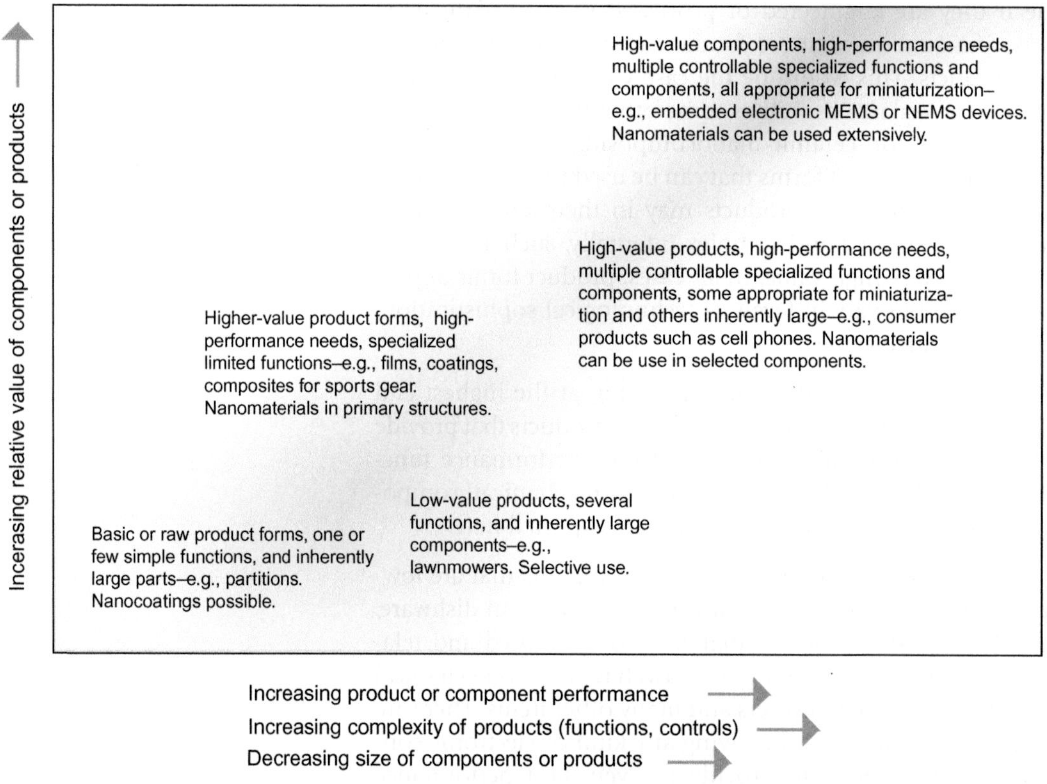

FIGURE 3.9

General trends in product types and forms in relation to nanomaterial and nanotechnology applicability.

Architecture, Building Engineering, and Civil Engineering

This section looks at the characteristics of the primary environments and systems that are present in architecture and civil engineering. These are individually huge fields, of course, with distinctly different design objectives and final products. Yet again there is a surprising amount of overlap in concerns, viewed from the perspective of introducing new materials into design vocabularies. These professions also necessarily come together and interact in complex design situations.

Any major building, for example, is designed by a whole team of designers from different professions who undertake different roles within the process. Initial forms and organizations may well be developed by the architectural team, with the primary responsibility for interfacing with the client and assuring that overall design objectives are met, but as design development occurs, structural, mechanical, and electrical engineers quickly become involved, as do a host of other consultants (acousticians, lighting experts, and so forth), depending on the nature of the building. These disciplines interact with one another, and decisions made by each influence design acts of the others. In other domains, structural engineers would normally take primary responsibility for civil works such as bridges, but again other disciplines are quickly involved. Other branches of civil engineering would assume primary responsibility for other kinds of infrastructure of projects, but even a casual reading of newspaper accounts of any big civil engineering infrastructure projects reveals how many other disciplines are involved (from planners and architects to lighting and telecommunications experts). We obviously cannot treat nuances in detail in this text but rather attempt to highlight some of the major issues that collectively face designers of projects such as these.

The Design and Development Context

Unlike products that are normally produced in quantity and sold to customers, most complex buildings are typically uniquely designed and are one-off constructions, albeit they incorporate many mass-produced products (e.g., doors) that are targeted for general use in buildings. Building projects may be sponsored by private or public groups. In the private arena, it is common for a single developer or development team to determine the general project intent and acquire a specific site for the building. The development team, often with the aid of an architect, also typically goes through various legal and community approval processes that are necessary to enable a project to go forward. Many feasibility studies are also naturally made at the time, since the high costs of any building generally necessitate that the costs be amortized over a long period of years. The development team typically will then engage an architectural firm to assemble a multidisciplinary project team to do the design and engineering work. This same team typically then facilitates making a contract with a construction firm to actually construct the building. Though many building components are bought as "products" (e.g., windows or air-handling units) and are simply installed by the contractor, a vast portion of the act of constructing a building is still idiosyncratic and occurs on site—for example, structural steel frames typically go up piece by piece. The design and construction team assembled to do a project might or might not ever work together again on future projects. In some cases the development team may retain ownership of the new building, but it is often the case that the development team sells it to other buyers. Whereas many exceptions can be identified, such as clients wanting "signature" buildings for one reason or another, it remains generally true that this prototypical process unfortunately does not naturally encourage exploration or risk taking with new designs—much less research that leads to a new design—that might use unproven materials, even if they might prove ultimately beneficial.

In situations where the building is ultimately to be owned and operated by the same group that develops it, prospects for innovation improve. Some ownership groups might well be interested in material innovations that have particular long-term economic benefits that are not easy for a private developer to justify, particularly in the area of energy sustainability. The employment of some new approach or new material might also enhance the image of a company in a positive way and thus increase the market appeal of its products and hence the company's ultimate revenue stream. Apple Computer stores in the United States, for example, have successfully promoted a particular kind of high-tech lifestyle image that is achieved via innovations in design and material choice—an image that ultimately improves the company's bottom line. Other examples include many stores that sell sophisticated sporting equipment, racing bicycles, fashion clothing, or other high-end goods. These are "high-value" buildings where high-end design is considered worth paying for. It is here that most interesting and innovative material explorations in architecture occur. The fact that a line of buildings is being built also helps defray additional design costs over several buildings rather than just one. In terms of exploring where nanomaterials might eventually find building owners immediately receptive to their use, it is indeed this kind of high-value building where the association with nanotechnologies could also serve to enhance the appeal of the company's products.

Another important aspect of the way buildings are designed and built is that their nature is highly dependent on not only the specific programmatic needs to which the building is responding but also to specific site locations and conditions. Climatic conditions associated with specific locations, for example, can dramatically influence external environmental conditions and must be taken into consideration and impact material choice. We can see obvious manifestations of the way that building shapes, forms, and material choices have evolved over time in indigenous building construction throughout the world in response to local climatic conditions. High-mass masonry walls such as adobe are commonly found in hot and arid parts of the world—a consequence not only of material availability but also because the thermal lag cycle associated with the way the material absorbs and emits heat is particularly useful in creating comfortable interior environments in areas where there are hot and dry days but where nights are cold. Heat is absorbed by the thermal mass during the day and reemitted at night, when it is needed. We can see other responses in hot and humid areas or in temperate zones. Even with today's seemingly advanced mechanical technologies, many interesting technologies and related material choices (including many of the interesting active double-layered curtain wall systems) that have proven so successful in energy-sustainable buildings in the temperate climate of Europe have not proven equally successful in the more extreme climates of other countries. Electrochromic glass that is now benefiting from developments using nanomaterials, for example, has been widely touted vis-à-vis its usefulness in meeting sustainable design objectives, but the heat gain associated with current types of electrochromic glass is dependent on sun angle and makes it most effectively used primarily in high-latitude zones. Many of these kinds of locational dependencies can ultimately affect material choice. There are, of course, many other locational dependencies related to urban or rural conditions, access to transportation, and so on that could be noted here as well. This kind of location dependency is normally less prevalent in product design.

The general processes noted here are quite different from those associated with thinking about developing a product. In the building field, there are indeed some cases in which a building is conceived, designed, built, and marketed much more like a product. A salient example here might be modular housing, which is now so ubiquitous. Multiple units are built in a factory and marketed and sold more like a product than a building. Specific sites are not considered, but broad locational dependencies remain important. The same is true for various kinds of premade "industrial" buildings. In current practice, however, these are usually fairly low-end products resistant to significant innovations, despite many attempts to the contrary.

Building Types

One of the most important characteristics of buildings to always keep in mind is that they must provide habitable spatial environments for living, work, or other common life activities that range from the cultural to entertainment; thus we can think about buildings in terms of different kinds of fundamental spatial environments. One way of looking at these environments is simply by building type—hospitals, laboratories, housing, and the like. A whole host of defining characteristics spring to mind when thinking about each, as do differences among types. Essential differences between hospitals and housing complexes are patently obvious and suggestive of not only uses but differences in overall forms and spatial organizations, types and uses of spaces, types of infrastructures (mechanical, electrical, telecommunications, etc.)—all the way down to images of specific material requirements or needs (as might be present in hospital operating rooms, for example, with their needs for surfaces that do not host bacteria and that are easily cleanable). Thus, a building types perspective can suggest a whole host of design requirements and subsequently stated design objectives that are useful for designers and engineers. The perspective becomes considerably less useful in other instances, particularly in cases where buildings serve multiple functions and must be designed to adapt to different uses over time. Building types also come and go; we have very few harness shops being built today. More generally, a types perspective can detrimentally lead to framing broad design questions in a restrictive way. Nonetheless, the approach is often used and it is valuable within the context of this book.

Other approaches to thinking about buildings are from the perspective of physical assemblies and systems (structural, mechanical, and so on). The following section briefly explores the idea of systems, particularly their functional definition.

3.3 ENVIRONMENTS, SYSTEMS, AND ASSEMBLIES

The design of any product or building is a complex undertaking. It requires both understanding and manipulating the whole form as well as orchestrating the many components and systems that make up the form and give it usability or functionality. As a way of understanding the design tasks at hand, it is useful to take a moment to first look at products and buildings via the nature of the involved physical environments, systems, and assemblies. *Environments* may be alternatively considered as either an integral part of a design, such as spaces within a building, or as defining the external context within which a product operates (see Figure 3.10). *Systems* are generally defined as physical components or parts that work interactively to provide some particular type of function. In architecture, systems of one type or another may serve to support building environments. In products they might provide the immediate function that characterizes the product itself, as in a lighting system. *Assembly* considerations spring largely from issues of manufacturability or constructability as well as from broader ideas about modularity and other assembly architectures.

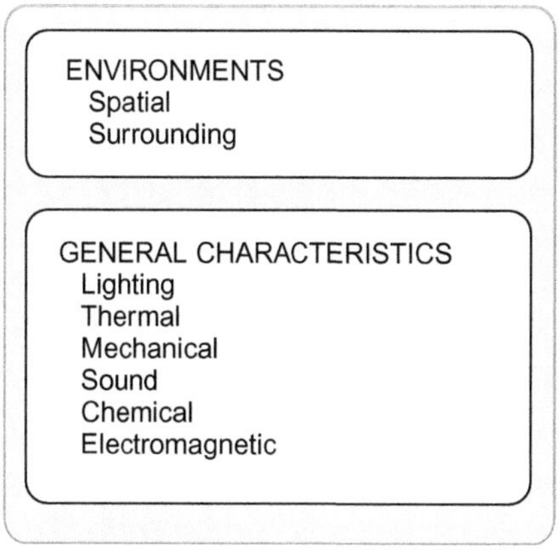

Environments

The notion of thinking of design in reference to various kinds of critical environments is a common approach. For buildings and some products, such as automobiles, we might think about the nature of the spatial environments that are *occupied* by humans. For products, we might think as well about the special characteristics of *surrounding* environments or those that are intrinsic within a product that might influence design approaches. In both products and buildings, broad considerations of *thermal, sound, lighting,* and *mechanical* issues are invariably important. In many situations, *chemical* factors might be a concern, as in designing a product for use within a surrounding environment that is corrosive or in considering the health effects of material outgassing in a closed habitable environment. *Electromagnetic* environments might be important in designing specific products. Common material properties play a fundamental role vis-à-vis these issues, and thus nanomaterials can be expected to play a role as well. Many other products can be described in terms of their *use* environments. Interface characteristics are particularly important here. Chapter 9 addresses these environments in detail. Here we review salient issues.

FIGURE 3.10

Basic kinds of design environments in buildings and products.

In a building, for example, the *spatial* environments where we live, work, or play are the *raison d'être* for the whole building. These spatial environments have specific configurations that are geometrically defined. They also have particular environmental qualities that support our well-being, make us comfortable, or support our tasks. We can consider these environmental qualities in terms of their primary *thermal, lighting,* and *sound* characteristics (Figure 3.10). For example, designers must provide an overall thermal environment that is comfortable for inhabitants. This, in turn, is normally accomplished via a complex set of passive material and ventilation elements and supporting mechanical systems (e.g., ubiquitous heating, ventilating, and air-conditioning systems) that function together to maintain the temperature of the room at prescribed levels. The geometric characteristics of the space, the nature of the surrounding walls, or the kinds of glass present in fenestrations, including their material characteristics, are all contributors to defining the thermal environment actually perceived by the occupant. Similar observations could be made about lighting environments. There are lighting systems that provide specific light-emission capabilities, but the physical characteristics of surrounding walls and windows contribute to the kind of luminous environment provided. The material reflectivity of wall surfaces or the amount and kind of light that passes through windows (which depends on the relative transparency of the window material) all play a fundamental role in defining the kind of lighting environment that's ultimately provided.

In these and similar situations, the *primary* evaluation metric is invariably associated with some type of measure that is related to the environmental parameter itself, such as final lighting levels at a particular task or living area within the space, rather than to a specific supporting component, such as emissions from a lighting system, albeit the latter's contribution to this measure can and should be assessed due to its integral contribution. In a product design situation, the performance of a thermos bottle with an internal volume designed to keep liquids hot or cold can be evaluated vis-à-vis the degree to which it can maintain the temperature of the liquid inside. Various kinds of mathematically based simulation tools can be used to aid in these kinds of performance predictions. Thus, in the thermos bottle, various analytical approaches or models can be developed that take into consideration temperature differentials between the inside and outside of the bottle and the thermal heat-transfer properties of the bottle material. The performance would be found to obviously depend on particular material properties but also on other parameters, such as the geometry of the bottle, that are not directly material dependent. The use of these kinds of computational models that play out the role of material properties in relation to other design parameters is intrinsically important in designing any object. Figure 3.11 suggests several typical examples of where and how nanomaterials are now or potentially can be applicable. Note that nanoscale coatings are expected to be used in many applications—lighting control or manipulation, surfaces that are self-cleaning, scratch-resistant, or anti microbial, and others.

Systems and Assemblies

Systems are normally defined as those sets of subsystems and components that perform some type of *functionally defined* role. A sophisticated design may consist of several *primary systems* that have both functional and physical characteristics but that are primarily thought about in functional terms. In this discussion, a *system* is generally defined as a series of lower-level subsystems and related functional components that collectively provide a larger function important to the use and operation of the whole configuration and that interact with one another. The provided function is intrinsic to the operation or existence of the whole. Several systems typically make up a complex design (e.g., a lifting system is part of a forklift assembly; a heating system is part of a building). Systems may have both functional and physical connotations (e.g., a power supply is defined primarily in functional terms, but the system is necessarily ultimately made up of a physical set of components that exist within a given layout). Primary systems in turn may consist of several subsystems that can each be considered as serving a specific function that contributes to the role of the system as a whole. Subsystems usually have functions that are quite well defined but of greater complexity than the lower-level functions served by components such as simple motors.

Assemblies are defined in terms of the sets of physical subassemblies and parts that are ultimately assembled as physical objects in a manufacturing or construction sense. This is a purely physical definition of the way individual parts are assembled into ever-larger groupings to eventually form a whole. At the lowest level there are *parts* that are more or less irreducible from an assembly hierarchy perspective—for example, basic cast housings—and that typically have limited and well-defined functions. The metrics defining the relative success or failure of a particular assembly design normally have less to do directly with product performance or use but everything to do with manufacturing assembly speed

GENERAL ENVIRONMENTAL
Air and water and other fluid purification and movement, odor control, other

LIGHTING ENVIRONMENTS
QLEDS, lasers, optics
Nanofilms and nanocoatings
  Reflection, transmission, absorption, refraction, wavelength tuning
    Antireflective, antiglare, brightness enhancing, other
Transparency/color-changing films and glasses
Electro-optical displays
Other

THERMAL ENVIRONMENTS
Nanoporous materials–insulation
  Nanofoams, aerogels, other
Nanocomposites–conduction
Thermal barriers, heat exchangers, heat spreaders
Heating/cooling devices
  Thermoelectric, other
Other

MECHANICAL ENVIRONMENTS
Nanocomposites for strength, stiffness, hardness, other
Nanocoatings–hardness, corrosion resistance, other
Vibration/damping control
Damage monitoring
Active/responsive structures
Other

SOUND ENVIRONMENTS
Speakers, microphones, damping, other

CHEMICAL ENVIRONMENTS
Fuel cells, catalytic converters, corrosion, purification, other

ELECTROMAGNETIC ENVIRONMENTS
Nanotechnology-based chips, magnetic storage, supercapacitors, batteries, shielding, solar cells, electro-optical displays, other

FIGURE 3.11

Basic design environments in buildings and products and typical examples of where nanomaterials or nanotechnologies can be applied (see Chapters 9 and 10).

and ease, assembly time, types of machines required, special conditions, and so forth.

In many cases, functional and assembly systems might be identical, but in other cases they might well be different; for example, various components of a functional system might be distributed in several different but interconnected subassembly units. However, constituents in a subassembly must still share a logic for being housed together. Reasons can vary: Constituents of a subassembly might be of similar sizes. They might all need to be assembled using a particular robotic device. They might all share a need to be protected from moisture or heat. Other criteria for creating subassemblies might be the stipulation that maximum external housing dimensions for all subassemblies be the same—a requirement that might literally force the placement of elements from various functional subsystems into different housings.

The basic distinctions we've drawn here are not always clear in real design situations, and terminology is by no means always agreed on (for example, what is one person's functional subsystem may be another's physical component). The use of multifunctional parts or subassemblies obviously complicates the making of simple characterizations. A simple connecting bolt may well serve to position multiple parts of a power supply system and a cooling system simultaneously and cannot easily be "assigned" to one larger system or another. Indeed, a major task in early design phases is to identify where parts or components serve *multifunctional* or *cross-functional* roles. Failure to do so can sometimes result in redundant and needlessly complex overall solutions.

There is not really any overarching way of describing or characterizing the kinds of systems and assemblies found in product design. As a simple gesture for discussion purposes, the diagram shown in Figure 3.12 suggests some basic characteristics of common products. Most sophisticated products contain control/logic systems or interfaces used in the direct operation of the product or in monitoring its functions. An energy system for power input, storage, conversion, or transfer is often present. Products exist for some purpose; hence an action system or process system is invariably involved that includes many supporting components. Many other elements are present as well, including structures or frameworks for which the function is to house other systems. Often many subsystems that have their own characteristics and functionalities are present. Obviously, not all products have all these characteristics; some might not be present at all (as in a simple table). In other cases, the basic

CONTROL/LOGIC COMPONENTS:
INTERFACES
*User interfaces–direct operation
and control, monitoring (sensors,
other), component/assembly
interfaces, other*

ENERGY SYSTEM COMPONENTS
*Energy or power input, storage,
conversion, and transfer components*

ACTION SYSTEM DEVICES AND
PROCESS COMPONENTS
*Sensors, control and logic components,
signal carriers, transducers and
modifiers, direct action or process
components
Mechanical/electrical/thermal/
chemical/optical devices
Sound/light and other devices*

STRUCTURES AND FRAMEWORKS

SURFACES, ENCLOSURES, AND
CONTAINERS
*Surface structures, coatings,
finishes, sealants, other*

ACCESS. HANDLING, AND OTHER
COMPONENTS
*Structures and surfaces
Mechanisms and other devices*

FUNCTIONAL SYSTEMS

GENERAL
Self-cleaning, self-repairing
HEALTH
*Antibacterial, antifungal, antimold,
antipollutants, and purification*
COLOR/LIGHT
*Reflective and antireflective,
emissivities, brightness enhancing,
other*
THERMAL
*Reflection and emissivities,
thermal barriers*
TRIBOLOGICAL FUNCTIONS
Hardness, wear, lubrication
ELECTRICAL
*Conductive/dialectric
Static dissipation, other*
OTHER

FUNCTIONAL SURFACES

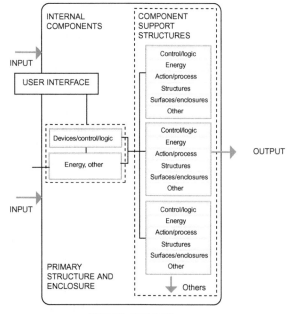

TYPICAL PRODUCT

Diagrammatic illustrations of common types of functionally defined systems and subsystems in an electronically complex product. Nanomaterials or nanotechnologies have applicability in most of the systems and subsystems illustrated, e.g., sensors, energy, enclosures.

Surfaces of products are always of crucial importance. The list to the left suggests some of the possible applications of nanomaterials and nanotechnologies for surfaces or surface treatments.

FIGURE 3.12

Diagrammatic illustrations of common types of functionally defined systems and subsystems in an electronically complex product. Obviously, not all products have this complexity or all the subsystem elements shown. For other products, few of the systems shown may apply; for example, a box is a structure and enclosure and nothing more. Nanomaterials or nanotechnologies have applicability in most of the illustrated systems and subsystems.

description presented is actually played out many times over in a highly complex or sophisticated product, with many components that are "products" in their own right becoming "constituent" components within a larger and more complex product.

Figure 3.12 suggests examples in which nanomaterials and nanotechnologies can play a role in systems found in products. To industrial designers, the role of the "surface" of a product is particularly important. There are many potential applications of nanomaterials in surfaces, such as self-cleaning, antimicrobial, improved abrasion or scratch resistance, self-healing, and others (Figure 3.12).

In buildings, the common description used for constituent systems includes structural systems; heating, ventilating, and cooling systems; lighting systems; sound systems; enclosure and fenestration systems (walls, doors, windows); transportation systems (elevators, escalators); fire-protection systems; various kinds of information and control systems; and others (see Figure 3.13). Basic ideas of hierarchical system/subsystem/component and assembly/subassembly/part relations are fundamental to the field. Similar observations can be made about systems and assemblies in bridges and other civil works, albeit bridges and similar constructions tend naturally to focus primarily on specific systems (e.g., structural). Many components that make up the assemblies and systems in buildings, such as telecommunication or lighting components, are in themselves complex and high-value "products" in every sense of the word.

On the other hand, the nature of building often requires many on-site construction activities that make massive use of relatively low-level "products" that are almost raw materials or marginally removed from them, such as concrete. Large sizes of components and low volume of use and production issues are even more present in these conditions than in more traditional products. Figure 3.13 suggests where nanomaterials and nanotechnologies might play a role in relation to building systems—for example, nanotechnologies in control systems or electro-optical displays, sensors, energy systems (including solar cells), films and coatings on enclosure panels or glass windows, water and air purification, and others. Figure 3.14 shows a brief selection of several buildings using some type of nanoscale material. As with products, surface characteristics of the type listed in Figure 3.12 are again extremely important. Many of these functionalities, such as self-cleaning or antimicrobial, are based on photocatalytic effects. Large surfaces with photocatalytic properties can also contribute to pollution reduction to a limited but useful extent (see Chapter 10).

CONTROL/INFORMATION/COMMUNICATION SYSTEMS

PRIMARY AND SECONDARY STRUCTURAL SYSTEMS

ENERGY SYSTEMS

LIGHTING SYSTEMS

THERMAL/AIR–HEATING, VENTILATING, AND AIR CONDITIONING

SOUND SYSTEMS

TRANSPORTATION SYSTEMS

WATER/WASTE SYSTEMS

FIRE PROTECTION SYSTEMS

ENCLOSURE AND FENESTRATION SYSTEMS
Interior and exterior walls
Windows
Other

ENVIRONMENTS

Indoor air-quality improvements
Air cleaning and purification
Water cleaning and purification
Thermal, lighting, sound
Wastes

COMPONENTS AND SYSTEMS

Surfaces and enclosures
Self-cleaning, self- repairing, antimicrobial, light and color, hardness, thermal, sound, other
Structural
Members, vibration damping and active response and control, damage, monitoring, other
Energy
Solar cells/films, fuel cells, batteries, sensors, controls
Communications, information, alerts
Electro-optical displays
sensors and control systems
Environmental, including biohazard alerts
Fire, other

FIGURE 3.13
A typical way systems are described in buildings is shown at left. The list on the right suggests examples of where nanomaterials or nanotechnologies might find applications in buildings. Many of the surface-related applications are based on hydrophobic, hydrophilic, and/or photocatalytic technologies (see Chapter 10).

Trends II

As we delve more deeply into nanomaterials in subsequent chapters, it will become increasingly clear that certain forms of nanomaterials or nanotechnologies can play a more effective role in some systems or assemblies than in others. Whole complex products consisting of multiple components will not normally be made exclusively of nano-based materials or technologies. Specific components might very well involve their use, depending on performance requirements. We can expect, for example, that nanotechnologies will play a central role in high-value electronic control and interface systems that support products, as well as in energy systems (generation, storage, distribution) or lighting systems. Nanocomposites might also find use in specific situations in which high performance is necessary, such as making high-value components lighter and stronger. Other nano-based products, such as films, coatings, and paints, are also expected to be used widely because of their interesting potential for self-cleaning, self-repairing, and other capabilities, discussed in Chapter 10. In general, widespread uses will initially be highly selective. In buildings, similar trends are expected. Developing nanomaterial use can be expected in control, monitoring (including sensors), and other systems with high-value components or where performance needs are particularly high. Applications in lighting are expected to be high. Nano-based films, coatings, paints, and insulation are also expected to quickly find wide applications. Nano-based paints and other surface treatments based on photocatalytic properties are expected to become widespread quickly. (Making photocatalytic surfaces is relatively inexpensive.) Nanoporous materials can benefit thermal insulation as well as air and water purification objectives. Chapters 10 and 11 discuss these expected trends in more detail.

Many optical films with nanoscale layers have light transmission and reflection properties that are angularly dependent and are thus affected by the relative positions of the object, the light source, and the viewer. The same object is shown in both images at right.

The Dragon's Lair in London uses a self-cleaning glass. The glass has a nanoscale coating but does not contain actual nanoparticles; see Chapter 10. (Courtesy of Pilkington.)

The Strucksbarg housing project by Renner Hainke Wirth Architekten uses a self-cleaning paint based on photocatalytic technologies. (Courtesy of Christof Gebler, Hamburg)

Concrete is poised to benefit from nanomaterial innovations in its mixing qualities, strengths, and other properties. Self-cleaning and antipollutant concrete has also been explored (see Chapter 10). The Dives Church in Misericordia in Rome, by Richard Meiers and Partners, shown here, uses self-cleaning concrete.

Textiles are being made stronger, tougher, and more abrasion resistant using nanocomposites, as well as imparting properties such as self-cleaning, antistaining, antimold, flame retardancy, and others. The Columbo Center in Lisbon is shown. (Courtesy of Birdair.)

FIGURE 3.14

Several examples of how nanoscale materials are currently being used. Applications include many optical films for light, color, or thermal control; light-emitting devices (QLEDs); surfaces that are self-cleaning, antibacterial, or self-healing and that aid pollution reduction; nanofoams and nanogels for thermal insulation; solar cells; and a host of applications related to the strength and stiffness of members (see Chapters 9 and 10).

Thermal insulation properties of enclosures can be improved through the use of nanofoams. Translucent nanogel-filled panels are shown in this view of the Four Points Hotel, Manchester, New Hampshire.

Unitary and Modular Design Approaches to Systems and Assemblies

Determining appropriate relationships between functionally driven and physically driven design responses is one of the major tasks of a designer. There are design choices that can be made about how specific functions may be assigned to particular physical parts or subassemblies, or, alternatively, distributed among several. Pursuing one or another direction has dramatic implications on the resulting qualities of the design and the way it is ultimately produced. Here we look at consequences of carrying one or the other of these approaches to an extreme. The broad distinction introduced here is between two general design approaches. In the first approach, holistic designs evolve from a systemic arrangement of discrete but interactive *modules* that house particular functions and/or that have particular standardized geometries. In the second approach, the overall design is carefully built up as a *single* whole and without subassemblies or components that are easily or necessarily distinguishable as such.

First, it is useful to consider what is meant by the term *module* in relation to any design activity. The term refers to a self-contained or distinct and identifiable unit that serves as a building block for an overall structure, but within the context of a prescribed or orchestrated set of interactions that relate to the performance of the whole object (see Figure 3.15). This broad meaning can have both physical and functional connotations. A general implication of a modular design approach is that all modules are replaceable *without* completely disassembling or changing the whole configuration.

There are many ways the term *module* is actually used. For example, the term *modular design* can assume a primarily functional meaning. In computer programming it generally refers to an approach in which separate logical tasks are reduced to "modules" of manageable sizes and programmed individually, albeit within the context of carefully prescribed interactions. In product design there are many examples of modules that are primarily functionally defined, such as a power module, although some physical shaping and dimensional implications are usually present as well. By contrast, the term *module* can also assume primarily a physical meaning. Here, it refers to a standardized unit of size. In its more extreme form, a "modular" physical design thus generally refers to an overall design based on the

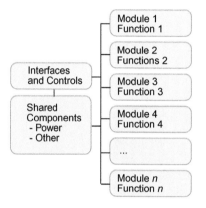

Modular design with functional subsystems housed within dedicated subassemblies; subsystems and subassemblies are identical.

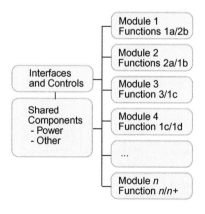

Modular design with functional subsystems distributed across several physical subassemblies; functional subsystems and physical subassemblies are not identical.

FIGURE 3.15
Modular design approaches.

use of a *prespecified dimensional array* (typically in the form of integral multiples of given dimensions) that serves to define the sizes and locations of all parts and subassemblies (locations of connection or interface points are often predefined as well). This latter physically based approach is widely used in many diverse fields. In building design, this approach is widely used in connection with prefabricated building elements, whereas in electronics the same conceptual approach is used to structure the sizes and shapes of electronic devices to go on larger circuit boards. This physically based notion of modular design is extremely important when it comes to manufacturing and assembly and can literally define the maximum "envelope" and location of "connectivity" points that a designer of any component must respect. In many physical approaches, there is often a literal supporting physical infrastructure (e.g., the board with the holes in largely modularly designed electronic devices) that provides a rationalized support system. Connective systems that are usually primarily functionally defined are used to link one module to another.

This modular approach clearly places a premium on the potential positive benefits associated with design ease and rationality, the potential for easy product variety and upgrades (vis-à-vis module interchangeability capabilities), direct maintenance (in the form of access and component replaceability), and straightforward manufacturing and assembly. The whole must obviously be made to work functionally. However, there is nothing implicit in the modular approach that *a priori* guarantees ideal part or component locations with respect to functional or use issues. In poor modular designs, there can be high redundancy, the need for additional support and infrastructure elements, and sometimes a loss in ideal performance.

These modular design techniques stand in marked contrast to single *holistic design* approaches. In these approaches, rarely are specific functions associated with specific distinguishable physical subassemblies (see Figure 3.16). Rather, the actual parts and elements making up various functionally defined systems or subsystems may coexist in the same internal spaces and be located throughout the whole design in an integral way. Single constituent entities often serve multiple or cross-functions. The driving force in this design approach is to focus on the final characteristics of the end product and to position elements *internally* such that they perform their respective roles in a collective way that

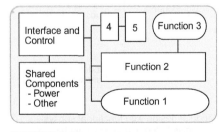

FIGURE 3.16

Holistic design approaches. All systems and subsystems are tightly housed within a single assembly fabric.

is most optimal for the performance of the whole design. Following this objective can result in tight designs with little redundancy and high performance. Shapes can become highly complex and derivative of the need to integrate all systems together (normally in a closely packed arrangement). These same holistic design strategies often make replacement of components difficult in a unitary design and correspondingly can potentially make upgrades and maintenance concerns much more difficult as well. Whole units may well have to be replaced for even simple upgrades (albeit this is already a common trend in microelectronics and other fields that has proven acceptable to the market).

For many products there have been increasing pushes toward smallness. This is particularly true for sensors, chips, and other electronic devices that are commonly integrated into some larger product form. This push has long been under way and has led to some remarkable technological developments in the form of microelectromechanical systems, or MEMS, devices. These devices are fabricated using the same technologies employed in the microchip industry, and hence they can be extremely small. They need not be solely electronic but can be small mechanical systems as well. Figure 3.17 illustrates several MEMS devices. Functionalities can be highly developed, such as "labs on chips" for performing various analyses (also see Chapter 11). These devices are invariably holistic designs. There is no notion of part exchangeability or subsequent repair. They is highly focused devices. The next step in the evolution of these devices is expected to be nanoelectromechanical systems (NEMS) that utilize nanotechnologies and synthesis methods.

A dust mite walking across gears in a MEMS device.

Courtesy Sandia National Laboratories, SUMMiTTM Technologies, www.mems.sandia.gov

A tiny gear train.

FIGURE 3.17

The recent push toward smallness is exemplified by MEMS, which are normally made using microchip technologies. NEMS are expected to be the next stage of evolution. (Courtesy of Sandia National Laboratories.)

Trends III

In terms of nanomaterial and nanotechnology use, the continuing affordances offered by nanomaterials and nanotechnologies in achieving ever greater size reduction and miniaturization of even highly complex functional subsystems are highly compatible with holistic or unitary design philosophies. Figure 3.18 suggests general trends. The adoption of nanotechnologies, with their small sizes and increased functionalities, will enable far more tightly integrated unitary designs than are now commonly available. Miniaturization trends already in progress via MEMS technologies will undoubtedly continue. MEMS devices

Continued

may well be enhanced via nanotechnologies, or whole new families based on nanoconcepts alone will emerge as NEMS devices. Many products that have complex functionalities may well appear as simple small objects with few or no visible distinguishable features. Still, the miniaturization trend will apply to certain products only, particularly those special-function devices that are intended for integration into other more complex product forms. Many components that require direct human interfaces will grow smaller, but interface needs will keep aspects of them larger. Other products might have large features because of certain of their functionalities, such as tennis racquets, which must exist at the macroscale in order for them to be useful. Even macroscale products, however, can be expected to become stiffer and lighter because of nanocomposites and have enhanced functionalities. Ideas of modularity and other design approaches will remain applicable with large products. At the building scale, modularity will remain a common design approach.

Increasing use of nanomaterials and nanotechnologies. Increased miniturization, increased functionalities relative to size, decreased weight, changes to precision manufacturing techniques necessary for nanotechnologies. Increasing use of unitary designs.

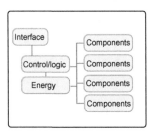

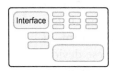

METER SCALE: Conventional products, large- and small-scale components. Modularity common.

CENTIMETER SCALE: Overall size reductions are limited by human interfaces. Process, mechanical, or light-emitting and sound components can remain large.

MILLIMETER SCALE: Products without human interfaces. Overall size decreases are easier. Any lighting/sound components remain fairly large.

MEMS SCALE: Miniaturization. Sophisticated functions.

NANOSCALE: Sophisticated special functions. Extreme miniturization.

FIGURE 3.18

Impact trends in technologically complex products as a consequence of introducing nanomaterials and nanotechnologies. Some products are expected to become very small, whereas others will have size limitations due to functionalities or interfaces.

3.4 THE DESIGN AND DEVELOPMENT PROCESS

The Overall Development Process

Typical product development processes may be generally thought of as occurring in several broad steps: definition of design needs or requirements, concept development, concept and preliminary design, design development, preproduction prototyping and evaluation, and production design. The design process does not necessarily march so literally from one step to another without a backward look. Good design work often comes from processes characterized as being the result of convergent back and forth iterations. Well-received design proposals are typically those that both meet functional requirements and have the kinds of design appeal with respect to both visual and user interface qualities that have long been known to characterize successful product designs. Of particu-

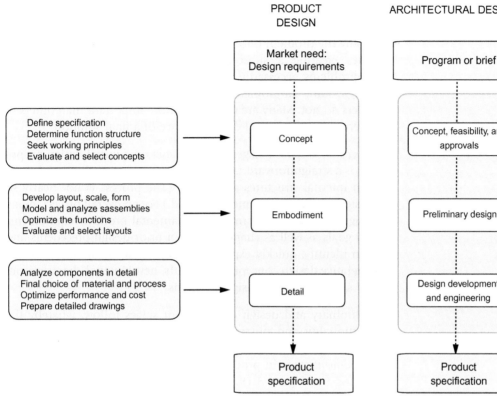

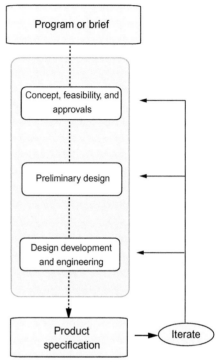

FIGURE 3.19
Design flowcharts for product design and architecture. Overall processes are similar, but terminology is different. In product design terminology, the design proceeds from the identification of a market need, clarified as a set of design requirements, through concept, embodiment, and detailed analysis to a product's specifications. Architectural processes are broadly similar.

lar importance in user-driven products are critical issues that center around classic design characteristics such as form, proportion, materiality, color, tactility, and so forth. Also critical are issues of what designers frequently call *design consistency*. Terms such *design coherency* or *design integrity* imply broadly similar concepts. These concepts generally refer to the extent to which a design collectively holds together with respect to a *collective* view of visual, interface, and functional issues.

Figure 3.19 suggests a broad view of the design flow process, from identification of market needs or a new idea to a final product specification. A first step is typically that of identifying what the product needs to do and what kind of market the product is addressing. What needs must be met or what requirements must the product fulfill? Are there specific constraints that the product must be designed to respect? More broadly, how is the design problem framed? Ultimately, *design requirements* are established. There are formal and informal methods for accomplishing this end (see the section on structured design processes that follows). In formal methodologies in product design, early steps often include decomposing the problem into as many subproblems as possible (sometimes creating a subsolution importance hierarchy). Not all subproblems necessarily have to be solved at once by a proposed design. Some subproblems may have preexisting solutions, whereas others do not. Many are ultimately more important than others in determining the critical design essence of a final solution.

Concept development stages follow. Generating initial design proposals is a straightforward task to a design professional; it involves its own internal structures and logics. (The process is not nearly so mysterious as is sometimes portrayed.) For an overall project, many designers typically internalize fundamental product requirements and goals as well as characteristics of prior design precedents and then identify, quickly explore, and rapidly delineate one or more generative design concepts. Frequently, new material characteristics will suggest original design solutions.

Preliminary and design development stages follow, whereby the functionality and physical form of the promising concepts are increasingly resolved and studied. (This is sometimes called the *embodiment* stage in product design or *preliminary design* in architecture; see Figure 3.19.) A series of geometric and functional modeling and technical evaluation studies are normally undertaken, as are initial cost studies. Subsequent steps include defining in detail the series of related systems and subsystems that provide a given function or set of functions. A result of the first part of this general process is to have a set of design requirements for each system or element detailing what it needs to do in unambiguous terms. These specifications are normally performance oriented but are much more technically directed than the more generic target requirements used in preliminary designs, which typically relate more to user need perspectives. For some subsystems, input and output values must be within specified ranges. Other examples might be those of specifications for a mechanism that require it to provide a given pushing force over a specified stroke distance. A pressure vessel might have to be able to contain a specified internal volume and pressure.

By this stage of the process, there should also be clear understanding of the maximum envelope dimensions of any particular assembly of subsystem components and where (in a physical layout sense) connections are made to related subsystems. The *detail* or *design development and engineering* phase of the development process builds explicit material solutions and/or physical devices. Specialized design and engineering teams usually undertake this process. Final material selections are determined at this point. Numerical simulation analyses (e.g., functional, structural, thermal, or other) are normally used to verify that an anticipated product meets original design requirements.

Once all systems and subsystems have been physically and functionally defined, the design team again addresses the overall design form that has been evolving in parallel. A major objective here is to make sure that there are no physical or functional system "conflicts." Quite often a system might simply not be able to fit in the intended space. There may be spatial conflicts between elements of different systems. (In building design, structural and mechanical system components often seem to have affinities for occupying the exact same physical spaces.) Or, in other situations, the proximity of one electronic system to another might require special shielding. The list goes on and on. All these events would require design alterations, either to the whole assembly or within its systems. The process is typically iterative. In product design, the term *systems approach* is often used to characterize these procedures (see Figure 3.20).

In subsequent stages, physical objects are also generally shaped in relation to anticipated *production processes*, that is, designing parts so that casting or molding procedures are ultimately possible. Specific design for manufacturability or design for assembly principles may be explicitly followed as well. Sometimes formal design for manufacturing and assembly methodologies with proprietary components can also be used.

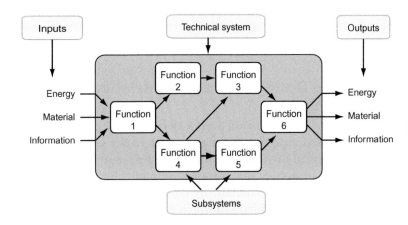

FIGURE 3.20

The systems approach to the analysis of a technical system, seen as transformation of energy, materials, and information (signals). This approach, when elaborated, helps structure thinking about alternative designs.

As part of the design efforts we've noted, various kinds of experimental prototypes may also be made during this stage. These prototypes serve to test one aspect or another of a design and help in the refinement and integration process. They may be functional prototypes that work as intended but are not yet physically packaged in their final anticipated physical form. These prototypes serve many broader roles as well, including fostering communication with all members of the design team (and with management and clients as well). Alpha prototypes are built using part geometry and intended final materials but are not necessarily fabricated in the same way intended for final production processes or by using the same machines. As part of this process, several competing final-use materials may still be tried for different parts. Actual material testing could occur, in which products are placed in intended-use environments that are known to be difficult (e.g., a high-temperature environment). The look, feel, and performance of the alpha prototype, however, must match its final production descendant.

Beta prototypes are further developed and normally use parts produced directly by the actual production processes intended for use. Individual parts, however, might be assembled differently. Beta testing by the user community within its own operational environments is widely acknowledged as a good way of debugging the final design and uncovering any reliability problems. Final design corrections are subsequently made.

The final stage of production design involves making initial runs of the product, normally done using the actual intended production process. Initial runs are quite carefully monitored for production quality. Some units may be again placed into a user context for further evaluation and testing. Normally production levels are quite small at first while process bugs are worked out and the workforce is trained. In combination with a marketing and distribution plan timetable, the product then goes into higher-volume production and is subsequently officially launched.

Structured Design Processes in Relation to Materials

Despite the complexities of overall development process issues we've discussed, specific material characteristics remain fundamental shapers of the design result. Some design problems can be very limited in scope and dictated primarily by the technical properties of various materials. By contrast, there are also more open-ended design problems in which many external factors might influence material choices, and there can be little or no determinism based solely on singular technical factors. The latter often also include multiple design objectives involving material choices that might even be in conflict with one another.

In many situations involving well-defined design problems, design processes including materials can be relatively straightforward and consist of a series of sequential steps. A structural engineer designing a beam that is part of a flooring system for a building, for example, typically assesses the external loads acting on the member and determines the forces and bending moments using well-developed structural analysis algorithms that relate the geometrical and material properties of a beam to the kinds of internal stresses and deformations that are developed. Cost functions could also be constructed. Depending on the design objective, such as minimum weight or minimum depth, the process can then be structured so that by specifying design criteria related to strength and stiffness, variables such as specific required material properties (minimum yield stress levels or needed modulus of elasticity) can be determined explicitly for the material to be used in a specified beam cross-section. (Alternatively, material properties could be specified and geometric properties of the cross-section determined.) Subsequently, different kinds of screening processes can be used to select real materials with the needed properties from existing databases of material properties. Design processes of this general type occur in many engineering fields literally by the thousands each day and form the working basis for many common design development activities.

In the scenario just illustrated, we clearly defined the function of the object to be designed. Clear design objectives, such as minimum weight or minimum cost, could also be easily specified. Constraints—for example, how much load the member has to support—could be specified as well. Design variables or parameters might then be either the geometric and dimensional characteristics of the beam for a given material or material properties themselves for a given beam geometry. Material selection procedures directly follow. More generally, the process of selecting a material includes the following steps: the *translation of design requirements, screening using constraints, ranking using objectives,* and *documentation* (see Figure 3.21). These processes are elaborated on in Chapter 5.

As design objectives become more complex, so do design processes. Meeting more complex design conditions might involve trying to achieve multiple design objectives simultaneously as well as meeting multiple constraints. These are invariably difficult goals

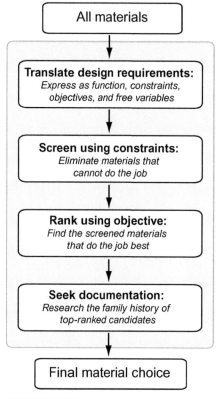

FIGURE 3.21

Material selection processes. Chapter 5 discusses in more detail structured approaches for selecting materials.

to accomplish. Optimally meeting some design objectives might necessitate a relaxing of others. In thinking yet again about simple thermos bottles, the function might be thought of as a need to hold and provide a transport container for hot or cold liquids. Dimensional constraints might exist—for example, such that the bottle could be carried in a typical bicycle bottle holder. An externally imposed constraint might initially be that the unit cannot exceed a certain cost (benchmarked from an analysis of a competitor's product). The choice of primary material might be open. Here the designer might want to optimally meet several design objectives: minimizing the weight of the container since it is for use by bicyclists, maximizing the length of time a contained liquid would remain at an initial temperature level, or minimizing the cost of the whole unit. Meeting all these design objectives simultaneously (i.e., optimizing all design objectives) is at best a difficult and sometimes unachievable task. Compromises are often made and tradeoffs normally occur.

Nonetheless, the designer can choose which of the design objectives is initially most important and which are less so (i.e., rank-order design objectives for optimization). For serious cyclists, weight might well be most important to the user and thus be the governing design objective for the designer. Thermal performance might well have to be compromised, and cost constraints may simply be ignored for all practical purposes. For everyday cyclists, costs and thermal performance might well be more important than absolute weight. Different design solutions will invariably result, depending on what objectives are most highly valued. Definition or prioritization of design objectives, therefore, often depends on the use context or other external driving forces.

Situations with design objectives that are inherently conflicting can, on one hand, try the souls of designers, but on the other hand they have provided both the spark plug and the setting for innumerable innovations in the materials world. Tradeoffs and value stipulations are invariably required to move forward.

Unstructured Design Processes

In the prior section, material-oriented design processes were seen to progress from a determination of design objectives and requirements to the determination of needed material properties and subsequent material type selection. Materials are seen as a factor to be chosen to implement a desired objective. Needs, objectives, and constraints are thus first established for a particular design approach, and a material is subsequently selected from among many on the basis of its material qualities best meeting specific design requirements. This is a common approach used in a great many design situations, particularly in engineering and many product designs with a technical orientation. Any of several materials might meet requirements, and a design process objective is to find the best among them. The process is certainly relevant to designing with nanomaterials, but it obviously demands that the useful mechanical, optical, thermal, and chemical properties of the nanomaterials be reliably known. At the current time, data of this type in a form useful to designers is often not readily available.

An alternative approach, often initially viewed as not rigorous and even unseemly by more scientifically oriented designers, flips the whole process. A designer becomes familiar, perhaps even fascinated, with a particular property or set of material properties and then quite literally imagines or identifies interesting design uses for the material. In this scenario, formal definitions of needs or design requirements do not come first. To be sure, the considered material does ultimately respond to some perception of needs or uses in a particular design context, but a material orientation is set first. This process is less deterministic and more opportunistic than the former. Its procedural steps are more difficult to formalize, but they are surely there for the experienced designer. It involves the systematic study and experimentation of the properties of a material by a designer, who internalizes these findings and who has a large portfolio of design problems to solve or design opportunities to explore. An internalized "fit" process occurs between material affordances and design opportunities. It is a time-honored and effective way of designing with materials. Interestingly, it appears to be one of the primary modalities that is operative in the nanomaterial world at the moment. Various nanomaterials have different technical properties that can be fascinating. What can we do with these kinds of nanomaterials? What applications can we imagine?

Many other approaches evoke different processes or are somehow combinations of the two extremes we've noted. Many designers' long preoccupation with objects that are "strong and light" is a case in point. In some cases there might be a strong technical or use reason for making something strong and light, as is the case with many handheld consumer products on one hand or aircraft components on the other hand, whereas in other cases there simply might not be (e.g., it is only marginally important that a park bench be really light in weight, and, indeed, extreme lightness could be problematical). The design value of "strong and light," however, might remain as an initial *generative* factor for a design, perhaps because the idea is symbolically associated with other designed objects that have captured our societal imaginations, such as aircraft. Why were many objects that were designed in the 1930s—including not only locomotives but vacuum cleaners and other common household goods—given "streamlined" appearances? It is, after all, rather hard to argue that the wind resistance provided by a floor vacuum cleaner is a particularly important functional generator of design form. The associative design value of "streamlining" and its related material suggestions of "sleek metallic," however, have long been with us and remain widespread as symbolic of newness and innovation. This is particularly true in many consumer-oriented products in which the ideas of "associative" properties, briefly discussed earlier, can be highly important.

3.5 A DESIGN VIGNETTE

As a way of entering into a discussion of design issues in relation to materials in architecture, let's first look at a highly particularized vignette: the design of a concert hall for listening to music. Here the material issues would seem to immediately devolve into classical technical problems of choosing materials with the right acoustical properties in relation to sound absorption or reflection. We, however, poke at this assertion here. This same vignette is used as a brief introduction to the way design processes occur.

Any music aficionado knows that there are great halls—Vienna's Grosser Musikvereinssaal, Amsterdam's Concertgebouw, Boston's Symphony Hall, and many others that are well known to performers and listeners alike (see Figures 3.22 and 3.23). By contrast, many others—some recent—have been designed with good intent but have simply failed to become known as good concert halls. Even the casual music listener knows that the quality musical experience an audience receives is very much dependent on the acoustics of the hall and that the seemingly odd shaping of a hall has something to do with this experience. The more careful observer would comment that in some places and not others, drapes are hung, and that many walls are covered with different kinds of soft materials in various places. Undoubtedly, our more knowledgeable observer would also know or suspect that these same materials might have something to do with absorbing or preventing unwanted sounds and their reflections from reaching the listener. But not only are the surfaces absorptive; hard surfaces can also be seen to be

FIGURE 3.22

Boston Symphony Hall. This concert hall is well known for its fine acoustics.

FIGURE 3.23

Vienna's Grosser Musikvereinssaal is another fine concert hall. Designing concert halls for premier acoustical performance is extremely difficult. Halls suitable for one music type are rarely optimum for other types.

suspended in selected locations—clearly to direct sound reflections without reducing them significantly.

To the designer, all these issues—geometric configurations, placements of different materials with different properties (which might indeed have different absorptive qualities appropriate for different locations), placements of reflectors, and many other factors—are surely things that someone in this situation (presumably an architect working with an acoustical engineer) has to decide. To the developer of a new material, including nanomaterials, who is seeking to develop application domains, the situation would be viewed as a design case where there is a need to have materials with different sound absorptive or reflective characteristics. The question might merely seem to be one of, "What values?"

In developing a design for a concert hall, experienced architects know that all the issues we've noted are interrelated and each affects the other; they would also know that there are no single solutions, and many variants might have similar performances. The design team charged with the task of developing a design would do many things, but thinking about specific material choices would come fairly late in the design process. The exact framing of the problem leading to these choices, however, is done quite early. In terms of steps, a series of immediately relevant *programmatic issues* are first identified (often through interviews and other techniques), and subsequently a series of stated *design requirements* are worked out: what kind of hall, for what type of music, for how many listeners, what ancillary facilities must be present, and so forth.

Within this context, specific *design objectives* would be established. Broad design objectives might be at a seemingly obvious level: to design a hall that provides a high level of musical experience for the audience (an objective that could presumably be subsequently defined more precisely in terms of various descriptors of the acoustical performance of the hall). A more precise objective might be to say that each listener seated somewhere in the hall must have the same high level of experience no matter where he or she is seated (something potentially much harder to accomplish).

These and other factors are usually defined in a *design brief* or *design program* of some sort. The designers would then generate a series of *initial conceptual design proposals* that impart an initial physical design sensibility to the programmatic requirements (e.g., initial layouts showing locations of the stage, seating arrangements, and so forth). This would have to be done with conscious consideration of the actual site involved (not only its area and shape but also its relation to streets and proximities to other buildings and elements of the urban infrastructure). Most of this work is primarily configurational and normally does not yet involve detailed material considerations. Initial internal configurations for the acoustical space are often based on the designer's own professional knowledge of what configurations might work and be worth further development or, as is often the case with concert halls, the shaping might come from a series of *precedent studies* of other concert halls known to perform well that are of a similar size and in a similar situation. Preliminary reviews and evaluations are then typically made with all stakeholders in the final hall. These stakeholders include client/ owners, representatives from the music world (perhaps the conductor or others who will eventually play there), and any other person or group with a vested interest in the hall.

During this process, broad design possibilities are invariably narrowed, and, ultimately, one is normally selected for further development. *Preliminary design proposals* are then developed that show the design in more detail. Here a better shaping of the internal acoustical volume is defined. At this point in current practice, the building is represented (via traditional means such as drawings or via a computer-generated model) in sufficient detail for all involved to understand the nature of the design. After further evaluations with stakeholders, a design proceeds to an intensive *design development and engineering process* to resolve the design even further. During this process, design teams involving various disciplines—acoustical engineering, lighting, structural, and so on—work together to develop the building. During this process, both two- and three-dimensional computer models of the whole building are commonly used to represent the design and to allow participants in the design process to share information with one another.

The design development and engineering step is normally where intensive studies are made of the potential acoustical performance of the space. To the material's engineer, the immediate and obvious point of interest is that the sound absorptive and reflective qualities of the materials undoubtedly play a pivotal role here. But exactly what role is this? What specific values are needed? If an acoustical engineer is asked what *single* metric is of primary importance in evaluating the acoustical performance of a space, it is undoubtedly not whether a measure of sound absorption in a particular material or set of materials is high or low, but it might well be the reverberation time of the whole acoustical volume. (Reverberation time is a measure of the length of time, in seconds, in which a listener hears an emanated sound as it arrives at various direct and reflected paths to his or her ear; see Figure 3.24.) Long rolling sounds can be heard, or quick sharp ones. Reverberation time is a measure that involves a consideration of material properties, but it is not limited to them and includes a measure of the geometric characteristics of the room. This idea of a primary evaluation metric being associated with the phenomena of interest rather than solely a material measure is common in all design fields. Materials may well play a crucial role, but the focus is invariably on the phenomena of interest.

The common way the role of material properties is understood with respect to a phenomenon such as sound is via the use of *analytical* or mathematical models that relate the involved variables to one another with the overall goal of performance prediction in relation to the phenomena of interest. Here various kinds of computer-based analysis or simulation models are typically used to scientifically model the way sound flows throughout the space. These models typically require the specification of configurations, materials, and all other design factors, including material qualities, that influence the way sound propagates through the space. The designer or acoustical engineer can then make various studies of how specific spatial shapings or material placements would affect the acoustical qualities of the space and attempt to ascertain whether original design objectives are met. By this time, broad criteria such as "musical experience" are no longer useful; more well-defined measures must be employed. Sometimes actual scale models of the space as well as mathematically based simulation models are constructed so that designers and acoustical engineers can use other mediums to gain this same kind of understanding. (The making of prototypes in the product design field is analogous.)

All through this complex procedure, other detailed architectural or engineering studies are being worked on as well. Some of these, such as structural design considerations or fire protection or egress requirements, might well have an impact on the acoustical qualities of the space and need to be taken into account. A typical problem encountered here is that many common materials currently used for sound control are poor choices when viewed from these other perspectives. The soft, sound-absorbent materials that have historically been used in this connection, for example, have also long been known to contribute to the spreading of flames and the development of toxic gases during fire circumstances. Special attention must be paid to make sure that focus is not solely on one design

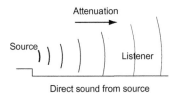
Direct sound from source

Direct plus multiple reflections

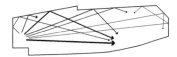
Sound arrives at different times, creating a reverberant field

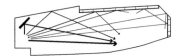
Sound reflectors and absorbent materials modify reverberant field and sound heard at a point

FIGURE 3.24

Reverberant sound fields in building design. The actual sounds from a source heard by a listener include both direct and reflected sounds that take different lengths of time to reach the hearer, thus leading to reverberation effects. Reflected sounds are affected by the absorption qualities of the materials used in reflective surfaces. Absorption qualities are often varied by using soft or hard materials, depending on how sound paths need to be controlled for optimum listening.

consideration at the expense of another equally important one. Common practice in this kind of situation is to be sure that the materials used meet certain minimum conditions with respect to each considered phenomenon. In many cases, design objectives for the materials are *directly conflicting*, and making material choices again often involves making considered compromises, as was the case with the closely related design optimization problem noted previously. This simple example brings up the need to consider materials with *multidimensional properties*, which we return to in Chapter 5. To the developer of nanomaterials or other new materials for use in a situation such as this, the problem is no longer one of how to impart only sound absorption qualities nor even one of what specific values are needed, but how to develop materials that meet multiple objectives.

Any number of other considerations might come into play as well. A designer intent on working with a company that might have developed a new nanomaterial coating that has seemingly attractive sound absorption or reflectance qualities may also have to face perfectly reasonable questions of environmental safety. An obvious question that arises is one of whether or not the coating poses a health hazard in any way. Does it contribute to indoor air pollution through offgassing, or are there particulates that might come from the application of the material? What test results are available to demonstrate that this is or is not a problem? If not already available, tests or experiments can only rarely be commissioned in the architectural world. Costs are high and would have to be borne by the project's owner. Unlike products for which costs can be distributed over multiple units, costs in a building are normally associated with a single product that is designed and produced only once—a problem that clearly militates against testing and experimentation. The list of other considerations goes on and on. The interior designer might want to have shiny surfaces in some places and matte surfaces in others. An owner would undoubtedly ask, "How long will it last?," "Can it be cleaned?," as well as the obvious "How much does it cost?"

As the design development progresses, stakeholder evaluations and inputs continue but necessarily become oriented to more detailed considerations. Ultimately, a decision is made to advance to *construction*—a process often requiring specific types of drawings and other representations of the design and involving yet other parties in a consideration of the design's buildability. Minor changes to the design typically occur all the way through the final construction phase.

Now that we have gone through the typical steps in a design process, let's look more critically at parts of it. Was the original design objective reasonable? Consider the objective "to design a hall that provides a high level of musical experience for the audience." Though it's seemingly reasonable, if we stopped at only this objective we could get a hall that is unexciting or uninteresting to musicians themselves—a feeling that would undoubtedly affect more general perceptions of the hall. It is well known, for example, that the qualities of not only the hall but the stage and other components affect how musicians play a hall. The stage, for example, can literally vibrate in a way such that it becomes rather like an orchestral instrument. Common experience also suggests that there is a relationship between the type of music played and the acoustical qualities of the performance space. Gregorian chants sound marvelous in spaces with long reverberation times (5 to 10 seconds)—times commonly found in spacious Medieval cathedrals. Secular Baroque music, with its clear articulated sounds and low fullness of tone, was originally played in relatively small rooms with hard surfaces—rooms with low reverberation times. Music of the Romantic period plays best in halls with longer reverberation times and low direct-to-reflected sound ratios that produce a fullness of tone and low definition. Music historians have long known that composers of these periods well knew the acoustical qualities of the halls in which their works were to be played. Richard Wagner is known to have composed his romantic *Parsifal* especially for the Festspielhaus in Bayreuth, Germany. In a like way, hall qualities evolved in response to the music of the day, which was in turn dependent on a whole realm of complex cultural conditions.

What is fascinating here, but problematic in design terms, is that, if this idea is carried to its logical conclusion, there is only one type of music that can be best played in a hall with a particular configuration and choice of materials qualities. Even if we speculate on the contributions that nanomaterials might make in terms of the optimization of selected material properties, such as absorption, the overall design for a concert hall can be truly optimized for only one music type as long as the property is a static one. The sound absorption coefficients of various materials that play so critical a role in the acoustics of the space are not actually constant for a given material but rather depend on the frequency of the impinging sound waves. Ultimately, this means that a set of materials can be chosen to optimize a single configuration and single type of music to be played within it but not necessarily for all types within the same hall! Choices come down to optimizing the hall for specific music types or using a series of suboptimal solutions that allow a wider variety of music to be played within it, albeit not in an optimized way. On a more general level, this same broad problem of *design optimization* with respect to various performance measures runs through most complex design problems and is not an easy one to address. We will return to this problem in Chapter 5.

These realizations make the overall design problem more acutely difficult, and the problem of there being no single solution that optimizes all performance needs necessarily requires that design goals and objectives be clarified. For many settings, a preferred aspiration might be to have a hall that works well for as many musical venues as possible, albeit not necessarily ever optimized for any one. Implications vis-à-vis specific design issues, such as material choice, ensue. Choices must be made.

A further assumption made in the overall design scenario we initially presented is quite simply that the discussion immediately assumed only natural acoustical music; no consideration was made of electronic assists in either amplification of direct sound or in its distribution through speakers located throughout the performance setting—an unthinkable assumption to many current musicians but natural to others. The impact on music of developments in electronics needs no comment here; it is all around us. This impact has been extended to concert halls. Why should the all-important reverberation time depend solely on geometrical and material characteristics? In this day and age, the reverberation time can be electronically altered and controlled through a careful orchestration of speakers and responsive manipulation and control systems. Many music settings of this type no longer even have any internal shaping designed to promote natural acoustics, and material implications follow. These interesting developments pose even further challenges to the designer. Electronic advocates would argue that electronic assists offer improved performance in general, and surely much current music is designed specifically to be played in "electronic" halls but would sound poor in even the best of the 19th century concert halls. Others would simply offer a pitying smile at assertions that equivalent quality exists. These interesting matters might not be solely the province of the architect and acoustical engineer to determine, but their resolution surely has an impact on the final nature of the building and the materials selected. Just as surely, these designers can play a fundamental role in helping the electronic vs. nonelectronic decision process by playing out design implications and providing performance simulations. The fundamental role of the design team as clarifier of design intent thus reappears.

Now that we have explored the seemingly endless chain of inter-dependent design issues that invariably affect material choice in a concert hall, let's go back just for a moment and look again at our original design assumptions: that our purpose was to design a hall that provides a high level of musical experience for the audience. Everything followed from this seemingly perfectly reasonable design objective. How could anyone argue? The history of architecture and other design fields, however, is literally filled with examples in which a metric associated with a seemingly unassailable function-ally driven design objective is overshadowed by some other, and perhaps less obvious, design objective. In designing for stage perfor-mances, for example, optimum views of the stage from the spectator boxes and seats as defined by clear sight lines and proximity would seemingly be a primary design determinant. Yet in the Renaissance period, we know that desirable seats were often those that had only adequate views of the stage but excellent views of the remainder of the theater. This reflected the understood social value at the time that seeing the audience was as important as seeing the stage. In many Baroque theaters, the royal box did have a good view of the stage, whereas other spectator seats often had poor orientations toward the stage. Here, the most desired seats were often the ones with the best view of the royal box, with it's vice versa implication ("to see and be seen"). Clear lines of sight and proximities to the stage were simply less important. Thus, the social interactions and conventions provided the dominant design context. We note this example simply to serve as a warning that design problems can often be too easily characterized. A good design team will spend a good amount of time being sure that the initial design question is cor-rectly framed in its broadest possible sense.

FURTHER READING

The language and approach in sections of this chapter dealing with characterizing products were largely adopted from sources such as Karl T. Ulrich and Steven Eppinger, *Product Design and Development*, McGraw-Hill, 1995, although terms such as *modular* have been used for years in many and various design contexts. The ideas are certainly not new to experienced designers.

Design process discussions were drawn from several sources, including the following books by Michael Ashby that elaborate in much more detail on many of the topics presented here. The book by Leo Baranek is an immensely interesting book on acoustics.

Michael F. Ashby, Materials selection in mechanical design, 3rd ed., Butterworth-Heinemann, Elsevier, 2005.

Michael Ashby and Kara Johnson, Materials and design: The art and science of material selection, Butterworth-Heinemann, 2002.

Leo Baranek, How they sound: Concert and opera halls, Acoustical Society of America, 1996.

Karl T. Ulrich and Steven Eppinger, Product Design and Development, McGraw-Hill, 1995.

Frederick Sports, Bayreuth, Yale University Press, 1994.

CHAPTER 4

Material Classes, Structure, and Properties

4.1 CLASSES OF MATERIALS

If you stop for a moment and look around you, you will notice a wide variety of materials, either artificially produced by humans or naturally existing in nature. Both types can be categorized in particular classes to provide a better understanding of their similarities and differences. In this book, we distinguish seven classes: metallic, ceramic, polymeric, composite, electronic, biomaterials, and nanomaterials. However, as you will note, some materials have characteristics across various classes.

Metallic Materials

Metallic materials consist principally of one or more metallic elements, although in some cases small additions of nonmetallic elements are present. Examples of metallic elements are copper, nickel, and aluminum, whereas examples of nonmetallic elements are carbon, silicon, and nitrogen. When a particular metallic element dissolves well in one or more additional elements, the mixture is called a *metallic alloy*. The best example of a metallic alloy is steel, which is composed of iron and carbon. Metallic materials exhibit metallic-type bonds and thus are good thermal and electrical conductors and are ductile, particularly at room temperature.

Ceramic Materials

Ceramic materials are composed of at least two different elements. Among the ceramic materials, we can distinguish those that are predominantly ionic in nature (these consist of a mixture of metallic elements and nonmetallic elements) and those that are covalent in nature (which consist mainly of a mixture of nonmetallic elements).

Examples of ceramic materials are glasses, bricks, stones, and porcelain. Because of their ionic and covalent types of bonds, ceramic materials are hard, brittle, and good insulators. In addition, they have very good corrosion resistance properties.

Polymeric Materials

Polymeric materials consist of long molecules composed of many organic molecule units, called *mer* (therefore the term *polymer*). Polymers are typically divided into natural polymers such as wood, rubber, and wool; biopolymers such as proteins, enzymes, and cellulose; and synthetic polymers such as Teflon and Kevlar. Among the synthetic polymers there are elastomers, which exhibit large elongations and low strength, and plastics, which exhibit large variations in properties. Polymeric materials are in general good insulators and have good corrosion resistance.

Composite Materials

Composite materials are formed of two or more materials with very distinctive properties, which act synergistically to create properties that cannot be achieved by each single material alone. Typically, one of the materials of the composite acts as a matrix, whereas the other materials act as reinforcing phases. Composite materials can be classified as metal-matrix, ceramic-matrix, or polymer-matrix. For each of these composite materials, the reinforcing phases can be a metal, a ceramic, or a polymer, depending on the targeted applications.

Electronic Materials

The electronic class of materials is a bit broader than the previous classes because electronic materials can encompass metals, ceramics, and polymers, such as the metal copper that is used as interconnects in most electronic chips, the ceramic silica that is used as optical fibers, and the polymer polyamides, which are used as a dielectric. However, the term *electronic material* is used to describe materials that exhibit semiconductor properties. The most important of these materials is silicon, which is used in practically all electronic components. Other materials such as germanium and gallium arsenide are also part of this class.

Biomaterials

The biomaterials class is related to any material, natural or synthetic, that is designed to mimic, augment, or replace a biological function. Biomaterials should be compatible with the human body and not induce rejection. This class of materials is rather broad and can comprise metals, ceramics, polymers, and composites. Typically these materials are used in prostheses, implants, and surgical instruments. Biomaterials should not be confused with bio-based materials, which are the material parts of our body, such as bone.

Nanomaterials

The nanomaterial class of materials is extremely broad because it can include all the previous classes of materials, provided they are composed of a structural component at the nanoscale or they exhibit one of the dimensions at the nanoscale. The prefix *nano* represents a billionth of a unit, so the nanoscale is normally considered to span from 1 to 100 nanometers. Nanomaterials are typically categorized as 0-D (nanoparticles), 1-D (nanowires, nanotubes, and nanorods), 2-D (nanofilms and nanocoatings), or 3-D (bulk), which represent the number of dimensions that are not at the nanoscale (see Chapter 6).

FIGURE 4.1
Relative dimensions of an atom, a strawberry, and the Earth.

4.2 THE INTERNAL STRUCTURE OF MATERIALS

Atomic Structure

What are things made of? Everything is made of atoms. How do we know? It is a hypothesis that has been confirmed in several ways. To illustrate the idea of how small an atom is, observe Figure 4.1. If a strawberry is magnified to the size of the Earth, the atoms in the strawberry are approximately the size of the original strawberry.

Let's now ask another question, which is: What are the properties of these entities called atoms? We shall divide the properties of atoms into two main categories, namely inertia and forces. First, the property of inertia: If a particle is moving, it keeps going in the same direction unless forces act on it. Second, the existence of short-range forces: They hold the atoms together in various combinations in a complicated way. What are these short-range forces? These are, of course, the electrical forces. What is it in the atom that can produce such an effect? To answer this question, consider the Bohr atomic model, depicted in Figure 4.2.

In the Bohr atomic model, there is a nucleus consisting of protons with a positive charge and a mass of 1.67×10^{-27} kg and neutrons with no charge but with the same mass as the protons. The nucleus is surrounded by electrons with a negative charge and a mass of

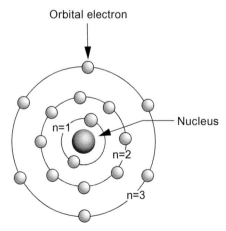

FIGURE 4.2
Schematic representation of the Bohr atom.

9.11 × 10^{-31} kg that revolve around the nucleus in discrete orbits. Thus, the nucleus is positively charged and very heavy and the electrons are negatively charged and light.

Let's now think about another aspect: What is particular about particles such as electrons, protons, neutrons, and photons? Newton thought that light was made up of particles and therefore should behave the way particles do. And in fact light does behave like particles. However, this is not the whole story. Light also behaves like a wave. How can particles such as electrons and light exhibit this dual behavior? We don't know. For the time being we need to accept this, keeping in mind that on a small scale the world behaves in a very different way. It is hard for us to imagine this because we have evolved in a different kind of world. However, we can still use our imagination. This is the field of quantum mechanics.

Let's explore the first idea of quantum mechanics. This idea claims that we are not allowed to know simultaneously the definite location and the definite speed of a particle. This is called the *Heinserberg Uncertainty Principle*. In other words, we can only say that there is a probability that a particle will have a position near some coordinate *x*. This is akin to watching Shaquille O'Neal throw a basketball to the basket. You can't say that he is going to hit the basket for sure! There is also a certain probability. This explains a very mysterious paradox, which is this: If the atoms are made of plus and minus charges, why don't the electrons get closer? Why are atoms so big? Why is the nucleus at the center with electrons around it? What keeps the electrons from simply falling in? The answer is that if the electrons were in the nucleus, we would know their position and then they would have to have a very high speed, which would lead to them breaking away from the nucleus.

So far, when we have been talking about atoms, we have considered their lowest possible energy configuration. But it turns out that electrons can exist in higher-energy configurations. Are those energies arbitrary? The answer is no. In fact, atoms interchange energy in a very particular away. An analogous idea is to have people exchange paper currency. Imagine that I want to buy a CD that costs $17 and that I only have $5 bills. Further imagine that the CD store only has $5 bills in the cash register. In this case, the CD I want to buy will cost either $15 or $20, depending on which party wants to assume the loss. Atoms are very similar. They can only exchange certain "dollar bills." For simplicity, let's look at the hydrogen atom (see Figure 4.3). As shown in this figure, the ground energy for the hydrogen atom is −13.6 eV (electron volts). Why is it negative?

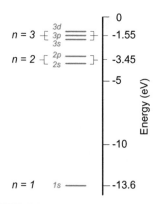

FIGURE 4.3

Electron energy states for a hydrogen atom.

The reason is because electrons have less energy when located in the atom than when outside the atom. Therefore, −13.6 eV is the energy required to remove the electron from the ground energy level. If an atom is in one of the excited states E_1, E_2, and so on, it does not remain in that state forever. Sooner or later it drops to a lower state and radiates energy in the form of light. The frequency of light v that is liberated in a transition, for example, from energy E_3 to energy E_1, is given by

$$v = (E_3 - E_1)/h \qquad (4.1)$$

where h is Planck's constant. Each energy level or shell is represented by the *principal quantum number n*, as shown in Figure 4.3. However, if we have equipment of very high resolution, we will see that what we thought was a single shell actually consists of several subshells close together in energy (Figure 4.3). In fact, for each value of n there are n possible subshells. In addition, in each subshell, it is possible that different energy states may coexist.

We are still left with one thing to worry about, and that is: How many electrons can we have in each state? To understand this problem, we should consider that electrons not only move around the nucleus but also spin while moving. In addition, we should consider a fundamental principle of atomic science, which is the *exclusion principle*. The exclusion principle says that two electrons cannot get into exactly the same energy state. In other words, it is not possible for two electrons to have the same momentum, be at the same location, and spin in the same direction. What is the consequence of this? Two electrons can occupy the same state if their spins are opposite. Where can we put a third electron? The third electron can't go near the place occupied by the other two, so it must take a special condition in a different kind of state farther away from the nucleus (see Figure 4.4). From this discussion, we can now realize that there is a spin moment associated with each electron, which must be oriented either up or down. Because every spinning electrical charge is magnetic, the electron acts as a tiny magnet. However, when two electrons are in the same orbital with opposite spins, the magnetic effect is counteracted and there is no magnetic effect.

With the ideas mentioned so far, we can now understand the periodic table. We should keep in mind that (1) the number of electrons in an electrically neutral atom depends on the number of protons in the nucleus, (2) an electron will enter the orbital possessing the least possible energy, and (3) only two electrons can fit into any one of the energy states.

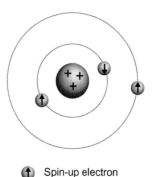

⬆ Spin-up electron

⬇ Spin-down electron

FIGURE 4.4

Atomic configurations for real spin one-half electrons.

All elements can be classified in which the position of the elements enables the prediction of the element's properties. The periodic table is subdivided into horizontal periods and vertical groups. The group indicates the number of electrons in its outermost shell. This property is very important because the outer electrons, being less tightly held to the nucleus than the inner ones, play the most important role in bonding. These are called the *valence electrons*. Thus, all atoms in Group I have one valence electron; those in Group II have two, and so on. On the other hand, the Period indicates the outermost energy shell containing electrons. Thus, all atoms in Period 3 have electrons in the third shell; atoms in Period 4 have electrons in the fourth, and so on.

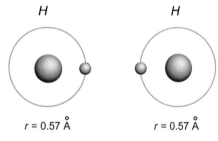

H   *H*

$r = 0.57$ Å $r = 0.57$ Å

FIGURE 4.5
Two atoms of hydrogen widely separated. The atomic radius is 0.57 Å.

Atomic Bonding

In the last section, we looked at the elements one by one. Now, based on the information we have acquired, we can imagine that it is possible to form a large variety of materials simply by combining different atoms with each other. How is that possible? If an atom is electrically neutral, why will an atom combine with other atoms? Remember what we have discussed before! When an electron is closest to the nucleus it will possess the least amount of energy. Hence, the closer it is to the nucleus, the more tightly it is held. In this context, we might now understand why atoms can combine with each other.

For simplification, look again at the hydrogen atom (see Figure 4.5). As shown in the figure, when two atoms of hydrogen are sufficiently apart, their atomic radius is 0.57 Å. Now let's see what happens when you bring the two atoms close enough (see Figure 4.6). The nucleus remains unchanged, but the electrons become attached to each nucleus and are brought closer to them. In this fashion, the electrons are shared by the two atoms, and as a consequence the system reduces its overall energy. Therefore, atomic bonding occurs because both electrons are now attached to both nuclei. This is why hydrogen usually occurs as the molecule H_2. The distance of 0.74 Å is then the equilibrium distance between the two atoms of hydrogen when forming the molecule, and thus the net force is zero. This can be expressed as

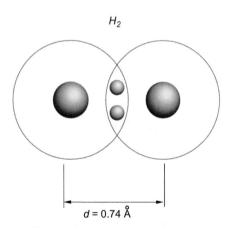

H_2

$d = 0.74$ Å

FIGURE 4.6
Sharing of electrons in the hydrogen molecule. The covalent radius r = 0.37 Å is now less than the atomic radius.

$$F_N = F_R + F_A = 0 \tag{4.2}$$

where F_N is the net force, F_A is the attractive force, and F_R is the repulsive force. This net force of zero corresponds to a minimum in energy. As a result, the system is in a state of less energy when the atoms are bonded together than when the energy used in each individual atom is added. In other words, the hydrogen molecule is more stable than each individual hydrogen atom. Thus, the hydrogen molecule is the result of overlapping electron orbitals. This overlapping of orbitals is called a *chemical bond*. Despite the fact that chemical bonds can be formed by atoms of the same kind, when atoms bond chemically, the properties are very different from those of the individual atom. For example, atoms of sodium (Na) and chlorine (Cl) bond to form NaCl, which is common table salt; however, individually, Na burns exposed to air, and Cl is a poisonous gas.

The reason we are so interested in chemical bonding is because the material properties and the processing of materials will depend strongly on the kind of chemical bonding that exists between the atoms. For example, materials that exhibit different bonding energies between their atoms will possess very different melting temperatures. Hence, high melting temperatures are associated with strong bonding energies. What about materials with high stiffness properties? By the same token, these materials also possess high bonding energies. The same applies for the coefficient of thermal expansion. It is now evident that materials with large bonding energies will possess smaller atomic vibrations and thus lower coefficients of thermal expansion.

Let's look at the types of chemical bonding that exist in nature. We start with perhaps the easiest type. This is *ionic bonding* (see Figure 4.7). It always happens in compounds composed of both metallic (Groups I and II A) and nonmetallic (Groups VI and VII A) elements. The role of the metallic element is to give its valence electrons, whereas the role of the nonmetallic element is to accept them. Again, the best example of this type of bonding is common table salt, NaCl (Figure 4.7). The sodium atom, with only one very loosely held electron in its outermost shell, bonds with one atom of chlorine, which has two paired orbits and one half-filled orbit in its outermost shell. When the sodium atom approaches the chlorine atom, overlapping takes place, resulting in bonding. However, the electron donated by the sodium atom spends most of its time near the chlorine atom, making the sodium atom a positively charged ion and the chlorine atom a negatively charged ion. As a result, the bond formed is *ionic*. In terms of properties, the ionic bonding is *isotropic*, which means that the bonds have the same characteristics in all directions and possess high bonding energies. As a result, ionic bonded materials exhibit higher melting temperatures, brittleness, and poor electrical conductivity.

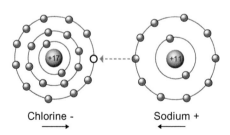

Chlorine -   Sodium +

FIGURE 4.7

Ionic bonding in sodium chloride—NaCl. The sodium atom "lends" one electron to the chlorine atom and becomes a positively charged ion, whereas the chlorine atom becomes a negatively charged ion.

One other type of chemical bonding is *covalent bonding*. In this type of bonding, there is a high probability of electrons spending most of their time in the region between atoms. Thus, for example, when two atoms of hydrogen form a molecule, the shared electrons spend most of their time in between the two hydrogen atoms. What would happen to the bonding if, for some reason, both electrons were on one side? The bonding would no longer be covalent. It would become an ionic bond because one atom would have an excess of electrons, whereas the other one would have a depletion of electrons. In general, although covalent bonds are very strong, materials bonded in this manner have poor ductility and poor electrical conductivity. Thus, for example, when silicon is bent, bonds must be broken for deformation to occur. This is not easy. In addition, for electrons to be mobile, bonds must be again broken. Therefore, covalent bonded materials are usually brittle and insulators.

Other materials are only partially covalent. Compounds, in general, are neither purely ionic nor purely covalent. Take, for example, sugar, which is composed of many units of the monomer $C_{12}H_{22}O_{11}$. In sugar, the atoms are tightly held by covalent bonds. However, between the various $C_{12}H_{22}O_{11}$ units, there are no covalent bonds and the interaction is weak. As a result, the crystals of sugar are easy to break. For a compound, the degree of either ionic or covalent bonding depends on the position of the elements in the periodic table. The greater the difference in electronegativity between the elements, the more ionic the bond, whereas the smaller the difference, the greater the degree of covalency.

Another important type of bonding is *metallic bonding*. We define that term as the ease with which atoms of elements lose valence electrons; it provides a basis for classifying atoms into two general groups, namely metals and nonmetals. In general, an element that has one, two, or three valence electrons in to its outermost shell will tend to lose the electrons and thus is considered a metal. What, then, is particular about metallic bonding? In fact, it is the idea that a metal is an aggregate of positively charged cores surrounded by a "sea" of electrons (see Figure 4.8). In the metallic structure, the electrons actually hold the ions in place; otherwise they would attract each other. One can think of this cohesion as a mass of solid balls bonded by a very strong liquid glue. As a result of this metallic structure, the path of an electron in the metallic bond is completely random around the aggregate of ions (see Figure 4.9). Therefore, metals are of course good conductors because electrons can easily move in any direction. Second, metals usually have a high

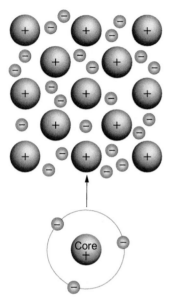

FIGURE 4.8

A metallic bond forms when atoms give up their valence electrons, which then form an electron sea.

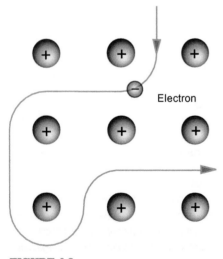

FIGURE 4.9

The path of an electron in a metallic structure.

degree of ductility because when the metal is bent, the electrons can easily rearrange their position, and thus bonds do not need to broken.

We are now left to discuss another type of chemical bonding, called *Van der Waals bonding* or *secondary bonding*. These are weak bonds that are typically electrostatic attractions. They arise from atomic or molecular *dipoles*, which are polar molecules (see Figure 4.10). Typically, some portions of the molecule tend to be positively charged, whereas other portions are negatively charged. A good example is water (see Figure 4.10). To show the polarity of water, a good experiment is to rub a comb against a piece of wool. As the wool becomes electrically charged, it attracts a thin stream of water.

Crystal Structure

In the last section we discussed the atomic structure and the various types of atomic bonding. We now examine the order in which atoms arrange themselves when they form a particular material. Argon, for example, is a gas, and thus it does not exhibit any atomic order. Water vapor is also a gas and hence behaves in a similar way to argon. However, water polar molecules can interact due to electrostatic forces. Therefore, when we compare the boiling temperatures of Ar $(-185\,°C)$ and H_2O $(100\,°C)$, it is clear that there is some degree of interaction between the water molecules, leading to stronger bonds. Now if we decrease sufficiently the temperature of the water vapor, there is less kinetic energy, and water vapor will transform into the liquid state. In this case there is some ordering but only at short distances. If we decrease the temperature even further, liquid water will turn into ice, a crystal that exhibits a high degree of order in three dimensions. This long-range order represented by a repeated array of atoms in three dimensions is called a *crystal*.

How do crystals form? In fact, it is really a statistical process. In general, nature is trying to come up with a structure that has the lowest possible energy configuration. But wait! How does an atom know how to recognize another atom? Actually, it uses a trial-and-error mechanism. What do we mean by this? Let's assume that we are talking about a metal in the liquid state. In this case, each atom in the liquid jumps about 10^{13} times/sec. In other words, a metal in this form has been testing over millions of years, at a rate of 10^{13} times/sec; this type of atoms would turn the compound more energetically favorable.

Let's now look in more detail at a crystal. The first thing we should know about a crystal is that it is composed of unit cells. A *unit cell* is a pattern that repeats itself in space (see Figure 4.11). Now

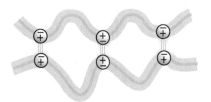

(a) Polymer molecules

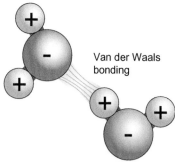

Van der Waals bonding

(b) Water molecules

FIGURE 4.10

A Van der Waals bond is formed due to polarization of molecules or groups of atoms. In water, electrons in the oxygen tend to concentrate away from the hydrogen. The resulting charge difference permits the molecule to be weakly bonded to other molecules.

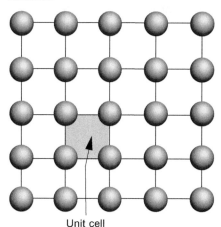

Unit cell

FIGURE 4.11

Concept of a unit cell. In this example the unit cell is formed by four solid circles arranged in a square array, which is repeated in a 2-D space. In a 3-D space, the unit cell could be for example a cube.

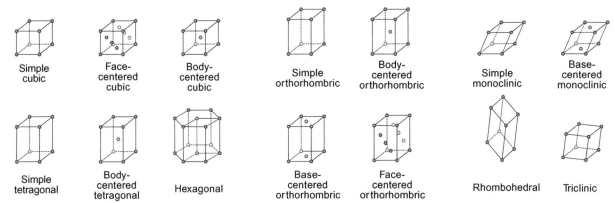

FIGURE 4.12

The 14 Bravais lattices.

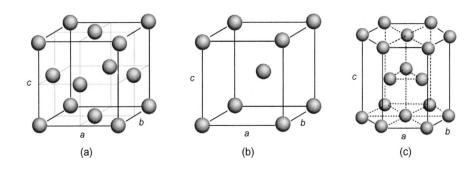

FIGURE 4.13

(a) Face-centered cubic unit cell, (b) body-centered cubic unit cell and (c) hexagonal closed packed unit cell.

that we know what a unit cell is, we can introduce another idea: the notion of a *crystal lattice*. This is the structure that arises from the repetition of a particular pattern or unit cell in a 3-D space. In nature, there are 14 possible unit cells, also called the *Bravais lattices* (see Figure 4.12). These 14 types of unit cells are grouped into seven possible crystal systems: cubic, tetragonal, orthorhombic, hexagonal, rhombohedral, monoclinic, and triclinic.

The simplest crystal structures typically belong to pure metals. The most common are the *face-centered cubic* (FCC), the *body-centered cubic* (BCC), and the *hexagonal closed packed* (HCP) crystal structures (see Figure 4.13). The FCC structure is composed of atoms located at each corner and at the center of each face. Some examples of elements that have this structure are the metals copper, silver, gold, nickel, and aluminum. The BCC unit cell has a configuration in which the atoms are located at each corner of the cube, with one atom at the center of the cube. Some examples of this crystal structure are iron, tungsten, and chromium. The HCP crystal structure has a configuration in which atoms are located in the corners of a hexagon, one atom in the center of the basal plane of the hexagon, and three atoms in the center of the cell, off-centered from the c-axis. Some examples of elements with this structure are the metals zinc, cobalt, and titanium.

One important aspect to point out is that some elements can assume more than one crystal structure. It all depends on certain thermodynamic conditions, such as temperature, pressure, stress, magnetic field, and electric field. These changes in crystal structure in a pure element are called *allotropic transformations*. A good example of this behavior is iron (Fe), an element that undergoes several changes in crystal structure. Iron is BCC at room temperature and atmospheric pressure, but it will change to an FCC crystal structure at around 750°C (see Figure 4.14). On the other hand, if we maintain the ambient temperature and increase the pressure to high levels (125 kbar), the BCC structure will transform to an HCP structure (Figure 4.14).

However, ceramic structures are considerably more complex. Most ceramic materials are compounds formed by metallic and nonmetallic elements. In addition, due to the fact that ceramics are predominantly ionic, they are usually described in terms of ions instead of atoms. Metallic ions are called *cations* (ions depleted in electrons), whereas nonmetallic ions are called *anions* (enriched in electrons). How can these ions affect the crystal structure? The answer is, by the magnitude of their electrical charge as well as the relative size of cations and anions. Regarding the first effect, the crystal will try as much as possible to be electrically neutral. With respect to the second effect, because cations are in general smaller than anions, the size ratio between the cations and anions will dictate the number of nearest neighbors surrounding each cation.

The most common ceramic structures are the rock salt structure, the cesium chloride structure, and the zinc blend structure. The rock salt structure derives it name from common table salt, which exhibits this type of crystal structure (see Figure 4.15). The structure is formed by two interpenetrating FCC lattices—one composed of cations, the other composed of anions. Other examples of ceramic materials with the rock salt structure include MgO, MnS, LiF, and FeO. The cesium chloride structure is a combination of two simple cubic structures, one formed by cations, the other formed by anions (Figure 4.15). The zinc blend structure is a combination of two FCC structures. Some examples of ceramics with zinc blend structures are ZnS, ZnTe, and SiC (a medium used in grinding paper). These three ceramic crystal structures have in common the fact that to fulfill the conditions of neutrality, there is one cation and one anion per unit cell.

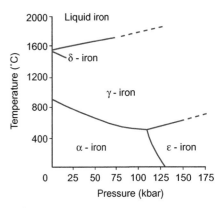

FIGURE 4.14
Pressure-temperature phase diagram for iron.

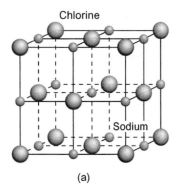
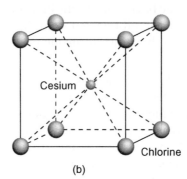
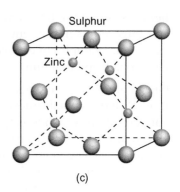

FIGURE 4.15

(a) Rock salt structure for NaCl, (b) cesium chloride structure, and (c) a zinc blend.

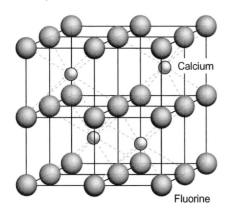

FIGURE 4.16

Calcium Fluorite structure (CaF₂).

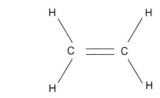

FIGURE 4.17

Mer unit of ethylene.

However, there are other types of ceramic structure for which the charges of cations and anions are not the same. An example is the calcium fluorite structure (CaF_2; see Figure 4.16). The unit cell is composed of eight cubes, for which the anions occupy the corner sites and the cations the center of the cubes. To achieve neutrality, there are half as many Ca^{2+} ions as F^- ions. As a result, only half the cubes exhibit the presence of a cation.

Molecular Structure

Normally, when we talk about metals or ceramics, we consider atoms or ions as single entities. However, in the case of polymers the situation is quite different; we normally think of molecules as the single entity. Polymer molecules consist of a repetition of a single unit called a *mer*, a term that derives from the Greek word *meros*, which means *part*. Therefore, *poly + mer* means *many parts*. The simplest type of polymer is called polyethylene, and the *mer* is ethylene. The ethylene molecule has the structure shown in Figure 4.17. The molecule is composed of carbon and hydrogen. As we have discussed, carbon has four valence electrons that can participate in bonding, whereas hydrogen has only one. Thus in the ethylene molecule, each carbon forms a single bond with each hydrogen atom (each atom is sharing one electron), whereas the two carbon atoms form a double bond between themselves (each atom is sharing two electrons).

Now let's imagine that we replace all hydrogen atoms with fluorine atoms. What will happen? The resulting polymer will be called polytetrafluroethylene (PTFE), or Teflon (see Figure 4.18). What about if we replace one hydrogen with one Cl atom? Then we obtain the polymer polyvinylchloride (PVC). What about doing the same thing but exchanging the hydrogen atom with a CH_3 methyl group? Then we obtain the polymer polypropylene (PP). We can further characterize the types of polymers by observing the repeating unit. Hence, when all the repeating units along a chain are of the same type, the resulting polymer is called a *homopolymer*. However, if chains are composed of two or more different units, the polymer is called a *copolymer*. In addition, if the different units are arranged in blocks, the polymer is called a *block copolymer* (see Figure 4.19).

The molecular structure of polymers can be further categorized as linear polymers, branched polymers, cross-linked polymers, and network polymers. In the case of linear polymers, the mer units are joined together end to end in single chains. The molecules will be spaghetti-like and characterized by many Van der Waals bonds between the chains (see Figure 4.20). Some examples are the polymers polyethylene, polystyrene, and nylon. In the case of branched polymers, the main chains are connected through branches. As a result, the density of the polymer is lowered (Figure 4.20). In cross-linked polymers, adjacent linear chains are joined one to another at various positions by covalent bonds. This process is called *cross-linking* and is accomplished by including additives that form the covalent bonds (Figure 4.20). Examples of this type of polymer are rubber elastic materials. Finally, network polymers form when mer units can bond in three dimensions (Figure 4.20). An example of this type of polymer is epoxy.

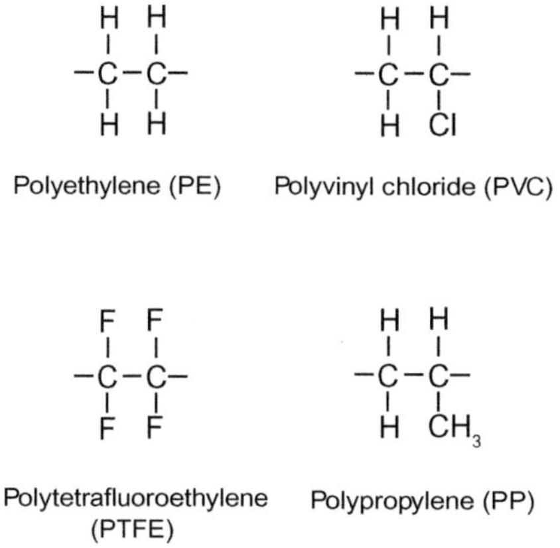

Defects

In the previous chapters we have always assumed a pure material or a compound with no defects in it. Thus the structures we have talked about were in "some way" ideal situations. However, in fact, 100% pure materials and with no defects do not exist, although some materials can have small amounts of impurities and/or defects. The origin of these defects is very diverse, ranging from atomic packing problems during processing to the formation of interfaces with poor atomic registry or the generation of defects during deformation.

On the basis of our discussion so far, you are probably thinking that the presence of defects is deleterious to materials. In some cases that's true, but in many cases defects are extremely beneficial. Some examples include:

- The presence of small amounts of carbon in iron (known as steel) makes possible the achievement of high strengths.

- The addition of 0.01% of arsenic can increase the conductivity of Si by 10,000 times.

- Some defects called *dislocations* are responsible for plastic deformation in materials.

Polyethylene (PE)

Polyvinyl chloride (PVC)

Polytetrafluoroethylene (PTFE)

Polypropylene (PP)

FIGURE 4.18

Mer structures of the more common polymeric materials.

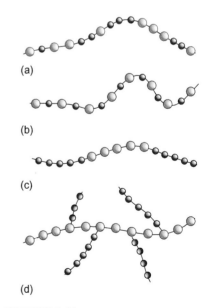

(a)

(b)

(c)

(d)

FIGURE 4.19

Schematic representations of (a) random, (b) alternating, (c) block, and (d) graft copolymers.

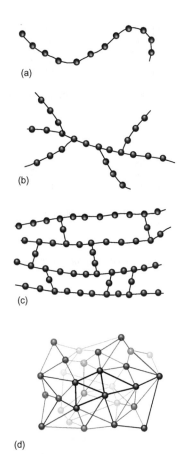

FIGURE 4.20

Schematic representation of (a) linear, (b) branched, (c) cross-linked, and (d) network polymer molecular structures.

In general, we can classify the defects as point defects, linear defects, planar defects (interfacial defects), or volume defects (bulk defects), for which the scale of each class is shown in Figure 4.21. Let's examine each class of defect in more detail.

The simplest point defect is called a *vacancy*, which is a lattice point from which an atom is missing (see Figure 4.22). Vacancies are introduced into a material during solidification or heat treatments or by radiation of atomic particles. In fact, in a nuclear power plant, where radiation is continuously being produced, the monitoring of the formation of vacancies is crucial for the safety of the plant. The presence of vacancies in a crystal is a necessity because the presence of vacancies will increase the entropy (randomness) of the crystal. In addition, the presence of a vacancy changes the stress field of the crystal. As shown in Figure 4.22, the vacancy induces a tensile stress field around the neighboring atoms. In addition, the equilibrium number of vacancies increases exponentially with temperature. Typically, at room temperature, there is one vacancy per 1 million atoms, whereas at the melting temperature, there are 1000 vacancies per million atoms.

A self-interstitial point defect can also form in materials. This occurs when an atom from the lattice goes into an interstitial position, a small space that is not usually occupied by any atom (see Figure 4.23). The self-interstitial atom creates large distortions in the lattice because the initial available space is smaller than the atom dimensions.

In ionic structures, such as ceramic materials, because of neutrality, there are two types of point defects: (1) the *Schottky defect*, which is a pair of defects formed by a vacancy of one cation and a vacancy of one anion (see Figure 4.24), and (2) the *Frenkel defect*, which is formed by a vacancy of one cation and a self-interstitial cation (Figure 4.24) or a vacancy of one anion and a self-interstitial anion.

So far we have only discussed defects in pure solids or compounds. However, as mentioned before, impurities must exist. Even if the metal is almost (99.9999%) pure, there are still 10^{23} impurities in 1 m^3. Therefore the following question arises: Where do the impurities go? That all depends on the impurity and the host material. The final result will be a consequence of (1) the kind of impurity, (2) the impurity concentration, and (3) the temperature and pressure. However, in general, impurities can go to a substitutional site, that is, a site occupied by the host atom. In this case they will be called *substitutional atoms*. Or impurities can go to an interstitial site and will be called *interstitial atoms* (different from self-interstitials). Whether the impurities go to a substitutional or interstitial site will depend on (1) atomic size, (2) crystal structure, (3) electronegativity, and (4) valence electrons. With respect to atomic size, interstitial atoms are small compared with host atoms.

In covalent bonded materials, substitutional atoms can create a unique imperfection in the electronic structure if the impurity atom is from a group in the periodic table other than that of the host atoms. An example is the addition of As or P (Group V) in Si (Group IV; see Figure 4.25). Only four of five valence electrons of these impurities can participate in the bonding, because there are only four possible bonds with neighboring atoms. The extra nonbonding electron is loosely bound to the region around the impurity atom in a weak electrostatic interaction. Thus the binding energy of this electron is relatively small, in which case it becomes a free or conducting electron.

In polymers, the addition of impurities can also have significant consequences. For example, natural rubber becomes cross-linked when small amounts of sulfur (5%) are added. As a result, the mechanical properties change dramatically. Natural rubber has a tensile strength of 300 psi, whereas vulcanized rubber (sulfur addition) has a tensile strength of 3000 psi.

Now let's discuss another class of defects called *linear defects.* These defects, also called *dislocations,* are the main mechanism in operation when a material is deformed plastically. Currently, several techniques are available for the direct observation of dislocations. The transmission electron microscope (TEM) is probably the most utilized in this respect.

Let's look at the simplest case, the *edge dislocation.* Imagine the following sequence of events: (1) take a perfect crystal, (2) make a cut in the crystal, (3) open the cut, and (4) insert an extra plane of atoms (see Figure 4.26). The end result is an edge dislocation. These dislocations are typically generated during processing or in service, if subjected to enough stress. The presence and motion of dislocations dictate whether materials are ductile or brittle. Because metals can easily generate and move dislocations, they are ductile. On the other hand, ceramic materials have a high difficulty in nucleating and moving dislocations due to the covalent/ionic character of the bonds, and they are therefore brittle. One of the important parameters related to dislocations is the knowledge of the amount of dislocation length per unit volume. This is called the *dislocation density* and is given by

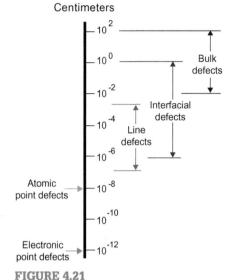

FIGURE 4.21
Range of scales for the various classes of defects.

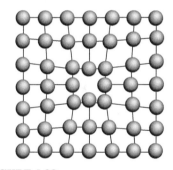

FIGURE 4.22
A vacancy defect.

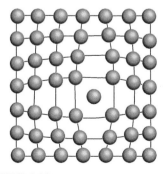

FIGURE 4.23
A self-interstitial defect.

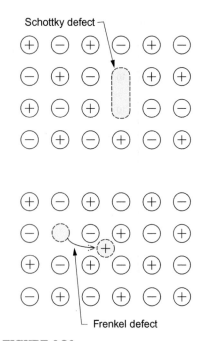

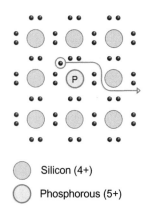

FIGURE 4.24
A Schottky defect and a Frenkel defect.

○ Silicon (4+)

○ Phosphorous (5+)

FIGURE 4.25
Addition of phosphorous, having five valence electrons, to silicon leads to an increase in electrical conductivity due to an extra bonding electron.

$$\rho_D = \frac{L_D}{V} \qquad (4.3)$$

where L_D is the total length of dislocations and V is the volume.

Another class of defects is that of *interfacial defects*. Among these, we'll first discuss the free surfaces. All solid materials have finite sizes. As a result, the atomic arrangement at the surface is different from within the bulk. Typically, surface atoms form the same crystal structure as in the bulk, but the unit cells have slightly larger lattice parameters. In addition, surface atoms have more freedom to move and thus have higher entropy. We can now understand that different crystal surfaces should have different energies, depending on the number of broken bonds.

The second type of interfacial defect is the *grain boundary*. This boundary separates regions of different crystallographic orientation. The simplest form of a grain boundary is called a *tilt boundary* because the misorientation is in the form of a simple tilt about an axis (see Figure 4.27).

We know that grain boundaries have also an interfacial energy due to the disruption of atomic periodicity. However, will they have higher or lower energies than free surfaces? The answer is lower energies. This is because grain boundaries exhibit some distortion in the type of bonds they form, but there are no absent bonds. Most of the materials utilized in our daily lives are polycrystalline materials. This means that the materials are composed of many crystals with different orientations, separated by grain boundaries (see Figure 4.28). It turns out that this network of grains significantly affects several of the material properties.

The last type of interfacial defect is the *interphase boundary*. These are the boundaries that separate regions of materials with different structure and/or composition. An example is a dentist's drill, which is a mixture of small crystals of tungsten carbide surrounded by a matrix of cobalt. The interfaces between the tungsten carbide and the cobalt matrix are interphase boundaries.

Finally, we are left to discuss bulk defects. Most of the time these volume defects are introduced during the production of materials. In some cases, impurities in a material combine with each other and form *inclusions*. These are second-phase particles that can considerably affect the mechanical properties of materials.

One other type of defect is the *casting defect*. These can be cavities and gas holes produced under certain conditions of temperature and pressure. A different type of volume defect is the cracks formed during heating and cooling cycles or during formability processes. Finally, welding defects can be formed during welding procedures due to the fact that the heat generated during welding is not uniform. As a consequence, a region affected by the heat is produced, where the properties change gradually away from the heat source.

4.3 MECHANICAL BEHAVIOR

Stress, Strain, Stiffness, and Strength

Stress is something that is applied to a material by loading it. *Strain*—a change of shape—is the material's response. It depends on the magnitude of the stress and the way it is applied—the *mode of loading*. Ties carry tension; often they are cables. Columns carry compression; tubes are more efficient as columns than are solid rods of similar areas because they don't buckle as easily. Beams carry bending moments. Shafts carry torsion. Pressure vessels contain a pressure. Often they are shells: curved, thin-walled structures.

Stiffness is the resistance to change of shape that is *elastic*, meaning that the material returns to its original shape when the stress is removed. *Strength* is its resistance to permanent distortion or total failure. Stress and strain are not material properties; they describe a stimulus and a response. Stiffness (measured by the elastic modulus E, defined in a moment) and strength (measured by the yield strength σ_y or tensile strength σ_{ts}) *are* material properties. Stiffness and strength are central to mechanical design.

The elastic moduli reflect the stiffness of the bonds that hold atoms together. There is not much you can do to change any of this, so the moduli of pure materials cannot be manipulated at all. If you want to control them you must either mix materials together, making *composites*, or disperse space within them, making *foams*.

Modes of loading

Most engineering components carry loads. Their elastic response depends on the way the loads are applied. Usually one mode dominates, and the component can be idealized as one of the simply loaded cases in Figure 4.29: *tie, column, beam, shaft,* or *shell*. Ties carry simple axial tension, shown in (a) in the figure; columns do the same in simple compression, as in (b). Bending of a beam (c) creates simple axial tension in elements above the neutral axis (the center line, for a beam with a symmetric cross-section) and simple

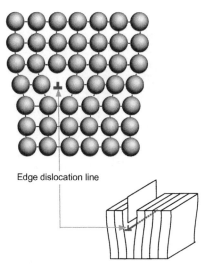

Edge dislocation line

FIGURE 4.26
Sequence of events leading to the formation of an edge dislocation.

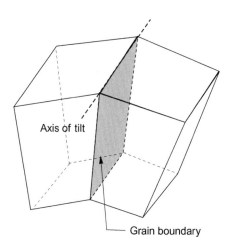

Axis of tilt

Grain boundary

FIGURE 4.27
Tilt boundary.

FIGURE 4.28

Transmission electron microscopy image of polycrystalline copper. These grains are separated by grain boundaries. (Courtesy of R. Calinas, University of Coimbra; M. Vieira, University of Coimbra and P.J. Ferreira, University of Texas at Austin.)

compression in those below. Shafts carry twisting or torsion (d), which generates shear rather than axial load. Pressure difference applied to a shell, such as the cylindrical tube shown at (e), generates biaxial tension or compression.

Stress

Consider a force F applied as normal to the face of an element of material, as in Figure 4.30 on the left of row (a) in the figure. The force is transmitted through the element and balanced by an equal but opposite force on the other side so that it is in equilibrium (it does not move). Every plane normal to F carries the force. If the area of such a plane is A, the *tensile stress* σ in the element (neglecting its self-weight) is

$$\sigma = \frac{F}{A} \tag{4.4}$$

If the sign of F is reversed, the stress is compressive and given a negative sign. Forces are measured in Newtons (N), so stress has the dimensions of N/m^2. But a stress of 1 N/m^2 is tiny—atmospheric pressure is 10^5 N/m^2—so the usual unit is MN/m^2 (10^6 N/m^2), called *megapascals*, symbol MPa.

If, instead, the force lies parallel to the face of the element, three other forces are needed to maintain equilibrium (Figure 4.30b). They create a state of shear in the element. The shaded plane, for instance, carries the *shear stress* τ of

$$\tau = \frac{F_s}{A} \tag{4.5}$$

The units, as before, are MPa.

One further state of multiaxial stress is useful in defining the elastic response of materials: that produced by applying equal tensile or compressive forces to all six faces of a cubic element, as in Figure 4.30c. *Any* plane in the cube now carries the same state of stress; it is equal to the force on a cube face divided by its area. The state of stress is one of *hydrostatic pressure*, symbol p, again with the units of MPa. There is an unfortunate convention here. Pressures are positive when they push—the reverse of the convention for simple tension and compression.

Engineering components can have complex shapes and can be loaded in many ways, creating complex distributions of stress. But no matter how complex, the stresses in any small element within the component can always be described by a combination of tension, compression, and shear.

Strain

Strain is the response of materials to stress (second column of Figure 4.30). A tensile stress σ applied to an element causes the element to stretch. If the element in Figure 4.30a, originally of side L_o, stretches by $\delta L = L - L_o$, the nominal *tensile strain* is

$$\varepsilon = \frac{\delta L}{L_o} \tag{4.6}$$

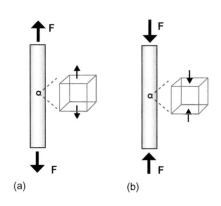
(a)   (b)

A compressive stress shortens the element; the nominal compressive strain (negative) is defined in the same way. Since strain is the ratio of two lengths, it is dimensionless.

A shear stress causes a *shear strain* γ (Figure 4.30b). If the element shears by a distance w, the shear strain

$$\tan(\gamma) = \frac{w}{L_o} \approx \gamma \tag{4.7}$$

In practice $\tan\gamma \approx \gamma$ because strains are almost always small. Finally, a hydrostatic pressure p causes an element of volume V to change in volume by δV. The volumetric strain, or *dilatation* (Figure 4.30c), is

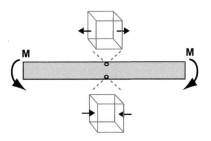
(c)

$$\Delta = \frac{\delta V}{V} \tag{4.8}$$

Stress-strain curves and moduli

Figures 4.31, 4.32, and 4.33 show typical tensile stress-strain curves for a metal, a polymer, and a ceramic, respectively; that for the polymer is shown at four different temperatures relative to its glass temperature, T_g. The initial part, up to the elastic limit σ_{el}, is approximately linear (Hooke's law), and it is elastic, meaning that the strain is recoverable; the material returns to its original shape when the stress is removed. Stresses above the elastic limit cause permanent deformation (ductile behavior) or brittle fracture.

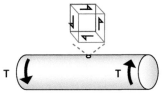
(d)

Within the linear elastic regime, strain is proportional to stress (Figure 4.30, third column). The tensile strain is proportional to the tensile stress:

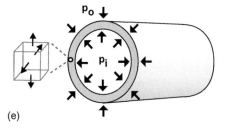
(e)

FIGURE 4.29

Modes of loading and states of stress: (a) tie, (b) column, (c) beam, (d) shaft, and (e) shell.

$$\sigma = E\varepsilon \tag{4.9}$$

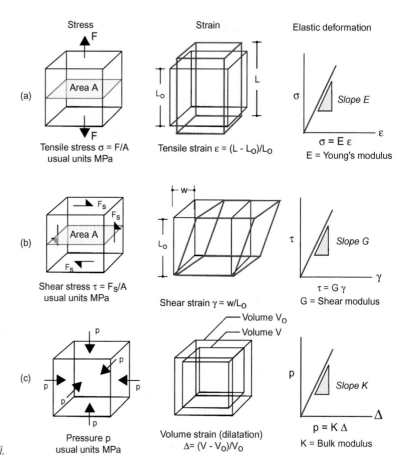

FIGURE 4.30

The definitions of stress, strain, and elastic moduli.

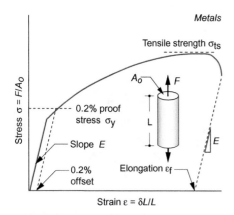

FIGURE 4.31

Stress-strain curve for a metal.

and the same is true in compression. The constant of proportionality, E, is called *Young's modulus*. Similarly, the shear strain γ is proportional to the shear stress τ

$$\tau = G\gamma \qquad (4.10)$$

and the dilatation Δ is proportional to the pressure p:

$$p = K\Delta \qquad (4.11)$$

where G is the *shear modulus* and K the *bulk modulus,* as illustrated in the third column of Figure 4.30. All three of these moduli have the same dimensions as stress, that of force per unit area (N/m^2 or Pa). As with stress it is convenient to use a larger unit, this time an even bigger one, that of 10^9 N/m^2, giga pascals, or GPa.

Strength and ductility

If a material is loaded above its yield strength, it deforms plastically or it fractures. The *yield strength* σ_y units, MPa or MN/m^2, require careful definition. For metals, the onset of plasticity is not always distinct, so we identify σ_y with the *0.2% proof stress*, that is, the stress at which the stress-strain curve for axial loading deviates by a strain of 0.2% from the linear-elastic line, as shown in Figure 4.31. When strained beyond the yield point, most metals *work harden*, causing the rising part of the curve, until a maximum, the *tensile strength*, is reached. This is followed in tension by nonuniform deformation (*necking*) and fracture.

For polymers, σ_y is identified as the stress at which the stress-strain curve becomes markedly nonlinear—typically, a strain of 1% (see Figure 4.32). The behavior beyond yield depends on the temperature relative to the glass temperature T_g. Well below T_g, most polymers are brittle. As T_g is approached, plasticity becomes possible until, at about T_g, thermoplastics exhibit *cold drawing*: large plastic extension at almost constant stress during which the molecules are pulled into alignment with the direction of straining, followed by hardening and fracture when alignment is complete. At still higher temperatures, thermoplastics become viscous and can be molded; thermosets become rubbery and finally decompose.

Ductility is a measure of how much plastic strain a material can tolerate. It is measured in standard tensile tests by the *elongation* ε_f (the tensile strain at break) expressed as a percent (Figures 4.31 and 4.32). Strictly speaking, ε_f is not a material property, because it depends on the sample dimensions—the values that are listed in handbooks and in the CES software are for a standard test geometry—but it remains useful as an indicator of a material's ability to be deformed.

Ceramics do not deform plastically in the way that metals and polymers do. It is still possible to speak of a strength or elastic limit, σ_{el} (Figure 4.33). Its value is larger in compression than in tension by a factor of about 12.

Hardness

Tensile and compression tests are not always convenient; you need a large sample and the test destroys it. The hardness test (see Figure 4.34) avoids these problems, although it has problems of its own. In it, a pyramidal diamond or a hardened steel ball is pressed into the surface of the material, leaving a tiny permanent indent, the size of which is measured with a microscope. The indent means

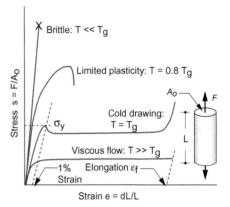

FIGURE 4.32
Stress-strain curve for a polymer.

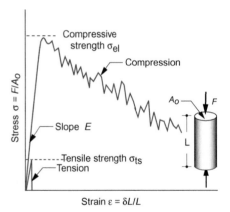

FIGURE 4.33
Stress-strain curve for a ceramic.

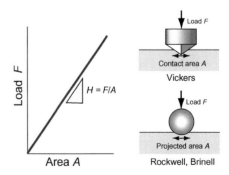

FIGURE 4.34
The hardness test. The Vickers test uses a diamond pyramid; the Rockwell and Brinell tests use a steel sphere.

that plasticity has occurred, and the resistance to it—a measure of strength—is the load F divided by the area A of the indent projected onto a plane perpendicular to the load

$$H = \frac{F}{A} \tag{4.12}$$

The indented region is surrounded by material that has not deformed, and this constrains it so that H is larger than the yield strength σ_y; in practice it is about $3\sigma_y$.

Strength, as we have seen, is measured in units of MPa, and since H is a strength it would be logical and proper to measure it in MPa, too. But things are not always logical and proper, and hardness scales are among those that are not. A commonly used scale, that of *Vickers*, symbol H_v, uses units of kg/mm², with the result that

$$H_v \approx \frac{\sigma_y}{3} \tag{4.13}$$

Figure 4.35 shows conversions to other scales.

The hardness test has the advantage of being nondestructive, so strength can be measured without destroying the component, and it requires only a tiny volume of material. But the information it provides is less accurate and less complete than the tensile test, so it is not used to provide critical design data.

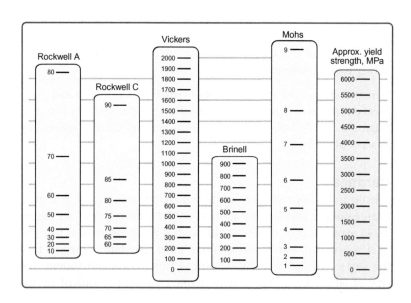

FIGURE 4.35
Common hardness scales compared with yield strength.

The Origins of Strength

In thinking of ways to make materials stronger, it is worth first asking: Is there an upper limiting strength that, for fundamental reasons, cannot be exceeded? If there is, then the proper measure of success in achieving high strength is proximity to this limit.

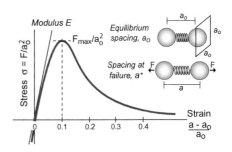

FIGURE 4.36
The stress-strain curve for a single atomic bond. (It is assumed that each atom occupies a cube of side a_o.)

Perfection: The ideal strength

Physicists calculate the greatest strength a material could, in theory, have from their understanding of the bonds that hold atoms together. The bonds in iron, titanium, and most other metals are strong. Those in diamond, silicon carbide, and other ceramics are even stronger. But those that attach one polymer molecule to another are weak—weaker by far than those of metals or ceramics. The most obvious consequence of this is in a material's stiffness: The elastic modulus, E, is a direct measure of the strength of the interatomic bonds. Physicists calculate that the greatest strength a material could have—the "ideal strength"—is also proportional to bond strength. The argument goes like this:

The bonds between atoms act like little springs, and like any other spring, they have a breaking point. Figure 4.36 shows a stress-strain curve for a single bond. Here an atom is assumed to occupy a cube of side a_o so that a force F acting on the cube is equivalent to a stress $\sigma = F/a_o^2$. The force stretches the bond from its initial length a_o to a new length a, giving a strain $\varepsilon = (a - a_o)/a_o$. The initial part of this curve is linear, with a slope equal to the modulus, E. Stretched further, the curve passes through a maximum and sinks to zero as the atoms lose communication. The peak is the bond strength; if you pull harder than this it will break. The same is true if you shear it rather than pull it.

The distance over which interatomic forces act is small; a bond is broken if it is stretched to more than about 10% of its original length. So the force needed to break a bond is roughly

$$F_{max} \approx S\frac{a_o}{10} \qquad (4.14)$$

where $S = F/(a - a_o) = Ea_o$ is the bond stiffness. On this basis, the *ideal strength* of a solid should therefore be roughly

$$\sigma_{ideal} \approx \frac{F_{max}}{a_o^2} = \frac{S}{10a_o} = \frac{E}{10}$$

or

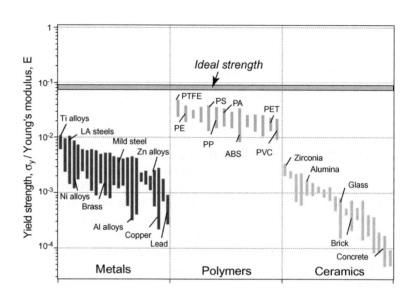

FIGURE 4.37

The ideal strength is predicted to be about E/15, where E is Young's modulus. The figure shows σ_y/E with a shaded band at the ideal strength.

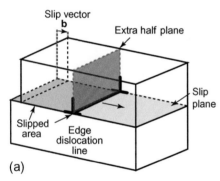

(a)

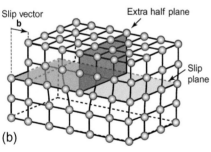

(b)

FIGURE 4.38

(a) Making a dislocation by cutting, slipping, and rejoining bonds across a slip plane. (b) The atom configuration at an edge dislocation in a simple cubic crystal. The configurations in other crystal structures are more complex, but the principle remains the same.

$$\frac{\sigma_{ideal}}{E} \approx \frac{1}{10} \tag{4.15}$$

This doesn't allow for the curvature of the force-distance curve; more refined calculations give a ratio of 1/15.

Figure 4.37 shows the ratio of the yield strength σ_y to the modulus E for metals, polymers, and ceramics. None achieves the ideal ratio of 1/15; most don't even come close. Why not? It's a familiar story: Like most things in life, materials are imperfect.

Crystalline imperfection: Dislocations and plasticity

Crystals contain imperfections of several kinds. The key player from a mechanical point of view is the *dislocation*, portrayed in Figure 4.38. *Dislocated* means *out of joint*, and this is not a bad description of what is happening here. The figure shows, on the left, how to make a dislocation. The crystal is cut along an atomic plane up to the line shown as ⊥ — ⊥, the top part is slid across the bottom by one full atom spacing, and the atoms are reattached across the cut plane to give the atom configuration shown on the right. There is now an extra half-plane of atoms with its lower edge along the ⊥ — ⊥ line, the *dislocation line*—the line separating the part of the plane that has slipped from the part that has not. Dislocations distort the lattice and so have elastic energy associated with them. If they cost energy, why are they there? To grow a perfect crystal just one cubic centimeter in volume from a liquid or vapor, about 10^{23} atoms have to find their proper sites on the perfect lattice, and the chance of this happening is just too small. Even with the greatest care in assembling them, all crystals contain dislocations (also see Section 4.2).

It is dislocation that makes metals soft and ductile. Recall that the strength of a perfect crystal computed from interatomic forces gives an "ideal strength" around $E/15$ but that the strengths of real engineering materials are much less. This was a mystery until halfway through the last century—a mere 60 years ago—when an Englishman, Geoffrey Taylor, and a Hungarian, Egon Orowan, realized that a "dislocated" crystal could deform at stresses far below the ideal strength. When a dislocation moves, it makes the material above the slip plane slide relative to that below, producing a shear strain. Figure 4.39 shows how this happens. At the top is a perfect crystal. In the central row a dislocation enters from the left, sweeps through the crystal, and exits on the right. By the end of the process the upper part has slipped by b, the slip vector (or *Burger's vector*) relative to the part below. The result is the shear strain γ shown at the bottom.

It is far easier to move a dislocation through a crystal, breaking and remaking bonds only along its line as it moves, than it is to simultaneously break all the bonds in the plane before remaking them. It is like moving a heavy carpet by pushing a fold across it rather than sliding the whole thing at one go. In real crystals it is easier to make and move dislocations on some planes than on others. The preferred planes are called *slip planes* and the preferred directions of slip in these planes are called *slip directions*. Slip displacements are tiny; one dislocation produces a displacement of about 10^{-10} m. But if large numbers of dislocations traverse a crystal, moving on many different planes, the shape of a material changes at the macroscopic length scale.

Why does a stress make a dislocation move?

A shear stress exerts a force f on a dislocation, pushing it across the slip plane. Crystals resist the motion of dislocations with a friction-like resistance f^* per unit length; we will examine its origins in a moment. For yielding to take place, the force f caused by the external stress must overcome the resistance f^*.

Imagine that one dislocation moves right across a slip plane, traveling the distance L_2, as in Figure 4.40. In doing so, it shifts the upper half of the crystal by a distance b relative to the lower half. The shear stress τ acts on an area L_1L_2, giving a shear force $F_s = \tau L_1 L_2$ on the surface of the block. If the displacement parallel to the block is b, the force does work:

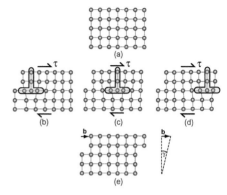

FIGURE 4.39

An initially perfect crystal is shown at (a). The passage of the dislocation across the slip plane, shown in the sequence (b), (c), and (d), shears the upper part of the crystal over the lower part by the slip vector b. When it leaves, the crystal has suffered a shear strain γ.

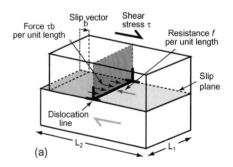

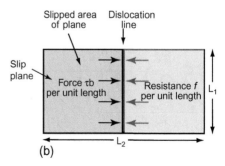

FIGURE 4.40

The force on a dislocation. (a) Perspective view, and (b) plan view of slip plane.

$$W = \tau L_1 L_2 b \qquad (4.16)$$

This work is done against the resistance f^* per unit length, or f^*L_1 on the length L_1, and it does so over a displacement L_2 (because the dislocation line moves this far against f^*), giving total work against f^* of $f^*L_1L_2$. Equating this to the work W done by the applied stress τ gives

$$\tau b = f^* \qquad (4.17)$$

So, provided the shear stress τ exceeds the value f^*/b, it will make dislocations move and cause the crystal to shear. The way to make it stronger is to increase this resisting force, and that is where the nanoscale comes in. We look first at the oldest of the nanostructuring schemes: *dispersion hardening*.

Strengthening mechanisms

As we have seen, dislocations distort the crystal locally, and the local distortion has an associated energy. This gives the dislocation a *line tension*—the equivalent, for a line, of the surface tension of a liquid. If the crystal is perfectly regular, a dislocation can remain straight as it sweeps across crystal planes. Dissolved impurities obstruct the motion a little (Figure 4.41a) by making the slip plane uneven; it is like dragging a carpet over a rough floor. Larger, stronger obstacles obstruct motion much more effectively—more like pinning the carpet down with tacks, though here the analogy is less good. To get real: It is because, when strong obstacles are placed in its path, the dislocation must bend between and around them, and in doing so its length increases (Figure 4.41b). Increase in length means increase in energy, and this energy increases the stress needed to deform the material, making it stronger. If the particles are far apart, the increases in line length and strength are small. But

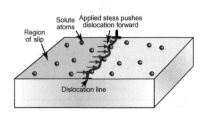
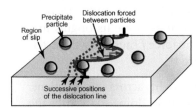
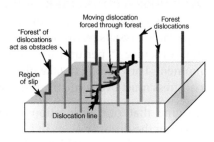

FIGURE 4.41

(a) Dissolved atoms obstruct dislocation motion, giving solution strengthening. *(b) Discrete obstacles are more effective in obstructing motion, provided their spacing is nanoscale, giving* dispersion hardening. *(c) Dislocation motion is obstructed by other dislocations introduced by plastic deformation, giving* work hardening.

when the size and spacing of the dispersed particles are of nano-dimensions, the increase in strength is dramatic. Finally, ductile materials become harder when deformed because of work hardening, shown in Figure 4.41c. Dislocations accumulate on many slip planes during plastic deformation; those on one set of slip planes intersect the slip planes of others, obstructing their motion.

4.4 THERMAL BEHAVIOR

Heat is atoms or molecules in motion. In gases, the molecules fly between occasional collisions with each other. In solids, by contrast, they vibrate about their mean positions; the higher the temperature, the greater the amplitude of vibrations. From this perception emerges all our understanding of the intrinsic thermal properties of solids: their heat capacity, expansion coefficient, conductivity, even melting.

Heat affects mechanical, electrical, and optical properties, too. As temperature rises, materials expand, the elastic modulus decreases, the strength falls, and the material starts to creep, deforming slowly with time at a rate that increases as the melting point is approached until, on melting, the solid loses all stiffness and strength. The electrical resistivity rises with temperature, the refractive index falls, color may change—all effects that can be exploited in design.

Intrinsic Thermal Properties

Two temperatures, the *melting temperature, T_m,* and the *glass temperature, T_g* (units for both: Kelvin, K, or Centigrade, C), are fundamental points of reference because they relate directly to the strength of the bonds in the solid. Crystalline solids have a sharp melting point, T_m. Noncrystalline solids do not; the glass temperature T_g characterizes the transition from true solid to very viscous liquid. It is helpful in engineering design to define two further temperatures: the *maximum* and *minimum service temperatures* T_{max} and T_{min} (units for both: K or C). The first tells us the highest temperature at which the material can be used continuously without oxidation, chemical change, or excessive distortion becoming a problem. The second is the temperature below which the material becomes brittle or otherwise unsafe to use.

It costs energy to heat a material. The energy to heat 1 kg of a material by $1°K$ is called the *heat capacity* or *specific heat,* and since the measurement is usually made at constant pressure (atmospheric

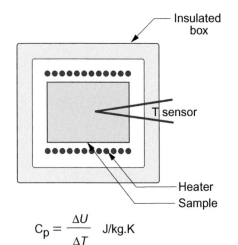

$$C_p = \frac{\Delta U}{\Delta T} \quad \text{J/kg.K}$$

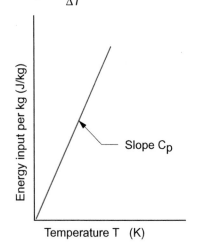

FIGURE 4.42

Measuring heat capacity, C_p. Its units are J/kg·K.

pressure), it is given the symbol C_p. Heat is measured in Joules, symbol J, so the units of specific heat are J/kg.K. In dealing with gases, it is more usual to measure the heat capacity at constant volume (symbol C_v), and for gases this differs from C_p. For solids the difference is so slight that it can be ignored, and we shall do so here. C_p is measured by the technique of calorimetry, which is also the standard way of measuring the glass temperature T_g. Figure 4.42 shows how, in principal, this is done. A measured quantity of energy (here, electrical energy) is pumped into a sample of material of known mass. The temperature rise is measured, allowing the energy/kg.K to be calculated. Real calorimeters are more elaborate than this, but the principle is the same.

Most materials expand when they are heated (see Figure 4.43). The thermal strain per degree of temperature change is measured by the *linear thermal-expansion coefficient, α*. It is defined by

$$\alpha = \frac{1}{L}\frac{dL}{dT} \qquad (4.18)$$

where L is a linear dimension of the body. If it is anisotropic, it expands differently in different directions, and two or more coefficients are required. Since strain is dimensionless, the units of α are K^{-1} or, more conveniently, "microstrain/K," that is, $10^{-6}\ K^{-1}$.

Power is measured in Watts; a Watt (W) is Joule/sec. The rate at which heat is conducted through a solid at the steady state (meaning that the temperature profile does not change with time) is measured by the *thermal conductivity, λ* (units: W/m.K). Figure 4.44 shows how it is measured: by recording the heat flux q (W/m²) flowing through the material from a surface at higher temperature T_1 to a lower one at T_2 separated by a distance x. The conductivity is calculated from Fourier's law:

$$q = -\lambda \frac{dT}{dx} = \lambda \frac{(T_1 - T_2)}{x} \qquad (4.19)$$

Thermal conductivity, as we have said, governs the flow of heat through a material at the steady state. The property governing transient heat flow (when temperature varies with time) is the *thermal diffusivity, a* (units: m²/s). The two are related by

$$a = \frac{\lambda}{\rho C_p} \qquad (4.20)$$

where ρ is the density and C_p is, as before, the heat capacity. The thermal diffusivity can be measured directly by measuring the time it takes for a temperature pulse to traverse a specimen of known thickness when a heat source is applied briefly to one side; or it can be calculated from λ and (ρC_p) via Equation 4.20.

The Physics of Thermal Properties

Heat capacity

Atoms in solids vibrate about their mean positions with an amplitude that increases with temperature. Atoms in solids can't vibrate independently of each other because they are coupled by their inter atomic bonds; the vibrations are like standing elastic waves. Some of these have short wavelengths and high energy, others long wavelengths and lower energy (see Figure 4.45). The shortest possible wavelength, λ_1, is just twice the atomic spacing; the other vibrations have wavelengths that are longer. In a solid with N atoms there are N discrete wavelengths, and each has a longitudinal mode and two transverse modes, $3N$ modes in all. Their amplitudes are such that, on average, each has energy $k_B T$ where k_B is Boltzmann's constant, 1.38×10^{-23} J/K. If the volume occupied by an atom is Ω, the number of atoms per unit volume is $N = 1/\Omega$ and the total thermal energy per unit volume in the material is $3k_B T/\Omega$. The heat capacity per unit volume, ρC_p, is the *change* in this energy per Kelvin change in temperature, giving

$$\rho C_p = \frac{3k_B}{\Omega} \, J/m^3 K \qquad (4.21)$$

The result matches well with measured values of the heat capacity.

Thermal expansion

If a solid expands when heated (and almost all do), it must be because the atoms are moving further apart. Figure 4.46 shows how this happens. The force-displacement curve is not quite straight; the bonds become stiffer when the atoms are pushed together and less stiff when they are pulled apart. Atoms vibrating in the way described earlier oscillate about a mean spacing that increases with the amplitude of oscillation and thus with increasing temperatures. So thermal expansion is a nonlinear effect; if the bonds between atoms were linear springs, there would be no expansion.

The stiffer the springs, the steeper the force-displacement curve and the narrower the energy well in which the atom sits, giving less scope for expansion. Thus materials with high modulus, E (stiff springs),

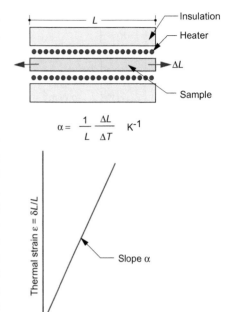

$$\alpha = \frac{1}{L}\frac{\Delta L}{\Delta T} \quad K^{-1}$$

FIGURE 4.43

Measuring the thermal expansion coefficient, α. Its units are 1/K or, more usually, $10^{-6}/K$ (microstrain/K).

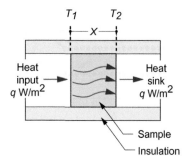

$$q = -\lambda \frac{\Delta T}{\Delta X} \quad \text{W/m}^2$$

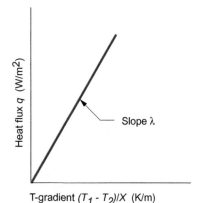

FIGURE 4.44
Measuring the thermal conductivity, λ. Its units are W/m·K.

have low expansion coefficient, α; those with low modulus (soft springs) have high expansion—indeed to a good approximation:

$$\alpha = \frac{1.6 \times 10^{-3}}{E} \qquad (44.22)$$

(E in GPa, α in K^{-1}). It is an empirical fact that all crystalline solids expand by about the same amount on heating from absolute zero to their melting point: about 2%. The expansion coefficient is the expansion per degree Kelvin, meaning that

$$\alpha \approx \frac{0.02}{T_m} \qquad (4.23)$$

For example, tungsten, with a melting point of around 3330°C (3600°K), has $\alpha = 5 \times 10^{-6}/C$, whereas lead, with a melting point of about 330°C (600°K, six times lower), expands six times more ($\alpha = 30 \times 10^{-6}/C$).

Thermal conductivity

Heat is transmitted through solids in three ways: by thermal vibrations, by the movement of free electrons in metals, and, if they are transparent, by radiation. Transmission by thermal vibrations involves the propagation of *elastic waves*. When a solid is heated, the heat enters as elastic wave packets, or *phonons*. The phonons travel through the material, and like any elastic wave, they move

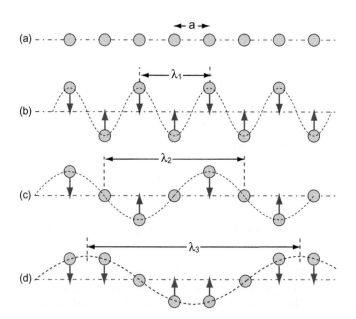

FIGURE 4.45
Thermal energy involves atom vibrations. There is one longitudinal mode of vibration and two transverse modes, one of which is shown here. The shortest meaningful wavelength, λ₁ = 2a, is shown at (b).

with the speed of sound, c_o ($c_o = \sqrt{E/\rho}$). If this is so, why does heat not diffuse at the same speed? It is because phonons travel only a short distance before they are scattered by the slightest irregularity in the lattice of atoms through which they move, even by other phonons. On average they travel a distance called the *mean-free path* ℓ_m before bouncing off something, and this path is short: typically less than 0.01 microns (10^{-8} m).

Phonon conduction can be understood using a *net flux model*, as suggested by Figure 4.47. Here a rod with a unit cross-section carries a uniform temperature gradient dT/dx between its ends. Phonons within it have 6 degrees of freedom of motion (they can travel in the $\pm x$, $\pm y$, and $\pm z$ directions). Focus on the midplane M-M. On average, 1/6 of the phonons are moving in the $+x$ direction; those within a distance ℓ_m of the plane will cross it from left to right before they are scattered, carrying with them an energy $\rho C_p(T + \Delta T)$ where T is the temperature at the plane M-M and $\Delta T = (dT/dx)\ell_m$. Another 1/6 of the phonons move in the $-x$ direction and cross M-M from right to left, carrying an energy $\rho C_p(T - \Delta T)$. Thus the energy flux q J/m^2.sec across unit area of M-M per second is

$$q = -\frac{1}{6}\rho C_p c_o\left(T + \frac{dT}{dx}\ell_m\right) + \frac{1}{6}\rho C_p c_o\left(T - \frac{dT}{dx}\ell_m\right)$$

$$= \frac{1}{3}\rho C_p \ell_m c_o \frac{dT}{dx}$$

Comparing this with the definition of thermal conductivity (Equation 4.19) we find the conductivity to be

$$\lambda = \frac{1}{3}\rho C_p \ell_m c_o \qquad (4.24)$$

Elastic waves contribute little to the conductivity of pure metals such as copper or aluminum because the heat is carried more rapidly by the free electrons. Equation 4.24 still applies, but now C_p, c_o, and ℓ_m become the thermal capacity, the velocity, and the mean-free path of the electrons. Free electrons also conduct electricity, with the result that metals with high electrical conductivity also have high thermal conductivity.

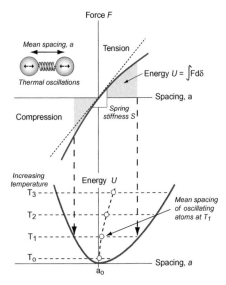

FIGURE 4.46
Thermal expansion results from the oscillation of atoms in an unsymmetrical energy well.

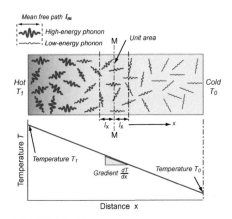

FIGURE 4.47
The transmission of heat by the motion of phonons.

4.5 ELECTRICAL BEHAVIOR

Electrical conduction (as in lightning conductors) and *insulation* (as in electric plug casings) are familiar properties. Dielectric behavior may be less so. A *dielectric* is an insulator. It is usual to

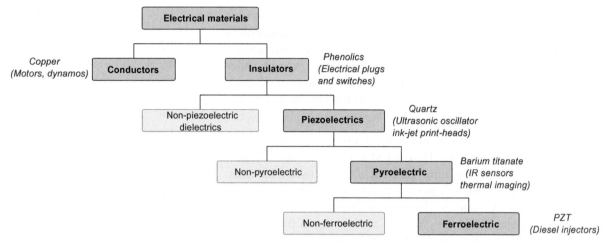

FIGURE 4.48

The hierarchy of electrical behavior. The interesting ones are in the darker colored boxes, with examples of materials and applications. Their nature and origins are described in this chapter.

use the word *insulator* in referring to its inability to conduct electricity and to use *dielectric* in referring to its behavior in an electric field.

Three properties are of importance here. The first, the *dielectric constant* (or *relative permittivity*), has to do with the way the material acquires a dipole moment (it *polarizes*) in an electric field. The second, the *dielectric loss factor*, measures the energy dissipated when radio-frequency waves pass through a material, the energy appearing as heat (the principle of microwave cooking). The third is the *dielectric breakdown potential*; lightning is dielectric breakdown, and it can be as damaging on a small scale—in a piece of electrical equipment, for example—as on a large one.

There are many kinds of electrical behavior, all of them useful. Figure 4.48 gives an overview, with examples of materials and applications.

Resistivity and Conductivity

The electrical resistance R (units: ohms, symbol Ω) of a rod of material is the potential drop V (volts) across it, divided by the current i (amps) passing through it, as in Figure 4.49. This relationship is Ohm's Law:

$$R = \frac{V}{i} \qquad (4.25)$$

The material property that determines resistance is the *electrical resistivity*, ρ_e. It is related to the resistance by

$$\rho_e = \frac{A}{L} R \qquad (4.26)$$

where A is the section and L the length of a test rod of material; think of it as the resistance of a unit cube of the material. Its units in the metric system are $\Omega \cdot m$, but it is commonly reported in units of $\mu\Omega \cdot cm$. It has an immense range, from a little more than 10^{-8} in units of $\Omega \cdot m$ for good conductors (equivalent to 1 $\mu\Omega \cdot cm$, which is why these units are still used) to more than 10^{16} $\Omega \cdot m$ (10^{24} $\mu\Omega \cdot cm$) for the best insulators. The electrical conductivity κ_e is simply the reciprocal of the resistivity. Its units are Siemens per meter (S/m or $(\Omega \cdot m)^{-1}$).

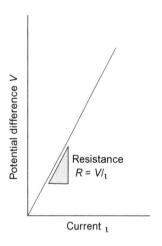

Dielectric Behavior

First, a reminder of what is meant by a field: It is a region of space in which objects experience forces if they have the right properties. Charge creates an *electric field*, E. The electric field strength between two oppositely charged plates separated by a distance t and with a potential difference V between them is

$$E = \frac{V}{t} \qquad (4.27)$$

and is independent of position except near the edge of the plates.

Two conducting plates separated by a dielectric make a capacitor (see Figure 4.50).

Capacitors (sometimes called *condensers*) store charge. The charge Q (coulombs) is directly proportional to the voltage difference between the plates, V (volts):

$$Q = CV \qquad (4.28)$$

where C (farads) is the capacitance. The capacitance of a parallel plate capacitor of area A, separated by empty space (or by air), is

$$C = \varepsilon_o \frac{A}{t} \qquad (4.29)$$

where ε_o is the *permittivity of free* space (8.85×10^{-12} F/m, where F is farads). If the empty space is replaced by a dielectric, capacitance increases. This is because the dielectric *polarizes*. The field created by the polarization opposes the field E, reducing the voltage difference V needed to support the charge. Thus the capacity of the condenser is increased to the new value

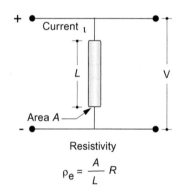

FIGURE 4.49

Electrical resistivity. Its value ranges from 1 to 10^{24} $\mu\Omega \cdot cm$.

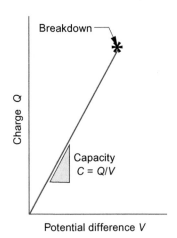

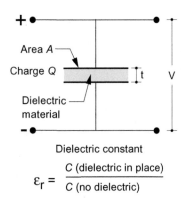

$$\varepsilon_r = \frac{C \text{ (dielectric in place)}}{C \text{ (no dielectric)}}$$

FIGURE 4.50

Dielectric constant and dielectric breakdown. The capacitance of a condenser is proportional to the dielectric constant, and the maximum charge it can hold is limited by breakdown.

$$C = \varepsilon \frac{A}{t} \tag{4.30}$$

where ε is the *permittivity of the dielectric* with the same units as ε_o. It is usual to cite not this but the *relative permittivity* or *dielectric constant*, ε_r:

$$\varepsilon_r = \frac{C_{\text{with dielectric}}}{C_{\text{no dielectric}}} = \frac{\varepsilon}{\varepsilon_o} \tag{4.31}$$

making the capacitance:

$$C = \varepsilon_r \varepsilon_o \frac{A}{t} \tag{4.32}$$

Being a ratio, ε_r is dimensionless. Its value for empty space and, for practical purposes, for most gases is 1. Most dielectrics have values between 2 and 20, though low-density foams approach the value 1 because they are largely air. Ferroelectrics are special: They have values of ε_r as high as 20,000.

Capacitance is one way to measure the dielectric constant of a material (Figure 4.50). The charge stored in the capacitor is measured by integrating the current that flows into it as the potential difference V is increased. The ratio Q/V is the capacitance. The dielectric constant ε_r is calculated from Equation 4.31.

Small capacitors, with capacitances measured in microfarads (μF) or picofarads (pF), are used in R-C circuits to tune oscillations and give controlled time delays. The time constant for charging or discharging a capacitor is

$$\tau = RC. \tag{4.33}$$

where R is the resistance of the circuit. When charged, the energy stored in a capacitor is

$$\frac{1}{2}QV = \frac{1}{2}CV^2 \tag{4.34}$$

and this can be large: "Supercapacitors" with capacitances measured in farads store enough energy to power a hybrid car.

The *breakdown potential* or *dielectric strength* (units: MV/m) is the electrical potential gradient at which an insulator breaks down and a damaging surge of current flows through it. It is measured by increasing, at a uniform rate, a 60 Hz alternating potential applied across the faces of a plate of the material in a configuration like that of Figure 4.50 until breakdown occurs, typically at a potential gradient of between 1 and 100 MV/m.

The *loss tangent* and the *loss factor* take a little more explanation. Polarization involves the small displacement of charge (either of electrons or of ions) or of molecules that carry a dipole moment when an electric field is applied to the material. An oscillating field drives the charge between two alternative configurations. This charge-motion is like an electric current that—if there were no losses—would be 90° out of phase with the voltage. In real dielectrics this current dissipates energy, just as a current in a resistor does, giving it a small phase shift, δ (see Figure 4.51). The *loss tangent, $\tan\delta$*, also called the *dissipation factor, D*, is the tangent of the loss angle. The *power factor, P_f*, is the sine of the loss angle. When δ is small, as it is for the materials of interest here, all three are essentially equivalent:

$$P_f \approx D \approx \tan\delta \approx \sin\delta \qquad (4.35)$$

More useful, for our purposes, is the *loss factor L*, which is the loss tangent times the dielectric constant:

$$L = \varepsilon_r \tan\delta \qquad (4.36)$$

It measures the energy dissipated by a dielectric when in an oscillating field. If you want to select materials to minimize or maximize dielectric loss, the measure you want is L.

When a dielectric material is placed in a cyclic electric field of amplitude E and frequency f, power P is dissipated and the field is correspondingly attenuated. The power dissipated per unit volume, in W/m³, is

$$P \approx f E^2 \varepsilon \tan\delta \approx f E^2 \varepsilon_0 \varepsilon_r \tan\delta \qquad (4.37)$$

where, as before, ε_r is the dielectric constant of the material and $\tan\delta$ is its loss tangent. This power appears as heat and is generated uniformly through the volume of the material. Thus the higher the frequency or the field strength and the greater the loss factor $\varepsilon_r\tan\delta$, the greater is the heating and energy loss. Sometimes this dielectric loss is exploited in processing—for example, in radio frequency welding of polymers.

All dielectrics change shape in an electric field, a consequence of the small shift in charge that allows them to polarize; the effect is called *electrostriction*. Electrostriction is a one-sided relationship in that an electric field causes deformation, but deformation does not produce an electric field. *Piezoelectric* materials, by contrast, display a two-sided relationship between polarization and deformation: A field induces deformation and deformation induces charge differences between its surfaces, thus creating a field. The piezoelectric

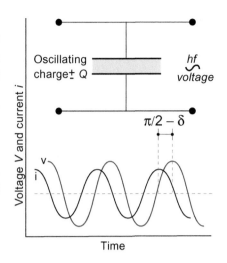

FIGURE 4.51
Dielectric loss. The greater the loss factor, the greater the microwave coupling and heating.

coefficient is the strain per unit of electric field, and although it is very small, it is a true linear effect, which makes it useful: When you want to position or move a probe with nanoscale precision, it is just what you need. *Pyroelectric* materials contain molecules with permanent dipole moments that, in a single crystal, are aligned, giving the crystal a permanent polarization. When the temperature is changed, the polarization changes, creating surface charges or, if the surfaces are connected electrically, a pyroelectric current—the principle of intruder-detection systems and of thermal imaging. *Ferroelectric* materials, too, have a natural dipole moment; they are polarized to start with, and the individual polarized molecules line up so that their dipole moments are parallel, like magnetic moments in a magnet. Their special feature is that the direction of polarization can be changed by applying an electric field, and the change causes a change of shape.

The Physics of Electrical Properties

Electrical conductivity

An electric field, E (volts/m), exerts a force Ee on a charged particle, where e is the charge it carries. Solids are made up of atoms containing electrons that carry a charge $-e$ and a nucleus containing protons, each with a positive charge $+e$. If charge carriers can move, the force Ee causes them to flow through the material—that is, it conducts. Metals are *electron conductors*, meaning that the charge carriers are the electrons. In ionic solids (which are composed of negatively and positively charged ions such as Na^+ and Cl^-), the diffusive motion of ions allows *ionic conduction*, but this is only possible at temperatures at which diffusion is rapid. Many materials have no mobile electrons, and at room temperature they are too cold to be ionic conductors. The charged particles they contain still feel a force in an electric field, and it is enough to displace the charges slightly, but they are unable to move more than a tiny fraction of the atom spacing. These are insulators; the small displacement of charge gives them dielectric properties.

How is it that some materials have mobile electrons and some do not? To explain this we need two of the stranger results of quantum mechanics. Briefly, the electrons of an atom occupy discrete energy states or orbits, arranged in shells (designated 1, 2, 3, and so on, from the innermost to the outermost); each shell is made up of subshells (designated s, p, d, and f), each of which contains 1, 3, 5, or 7 orbits, respectively. The electrons fill the shells with the lowest energy, two electrons of opposite spin in each orbit; the *Pauli exclusion principle* prohibits an energy state with more than two. When *n* atoms (a large number) are brought together to form a solid, the inner electrons remain the property of the atom on which they started, but the outer ones interact. Each atom now sits in the field created by the charges of its neighbors. This has the effect of decreasing slightly the energy levels of electrons spinning in a direction favored by the field of its neighbors and raising that of those with spins in the opposite direction, splitting each energy level. Thus the discrete levels of an isolated atom broaden, in the solid, into *bands* of very closely spaced levels. The number of electrons per atom that have to be accommodated depends only on the atomic number of the atoms. These electrons fill the bands from the bottom, lowest-energy slot on up, until all are on board, so to speak. The topmost filled energy level is called the *Fermi level*. An electron in this level still has an energy that is lower than it would have been if it were isolated in a vacuum far from the atoms.

Whether the material is a conductor or an insulator depends on how full the bands are and whether or not they overlap. In Figure 4.52 the central column describes an isolated atom and the outer ones illustrate the possibilities created by bringing atoms together into an array, with the energies spread into energy bands. Conductors such as copper, shown on the left, have an unfilled outer band; there are many very closely spaced levels just above the last full one, and, when accelerated by a field, electrons can use these levels to move freely through the material. In insulators, shown on the right, the outermost band with electrons in it is full, and the nearest empty band is separated from it in energy by a wide *band gap*. Semiconductors, too, have a band gap, but it is narrower—narrow enough that thermal energy can pop a few electrons into the empty band, where they conduct. Deliberate doping (adding trace levels of impurities) creates new levels in the band gap, reducing the energy

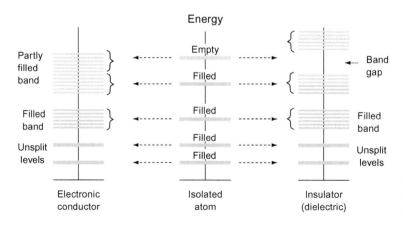

FIGURE 4.52

Conductors, on the left, have a partly filled outer band; electrons in the band can move easily. Insulators, on the right, have an outer filled band, separated from the nearest infilled band by a band gap.

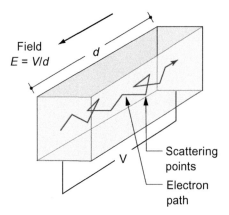

FIGURE 4.53

An electron, accelerated by the field E, is scattered by imperfections that create a resistance to its motion.

barrier to entering the empty states and thus allowing more carriers to become mobile.

Electrical resistance

If a field E exerts a force Ee on an electron, why does it not accelerate forever, giving a current that continuously increases with time? This is not what happens; instead, switching on a field causes a current that almost immediately reaches a steady value. Referring back to Equations 4.25 and 4.26, the current density i/A is proportional to the field E

$$\frac{i}{A} = \frac{E}{\rho_e} = \kappa_e E \tag{4.38}$$

where ρ_e is the resistivity and κ_e, its reciprocal, is the electrical conductivity.

Broadly speaking, the picture is this: Conduction electrons are free to move through the solid. Their thermal energy $k_B T$ (k_B = Boltzmann's constant, T = absolute temperature) causes them to move like gas atoms in all directions. In doing this they collide with *scattering centers*, bouncing off in a new direction. Impurity or solute atoms are particularly effective scattering centers (which is why alloys always have a higher resistivity than pure metals), but electrons are scattered also by imperfections such as dislocations and by the thermal vibration of the atoms themselves. When there is no field, there is no *net* transfer of charge in any direction, even though all the conduction electrons are moving freely. A field imposes a drift velocity $v_d = \mu_e E$ on the electrons, where μ_e is the electron mobility, and it is this that gives the current (see Figure 4.53). The greater the number of scattering centers, the shorter is the mean-free path, λ_{mfp}, of the electrons between collisions, and the slower, on average, they move. Just as with thermal conductivity, the electrical conductivity depends on the mean-free path, on the density of carriers (the number n_v of mobile electrons per unit volume), and the charge they carry. Thus the current density, i/A, is given by

$$\frac{i}{A} = n_v e v_d = n_v e \mu_e E$$

Comparing this with Equation 4.38 gives the conductivity:

$$\kappa_e = n_v e \mu_e \tag{4.39}$$

Thus the conductivity is proportional to the density of free electrons and to the drift velocity, and this is directly proportional to the mean-free path. The purer and more perfect the conductor, the higher the conductivity.

The resistivity of metals increases with temperature because thermal vibration scatters electrons. Resistance decreases as temperature falls, which is why very high-powered electromagnets are precooled in liquid nitrogen. As absolute zero is approached, most metals retain some resistivity, but a few, at between 0 and 4°K, suddenly lose all resistance and become superconducting.

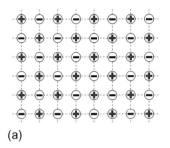

(a)

Dielectric behavior

Dielectrics are insulators. So what happens to their electrons when a field E is applied? In zero field, the electrons and protons in most dielectrics are symmetrically distributed, so the material carries no net charge or dipole moment. A field exerts a force on a charge, pushing positive charges in the direction of the field and negative charges in the opposite direction. The effect is easiest to see in ionic crystals, since here neighboring ions carry opposite charges, as on the left of Figure 4.54. Switch on the field and the positive ions (charge $+q$) are pulled in the field direction, the negative ones (charge $-q$) in the reverse, until the restoring force of the interatomic bonds simply balances the force due to the field at a displacement of Δx, as on the right of the figure. Two charges $\pm q$ separated by a distance Δx create a dipole with dipole moment, d, given by

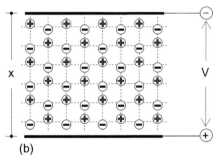

(b)

FIGURE 4.54

An ionic crystal (a) in zero applied field, and (b) when a field V/x is applied. The electric field displaces charge, causing the material to acquire a dipole moment.

$$d = q\Delta x \qquad (4.40)$$

The polarization of the material, P, is the volume-average of all the dipole moments it contains:

$$P = \frac{\Sigma d}{Volume}$$

Even in materials that are not ionic, such as silicon, a field produces a dipole moment because the nucleus of each atom is displaced a tiny distance in the direction of the field and its surrounding electrons are displaced in the opposite direction. The resulting dipole moment depends on the magnitude of the displacement and the number of charges per unit volume, and it is this that determines the dielectric constant. The bigger the shift, the bigger the dielectric constant. Thus compounds with ionic bonds and polymers that contain polar groups such as $-OH^-$ and $-NH^-$ (nylon, for example) have larger dielectric constants than those that do not.

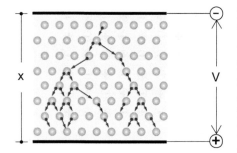

FIGURE 4.55
Breakdown involves a cascade of electrons like a lightning strike.

Dielectric loss

Think now of polarization in an alternating electric field. When the upper plate of Figure 4.54b is negative, the displacements are in the direction shown in the figure. When its polarity is reversed, it is the negative ions that are displaced upward, the positive ions downward; in an oscillating field, the ions oscillate. If their oscillations were exactly in phase with the field, no energy would be lost, but this is never exactly true, and often the phase shift is considerable.

Materials with high dielectric loss usually contain awkwardly shaped molecules that themselves have a dipole moment; a water molecule is an example. These respond to the oscillating field by rotating, but because of their shape they interfere with each other (you could think of it as molecular friction), and this dissipates energy that appears as heat; that is how microwave heating works. As Equation 4.38 shows, the energy that is dissipated depends on the frequency of the electric field; generally speaking, the higher the frequency, the greater the power dissipated (because power is work-per-second, and the more times the molecules shuttle, the more energy is lost), but there are peaks at certain frequencies that are characteristic of the material structure.

Dielectric breakdown

In metals, as we have seen, even the smallest field causes electrons to flow. In insulators they can't, because of the band gap. But if, at some weak spot, one electron is torn free from its parent atom, the force Ee exerted by the field E accelerates it, giving it kinetic energy; it continues to accelerate until it collides with another atom. A sufficiently large field can give the electron so much kinetic energy that, in the collision, it kicks one or more new electrons out of the atom it hits, and they, in turn, are accelerated and gain energy. The result, sketched in Figure 4.55, is a cascade—an avalanche of charge. It is sufficiently violent that it can damage the material permanently.

The critical field strength to make this happen, called the *breakdown potential*, is hard to calculate: It is that at which the first electron breaks free at the weak spot—a defect in the material such as a tiny crack, void, or inclusion that locally concentrates the field. The necessary fields are large, typically 1–15 MV/m. That sounds a lot, but such fields are found in two very different circumstances: when the voltage is very high or when the distances are very small. In power transmission, the voltages are sufficiently high—20,000 volts or so—that breakdown can occur at the insulators that support the line, whereas in microcircuits and thin-film devices the distances between components are very small: a 1 volt difference across a distance of 1 micron gives a field of 1 MV/m.

4.6 MAGNETIC BEHAVIOR

Magnetic fields are created by moving electric charge—electric current in electromagnets, electron spin in atoms of magnetic materials. This section is about magnetic materials: how they are characterized and where their properties come from.

Magnetic Fields in a Vacuum

When a current i passes through a long, empty coil of n turns and length L, as in Figure 4.56, a magnetic field is generated. The magnitude of the field, H, is given by Ampère's law as

$$H = \frac{ni}{L} \qquad (4.41)$$

and thus has units of amps/meter (A/m). The field has both magnitude and direction; it is a vector field.

Magnetic fields exert forces on a wire carrying an electric current. A current i flowing in a single loop of area S generates a dipole moment m where

$$m = iS \qquad (4.42)$$

with units $A \cdot m^2$, and it too is a vector with a direction normal to the plane of S (Figure 4.57). If the loop is placed at right angles to the field H, it feels a torque T (units: Newton $\cdot$ meter, or $N \cdot m$) of

$$T = \mu_o m H \qquad (4.43)$$

where μ_o is called the *permeability of vacuum*, $\mu_o = 4\pi x 10^{-7}$ henry/meter (H/m). To link these we define a second measure of the magnetic field, one that relates directly to the torque it exerts on a unit magnetic moment. It is called the *magnetic induction* or *flux density*, B, and for vacuum or nonmagnetic materials it is

$$B = \mu_o H \qquad (4.44)$$

Its units are *tesla*, so a tesla is 1 HA/m^2. A magnetic induction B of 1 tesla exerts a torque of 1 N $\cdot$ m on a unit dipole at right angles to the field H.

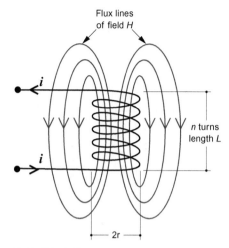

FIGURE 4.56
A solenoid creates a magnetic field H; the flux lines indicate the field strength.

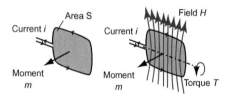

FIGURE 4.57
Definition of magnetic moment and moment-field interaction.

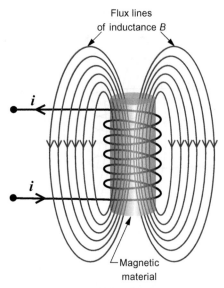

FIGURE 4.58

A magnetic material exposed to a field H becomes magnetized, concentrating the flux lines.

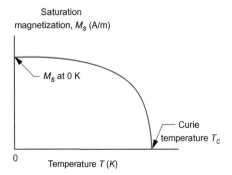

FIGURE 4.59

Saturation magnetization decreases with temperature, falling to zero at the Curie temperature T_c.

Magnetic Fields in Materials

If the space inside the coil of Figure 4.56 is filled with a material, as in Figure 4.58, the induction within it changes. This is because its atoms respond to the field by forming little magnetic dipoles. The material acquires a macroscopic dipole moment or *magnetization*, M (its units are A/m, like H). The induction becomes

$$B = \mu_o(H + M) \tag{4.45}$$

The simplicity of this equation is misleading, since it suggests that M and H are independent; in reality M is the response of the material to H, so the two are coupled. If the material of the core is ferromagnetic, the response is a very strong one and it is nonlinear, as we shall see in a moment. It is usual to rewrite Equation 4.45 in the form

$$B = \mu_R \mu_o H$$

where μ_R is called the *relative permeability*. The magnetization, M, is thus

$$M = (\mu_R - 1)H = \chi H \tag{4.46}$$

where χ is the *magnetic susceptibility*.

Nearly all materials respond to a magnetic field by becoming magnetized, but most are paramagnetic with a response so faint that it is of no practical use. A few, however, contain atoms that have large dipole moments and the ability to spontaneously magnetize—to align their dipoles in parallel. These are called *ferromagnetic* and *ferrimagnetic* materials (the second one is called *ferrites* for short), and it is these that are of real practical utility.

Magnetization decreases with increasing temperature. There is a temperature, the Curie temperature T_c, above which it disappears, as in Figure 4.59. Its value for most ferromagnetic materials is well above room temperature (typically 300–500°C).

Measuring Magnetic Properties

Magnetic properties are measured by plotting an M–H curve. It looks like Figure 4.60. If an increasing field H is applied to a previously demagnetized sample, starting at Point A on the figure, its magnetization increases, slowly at first and then faster, following the broken line until it finally tails off to a maximum, the *saturation magnetization M_s* at Point B. If the field is now backed off, M does not retrace its original path but retains some of its magnetization so that when H has reached zero, at Point C, some magnetiza-

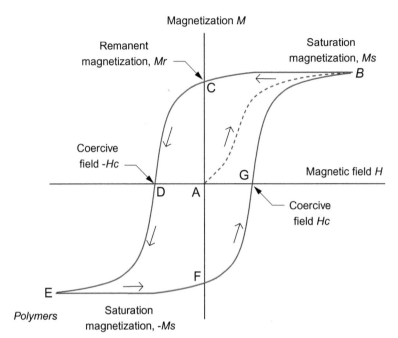

FIGURE 4.60
A hysteresis curve, showing the important magnetic properties.

tion remains; it is called the *remanent magnetization* or *remanence*, M_R, and is usually only a little less than M_s. To decrease M further we must increase the field in the opposite direction until M finally passes through zero at Point D when the field is $-H_c$, which is the *coercive field*, a measure of the resistance to demagnetization. Some applications require H_c to be as high as possible, others as low as possible. Beyond Point D the magnetization M starts to increase in the opposite direction, eventually reaching saturation again at Point E. If the field is now decreased again M follow the curve through F and G back to full forward magnetic saturation again at B to form a closed $M–H$ circuit, called the *hysteresis loop*.

Magnetic materials are characterized by the size and shape of their hysteresis loops. The initial segment AB is called the *initial magnetization curve*, and its average slope (or sometime its steepest slope) is the magnetic susceptibility, χ. The other key properties—the saturation magnetization M_s, the remanence M_R, and the coercive field H_c—have already been defined. Each full cycle of the hysteresis loop dissipates an energy per unit volume equal to the area of the loop multiplied by μ_o, the permeability of a vacuum. This energy appears as heat; it is like magnetic friction.

Magnetic materials differ greatly in the shape of an area of their hysteresis loop, the greatest difference being that between *soft*

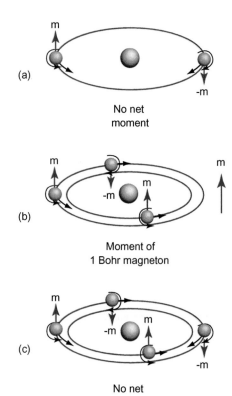

FIGURE 4.61

Orbital and electron spins create a magnetic dipole. Even numbers of electrons filling energy levels in pairs have moments that cancel, as at (a) and (c). An unpaired electron gives the atom a permanent magnetic moment, as at (b).

magnets, which have thin loops, and *hard magnets*, which have fat ones. The coercive field H_c (which determines the width of the loop) of hard magnetic materials such as Alnico is greater by a factor of about 10^5 than that of soft magnetic materials such as silicon-iron.

The Physics of Magnetic Behavior

Ferromagnetic atoms

The classical picture of an atom is that of a nucleus around which swing electrons, as in Figure 4.61. Moving charge implies an electric current, and an electric current flowing in a loop creates a magnetic dipole, as in Figure 4.57. There is, therefore, a magnetic dipole associated with each orbiting electron. That is not all. Each electron has an additional moment of its own: its spin moment. A proper explanation of this phenomenon requires quantum mechanics, but a way of envisaging its origin is to think of an electron not as a point charge but as slightly spread out and spinning on its own axis, again creating rotating charge and a dipole moment—and this turns out to be large. The total moment of the atom is the sum of the whole lot.

A simple atom like that of helium has two electrons per orbit, and they configure themselves such that the moment of one exactly cancels the moment of the other, as in Figure 4.61a and 4.61c, leaving no net moment. But now think of an atom with three, not two, electrons, as in Figure 4.61b. The moments of two may cancel, but there remains the third, leaving the atom with a net moment represented by the red arrow at the right of Figure 4.61b. Thus atoms with electron moments that cancel are nonmagnetic; those with electron moments that don't cancel carry a magnetic dipole. Simplifying a little, one unpaired electron gives a magnetic moment of $9.3 \times 10^{-24} \, \text{A} \cdot \text{m}^2$, called a *Bohr magneton*; two unpaired electrons give 2 Bohr magnetons, three give three, and so on.

Think now of the magnetic atoms assembled into a crystal. In most materials the atomic moments interact so weakly that thermal motion is enough to randomize their directions, as in Figure 4.62a. Despite their magnetic atoms, the structure as a whole has no magnetic moment; these materials are *paramagnetic*. In a few materials, though, something quite different happens. The fields of neighboring atoms interact such that their energy is reduced if their magnetic moments line up. This drop in energy is called the *exchange energy*, and it is strong enough that it beats the randomizing effect of thermal energy so long as the temperature is not too high. (The shape of the Curie curve of Figure 4.59 shows how thermal energy overwhelms the exchange energy as the Curie temperature is approached.) They may line up antiparallel, head to tail, so to speak, as in Figure 4.62b, and we still have no net moment; such materials are called *antiferromagnets*. But in a few elements, notably iron, nickel, and cobalt, exactly the opposite happens: The moments spontaneously align so that, if all are parallel, the structure has a net moment that is the sum of those of all the atoms it contains. These materials are *ferromagnetic*. Iron has three unpaired electrons per atom, nickel has two, and cobalt just one, so the net moment if all the spins are aligned (the saturation magnetization M_s) is greatest for iron, less for nickel, and still less for cobalt.

Compounds give a fourth possibility. The materials we refer to as ferrites are oxides; one class of them has the formula $M\,Fe_2O_4$, where M is also a magnetic atom, such as cobalt, Co. Both the Fe and the Co atoms have dipoles, but they differ in strength. They line up in the antiparallel configuration, but because of the difference, the moment the cancellation is incomplete, leaving a net moment M; these are *ferrimagnets*, or *ferrites* for short. The partial cancellation and the smaller number of magnetic atoms per unit volume mean that they have lower saturation magnetization than, say, iron, but they have other advantages, notably that, being oxides, they are electrical insulators.

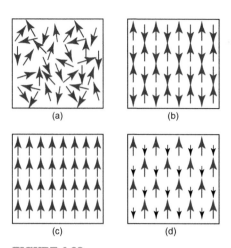

FIGURE 4.62

Types of magnetic behavior: (a) paramagnetic, (b) antiferromagnetic, (c) ferromagnetic, and (d) ferrimagnetic.

Domains

If atomic moments line up, shouldn't every piece of iron, nickel, or cobalt be a permanent magnet? Magnetic materials they are; magnets, in general, they are not. Why not?

A uniformly magnetized rod creates a magnetic field, H, like that of a solenoid. The field has an energy associated with it. The smaller the field and the smaller the volume that it invades, the smaller the energy. If the structure can arrange its moments to minimize its H or get rid of it entirely (remembering that the exchange energy wants neighboring atom moments to stay parallel), it will try to do so.

Figure 4.63 illustrates how this can be done. The material splits into *domains* within which the atomic moments are parallel but with a switch of direction between mating domains to minimize the external field. The domains meet at *domain walls*, regions a few atoms thick in which the moments swing from the orientation of one domain to that of the other. Splitting into parallel domains of opposite magnetization, as at (b) and (c), reduces the field a good deal; adding caps magnetized perpendicular to both, as in (d), kills

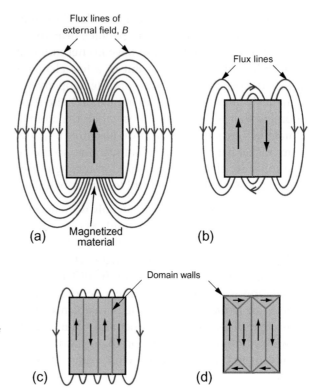

FIGURE 4.63

Domains allow a compromise in the cancellation of the external field while retaining magnetization of the material itself. The arrows show the direction of magnetization.

it almost completely. The result is that most magnetic materials, unless manipulated in some way, adopt a domain structure with minimal external field; that's the same as saying that, though magnetic, they are not magnets.

How can they be magnetized? Placed in a magnetic field, created, say, with a solenoid, the domains already aligned with the field have lower energy than those aligned against it. The domain wall separating them feels a force, the *Lorentz force*, pushing it in a direction to make the favorably oriented domains grow at the expense of the others. As they grow, the magnetization of the material increases, moving up the $M–H$ curve, finally saturating at M_s, when the whole sample is one domain oriented parallel to the field.

4.7 OPTICAL BEHAVIOR

Electromagnetic (e-m) radiation permeates the entire universe. Observe the sky with your eye and you see the visible spectrum, the range of wavelengths we call *light* (0.40–0.77 μm). Observe it with a detector of X-rays or γ-rays and you see radiation with far shorter wavelengths (as short as 10^{-4} nm, one thousandth the size of an atom). Observe it instead with a radio telescope and you pick up radiation with wavelengths measured in millimeters, meters, or even kilometers, known as *radio waves* and *microwaves*. The range of wavelengths of radiation are vast, spanning 18 orders of magnitude (see Figure 4.64). The visible part of this spectrum is only a tiny part of it—but even that has entrancing variety, giving us colors ranging from deep purple through blue, green, and yellow to deep red.

When e-m radiation strikes materials, things can happen. Materials interact with radiation by reflecting it, absorbing it, transmitting it, and refracting it. This chapter is about these interactions, the materials that do them best, and the ways we use them.

The Interaction of Materials and Radiation

The intensity I of an e-m wave, proportional to the square of its amplitude, is a measure of the energy it carries. When radiation with intensity I_o strikes a material, a part I_R of it is reflected, a part I_A absorbed, and a part I_T may be transmitted. Conservation of energy requires that

$$\frac{I_R}{I_o} + \frac{I_A}{I_o} + \frac{I_T}{I_o} = 1 \qquad (4.47)$$

The first term is called the *reflectivity* of the material, the second the *absorptivity*, and the last the *transmittability* (all dimensionless). Each depends on the wavelength of the radiation, on the nature of the material, and on the state of its surfaces. They can be thought of as properties of the material in a given state of surface polish, smoothness, or roughness.

In optics we are concerned with wavelengths in the visible spectrum. Materials that reflect or absorb all visible light, transmitting none, are called *opaque*, even though they may transmit in the near visible (infrared or ultraviolet). Those that transmit a little diffuse light are called *translucent*. Those that transmit light sufficiently well that you can see through them are called *transparent*; and a subset of these that transmit almost perfectly, making them suitable for lenses, light guides, and optical fibers, are given the additional title of *optical quality*. Metals are opaque. To be transparent, a material must be a dielectric.

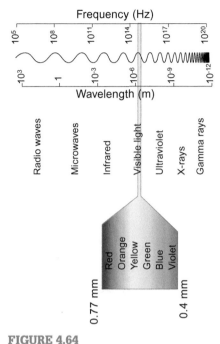

FIGURE 4.64

The spectrum of electromagnetic (e-m) waves. The visible spectrum lies between the wavelengths 0.4 and 0.77 microns.

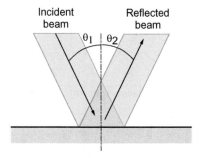

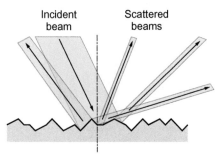

FIGURE 4.65
Perfectly flat, reflective surfaces give specular reflection, such that θ₁ = θ₂. The angles of incidence and reflection are always equal, but the rough surface gives diffuse reflection, even though the angles of incidence and reflection are still, locally, equal.

Specular and Diffuse Reflection

Metals reflect almost all the light that strikes them; none is transmitted and little is absorbed. When light strikes a reflecting surface at an incident angle θ_1, part of it is reflected, leaving the surface with an angle of reflection θ_2 such that

$$\theta_1 = \theta_2 \tag{4.48}$$

Specular surfaces are microscopically smooth and flat; a beam striking such a surface suffers specular reflection, meaning that it is reflected as a beam, as on the left in Figure 4.65. *Diffuse* surfaces are irregular; the law of reflection (Equation 4.48) still holds locally, but the incident beam is reflected in many different directions because of the irregularities, as on the right in the figure.

Absorption

If radiation can penetrate a material, some is absorbed. The greater the thickness x through which the radiation passes, the greater the absorption. The intensity I, starting with the initial value I_o, decreases such that

$$I = I_o \exp{-\beta x} \tag{4.49}$$

where β is the absorption coefficient, with dimensions of m^{-1} (or, more conveniently, mm^{-1}). The absorption coefficient depends on wavelength with the result that white light passing through a material may emerge with a color corresponding to the wavelength that is least absorbed; that is why a thick slab of ice looks blue.

Transmission

By the time a beam of light has passed completely through a slab of material, it has lost some intensity through reflection at the surface at which it entered, some in reflection at the surface at which it leaves, and some by absorption in between. Its intensity is

$$I = I_o\left(1 - \frac{I_R}{I_o}\right)^2 \exp{-\beta x} \tag{4.50}$$

The term $(1 - I_R/I_o)$ occurs to the second power because intensity is lost through reflection at both surfaces.

Refraction

The velocity of light in vacuum, $c_o = 3 \times 10^{23}$ m/s, is as fast as it ever goes. When it (or any other electromagnetic radiation) enters a material, it slows down. The *index of refraction, n,* is the ratio of its velocity in vacuum, c_o, to that in the material, c:

$$n = \frac{c_o}{c} \qquad (4.51)$$

This retardation makes a beam of light bend, or *refract,* when it enters a material of different refractive index. When a beam passes from a material *1* of refractive index n_1 into a material *2* of index n_2 with an angle of incidence θ_1, it deflects to an angle θ_2, such that

$$\frac{\sin \theta_1}{\sin \theta_2} = \frac{n_2}{n_1} \qquad (4.52)$$

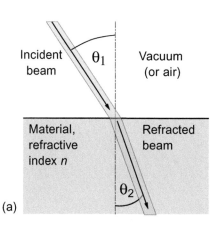
(a)

as in Figure 4.66a; the equation is known as *Snell's law.* The refractive index depends on wavelength, so each of the colors that make up white light is diffracted through a slightly different angle, producing a spectrum when light passes through a prism. When material 1 is vacuum or air, for which $n_1 = 1$, the equation reduces to

$$\frac{\sin \theta_1}{\sin \theta_2} = n_2$$

Equation 4.52 says that light passes from a material with index $n_1 = n$ into air with $n_2 = 1$, it is bent away from the normal to the surface, like that in Figure 4.66b. If the incident angle, here θ_1, is slowly increased, the emerging beam tips down until it becomes parallel with the surface when $\theta_2 = 90°$ and $\sin \theta_2 = 1$. This occurs at an incident angle, from Equation 4.52, of

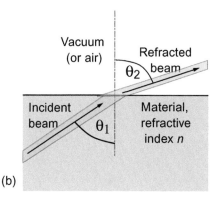
(b)

$$\sin \theta_1 = \frac{1}{n} \qquad (4.53)$$

For values of θ_1 greater than this, the equation predicts values of $\sin \theta_2$ that are greater than 1, and that can't be. Something strange has to happen, and it does: The ray is totally reflected back into the material, as in Figure 4.66c. This *total internal reflection* has many uses, one being to bend the path of light in prismatic binoculars and reflex cameras. Another is to trap light within an optical fiber, an application that has changed the way we communicate.

Reflection is related to refraction. When light traveling in a material of refractive index n_1 is incident normal to the surface of a second material with a refractive index n_2, the reflectivity is

$$R = \frac{I_R}{I_o} = \left(\frac{n_2 - n_1}{n_2 + n_1} \right)^2 \qquad (4.54a)$$

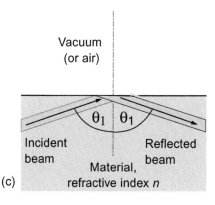
(c)

$$n = \frac{\sin \theta_1}{\sin \theta_2}$$

FIGURE 4.66

Refraction and total internal reflection.

If the incident beam is in air, with refractive index 1, this becomes

$$R = \frac{I_R}{I_o} = \left(\frac{n-1}{n+1}\right)^2. \tag{4.54b}$$

Thus materials with high refractive index have high reflectivity.

Before leaving Snell and his law, ponder for a moment on the odd fact that the beam is bent at all. Light entering a dielectric, as we have said, slows down. If that is so, why does it not continue in a straight line, only more slowly? Why should it bend? To understand this concept we need Fresnel's construction. We think of light as advancing via a series of wave fronts. Every point on a wave front acts as a source so that, in a time Δt, the front advances by $v\Delta t$, where v is the velocity of light in the medium through which it is passing ($v = c$ in vacuum or air), as shown in Figure 4.67 with *wave-front 1* advancing $c\Delta t$. When the wave enters a medium of higher refractive index it slows down so that the advance of the wave front within the medium is less than that outside, as shown with *wave-front 2* advancing $v\Delta t$ in the figure. If the angle of incidence θ_1 is not zero, the wave front enters the second medium progressively, causing it to bend, so that when it is fully in the material it is traveling in a new direction, characterized by the angle of refraction, θ_2. Simple geometry then gives Equation 4.52.

The Physics of Optical Properties

Light, like all radiation, is an electromagnetic (e-m) wave. The coupled fields are sketched in Figure 4.67. The electric part fluctuates with a frequency v that determines where it lies in the spectrum of Figure 4.64. A fluctuating electric field induces a fluctuating magnetic field that is exactly $\pi/2$ out of phase with the electric one because the induction is at its maximum when the electric field is changing most rapidly. A plane-polarized beam looks like this one: The electric and magnetic fields lie in fixed planes. Natural light is not polarized; then the wave also rotates so that the plane containing each wave continuously changes.

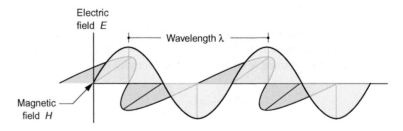

Electric field E

Wavelength λ

Magnetic field H

FIGURE 4.67

An electromagnetic wave. The electric component is $\pi/2$ out of phase with the magnetic.

Many aspects of radiation are most easily understood by thinking of radiation as a wave. Others need a different picture—that of radiation as discrete packets of energy, *photons*. The idea of a wave that is also discrete energy units is not intuitive; it is another of the results of quantum theory that do not correspond to ordinary experience. The energy E_{ph} of a photon of radiation of frequency v or wavelength λ is

$$E_{ph} = hv = \frac{hc}{\lambda} \qquad (4.55)$$

where h is Planck's constant (6.626×10^{-34} J·s) and c is the speed of the radiation. Thus radiation of a given frequency has photons of fixed energy, regardless of its intensity; an intense beam simply has more of them. This is the key to understanding reflection, absorption, and transmission.

Why aren't metals transparent?

Electrons in materials circle their parent atom in orbits with discrete energy levels, and only two can occupy the same level. Metals have an enormous number of very closely spaced levels in their conduction band; the electrons in the metal only fill part of this number. Filling the levels in a metal is like pouring water into a container until it is part full—its surface is the Fermi level—and levels above it are empty. If you "excite" the water—say, by shaking the container—some of it can slosh to a higher level. If you stop sloshing, it will return to its Fermi level.

Radiation excites electrons, and in metals there are plenty of empty levels in the conduction band into which they can be excited. But here quantum effects cut in. A photon with energy hv can excite an electron only if there is an energy level that is exactly hv above the Fermi level—and in metals there is. So all the photons of a light beam are captured by electrons of a metal, regardless of their wavelength. Figure 4.68 shows, on the left, what happens to just one.

What next? Shaken water settles back, and electrons do the same. In doing so they release a photon with exactly the same energy that excited them in the first place, but in a random direction. Any photons moving into the material are immediately recaptured, so none makes it more than about 0.01 µm (about 30 atom diameters) below the surface. All, ultimately, reemerge from the metal surface—that is, they are reflected. Many metals—silver, aluminum, and stainless steel are examples—reflect all wavelengths almost equally

FIGURE 4.68

Metals absorb photons, capturing their energy by promoting an electron from the filled part of the conduction band into a higher, empty level. When the electron falls back, a photon is reemitted.

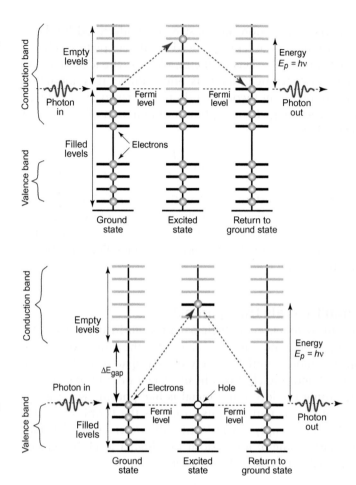

FIGURE 4.69

Dielectrics have a full band, separated from the empty conduction band by an energy gap. The material cannot capture photons with energy less than ΔE_{gap}, meaning that, for those frequencies, the material is transparent. Photons with energy greater than ΔE_{gap} are absorbed, as illustrated here.

well, so if exposed to white light they appear silver. Others—copper, brass, bronze—reflect some colors better than others and appear colored because, in penetrating this tiny distance into the surface, some wavelengths are slightly absorbed.

Reflection by metals, then, has to do with electrons at the top of the conduction band. These same electrons provide electrical conduction. For this reason, the best metallic reflectors are the metals with the highest electrical conductivities—the reason that high-quality mirrors use silver and cheaper ones use aluminum.

How does light get through dielectrics?

The reason that radiation of certain wavelengths can enter a dielectric is that its Fermi level lies at the top of the valence band, just below a band gap (see Figure 4.69). The conduction band with its vast number of empty levels lies above it. To excite an electron across the gap requires a photon with an energy at least as great as the width of the gap, ΔE_{gap}. Thus radiation with photon energy less than ΔE_{gap} cannot excite electrons; there are no energy states within the gap for the electron to be excited into. The radiation sees the material as transparent, offering no interaction of any sort, so it goes straight through.

Electrons are, however, excited by radiation with photons that have energies greater than ΔE_{gap} (i.e., higher frequency, shorter wavelength). These have enough energy to pop electrons into the conduction band, leaving a "hole" in the valence band from which they came. When they jump back, filling the hole, they emit radiation of the same wavelength that first excited them, and for these wavelengths the material is not transparent (Figure 4.69, right side). The critical frequency v_{crit}, above which interaction starts, is given by

$$h v_{crit} = \Delta E_{gap} \qquad (4.56)$$

The material is opaque to frequencies higher than this. Thus Bakelite is transparent to infrared light because its frequency is too low and its photons too feeble to kick electrons across the band gap, but the visible spectrum has higher frequencies with more energetic photons, exceeding the band-gap energy; they are captured and reflected.

Although dielectrics can't absorb radiation with photons of energy less than that of the band gap, they are not all transparent. Most are polycrystalline and have a refractive index that depends on direction; then light is *scattered* as it passes from one crystal to another. Imperfections, particularly porosity, do the same. Scattering explains why some polymers are translucent or opaque: Their microstructure is a mix of crystalline and amorphous regions with different refractive indices. It explains, too, why some go white when you bend them; it is because light is scattered from internal microcracks, or *crazes*.

Color

If a material has a band gap with an energy ΔE_{gap} that lies within the visible spectrum, the wavelengths with energy greater than this are absorbed, and those with energy that is less are not. The absorbed radiation is reemitted when the excited electron drops back into a lower energy state, but this might not be the one it started from, so the photon it emits has a different wavelength than the one that

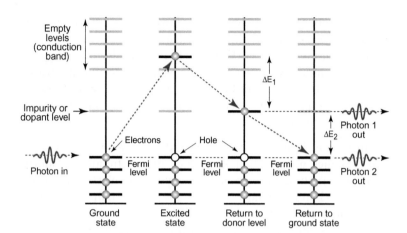

FIGURE 4.70

Impurities or dopants create energy levels in the band gap. Electron transitions to and from the dopant level emit photons of specific frequency and color.

was originally absorbed. The light emitted from the material is a mix of the transmitted and the reemitted wavelengths, and it is this that gives it a characteristic color.

More specific control of color is possible by *doping*—the deliberate introduction of impurities that create a new energy level in the band gap, as in Figure 4.70. Radiation is absorbed as before, but it is now reemitted in two discrete steps as the electrons drop first into the dopant level, emitting a photon of frequency $v_1 = \Delta E_1/h$, and from there back into the valence band, emitting a second photon of energy $v_2 = \Delta E_2/h$. Particularly pure colors are created when glasses are doped with metal ions: copper gives blue; cobalt, violet; chromium, green; manganese, yellow.

Fluorescence, phosphorescence, and electroluminescence

Electrons can be excited into higher energy levels by incident photons, provided they have sufficient energy. Energetic electrons, like those of the electron beam of a cathode-ray tube, do the same. In most materials the time delay before they drop back into lower levels, reemitting the energy they captured, is extremely short, but in some there is a delay. If, on dropping back, the photon they emit is in the visible spectrum, the effect is that of luminescence; the material continues to glow even when the incident beam is removed. When the time delay is fractions of seconds, it is called *fluorescence*, used in fluorescent lighting, where it is excited by ultraviolet from a gas discharge, and in TV tubes, where it is excited by the scanning electron beam. When the time delay is longer, it is called *phosphorescence*; it is no longer useful for creating moving images but is used instead for static displays, such as watch faces, where it is excited by electrons (β-particles) released by a mildly radioactive ingredient in the paint.

4.8 ACOUSTIC BEHAVIOR

The human ear responds to frequencies from 20 Hz to about 20,000 Hz, corresponding to wavelengths in air between 17 m and 17 mm. The range of acoustic frequency is far greater than this (see Figure 4.71), with wavelengths that extend into the nanorange. The vibrations that cause sound produce a change of air pressure in the range 10^{-4} Pa (low-amplitude sound) to 10 Pa (the threshold of pain). It is usual to measure this on a relative, logarithmic scale, with units of *decibels* (dB). The decibel scale compares two sound intensities using the *threshold of hearing* as the reference level (0 dB).

Sound Velocity and Wavelength

The sound velocity in a long rod of a solid material (such that the thickness is small compared with the wavelength) is

$$v_1 = \sqrt{\frac{E}{\rho}} \tag{4.57}$$

where E is Young's modulus and p is the density. If the thickness of the rod is large compared with the wavelength, the velocity instead is

$$v_B = \sqrt{\frac{E(1-v)}{(1-v-2v^2)\rho}}$$

where v is Poisson's ratio. These two sound velocity values don't differ much—a maximum of 25% at most.

Elastically anisotropic solids (fiber composites such as CFRP and all woods are anisotropic) have sound velocities that depend on direction. If $E_\parallel$ is the modulus along the direction of the fibers or grain and $E_\perp$ is that perpendicular to it, the ratio of the two velocities, as given by Equation 4.57, is

$$\frac{v_\parallel}{v_\perp} = \sqrt{\frac{E_\parallel}{E_\perp}} \tag{4.58}$$

Sound Management

The mechanism for reducing sound intensity within a given enclosed space (a room, for instance) depends on where the sound

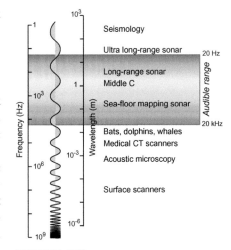

FIGURE 4.71

The acoustic spectrum from the extreme ultrasonic to the seismic, a range of 10^8. The audible part is a thick slice of this range. The figure is based on acoustic waves in air, velocity 350 m/s; for water, shift the wavelength scale up relative to the frequency scale by a factor of 3; for rock, shift it up by a factor of 10.

comes from. If it is generated within the room, one seeks to *absorb* the sound. If it is airborne and comes from outside, one seeks to *insulate* the space to keep the sound out. And if it is transmitted through the frame of the structure itself, one seeks to *isolate* the structure from the source of vibration. Cellular, porous solids are good for absorbing sound, and, in combination with other materials, they can help in isolation. They are very poor at providing insulation.

Soft, porous materials absorb incident, airborne sound waves, converting them into heat. Sound power, even for very loud noise, is small, so the temperature rise is negligible. Porous or highly flexible materials—for example, low-density polymeric foams such as plaster and fiberglass—absorb well; so do woven polymers in the form of carpets and curtains. Several mechanisms of absorption are at work here. First, there is the viscous loss as air is pumped into and out of the open, porous structures; sealing the surface (with a paint film, for instance) greatly reduces the absorption. Second, there is the intrinsic damping in the material. It measures the fractional loss of energy of a wave, per cycle, as it propagates within the material itself. Intrinsic damping in most metals and ceramics is low, typically 10^{-6} to 10^{-4}; in polymers and foams made from them it is high, in the range of 0.01 to 0.2. The proportion of sound absorbed by a surface is called the *sound-absorption coefficient*. A material with a coefficient of 0.8 absorbs 80% of the sound that is incident on it; material of coefficient 0.03 absorbs only 3% of the sound, reflecting 97%.

Sound insulation requires materials of a completely different sort. The degree of insulation is proportional to the mass of the wall, floor, or roof through which sound has to pass. This is known as the *mass law*: the heavier the material, the better it insulates. The lightweight walls of modern buildings, designed for good thermal insulation, do not in general provide good sound insulation. Foam plaster or fiberglass, for obvious reasons, are not a good choice, either; instead, the practice is to add mass to the wall or floor with an extra layer of brick, concrete, or a lead cladding.

Impact noise is transmitted directly into the structure or fabric of a building. Elastic materials, particularly a steel frame if there is one, transmit vibrations throughout the building. Unlike airborne sound, noise of this sort is not attenuated by additional mass. Since it is transmitted by the continuous solid part of the structure, it can be reduced by interrupting the sound path by floating the floor or building foundation on a resilient material. Here low modulus materials can be useful. The impact sound of footsteps can be suppressed using cork, porous rubber, or plastic tiles. On a larger scale, buildings are isolated by setting the entire structure on resilient pads such as a composite of rubber filled with cork particles. The low shear modulus of the rubber isolates the building from shear waves, and the compressibility of the cork adds a high impedance to compressive waves.

Sound Wave Impedance and Radiation of Sound Energy

If a sound-transmitting material is interfaced with a second one with different properties, part of the sound wave is transmitted across the interface, and part is reflected back into the first material. The transmission and reflection factors are determined by the relative impedances of the two materials. The impedance is defined by

$$Z = \sqrt{\rho E} \qquad (4.59)$$

where E is the appropriate modulus and ρ the density. The reflection and transmission coefficients between Material 1 and Material 2 are given by

$$R = \frac{Z_1 - Z_2}{Z_1 + Z_2} \quad \text{and} \quad T = 1 - R = \frac{2Z_2}{Z_1 + Z_2} \qquad (4.60)$$

Thus if the two impedances are about equal, most of the sound is transmitted, but if the impedances differ greatly, most is reflected. This is the origin of the mass law cited earlier: A heavy wall gives a large impedance mismatch with air so that most of the sound is reflected and does not penetrate to the neighboring room.

Table 4.1 lists typical values of the acoustic impedance. In the design of sound boards (the front plate of a violin, the sound board of a harpsichord, the panel of a loudspeaker), the intensity of sound radiation is an important design parameter. Fletcher and Rossing (1991) and Meyer (1995) demonstrate that the intensity, I, is proportional to the surface velocity and that for a given driving function, this scales with modulus and density as:

$$I \, \alpha \sqrt{\frac{E}{\rho^3}} \qquad (4.61)$$

A high value of the combination of properties $\sqrt{E/\rho^3}$, called the *radiation factor*, is used by instrument makers to select materials for sound boards. When the material is elastically anisotropic (as is wood), E is replaced by $E = (E_\parallel \cdot E_\perp)$. The last column of Table 4.1

Table 4.1 Acoustic Impedances and Radiation Factors*

Material	Density ρ (Mg/m³)	Young's Modulus E^* (GPa)	Acoustic Impedance, Calculated $\sqrt{\rho E}$ (GN s/m³)	Radiation Factor, Calculated $\sqrt{E/\rho^3}$ (m⁴/kgs)
PS foam, low density	0.03	0.01	0.01	19.2
PS foam, high density	0.78	1.6	1.1	2.0
Solid PS	1.04	3.6	1.9	1.8
Nylon 6/6	1.14	2.4	1.7	1.3
Spruce, ‖ to grain	0.38	10.8	1.2	8.6
Maple, ‖ to grain	0.45	10.8	1.7	5.4
Aluminum	2.7	69	13.6	1.87
Steel	7.9	210	40.7	0.65
Silver	10.4	75	27.9	0.26
Glass	2.24	46	10.1	2.0

Data from American Institute of Physics Handbook, 1972, and Meyer, 1995.

gives some representative values for the radiation factor. Spruce, widely used for the front plates of violins, has a particularly high value—nearly twice that of maple, which is used for the back plate, the function of which is to reflect, not radiate.

FURTHER READING

The Internal Structure of Materials

W. D. Callister, Jr., Materials science and engineering, an introduction, 7th ed., Wiley, 2007, ISBN 0-471-73696-1.

W. F. Smith, and J. Hashemi, Foundations of materials science and engineering, 4th ed., McGraw Hill, 2005.

J. F. Shackelford, Introduction to materials science for engineers, 6th ed., Prentice Hall, 2004.

D. R. Askeland, and P. P. Phule, The Science and Engineering of Materials, 4th ed., Thomson, Brooks/Cole, 2002.

Mechanical Properties

M. F. Ashby and D. R. H. Jones, Engineering materials 1, 3rd ed., Elsevier Butterworth Heinemann, 2006, ISBN 0-7506-6380-2.

M. F. Ashby, H. R. Shercliff, and D. Cebon, Materials: engineering, science, processing and design, Butterworth Heinemann, 2007, ISBN-13: 978-0-7506-8391-3.

D. R. Askeland and P. P. Phule, The science of engineering materials, 5th ed., Thomson Publishing, 2006.

W. D. Callister, Jr., Materials science and engineering, an introduction, 7th ed., Wiley, 2007, ISBN 0-471-73696-1.

Thermal Properties

A. H. Cottrell, The mechanical properties of matter, Wiley, 1964, Library of Congress no. 64-14262.

J. P. Hollman, Heat transfer, 5th ed., McGraw Hill, 1981, ISBN 0-07-029618-9.

Electrical Properties

N. Braithwaite and G. Weaver, Electronic materials, The Open University and Butterworth Heinemann, 1990, ISBN 0-408-02840-8.

D. Jiles, Introduction to the electronic properties of materials, 2nd ed., Nelson Thornes Ltd., 2001, ISBN 0-7487-6042-3.

L. Solymar and D. Walsh, Electrical properties of materials, 7th ed., Oxford University Press, 2004, ISBN 0-19-926793-6.

Magnetic Behavior

N. Braithwaite and G. Weaver, Electronic materials, The Open University and Butterworth Heinemann, 1990, ISBN 0-408-02840-8.

P. Campbell, Permanent magnetic materials and their applications, Cambridge University Press, 1994.

W. D. Douglas, Magnetically soft materials, in ASM Metals Handbook, 9th ed., Vol. 2, Properties and selection of nonferrous alloys and special purpose materials, pp. 761–781, ASM, 1995.

J. W. Fiepke, Permanent magnet materials, in ASM Metals Handbook, 9th ed., Vol. 2, Properties and selection of nonferrous alloys and special purpose materials, pp. 782–803, ASM, 1995.

J. P. Jakubovics, Magnetism and magnetic materials, 2nd ed., The Institute of Materials, 1994, ISBN 0-901716-54-5.

Optical Behavior

W. D. Callister, Jr., Materials science and engineering, an introduction, 6th ed., John Wiley and Sons, 2003, ISBN 0-471-13576-3.

D. Jiles, Introduction to the electronic properties of materials, Nelson Thomson Ltd., 2001, ISBN 0-7487-6042-3.

Acoustic Behavior

E. E. Gray (ed.), American Institute of Physics, Handbook, 3rd ed., McGraw-Hill, 1972.

L. L. Beranek (ed.), Noise reduction, McGraw-Hill, pp. 349–395, 1960.

N. H. Fletcher, and T. D. Rossing, The physics of musical instruments, Springer-Verlag, 1991.

W. Lauriks, Acoustic characteristics of low density foams, in: Low density cellular plastics, N. C. Hilyard and A. Cunningham (eds.), Ch. 10, Chapman and Hall, 1994.

H. G. Meyer, Catgut acoust. soc. J., 2, No 7, 9–12, 1995.

P. H. Parkin, J. R. Humphreys, and J. R. Cowell, Acoustics, noise and buildings, 4th ed., Faber, pp. 279–83, 1979.

CHAPTER 5

Material Property Charts and Their Uses

5.1 MATERIAL PROPERTY CHARTS

Data sheets for materials list their mechanical, thermal, electrical, and optical properties, but they give no perspective and present no comparisons. The way to achieve these is to plot *material property charts*. Such charts display material properties in ways that allow ease of comparison and, in conjunction with *material indices*, enable optimized selection to meet specified design requirements. This chapter introduces the charts, selection methods, and optimization techniques.

The Modulus Bar Chart and the Modulus–Density Bubble Chart

Property charts are of two types: bar charts and bubble charts. A *bar chart* is simply a plot of one property. Figure 5.1 shows an example; it is a bar chart for Young's modulus, *E*. The largest is more than 10 million times greater than the smallest (many other properties have similarly large ranges) so it makes sense to plot them on logarithmic,[1] not linear, scales, as here. The length of each bar shows the range of the property for each of the materials, here segregated by family. The differences between the families now become apparent. Metals and ceramics have high moduli. Those of polymers are smaller, by a factor of about 50, than those of metals. Those of elastomers are some 500 times smaller still.

---

[1] *Logarithmic* means that the scale goes up in constant multiples, usually of 10. We live in a logarithmic world; our senses, for instance, all respond in that way.

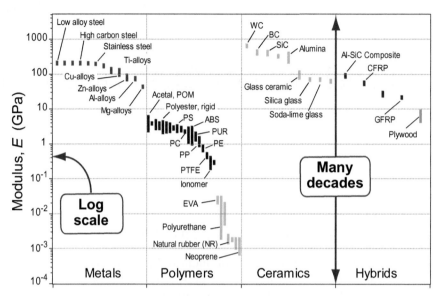

FIGURE 5.1

A bar chart of modulus. The chart reveals the difference in stiffness between the families.

More information is packed into the picture if two properties are plotted to give a *bubble chart*, as in Figure 5.2, here showing modulus E and density ρ. As before, the scales are logarithmic. Now families are more distinctly separated. Ceramics lie in the yellow envelope at the very top: They have moduli as high as 1000 GPa. Metals lie in the reddish zone near the top right; they, too, have high moduli, but they are heavy. Polymers lie in the dark blue envelope in the center, elastomers in the lighter blue envelope below, with moduli as low as 0.0001 GPa. Materials with a lower density than polymers are porous; examples are manmade foams and natural cellular structures such as wood and cork. Each family occupies a distinct, characteristic field.

The property chart thus gives an overview, showing where families and their members lie in $E - \rho$ space. It helps in the common problem of material selection for stiffness-limited applications in which weight must be minimized. Material property charts like these are a core tool, used throughout this book for the following reasons:

- They give an overview of the physical, mechanical, and functional properties of materials, presenting the information about them in a compact way.

- They reveal aspects of the physical origins of properties, helpful in understanding the underlying science.

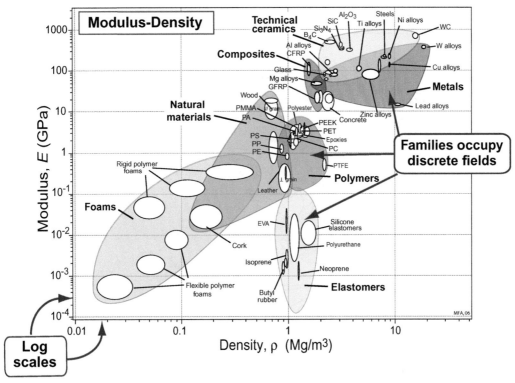

FIGURE 5.2

A bubble chart of modulus and density. Families occupy discrete areas of the chart.

■ They become a tool for optimized selection of materials to meet given design requirements, and they help us understand the use of materials in existing products.

■ They allow the properties of new materials, such as those with nano or amorphous structures, to be displayed and compared with those of conventional materials, bringing out their novel characteristics and suggesting possible applications.

To introduce the charts a little further, we briefly describe five more: the modulus–relative cost chart, the strength–density chart, the thermal expansion/thermal conductivity chart, and two that relate to the tactile and the acoustic properties of materials. Fuller descriptions of a wider range of charts can be found in Ashby (2005) and Ashby et al. (2007).

The Modulus–Relative Cost Chart

Often it is minimizing cost, not weight, that is the overriding objective of a design. The chart of Figure 5.3 shows, on the x axis, the

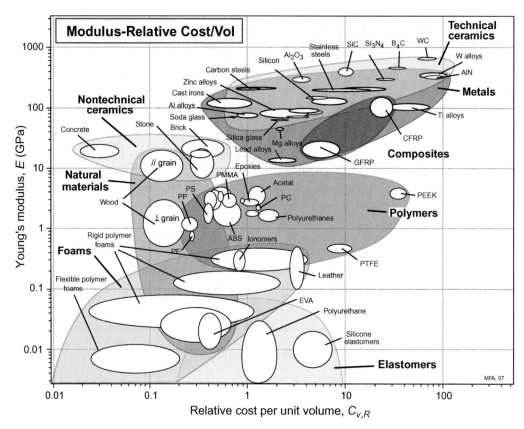

FIGURE 5.3

The modulus–relative cost chart. (The CES software contains material prices, regularly updated.)

relative prices per unit volume of materials, normalized to that of the metal used in larger quantities than any other: mild steel. Concrete and wood are among the cheapest; polymers, steels, and aluminum alloys come next; special metals such as titanium, most technical ceramics, and a few polymers such as PTFE and PEEK are expensive. The chart allows the selection of materials that are stiff and cheap.

The Strength–Density Chart

Figure 5.4 shows the yield strength (or elastic limit) σ_y plotted against density ρ. The range of strength for engineering materials, like that of the modulus, spans about six decades: from less than 0.01 MPa for foams, used in packaging and energy-absorbing systems, to 10^4 MPa for diamond, exploited in diamond tooling for machining and as the indenter of the Vickers hardness test. Members of each family again cluster together and can be enclosed in envelopes, each of which occupies a characteristic part of the chart.

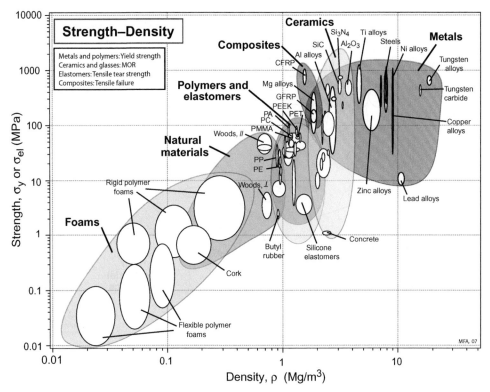

FIGURE 5.4
The strength–density chart.

Comparison with the modulus–density chart (Figure 5.2) reveals some marked differences. The modulus of a solid is a well-defined quantity with a narrow range of values. The yield strength is not. The strength range for a given class of metals, such as stainless steels, can span a factor of 10 or more, depending on composition and treatment, whereas the spread in stiffness is at most 10%. Since density varies very little, 'the strength bubbles for metals are long and thin. The wide ranges for metals reflect the underlying physics of yielding and present designers with an opportunity to manipulate the strength by varying composition and process history.

Polymers cluster together with strengths between 10 and 100 MPa. The composites CFRP and GFRP have strengths that lie between those of polymers and ceramics, as one might expect since they are mixtures of the two.

Thermal Expansion, α, and Thermal Conductivity, λ

Figure 5.5 maps thermal expansion, α, and thermal conductivity, λ. Metals and technical ceramics lie toward the lower right; they

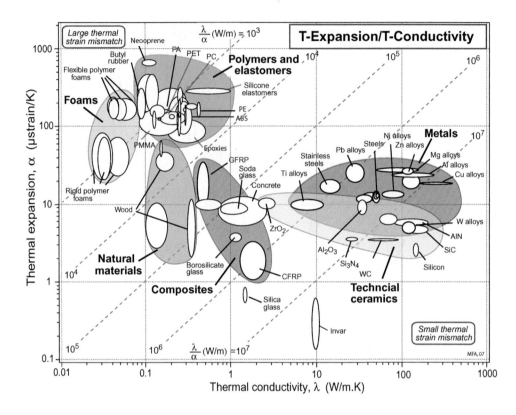

FIGURE 5.5

The linear expansion coefficient, α, plotted against the thermal conductivity, λ. The contours show the thermal distortion parameter λ/α.

have high conductivities and modest expansion coefficients. Polymers and elastomers lie at the upper left; their conductivities are 100 times less and their expansion coefficients 10 times greater than those of metals. The chart shows contours of λ/α, a quantity important in designing against thermal distortion. An extra material, Invar (a nickel alloy), has been added to the chart because of its uniquely low expansion coefficient at and near room temperature, a consequence of a tradeoff between normal expansion and a contraction associated with a magnetic transformation.

Feel: Tactile Attributes

Steel is "hard"; so is glass; diamond is harder than either of them. Hard materials do not scratch easily; indeed, they can be used to scratch other materials. They generally accept a high polish, resist distortion, and are durable. The impression that a material is hard is directly related to the material property "hardness" measured by materials engineers and tabulated in the handbook. "Soft" sounds like the opposite of "hard" but, in engineering terms, it is not; there

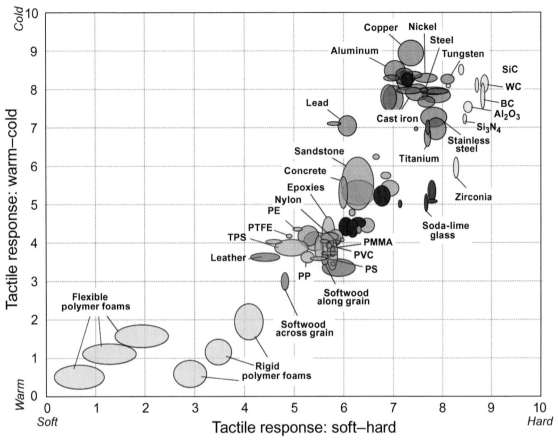

FIGURE 5.6

A chart for the tactile properties of materials.

is no engineering property called softness. A soft material deflects when handled, it gives a little, it is squashy, but when it is released it returns to its original shape. This is elastic (or visco-elastic) behavior and the material property that has most influence on it is the modulus, not the hardness. Elastomers (rubbers) feel soft; so do polymer films; both classes of materials have moduli that are 100 to 10,000 times lower than ordinary "hard" solids. It is this that makes them feel soft. Hard materials can be made soft by forming them into shapes in which they bend or twist—hard steel shaped into soft springs or glass drawn into fibers and woven into cloth, for instance. To compare the intrinsic softness of materials (as opposed to the softness acquired by shape), materials must be compared in the same shape, and then it is the modulus that is the key property. An appropriate measure of hardness and softness is the product EH. It is used (appropriately normalized) as one axis of Figure 5.6.

A material feels "cold" to the touch if it conducts heat away from the finger quickly; it is "warm" if it does not. This has something to do with the technical attribute thermal conductivity, but there is more to it than that; it also depends on specific heat. A measure of this perceived coldness or warmth of a material (in the sense of heat, not of color) is the quantity $\sqrt{\lambda \cdot \rho \cdot C_p}$, where λ is the thermal conductivity, ρ is the density, and C_p is the specific heat. It is shown as the other axis of Figure 5.6. The figure nicely displays the tactile properties of materials. Polymer films and low-density woods are warm and soft. Ceramics, stone, and metals are cold and hard. Polymers and composites lie in between.

Hearing: Acoustic Attributes

The frequency of sound (pitch) emitted when an object is struck relates to two material properties: modulus and density. A measure of this pitch is used as one axis of Figure 5.7. Frequency is not the only aspect of acoustic response; another one has to do with damping. A highly damp material sounds dull and muffled; one with low damping rings. Acoustic brightness—the inverse of damping—is used as the other axis of Figure 5.7. It groups materials that have similar acoustic behavior.

Bronze, glass, and steel ring when struck, and the sound they emit has, on a relative scale, a high pitch. These materials are used to make bells. Alumina, on this ranking, has the same bell-like qualities. Rubber, foam, and many polymers sound dull, and, relative to metals, they vibrate at low frequencies; they are used for sound damping. Lead, too, is dull and low-pitched; it is used to clad buildings for sound insulation.

5.2   USING CHARTS TO SELECT TRANSLATION, SCREENING, RANKING, AND DOCUMENTATION

Selection involves seeking the best match between the attribute profiles of the materials—bearing in mind that these must be mutually compatible—and those required by the design. The strategy, applied to materials, is sketched in Figure 5.8. The first task is that of *translation*: converting the design requirements into a prescription for selecting a material. This proceeds by identifying the *constraints* that the material must meet and the *objectives* that the design must fulfill. These become the filters; materials that meet the constraints

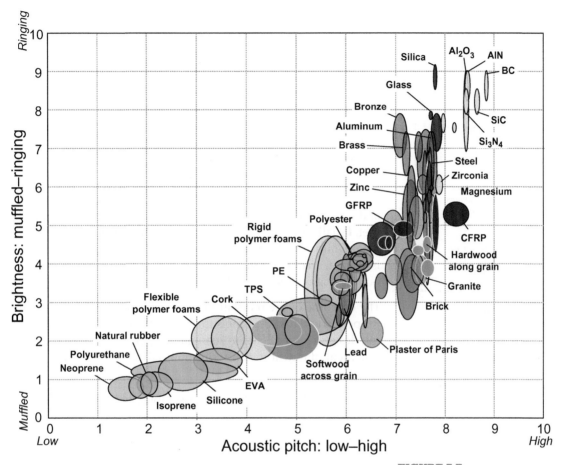

FIGURE 5.7
A chart for the acoustic properties of materials.

and rank highly in their ability to fulfill the objectives are potential candidates for the design. The second task, then, is that of *screening*: eliminating the material that cannot meet the constraints. This is followed by the *ranking* step, ordering the survivors by their ability to meet a criterion of excellence, such as that of minimizing cost. The final task is to explore the most promising candidates in depth, examining how they are used at present, case histories of failures, and how best to design with them, a step we call *documentation*.

Translation

Any engineering component has one or more functions: to support a load, to contain a pressure, to transmit heat, and so forth. This must be achieved subject to constraints: that certain dimensions are fixed, that the component must carry the design loads without

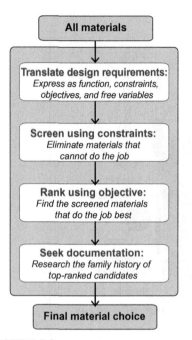

FIGURE 5.8

The strategy. There are four steps: translation, screening, ranking, and supporting information. All can be implemented in software, allowing large populations of materials to be investigated.

failure, the need to insulate against or to conduct heat or electricity, that it can function in a certain range of temperature and in a given environment, and many more. In designing the component, the designer has one or more objectives: to make it as cheap as possible, perhaps, or as light, or as safe, or some combination of these. Certain parameters can be adjusted to optimize the objective; the designer is free to vary dimensions that are not constrained by design requirements and, most important, is free to choose the material for the component. We refer to these as *free variables*. Constraints, objectives, and free variables (see Table 5.1) define the boundary conditions for selecting a material and, in the case of load-bearing components, a shape for its cross-section.

It is important to be clear about the distinction between constraints and objectives. A *constraint* is an essential condition that must be met, usually expressed as a limit on a material attribute. An *objective* is a quantity for which an extreme value (a maximum or minimum) is sought, frequently the minimization of cost, mass, or volume, but there are others (see Table 5.2).

The outcome of the translation step is a list of the design-limiting properties and the constraints they must meet. The first step in relating design requirements to material properties is therefore a clear statement of function, constraints, objectives, and free variables.

Table 5.1   Function, Constraints, Objectives, and Free Variables

Function	What does the component do?
Constraints	What nonnegotiable conditions must be met?
Objective	What is to be maximized or minimized?
Free variables	What parameters of the problem is the designer free to change?

Table 5.2   Common Constraints and Objectives

Common Constraints	Common Objectives
Meet a target value of:	**Minimize:**
Stiffness	Cost
Strength	Mass
Fracture toughness	Volume
Thermal conductivity	Impact on the environment
Electrical resistivity	Heat loss
Magnetic remanence	**Maximize:**
Optical transparency	Energy storage
Cost	Heat flow
Mass	

Screening

Constraints are gates: Meet the constraint and you pass through the gate; fail to meet it and you are out. Screening (see Figure 5.8) does just that; it eliminates candidates that cannot do the job at all because one or more of their attributes lie outside the limits set by the constraints. As examples, the requirement that "the component must function in boiling water" or that "the component must be transparent" imposes obvious limits on the attributes of maximum service temperature and optical transparency that successful candidates must meet. We refer to these as *attribute limits*.

Ranking: Material Indices

To rank the materials that survive the screening step, we need criteria of excellence. They are found in the material indices, introduced here, which measure how well a candidate that has passed the screening step can do the job (Figure 5.8). Performance is sometimes limited by a single property, sometimes by a combination of them. Thus the best materials for buoyancy are those with the lowest density, ρ; those best for thermal insulation are the ones with the smallest values of the thermal conductivity, λ—provided, of course, that they also meet all other constraints imposed by the design. Here maximizing or minimizing a single property maximizes performance. Often, though, it is not one but a group of properties that are relevant. Thus the best materials for a light, stiff tie rod are those with the greatest value of the specific stiffness, E/ρ, where E is Young's modulus. The best materials for a spring are those with the greatest value of σ_y^2/E, where σ_y is the yield strength. The property or property group that maximizes performance for a given design is called its *material index*. There are many such indices, each associated with maximizing some aspect of performance. They provide criteria of excellence that allow ranking of materials by their ability to perform well in the given application. The appendix to this chapter lists some of the more common ones. A further discussion of material indices, including their derivations and applications, is found in Ashby (2005) and Ashby, et al. (2007).

To summarize: *Screening* isolates candidates that are capable of doing the job; *ranking* identifies those among them that can do the job best.

Documentation

The outcome of the steps so far is a ranked shortlist of candidates that meet the constraints and that maximize or minimize the criterion of excellence, whichever is required. You could simply choose the top-ranked candidate, but what hidden weaknesses might it have? What is its reputation? Has it a good track record? To proceed further, we seek a detailed profile of each: its documentation (Figure 5.8, bottom).

What form does documentation take? Typically it is descriptive, graphical, or pictorial: case studies of previous uses of the material, details of its corrosion behavior in particular environments, of its availability and pricing, warnings of its environmental impact or toxicity. Such information is found in handbooks, suppliers' data sheets, CD-based data sources, and high-quality Websites. Documentation helps narrow the shortlist to a final choice, allowing a definitive match to be made between design requirements and material attributes.

Why are all these steps necessary? Without screening and ranking, the candidate pool is enormous and the volume of documentation is overwhelming. Dipping into it, hoping to stumble on a good material, gets you nowhere. But once a small number of potential candidates have been identified by the screening/ranking steps, detailed documentation can be sought for these few alone, and the task becomes viable.

5.3 PLOTTING LIMITS AND INDICES ON CHARTS

Screening: Constraints on Charts

Any design imposes certain nonnegotiable demands ("constraints") on the material of which it is made. These limits can be plotted as horizontal or vertical lines on material property charts, as illustrated in Figure 5.9. The figure shows a schematic of the E–relative cost chart shown earlier. We suppose that the design imposes limits on these of $E > 10$ GPa and relative cost <3, shown in the figure. All materials in the window defined by the limits, labeled "Search region," meet both constraints.

Charts exist for many properties; others can be generated using the CES software.[2] They allow limits to be imposed on other properties.

---

[2] The CES Edu software (www.Grantadesign.com). It was used to create all the property charts shown in this and other chapters.

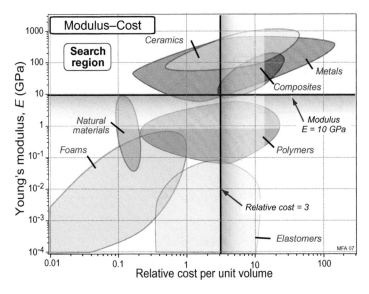

FIGURE 5.9

A schematic E-relative cost chart showing a lower limit for E and an upper one for relative cost.

Ranking: Indices on Charts

The next step is to seek, from the subset of materials that meet the property limits, those that maximize the performance of the component. We use the design of light, stiff components as examples; the other material indices are used in a similar way.

Figure 5.10 shows a schematic of the $E - \rho$ chart shown earlier. The logarithmic scales allow all three of the indices E/ρ, $E^{1/3}/\rho$, and $E^{1/2}/\rho$, listed in Table A.2 in the Appendix, to be plotted onto it. Consider the condition

$$M = \frac{E}{\rho} = cons\,tant, C \tag{5.1}$$

that is, a particular value of the specific stiffness. Taking logs

$$\log(E) = \log(\rho) + \log(C) \tag{5.2}$$

This is the equation of a straight line of slope 1 on a plot of $\log(E)$ against $\log(\rho)$, as shown on the figure. Similarly, the condition

$$M = \frac{E^{1/3}}{\rho} = cons\,tant, C \tag{5.3}$$

becomes, on taking logs,

$$\log(E) = 3\log(\rho) + 3\log(C) \tag{5.4}$$

This is another straight line, this time with a slope of 3, also shown. And by inspection, the third index $E^{1/2}/\rho$ will plot as a line of slope

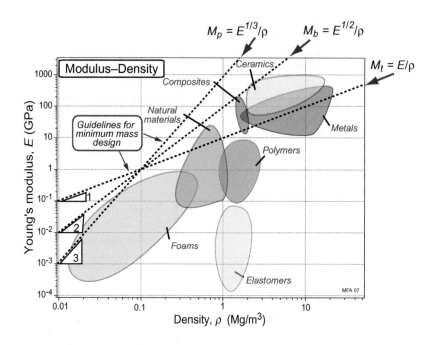

$$M_p = E^{1/3}/\rho \qquad M_b = E^{1/2}/\rho$$

$$M_t = E/\rho$$

FIGURE 5.10

A schematic E – ρ chart showing guidelines for three material indices for stiff, lightweight structures.

2. We refer to these lines as *selection guidelines*. They give the slope of the family of parallel lines belonging to that index.

It is now easy to read off the subset of materials that maximize performance for each loading geometry. For example, all the materials that lie on a line of constant $M = E^{1/3}/\rho$ perform equally well as a light, stiff panel; those above the line perform better, those below, less well. Figure 5.11 shows a grid of lines corresponding to values of $M = E^{1/3}/\rho$ from $M = 0.22$ to $M = 4.6$ in units of $GPa^{1/3}/(Mg \cdot m^{-3})$. A material with $M = 3$ in these units gives a panel that has one tenth the weight of one with $M = 0.3$. Ashby (2005) and Ashby et al. (2007) develop numerous case studies illustrating the use of the method.

Computer-Aided Selection

The charts give an overview, but the number of materials that can be shown on any one of them is obviously limited. Selection using them is practical when there are very few constraints, but when there are many, as there usually are, checking that a given material meets them all is cumbersome. Both problems are overcome by a computer implementation of the method.

The CES material selection software is an example of such an implementation. Its database contains records for materials, organized

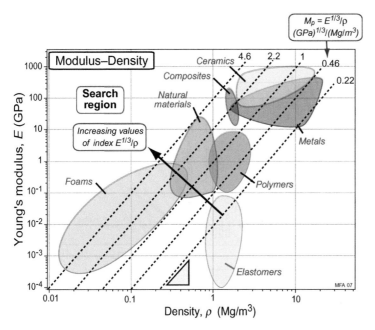

FIGURE 5.11

A schematic E − ρ chart showing a grid of lines for the index $E^{1/3}/\rho$. The units are $(GPa)^{1/3}/(Mg/m^3)$.

in the hierarchical manner. Each record contains property data for a material, each property stored as a range spanning its typical (or, often, permitted) values. It also contains limited documentation in the form of text, images, and references to sources of information about the material. The data are interrogated by a search engine that offers the search interfaces shown schematically in Figure 5.12. On the left is a simple query interface for screening on single attributes. The desired upper or lower limits for constrained properties are entered; the search engine rejects all materials with attributes that lie outside the limits. In the center is shown a second way of interrogating the data: a bar chart, constructed by the software, for any numeric property in the database. It and the bubble chart shown on the right are ways of both applying constraints and ranking. For screening, a selection line or box is superimposed on the charts, with edges that lie at the constrained values of the property (bar chart) or properties (bubble chart). This eliminates the materials in the shaded areas and retains the materials that meet the constraints. If, instead, ranking is sought (having already applied all necessary constraints), an index line like that shown in Figure 5.11 is positioned so that a small number of materials—say, 10—are left in the selected area; these are the top-ranked candidates. The software delivers a ranked list of the top-ranked materials that meet all the constraints.

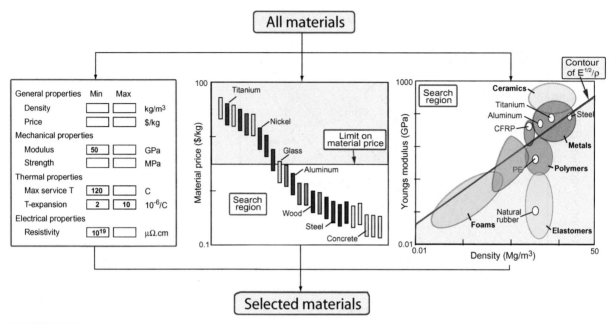

FIGURE 5.12

Computer-aided selection using the CES software. The schematic shows the three types of selection window. They can be used in any order and any combination. The selection engine isolates the subset of materials that pass all the selection stages.

5.4 RESOLVING CONFLICTING OBJECTIVES

The choice of material that best meets one objective will not usually be that which best meets the others. The designer charged with selecting a material for a portable building that must be both as light *and* as cheap as possible faces an obvious difficulty: The lightest material will certainly not be the cheapest, and vice versa. To make any progress, the designer needs a way of trading weight against cost. This section describes ways of resolving this and other conflicts of objective.

Suppose you want a component with a required stiffness (Constraint 1) and strength (Constraint 2) that must be as light as possible (one objective). You could choose materials with high modulus E for stiffness, and then the subset of these that have high elastic limits σ_y for strength, and the subset of those that have low density ρ for light weight. Then again, if you wanted a material with a required stiffness (one constraint) that was simultaneously as light (Objective 1) and as cheap (Objective 2) as possible, you could apply the constraint and then find the subset of survivors that were light and the subset of *those* survivors that were cheap. Some selection systems work that way, but it is not a good idea because there is no guidance in deciding the *relative* importance of the limits on stiffness, strength, weight, and cost. This is not a trivial difficulty; it is exactly this relative importance that makes aluminum the prime structural material for aerospace and steel that for ground-based structures.

These problems of relative importance are old; engineers have sought methods to overcome them for at least a century. The traditional approach is that of assigning *weight factors* to each constraint and objective and using them to guide choice in ways that are summarized in the following discussion. Experienced engineers can be good at assessing relative weights, but the method nonetheless relies on judgment, and judgments can differ. For this reason, this section focuses on systematic methods. But one should know about the traditional methods because they are still widely used. We start with them.

The Method of Weight Factors

Weight factors seek to quantify judgment. The method works like this: The key properties are identified and their values M_i are tabulated for promising candidates. Since their absolute values can differ widely and depend on the units in which they are measured, each is first scaled by dividing it by the largest value of its group, $(M_i)_{max}$, so that the largest, after scaling, has the value 1. Each is then multiplied by a weight factor, w_i, with a value between 0 and 1, expressing its relative importance for the performance of the component. This give a weighted index W_i:

$$W_i = w_i \frac{M_i}{(M_i)_{min}} \qquad (5.5)$$

For properties that are not readily expressed as numerical values, such as weldability or wear resistance, rankings such as A to E are expressed instead by a numeric rating, $A = 5$ (very good) to $E = 1$ (very bad), then dividing by the highest rating value as before. For properties that are to be minimized, such as corrosion rate, scaling uses the minimum value $(M_i)_{min}$, expressed in the form

$$W_i = w_i \frac{(M_i)_{min}}{M_i} \qquad (5.6)$$

The weight factors w_i are chosen such that they add up to 1, that is, $w_i < 1$ and $\Sigma w_i = 1$. There are numerous schemes for assigning their values (see Further Reading); all require, in varying degrees, the use of judgment. The most important property is given the largest w; the second most important, the second largest, and so on. The $W_i's$ are calculated from Equation 5.5 and 5.6 and summed. The best choice is the material with the largest value of the sum

$$W = \overline{\iota}_i W_i \qquad (5.7)$$

Sounds simple, but there are problems, the most obvious of which is the subjectivity in assigning the weights.

Systematic Methods: Penalty Functions and Exchange Constants

Real-life materials selection almost always requires that a compromise be reached between conflicting objectives. Three crop up all the time. They are:

- *Minimizing mass.* A common target in designing things that move or have to move, or oscillate or that must respond quickly to a limited force (think of aerospace and ground transport systems).

- *Minimizing volume.* Because less material is used and because space is increasingly precious (think of the drive for ever-thinner, smaller mobile phones, portable computers, MP3 players ... and the need to pack more and more functionality into a fixed volume).

- *Minimizing cost.* Profitability depends on the difference between cost and value; the most obvious way to increase this difference is to reduce cost.

To this we must now add a fourth objective:

- *Minimizing environmental impact.* The damage to our surroundings caused by product manufacture and use.

There are, of course, other objectives specific to particular applications. Some are just one of the four previously described, expressed in different words. The objective of maximizing *power-to-weight ratio*, for example, translates into minimizing mass for a given power output. Some can be quantified in engineering terms, such as maximizing *reliability* (although this can translate into achieving a given reliability at a minimum cost) and others cannot, such as maximizing *consumer appeal*, an amalgam of performance, styling, image, and marketing.

So we have four common objectives, each characterized by a performance metric P_i. At least two are involved in the design of almost any product. Conflict arises because the choice that optimizes one objective will not, in general, do the same for the others; then the best choice is a compromise, optimizing none but pushing all as close to their optima as their interdependence allows. This highlights the central problem: How is mass to be compared with cost, or volume with environmental impact? Each is measured in different units; they are incommensurate. We need a way of expressing both in a common currency. This comes in a moment. First, some definitions.

Tradeoff Strategies

Consider the choice of material to minimize both mass (performance metric P_1) and cost (performance metric P_2) while also meeting a set of constraints such as a required strength or durability in a certain environment. Following the standard terminology of optimizations theory, we define a *solution* as a viable choice of material, meeting all the constraints but not necessarily optimal by either of the objectives. Figure 5.13 is a plot of P_1 against P_2 for alternative solutions, each bubble describing a solution. The solutions that minimize P_1 do not minimize P_2, and vice versa. Some solutions, such as that at A, are far from optimal; all the solutions in the box attached to it have lower values of both P_1 and P_2. Solutions like A are said to be *dominated* by others. Solu-

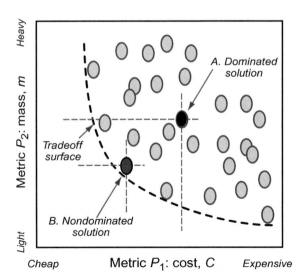

FIGURE 5.13

Multiple objectives. Mass and cost for a component made from alternative material choices. The tradeoff surface links nondominated solutions.

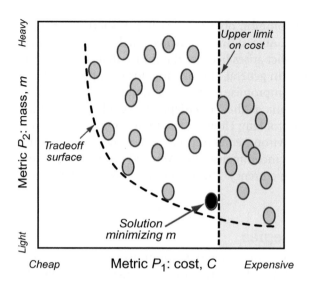

FIGURE 5.14

The tradeoff plot with a simple constraint imposed on cost. The solution with the lowest mass can now be read off, but it is not a true optimization.

tions like those at *B* have the characteristic that no other solutions exist with lower values of both P_1 and P_2. These are said to be *nondominated* solutions. The line or surface on which they lie is called the nondominated or optimal *tradeoff surface*. The values of P_1 and P_2 corresponding to the nondominated set of solutions are called the *Pareto set*.

There are three strategies for progressing further. The solutions on or near the tradeoff surface offer the best compromise; the rest can be rejected. Often this is enough to identify a shortlist, using intuition to rank them (Strategy 1). Alternatively (Strategy 2), one objective can be reformulated as a constraint, as illustrated in Figure 5.14. Here an upper limit is set on cost; the solution that minimizes the other constraint can then be read off. But this is cheating; it is not a true optimization. To achieve that, we need Strategy 3: that of *penalty functions*.

Penalty Functions

The tradeoff surface identifies the subset of solutions that offer the best compromises between the objectives. Ultimately, though, we want a single solution. One way to do this is to aggregate the various objectives into a single objective function, formulated such that its minimum defines the most preferable solution.

Consider the case in which one of the objectives to be minimized is cost, C (units: $), and the other is mass, m (units: kg). It makes

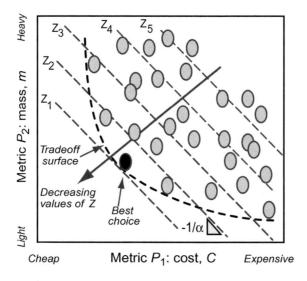

FIGURE 5.15
The penalty function Z superimposed on the tradeoff plot. The contours of Z have a slope of $-1/\alpha$. The contour that is tangent to the tradeoff surface identifies the optimum solution.

sense to measure penalty Z in the same units as that of cost ($). We then define a locally linear penalty function[3] Z:

$$Z = C + \alpha m \qquad (5.8)$$

or

$$m = -\frac{1}{\alpha}C + \frac{1}{\alpha}Z \qquad (5.9)$$

Here α is the change in Z associated with unit increase in m and has the units of $/kg. Equation 5.9 defines a linear a relationship between m and C. This plots as a family of parallel penalty lines, each for a given value of Z, as shown in Figure 5.15. The slope of the lines is the negative reciprocal of the exchange constant, $-1/\alpha$. The value of Z decreases toward the bottom left; the best choices lie there. The optimum solution is the one nearest the point at which a penalty line is tangential to the tradeoff surface, since it is the one with the smallest value of Z.

Values for the Exchange Constants, α

An exchange constant is the value or *utility* of a unit change in a performance metric. Its magnitude and sign depend on the

---

[3] Also called a *value function* or *utility function*. The method allows a local minimum to be found. When the search space is large, it is necessary to recognize that the values of the exchange constants α_i may themselves depend on the values of the performance metrics P_i.

application. Thus the utility of weight saving in a family car is small, though significant; in aerospace it is much larger. The utility of heat transfer in house insulation is directly related to the cost of the energy used to heat the house; that in a heat exchanger for power electronics can be much higher because high heat transfer means faster computing. The utility can be real, meaning that it measures a true saving of cost. But it can also sometimes be perceived, meaning that the consumer, influenced by scarcity, advertising, or fashion, will pay more or less than the true value of the metric.

In many engineering applications the exchange constants can be derived approximately from technical models for the life cost of a system. Thus the utility of weight saving in transport systems is derived from the value of the fuel saved or that of the increased payload, evaluated over the life of the system. Table 5.3 gives approximate values for α for various modes of transport. The most striking thing about them is the enormous range: The exchange constant depends in a dramatic way on the application in which the material will be used. It is this that lies behind the difficulty in adopting aluminum alloys for cars, despite their universal use in aircraft; it explains the much greater use of titanium alloys in military than in civil aircraft, and it underlies the restriction of beryllium (a very expensive metal) to use in space vehicles.

Exchange constants can be estimated in various ways. The cost of launching a payload into space lies in the range of $3000 to $10,000/kg; a reduction of 1 kg in the weight of the launch structure would allow a corresponding increase in payload, giving the ranges of α shown in the table. Similar arguments based on increased payload or decreased fuel consumption give the values shown for civil aircraft, commercial trucks, and automobiles. The values change with time, reflecting changes in fuel costs, legislation to increase fuel economy, and so on.

Table 5.3 Exchange Constants α for the Mass–Cost Tradeoff for Transport Systems

Sector: Transport Systems	Basis of Estimate	Exchange Constant, α US$/kg
Family car	Fuel saving	1–2
Truck	Payload	5–20
Civil aircraft	Payload	100–500
Military aircraft	Payload, performance	500–1000
Space vehicle	Payload	3000–10,000

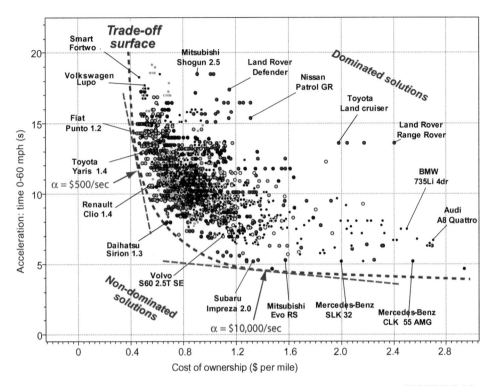

FIGURE 5.16
A tradeoff plot for selecting cars that have high performance but are cheap to run. The best choices are those that lie on or near the tradeoff surface.

These values for the exchange constant are based on engineering criteria. More difficult to assess are those based on perceived value. That for the performance/cost tradeoff for cars is an example. To the enthusiast, a car that is able to accelerate rapidly is alluring. He (or she) is prepared to pay more to go from 0 mph to 60 mph in 5 seconds than to wait around for 10 seconds. Figure 5.16 shows just how much. It is a tradeoff plot of 0–60 acceleration time t_a against all-inclusive running cost per mile, C_r (the total of fuel, depreciation, insurance, etc.), for about 2500 cars. Dominated solutions, some labeled, lie far from the tradeoff surface. Nondominated solutions, some of these, too, labeled, lie on or near it. They are the ones with the most attractive combinations of acceleration and running cost.

To refine the selection further, it is necessary to assign a value to acceleration. Suppose that the car will be driven 10,000 miles per year. Then the penalty function describing the way the tradeoff between acceleration and annual running cost takes the form:

$$Z = 10,000 \, C_r + \alpha t_a \qquad (5.10)$$

Here α describes the (perceived) value of acceleration. Two penalty lines are shown, both tangent to the tradeoff surface. The first is for a value of α of \$500/sec. The second is for a value of α = \$10,000/

sec. If acceleration is not important to you, cars near the tangent point of the first penalty line are the best choice: the Renault Clio 1.4 or the Toyota Yaris 1.4. If instead acceleration is what you want, cars near the tangent point of the second penalty line are the best bet: the Subaru Impreza 2.0 or the Mitsubishi Evo R3. But you should be aware that your running costs are inflated, even with the optimizing of cost we have just performed, by a factor of almost 3. Acceleration does not come cheap.

This is all about *perceived value*. Advertising aims to increase the perceived value of a product, increasing its value without increasing its cost. It influences the exchange constants for family cars, and it is the driver for the development of titanium watches, carbon fiber spectacle frames, and a great deal of sports equipment.

There are other circumstances in which establishing the exchange constant can be more difficult. An example is that of *environmental impact*—the damage to the environment caused by manufacture, use, or disposal of a given product. Minimizing environmental impact has now become an important objective, almost as important as minimizing cost. Ingenious design can reduce the first without driving the second up too much. But how much is unit decrease in impact worth? Until an exchange constant is agreed or imposed, it is difficult for the designer to respond.

FURTHER READING

M. F. Ashby, Materials selection in mechanical design, 3rd ed., Chapter 4, Butterworth Heinemann, 2005, ISBN 0-7506-6168-2.

M. F. Ashby, Multi-objective optimization in material design and selection, Acta Mater., 2008, Vol. 48, pp. 359–369.

M. F. Ashby and K. Johnson, Materials and design: the art and science of material selection in product design, Butterworth Heinemann, 2002, ISBN 0-7506-5554-2.

M. F. Ashby, H. R. Shercliff, and D. Cebon, Materials: engineering, science, processing and design, Butterworth Heinemann, 2007, ISBN-13: 978-0-7506-8391-3.

J. P. Clark, R. Roth, and F. R. Field, Techno-economic issues in material science, ASM Handbook, Vol. 20, Materials Selection and Design, G. E. Dieter (ed.), ASM International, 1997, pp. 255–265, ISBN 0-87170-386-6.

G. E. Dieter, Engineering design, a materials and processing approach, 3rd ed., McGraw-Hill, 2000, pp. 150–153 and 255–257, ISBN 0-07-366136-8.

F. R. Field and R. de Neufville, Material selection – maximizing overall utility, Metals and Materials, June 1998, pp. 378–382.

A. Goicoechea, D. R. Hansen, and L. Druckstein, Multi-objective decision analysis with engineering and business applications, Wiley, 1982.

Appendix: Material Indices

The performance, P, of a component is characterized by a performance equation. The performance equation contains groups of material properties. These groups are the material indices. Sometimes the "group" is a single property; thus if the performance of a beam is measured by its stiffness, the performance equation contains only one property, the elastic modulus E. It is the material index for this problem. More commonly the performance equation contains a group of two or more properties. Familiar examples are the specific stiffness, E/ρ, and the specific strength, σ_y/ρ (where σ_y is the yield strength or elastic limit, and ρ is the density), but there are many others. They are a key to the optimal selection of materials. Details of the method, with numerous examples, are given in the texts cited under Further Reading. This appendix compiles indices for a range of common applications. The symbols used in the tables are defined in the first of them. A simple example is shown here to illustrate the general process used in determining an index for a particular situation.

EXAMPLE: DETERMINATION OF A MATERIAL INDEX FOR A LIGHT, STRONG TIE ROD

Minimizing Weight

Choose a material for a cylindrical tie rod that has sufficient strength to carry a specified load and is as light as possible (see Figure 5.17).

The mass of any tension rod is given by

$$M_{mass} = \rho V = \rho(AL)$$

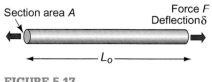

FIGURE 5.17

A simple tie-rod in tension.

where A is the cross-sectional area and L is the length of the member, V is the volume of the member, and ρ is the unit weight of the material. The stress in any tension member is given by:

$$Stress = Force/Area \text{ or } \sigma = F/A$$

For a given material stress level, σ_y, the required cross-sectional area becomes:

$$A_{required} = F/\sigma_y$$

From the mass expression:

$$M_{mass} = \rho A_r L = \rho L\left(F/\sigma_y\right)$$

If we put all the material-related parameters into one place, the mass becomes:

$$M_{mass} = FL\left(\rho/\sigma_y\right) \text{ or } M_{mass} = \rho A_r L = FL\left(\rho/\sigma_y\right)$$

Note that if we minimize the factor (ρ/σ_y) or maximize its inverse (σ_y/ρ), we will get a minimum weight tension member. The latter factor (σ_y/ρ) is typically called a *material index* or *performance index*, in this case for tension:

$$Material\ index\ for\ tension\ member = m_t = \left(\sigma_y/\rho\right)$$

To find what material should be used to minimize the weight of the tension member, all we have to do is find a material with a maximum value of (σ_y/ρ). As described in the text, this measure can be used in conjunction with material property charts to select an appropriate material.

Material indices for other design situations can be similarly determined (see Further Reading). Relevant indices are listed in the tables.

Table A.1 The Symbols

Class	Property	Symbol and Units
General	Density	ρ (kg/m^3 or Mg/m^3)
	Price	C_m ($/kg)
Mechanical	Elastic moduli (Young's, shear, bulk)	E, G, K (GPa)
	Poisson's ratio	v (—)
	Failure strength (yield, fracture)	σ_f (MPa)
	Fatigue strength	σ_e (MPa)
	Hardness	H (Vickers)
	Fracture toughness	K_{Ic} (MPa·m$^{1/2}$)
	Loss coefficient (damping capacity)	η (—)
Thermal	Thermal conductivity	λ (W/m·K)
	Thermal diffusivity	a (m^2/s)
	Specific heat	C_p (J/kg·K)
	Thermal expansion coefficient	α (°K^{-1})
	Difference in thermal conductivity	$\Delta\alpha$ (°K^{-1})
Electrical	Electrical resistivity	ρ_e ($\mu\Omega$·cm)
Ecoproperties	Energy/kg to produce material per kg	E_p (MJ/kg)
	CO_2/kg burden of material production per kg	CO_2 (kg/kg)

Table A.2 Stiffness-Limited Design at Minimum Mass (Cost, Energy, Environmental Impact)

Function and Constraints	Maximize
Tie (tensile strut) Stiffness, length specified; section area free	E/ρ
Shaft (loaded in torsion) Stiffness, length, shape specified, section area free Stiffness, length, outer radius specified; wall thickness free Stiffness, length, wall thickness specified; outer radius free	$G^{1/2}/\rho$ G/ρ $G^{1/3}/\rho$
Beam (loaded in bending) Stiffness, length, shape specified; section area free Stiffness, length, height specified; width free Stiffness, length, width specified; height free	$E^{1/2}/\rho$ E/ρ $E^{1/3}/\rho$
Column (compression strut, failure by elastic buckling) Buckling load, length, shape specified; section area free	$E^{1/2}/\rho$
Panel (flat plate, loaded in bending) Stiffness, length, width specified, thickness free	$E^{1/3}/\rho$
Plate (flat plate, compressed in-plane, buckling failure) Collapse load, length and width specified, thickness free	$E^{1/3}/\rho$
Cylinder with internal pressure Elastic distortion, pressure and radius specified; wall thickness free	E/ρ
Spherical shell with internal pressure Elastic distortion, pressure and radius specified, wall thickness free	$E/(1-v)\rho$

Table A.3 Strength-Limited Design at Minimum Mass (Cost, Energy, Environmental Impact)

Functions and Constraints	Maximize
Tie (tensile strut) Stiffness, length specified; section area free	σ_f/ρ
Shaft (loaded in torsion) Load, length, shape specified, section area free Load, length, outer radius specified; wall thickness free Load, length, wall thickness specified; outer radius free	$\sigma_f^{2/3}/\rho$ σ_f/ρ $\sigma_f^{1/2}/\rho$
Beam (loaded in bending) Load, length, shape specified; section area free Load length, height specified; width free Load, length, width specified; height free	$\sigma_f^{2/3}/\rho$ σ_f/ρ $\sigma_f^{1/2}/\rho$
Column (compression strut) Load, length, shape specified; section area free	σ_f/ρ
Panel (flat plate, loaded in bending) Stiffness, length, width specified, thickness free	$\sigma_f^{1/2}/\rho$
Plate (flat plate, compressed in-plane, buckling failure) Collapse load, length and width specified, thickness free	$\sigma_f^{1/2}/\rho$
Cylinder with internal pressure Elastic distortion, pressure and radius specified; wall thickness free	σ_f/ρ
Spherical shell with internal pressure Elastic distortion, pressure and radius specified, wall thickness free	σ_f/ρ
Flywheels, rotating disks Maximum energy storage per unit volume; given velocity Maximum energy storage per unit mass; no failure	ρ σ_f/ρ

Table A.4 Strength-Limited Design: Springs, Hinges, Etc. for Maximum Performance

Functions and Constraints	Maximize
Springs Maximum stored elastic energy per unit volume; no failure Maximum stored elastic energy per unit mass; no failure	σ_f^2/E $\sigma_f^2/E\rho$
Elastic hinges Radius of bend to be minimized (max flexibility without failure)	σ_f/E
Knife edges, pivots Minimum contact area, maximum bearing load	σ_f^3/E^2 and H
Compression seals and gaskets Maximum conformability; limit on contact pressure	$\sigma_f^{3/2}/E$ and $1/E$
Diaphragms Maximum deflection under specified pressure or force	$\sigma_f^{3/2}/E$
Rotating drums and centrifuges Maximum angular velocity; radius fixed; wall thickness free	σ_f/ρ

Table A.5 Vibration-Limited Design

Functions and Constraints	Maximize
Ties, columns	
Maximum longitudinal vibration frequencies	E/ρ
Beams, all dimensions prescribed Maximum flexural vibration frequencies Beams, length and stiffness prescribed Maximum flexural vibration frequencies	E/ρ $E^{1/2}/\rho$
Panels, all dimensions prescribed Maximum flexural vibration frequencies Panels, length, width and stiffness prescribed Maximum flexural vibration frequencies	E/ρ $E^{1/3}/\rho$
Ties, columns, beams, panels, stiffness prescribed Minimum longitudinal excitation from external drivers, ties Minimum flexural excitation from external drivers, beams Minimum flexural excitation from external drivers, panels	$\eta E/\rho$ $\eta E^{1/2}/\rho$ $\eta E^{1/3}/\rho$

Table A.6 Damage-Tolerant Design

Functions and Constraints	Maximize
Ties (tensile member)	
Maximum flaw tolerance and strength, load-controlled design	K_{Ic} and σ_f
Maximum flaw tolerance and strength, displacement control	K_{Ic}/E and σ_f
Maximum flaw tolerance and strength, energy control	K_{1c}^2/E and σ_f
Shafts (loaded in torsion)	
Maximum flaw tolerance and strength, load-controlled design	K_{Ic} and σ_f
Maximum flaw tolerance and strength, displacement control	K_{Ic}/E and σ_f
Maximum flaw tolerance and strength, energy-control	K_{1c}^2/E and σ_f
Beams (loaded in bending)	
Maximum flaw tolerance and strength, load-controlled design	K_{Ic} and σ_f
Maximum flaw tolerance and strength, displacement control	K_{Ic}/E and σ_f
Maximum flaw tolerance and strength, energy control	K_{1c}^2/E and σ_f
Pressure vessel Yield before break Leak before break	K_{Ic}/σ_f K_{1c}^2/σ_f

Table A.7 Electromechanical Design

Functions and Constraints	Maximize
Bus bars Minimum life cost; high current conductor	$1/\rho_e \rho C_m$
Electromagnet windings Maximum short-pulse field; no mechanical failure Maximize field and pulse length, limit on temperature rise	σ_f $C_p \rho/\rho_e$
Windings, high-speed electric motors Maximum rotational speed; no fatigue failure Minimum ohmic losses; no fatigue failure	σ_e/ρ_e $1/\rho_e$
Relay arms Minimum response time; no fatigue failure Minimum ohmic losses; no fatigue failure	$\sigma_e/E\rho_e$ $\sigma_e^2/E\rho_e$

Table A.8 Thermal and Thermomechanical Design

Functions and Constraints	Maximize
Thermal insulation materials Minimum heat flux at steady state; thickness specified Minimum temp rise in specified time; thickness specified Minimize total energy consumed in thermal cycle (kilns, etc.)	$1/\lambda$ $1/a = \rho C_p/\lambda$ $\sqrt{a}/\lambda = \sqrt{1/\lambda \rho C_p}$
Thermal storage materials Maximum energy stored/unit material cost (storage heaters) Maximize energy stored for given temperature rise and time	C_p/C_m $\lambda/\sqrt{a} = \sqrt{\lambda \rho C_p}$
Precision devices Minimize thermal distortion for given heat flux	λ/a
Thermal shock resistance Maximum change in surface temperature; no failure	$\sigma_f/E\alpha$
Heat sinks Maximum heat flux per unit volume; expansion limited Maximum heat flux per unit mass; expansion limited	$\lambda/\Delta a$ $\lambda/\rho\Delta a$
Heat exchangers (pressure-limited) Maximum heat flux per unit area; no failure under Δp Maximum heat flux per unit mass; no failure under Δp	$\lambda \sigma_f$ $\lambda \sigma_f/\rho$

CHAPTER 6

Nanomaterials: Classes and Fundamentals

6.1 CLASSIFICATION OF NANOMATERIALS

To properly understand and appreciate the diversity of nanomaterials, some form of categorization is required. Currently, the most typical way of classifying nanomaterials is to identify them according to their dimensions. As shown in Figure 6.1, nanomaterials can be classified as (1) zero-dimensional (0-D), (2) one-dimensional (1-D), (3) two-dimensional (2-D), and (4) three-dimensional (3-D). This classification is based on the number of dimensions, which are not confined to the nanoscale range (<100 nm). As we will see in what follows, as these categories of nanomaterials move from the 0-D to the 3-D configuration, categorization becomes more and more difficult to define as well.

Beginning with the most clearly defined category, zero-dimensional nanomaterials are materials wherein all the dimensions are measured within the nanoscale (no dimensions, or 0-D, are larger than 100 nm). The most common representation of zero-dimensional nanomaterials are nanoparticles (see Figure 6.2). These nanoparticles can:

- Be amorphous or crystalline

- Be single crystalline or polycrystalline

- Be composed of single or multichemical elements

- Exhibit various shapes and forms

- Exist individually or incorporated in a matrix

- Be metallic, ceramic, or polymeric

0-D

All dimensions (x,y,z) at nanoscale

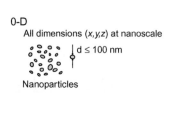

d ≤ 100 nm

Nanoparticles

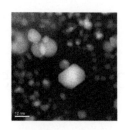

1-D

Two dimensions (x,y) at nanoscale, other dimension (L) is not

d ≤ 100 nm   L

y
x

Nanowires, nanorods, and nanotubes

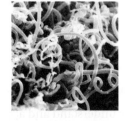

2-D

One dimension (t) at nanoscale, other two dimensions- (L_x, L_y) are not

L_x L_y

t ≤ 100 nm

Nanocoatings and nanofilms

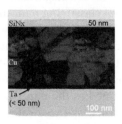

3-D

No bulk dimension at nanoscale

L_x L_y

L_z

Nanocrystalline and nanocomposite materials

FIGURE 6.1

Classification of nanomaterials according to 0-D, 1-D, 2-D, and 3-D.

On the other hand, 1-D nanomaterials differ from 0-D nanomaterials in that the former have one dimension that is outside the nanoscale. This difference in material dimensions leads to needle-like-shaped nanomaterials (see Figure 6.3). One-dimensional nanomaterials include nanotubes, nanorods, and nanowires. Yet, as was the case with 0-D nanomaterials, these 1-D nanomaterials can be:

- Amorphous or crystalline

- Single crystalline or polycrystalline

- Chemically pure or impure

- Standalone materials or embedded in within another medium

- Metallic, ceramic, or polymeric

Two-dimensional nanomaterials are somewhat more difficult to classify, as we will see later. However, assuming for the time being the aforementioned definitions for 0-D and 1-D nanomaterials, 2-D nanomaterials are materials in which two of the dimensions are not confined to the nanoscale. As a result, 2-D nanomaterials exhibit platelike shapes (see Figure 6.4). Two-dimensional nanomaterials include nanofilms, nanolayers, and nanocoatings. These nanomaterials can be:

- Amorphous or crystalline

- Made up of various chemical compositions

- Used as a single layer or as multilayer structures

- Deposited on a substrate

- Integrated in a surrounding matrix material

- Metallic, ceramic, or polymeric

Three-dimensional nanomaterials, also known as *bulk nanomaterials*, are relatively difficult to classify, as we discuss shortly. However, in keeping with the dimensional parameters we've established so far, it is true to say that bulk nanomaterials are materials that are not confined to the nanoscale in any dimension. These materials are thus characterized by having three arbitrarily dimensions above 100 nm (see Figure 6.5). With the introduction of these arbitrary dimensions, we could certainly ask why these materials are called nanomaterials at all. The reason for the continued classification of these materials as nanomaterials is this: Despite their nanoscale dimensions, these materials possess a nanocrystalline structure or involve the presence of features at the nanoscale. In terms of nanocrystalline structure, bulk nanomaterials can be composed of a multiple arrangement of nanosize crystals, most typically in different orientations. With respect to the presence of features at the nanoscale, 3-D nanomaterials can contain dispersions of nanoparticles,

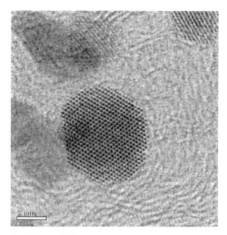

FIGURE 6.2
Bright-field scanning transmission electron microscopy image of a platinum-alloy nanoparticle. The nanoparticle is classified as a 0-D nanomaterial because all its dimensions are at the nanoscale. (Courtesy of P. J. Ferreira, University of Texas at Austin; L. F. Allard, Oak Ridge National Laboratory; and Y. Shao-Horn, MIT.)

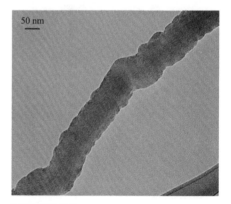

FIGURE 6.3
Transmission electron microscopy image of a carbon nanorod, which can be classified as 1-D nanomaterial because the cross-sectional dimensions are at the nanoscale, whereas the long axis of the tube is not. (Courtesy of P. J. Ferreira, University of Texas at Austin; J. B. Vander Sande, MIT; Peter Szakalos, Royal Institute of Technology, Sweden.)

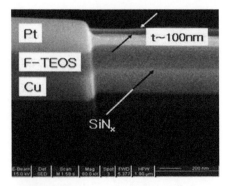

FIGURE 6.4

Scanning electron microscopy image of a multilayered structure. The top layer is a nanocoating of Platinum (Pt), which be classified as a 2-D nanomaterial because one of its dimensions (the thickness) is at the nanoscale and the other two dimensions are not. (Courtesy of Jin An and P. J. Ferreira, University of Texas at Austin.)

FIGURE 6.5

Transmission electron microscopy image showing the nanocrystalline structure of bulk copper, which can be classified as a 3-D nanomaterial. Although the grains are at the nanoscale, the material dimensions can be at the micro or macro scale. (Courtesy of R. Calinas, University of Coimbra; T. Vieira, University of Coimbra; S. Simoes, University of Porto; M. Vieira, University of Porto; P. J. Ferreira, University of Texas at Austin.)

bundles of nanowires, and nanotubes as well as multinanolayers. Three-dimensional nanomaterials can be:

- Amorphous or crystalline
- Chemically pure or impure
- Composite materials
- Composed of multinanolayers
- Metallic, ceramic, or polymeric

This procedure of classification by dimensions allows nanomaterials to be identified and classified in a 3-D space, as shown in Figure 6.6. The distances x, y, and z represent dimensions below 100 nm. As we look in more detail at the aforementioned categories, the straightforward nature of 0-D and 1-D nanomaterials speak for themselves, and we will look at their synthesis, characterization, properties, and applications in further detail in Chapters 7 and 8. Yet for us to begin thinking in more detail about 2-D and 3-D nanomaterials, we need a stronger understanding of their classification. With that in mind, we start by discussing 2-D nanomaterials.

The simplest case of a 2-D nanomaterial looks like the example given in Figure 6.1. Here the assumption is that the 2-D nanomaterial is a single-layer material, with a thickness below 100 nm and length and width that exceed nanometer dimensions. However, as discussed, a material may be categorized as a nanomaterial simply on the basis of its internal structural dimensions, regardless of its exterior material dimensions. The inclusion of these internal structural qualifications is part of what makes the classification of 2-D nanomaterials more complex. In this regard, look at Figure 6.7 for an example. Here, a 2-D nanomaterial is shown with a particular internal structure, composed of crystals (or grains) with nanoscale dimension. This 2-D nanomaterial may be called a nanocrystalline film because of two features: (1) the material exhibits an overall exterior thickness with nanoscale dimensions, and (2) its internal structure is also at the nanoscale. Though this example helps illustrate two possible ways of categorizing of 2-D nanomaterials, both these restrictions do not need to be in place for the material to be considered a nanomaterial. In fact, as we see in Figure 6.8, if the exterior thickness remains at the nanoscale, it is possible for the same film shown in Figure 6.1 to have a larger (above 100 nm) internal grain structure and still qualify the entire material as a nanoscale material. These examples help point out how the

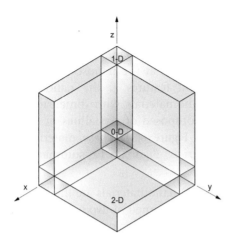

0-D: All dimensions at the nanoscale

1-D: Two dimensions at the nanoscale, one dimension at the macroscale

2-D: One dimension at the nanoscale, two dimensions at the macroscale

3-D: No dimensions at the nanoscale, all dimensions at the macroscale

FIGURE 6.6
Three-dimensional space showing the relationships among 0-D, 1-D, 2-D, and 3-D nanomaterials.

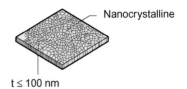

$t \leq 100$ nm

FIGURE 6.7
Two-dimensional nanomaterial with thickness and internal structure at the nanoscale.

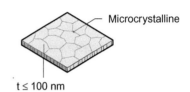

$t \leq 100$ nm

FIGURE 6.8
Two-dimensional nanomaterial with thickness at the nanoscale and internal structure at the microscale.

internal structural dimensions and external surface dimensions are independent variables for the categorization of 2-D nanomaterials.

The way 2-D nanomaterials are produced adds to the complexity of their categorization. Generally, 2-D nanomaterials, like the one shown in Figure 6.1, are deposited on a substrate or support with typical dimensions above the nanoscale. In these cases, the overall sample thickness dimensions become a summation of the film's and substrate's thickness. When this occurs, the 2-D nanomaterial can be considered a nanocoating (see Figure 6.9). Yet at times when the substrate thickness does have nanoscale dimensions or when multiple layers with thicknesses at the nanoscale are deposited sequentially, the 2-D nanomaterial can be classified as a multilayer 2-D nanomaterial (see Figure 6.10). Within each layer, the internal structure can be at the nanoscale or above it (Figure 6.10).

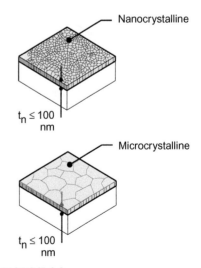

$t_n \leq 100$ nm

$t_n \leq 100$ nm

FIGURE 6.9
Two-dimensional nanomaterials with thickness at the nanoscale, internal structure at the nanoscale/microscale, and deposited as a nanocoating on a substrate of any dimension.

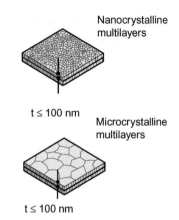

Nanocrystalline multilayers

t ≤ 100 nm

Microcrystalline multilayers

t ≤ 100 nm

FIGURE 6.10

Two-dimensional nanocrystalline and microcrystalline multilayered nanomaterials.

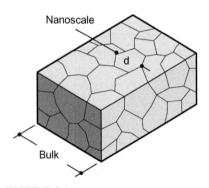

Nanoscale

d

Bulk

FIGURE 6.11

Three-dimensional nanocrystalline nanomaterial in bulk form.

Since we now have several working models for the categorization of 2-D nanomaterials, let's move on to 3-D nanomaterials. Following our previous definition, bulk nanomaterials are materials that do not have any dimension at the nanoscale. However, bulk nanomaterials still exhibit features at the nanoscale. As previously discussed, bulk nanomaterials with dimensions larger than the nanoscale can be composed of crystallites or grains at the nanoscale, as shown in Figure 6.11. These materials are then called *nanocrystalline materials*. Figure 6.12 summarizes 2-D and 3-D crystalline structures.

Another group of 3-D nanomaterials are the so-called *nanocomposites*. These materials are formed of two or more materials with very distinctive properties that act synergistically to create properties that cannot be achieved by each single material alone. The matrix of the nanocomposite, which can be polymeric, metallic, or ceramic, has dimensions larger than the nanoscale, whereas the reinforcing phase is commonly at the nanoscale. Examples of this type of 3-D nanomaterial are shown in Figures 6.13 and 6.14, where various nanocomposites are shown. Distinctions are based on the types of reinforcing nanomaterials added, such as nanoparticles, nanowires, nanotubes, or nanolayers. Within the nanocomposite classification, we should also consider materials with multinanolayers composed of various materials or sandwiches of nanolayers bonded to a matrix core.

Many applications, especially in nanoelectronics, require the use of various kinds of physical features, such as channels, grooves, and raised lines, that are at the nanoscale (see Figure 6.15). A typical copper interconnect is shown in Figure 6.16. Nanofilms, nanocoatings, and multilayer 2-D nanomaterials can be patterned with various features at various scales. In the case of multilayered nanomaterials, the patterns can be made on any layer. These patterns can have different geometries and dimensions at the nanoscale or at larger scales. Most electronic materials fall into the category of patterned 2-D nanomaterials. Figure 6.17 broadly summarizes types of nanomaterials in relation to their dimensionalities.

6.2 SIZE EFFECTS

Surface-to-Volume Ratio Versus Shape

One of the most fundamental differences between nanomaterials and larger-scale materials is that nanoscale materials have an extraordinary ratio of surface area to volume. Though the properties of traditional large-scale materials are often determined entirely by

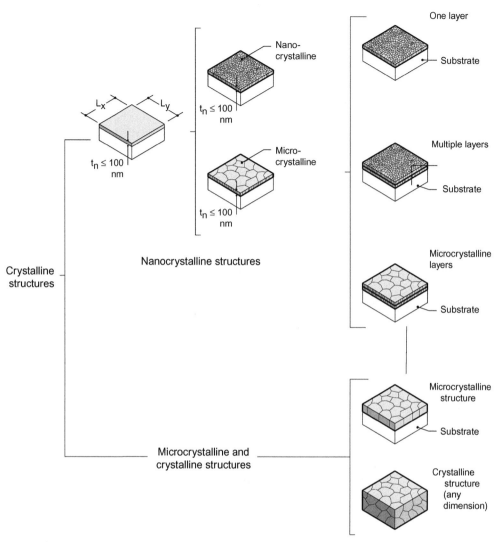

Large-Scale Forms

One layer

Substrate

Nano-
crystalline

$t_n \leq 100$ nm

Micro-
crystalline

$t_n \leq 100$ nm

Nanocrystalline structures

L_x L_y

$t_n \leq 100$ nm

Crystalline
structures

Multiple layers

Substrate

Microcrystalline
layers

Substrate

Microcrystalline
structure

Substrate

Crystalline
structure
(any
dimension)

Microcrystalline and
crystalline structures

FIGURE 6.12

Summary of 2-D and 3-D crystalline structures.

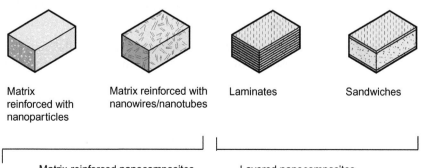

Matrix
reinforced with
nanoparticles

Matrix reinforced with
nanowires/nanotubes

Laminates

Sandwiches

Matrix-reinforced nanocomposites   Layered nanocomposites

FIGURE 6.13

Matrix-reinforced and layered nanocomposites.

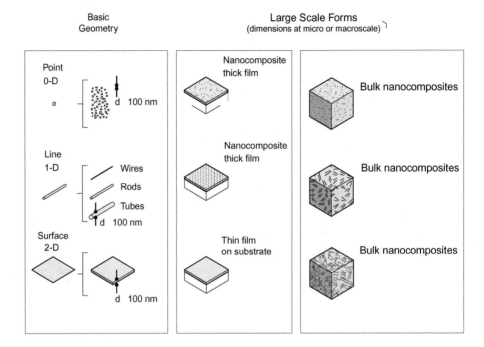

FIGURE 6.14

Basic types of large-scale nanomaterials bulk forms. The filler materials, whether 0-D, 1-D, or 2-D nanomaterials are used to make film and bulk nanocomposites.

the properties of their bulk, due to the relatively small contribution of a small surface area, for nanomaterials this surface-to-volume ratio is inverted, as we will see shortly. As a result, the larger surface area of nanomaterials (compared to their volume) plays a larger role in dictating these materials' important properties. This inverted ratio and its effects on nanomaterials properties is a key feature of nanoscience and nanotechnology.

For these reasons, a nanomaterial's shape is of great interest because various shapes will produce distinct surface-to-volume ratios and therefore different properties. The expressions that follow can be used to calculate the surface-to-volume ratios in nanomaterials with different shapes and to illustrate the effects of their diversity.

We start with a sphere of radius r. This is typically the shape of nanoparticles used in many applications. In this case, the surface area is given by

$$A = 4\pi r^2 \tag{6.1}$$

whereas the volume of a sphere is given by

$$V = \frac{4\pi r^3}{3} \tag{6.2}$$

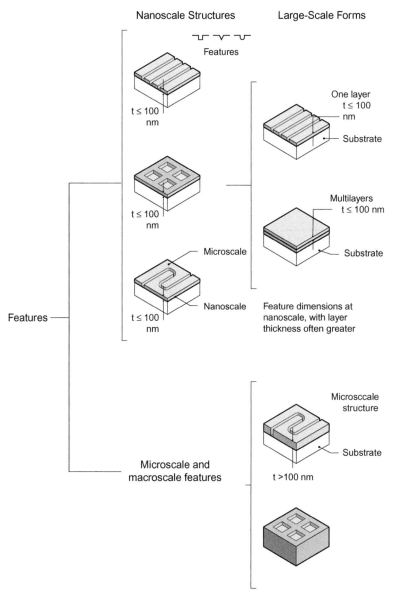

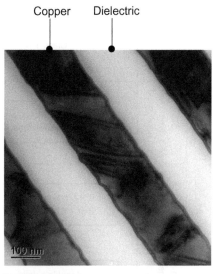

FIGURE 6.16

Nanocopper interconnects used in electronic devices. The copper lines were produced by electrodeposition of copper on previously patterned channels existent in the dielectric material. (Courtesy of Jin An and P. J. Ferreira, University of Texas at Austin.)

FIGURE 6.15

Two-dimensional nanomaterials containing patterns of features (e.g., channels, holes).

			Classes		
			Class 1 Discrete nano- objects	Class 2 Surface nano- featured materials	Class 3 Bulk nano- structured materials
Dimensionality	0-D All 3 dimensions on nano scale		Nanoparticles (smoke, diesel fumes)	Nanocrystalline films	Nanocrystalline materials Nanoparticle composites
	1-D 2 dimensions on nano scale		Nanorods and tubes (carbon nano tubes)	Nano interconnects	Nanotube-reinforced composites
	2-D 1 dimension on nano scale		Nanofilms, foils (gilding foil)	Nano surface layers	Multilayer structures

FIGURE 6.17

General characteristics of nanomaterial classes and their dimensionality.

Thus, the surface-to-volume ratio of a sphere is given by

$$\frac{A}{V} = \frac{4\pi r^2}{\frac{4\pi r^3}{3}} = \frac{3}{r} \tag{6.3}$$

On the basis of Equation 6.3, the results for various radii are shown in Figure 6.18. Clearly, as the radius is decreased below a certain value, there is a dramatic increase in surface-to-volume ratio. Next, consider a cylinder of radius r and height H—for example, a nanowire. In this case, the volume $V = \pi r^2 H$, whereas the surface area $A = 2\pi rH$. Thus, the surface-to-volume ratio is given by

$$\frac{A}{V} = \frac{\pi r^2 H}{2\pi rH} = \frac{2}{r} \tag{6.4}$$

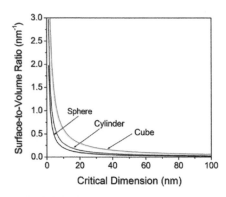

FIGURE 6.18

Surface-to-volume ratios for a sphere, cube, and cylinder as a function of critical dimensions. Nanoscale materials have extremely high surface-to-area ratios as compared to larger-scale materials.

The ratios of surface-to-volume as a function of critical dimension for the cylinder case are shown in Figure 6.18. The trend is similar to the sphere case, although the severe increase in surface-to-volume ratio occurs at larger critical dimensions. Let's now turn to a cube of side L. In this case, the volume and surface area of the cube are given by $V = L^3$ and $A = 6L^2$, respectively. Therefore the surface-to-volume ratio of a cube is given by

$$\frac{A}{V} = \frac{6L^2}{L^3} = \frac{6}{L} \tag{6.5}$$

As shown in Figure 6.18, the overall trend remains for the case of the cube, but the significant variation in surface-to-volume ratio is observed at larger critical dimensions compared with the sphere and cylinder cases.

After stressing the importance of the increase in surface area in nanomaterials relative to traditional larger-scale materials, let's put this information into context. With the help of a few simple calculations, we can determine how much of an increase in surface area will result—for example, from a spherical particle of 10 μm to be reduced to a group of particles with 10 nanometers, assuming that the volume remains constant. To do this, first we calculate the volume of a sphere with 10 microns. Following Equation 6.2 gives V (10 μm) = 5.23×10^{11} nm³. We then calculate the volume of a sphere with 10 nm. Again, with the help of Equation 6.2, we get V (10 nm) = 523 nm³. Because the mass of the 10 micron particle is converted to a group of nanosized particles, the total volume remains the same. Therefore, to calculate the number of nanosized particles generated by the 10 micron particle, we simply need to divide V (10 μm) by V (10 nm) in the form:

$$N = \frac{V(10\mu m)}{V(10 nm)} = \frac{5.23 \times 10^{11}}{523} = 1 \times 10^{9} \textbf{ particles} \qquad (6.6)$$

Hence, so far we can conclude that one single particle with 10 microns can generate 1 billion nanosized particles with a diameter of 10 nm, whereas the total volume remains the same.

We are thus left with the task of finding the increase in surface area in going from one particle to 1 billion particles. This can be done by first calculating the surface area of the 10 micron particle. Following Equation 6.1 gives A (10 μm) = 3.14×10^{8} nm². On the other hand, for the case of the 10 nm particle A (10 nm) = 314 nm². However, since we have 1 billion 10 nm particles, the surface area of all these particles amounts to 3.14×10^{11} nm². This means an increase in surface area by a factor of 1000.

Magic Numbers

As discussed, for a decrease in particle radius, the surface-to-volume ratio increases. Therefore the fraction of surface atoms increases as the particle size goes down. In general, for a sphere, we can relate the number of surface and bulk atoms according to the expressions

$$V = \frac{4\pi}{3} r_A^3 n \qquad (6.7)$$

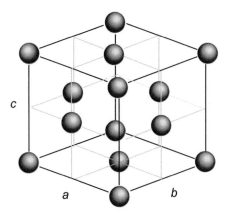

FIGURE 6.19

Face-centered cubic (FCC) structure. All 14 atoms are on the surface.

$$A = 4\pi r_A^2 n^{2/3} \qquad (6.8)$$

where V is the volume of the nanoparticle, A is the surface area of the nanoparticle, r_A is the atomic radius, and n is the number of atoms. On this basis, the fraction of atoms F_A on the surface of a spherical nanoparticle can be given by

$$F_A = \frac{3}{r_A n^{1/3}} \qquad (6.9)$$

We now consider a crystalline nanoparticle. In this case, in addition to the shape of the particle, we have to take into consideration the crystal structure. For illustration purposes, we assume a nanoparticle with a face-centered cubic (FCC) structure. This crystal structure is of practical importance because nanoparticles of gold (Au), silver (Ag), nickel (Ni), aluminum (Al), copper (Cu) and platinum (Pt) exhibit such a structure. We start with the FCC crystal structure shown in Figure 6.19. Clearly, the 14 atoms are all surface atoms. If another layer of atoms is added so that the crystal structure is maintained, a specific number of atoms must be introduced. In general, for n layers of atoms added, the total number of surface atoms can be given by

$$N_{Total}^{S} = 12n^2 + 2 \qquad (6.10)$$

On the other hand, the total number of bulk (interior) atoms can be given by

$$N_{Total}^{B} = 4n^3 - 6n^2 + 3n - 1 \qquad (6.11)$$

Thus, Equations 6.10 and 6.11 relate the number of surface and bulk atoms as a function of the number of layers. These numbers, so-called *structural magic numbers*, are shown in Table 6.1.

The assumption so far has been that a nanoparticle would exhibit a cube-type shape. However, from a thermodynamic point of view, the equilibrium shape of nanocrystalline particles is determined by

$$\sum A_i \gamma_i = \text{minimum} \qquad (6.12)$$

where γ_i is the surface energy per unit area A_i of exposed surfaces, if edge and curvature effects are negligible. For ideal FCC metals, the surface energy of atomic planes with high symmetry should follow the order $\gamma\{111\}_{Pt} < \gamma\{100\}_{Pt} < \gamma\{110\}_{Pt}$ due to surface atomic density. On the basis of calculated surface energies, the equilibrium crystal shape can be created. Among the possible shapes, the smallest *FCC* nanoparticle that can exist is a cubo-octahedron, which is a 14-sided polyhedron (see Figure 6.20).

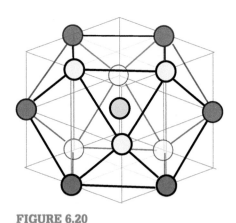

FIGURE 6.20

The smallest FCC nanoparticle that can exist: a cubo-octahedron. A bulk atom is at the center. Others are surface atoms..

Table 6.1 Structural Magic Numbers for a Cube-Type FCC Nanoparticle

n	Surface Atoms	Bulk Atoms	Surface/Bulk Ratio	Surface Atoms (%)
1	14	0	—	100
2	50	13	3.85	79.3
3	110	62	1.78	63.9
4	194	171	1.13	53.1
5	302	364	0.83	45.3
6	434	665	0.655	39.4
7	590	1098	0.535	34.9
8	770	1687	0.455	31.3
9	974	2456	0.395	28.3
10	1202	3429	0.350	25.9
11	1454	4630	0.314	23.8
12	1730	6083	0.284	22.1
100	120,002	3,940,299	0.0304	2.9

This nanoparticle has 12 surface atoms and one bulk atom. If additional layers of atoms are added to the cubo-octahedral nanoparticle such that the shape and crystal structure of the particle are maintained, a series of structural magic numbers can be found. In particular, for n layers of atoms added, the total number of surface atoms can be given by

$$N^S_{Total} = 10n^2 - 20n + 12 \qquad (6.13)$$

whereas the total number of bulk (interior) atoms can be given by

$$N^B_{Total} = \frac{1}{3}\left(10n^3 - 15n^2 + 11n - 3\right) \qquad (6.14)$$

Table 6.2 shows the number of surface and bulk atoms for each value of n as well as the ratio of surface-to-volume atoms.

These structural magic numbers do not take into account the electronic structure of the atoms in the nanoparticle. However, sometimes the dominant factor in determining the minimum in energy of nanoparticles is the interaction of the valence electrons of the atoms with an averaged molecular potential. In this case, electronic magic numbers, representing special electronic configuration may

Table 6.2 Structural Magic Numbers for a Cubo-Octahedral FCC Nanoparticle

n	Surface Atoms	Bulk Atoms	Surface/Bulk Ratio	Surface Atoms (%)
2	12	1	12	92.3
3	42	13	3.2	76.4
4	92	55	1.6	62.6
5	162	147	1.1	52.4
6	252	309	0.8	44.9
7	362	561	0.6	39.2
8	492	923	0.5	34.8
9	642	1415	0.4	31.2
10	812	2057	0.39	28.3
11	1002	2869	0.34	25.9
12	1212	3871	0.31	23.8
100	98,000	3,280,000	0.029	3.0

occur for certain cluster sizes. For example, potassium clusters produced in a supersonic jet beam and composed of 8, 20, 40, 58, and 92 atoms occur frequently. This is because potassium has the 4s orbital (outermost shell) occupied and thus clusters for which the total number of valence electrons fill an electronic shell are especially stable. Thus, in general, electronic magic numbers correspond to main electronic shell closings.

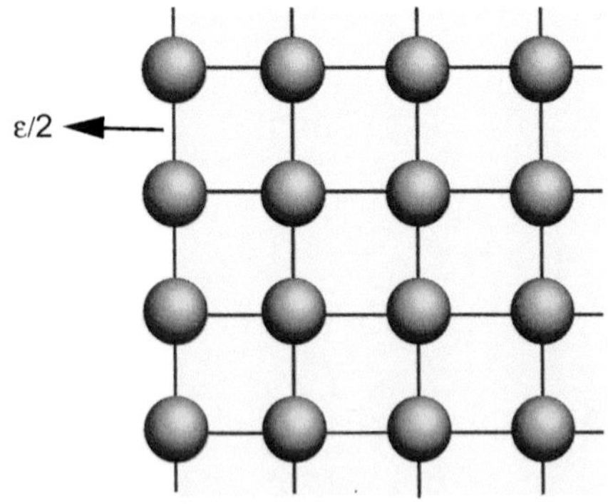

Surface Curvature

All solid materials have finite sizes. As a result, the atomic arrangement at the surface is different from that within the bulk. As shown in Figure 6.21, the surface atoms are not bonded in the direction normal to the surface plane. Hence if the energy of each bond is $\varepsilon/2$ (the energy is divided by 2 because each bond is shared by two atoms), then for each surface atom not bonded there is an excess internal energy of $\varepsilon/2$ over that of the atoms in the bulk. In addition, surface atoms will have more freedom to move and thus higher entropy. These two conditions are the origin of the surface free energy of materials. For a pure material, the surface free energy γ can be expressed as

FIGURE 6.21

For each surface atom there is an excess internal energy of ε/2 due to the absence of bonds.

$$\gamma = E^S - TS^S \tag{6.15}$$

where E^S is the internal energy, T is the temperature, and S^S is the surface thermal entropy. Equally important is the fact that the geometry of the surface, specifically its local curvature, will cause a change in the system's pressure. These effects are normally called *capillarity effects* due to the fact that the initial studies were done in fine glass tubes called *capillaries*. To introduce the concept of surface curvature, consider the 2-D curve shown in Figure 6.22. A circle of radius r just touches the curve at point C. The radius r is called the radius of curvature at C, whereas the reciprocal of the radius

$$k = 1/r \tag{6.16}$$

is called the local curvature of the curve at C. As shown in Figure 6.22, the local curvature may vary along the curve. By convention, the local curvature is defined as positive if the surface is convex and negative if concave (see Figure 6.23). As the total energy (Gibbs free energy) of a system is affected by changes in pressure, variations in surface curvature will result in changes in the Gibbs free energy given by

$$\Delta G = \Delta PV = \frac{2\gamma V}{r} \tag{6.17}$$

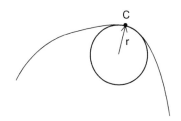

FIGURE 6.22
Surface curvature in two dimensions.

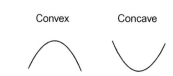

FIGURE 6.23
Concave and convex surface curvatures.

On the basis of Equation 6.17, the magnitude of the pressure difference increases as the particle size decreases, that is, as the local curvature increases. Therefore, at the nanoscale, this effect is very significant. In addition, because the sign for the local curvature depends on whether the surface is convex or concave, the pressure inside the particle can be higher or lower than outside. For example, if a nanoparticle is under atmospheric pressure, it will be subject to an extra pressure ΔP due to the positive curvature of the nanoparticle's surface, described in Equation 6.17.

Another important property that is significantly altered by the curvature effect is the equilibrium number of vacancies (see the section on crystalline defects in Chapter 4). In general, the total Gibbs free energy change for the formation of vacancies in a nanoparticle can be expressed by

$$\Delta G_v^{Total} = \Delta G_v^{bulk} + \Delta G_v^{excess} \tag{6.18}$$

where ΔG_v^{bulk} is the equilibrium Gibbs free energy change for the formation of vacancies in the bulk and ΔG_v^{excess} is the excess Gibbs free energy change for vacancy formation due to curvature effects. Assuming no surface stress, $\Delta G_v^{excess} = \frac{\Omega \gamma}{r}$, where Ω is the atomic volume, γ the surface energy, and r the radius of curvature, Equation 6.18 can be rewritten as:

$$\Delta G_v^{Total} = \Delta G_v^{bulk} + \frac{\Omega\gamma}{r} \tag{6.19}$$

Therefore, the total equilibrium vacancy concentration in a nanoparticle can be given by

$$X_v^{Total} = \exp\left(-\frac{\Delta G_v^{Total}}{k_B T}\right) \tag{6.20}$$

where k_B is the Boltzmann constant and T the temperature. Inserting Equation 6.19 into Equation 6.20 yields

$$X_v^{Total} = \exp\left(-\frac{\Delta G_v^{bulk}}{k_B T}\right)\exp\left(-\frac{\Omega\gamma}{r k_B T}\right) \tag{6.21}$$

For the bulk case, where curvature effects can be neglected, the concentration of vacancies can be expressed as

$$X_v^{bulk} = \exp\left(-\frac{\Delta G_v^{bulk}}{k_B T}\right) \tag{6.22}$$

However for a nanoparticle, the concentration of vacancies can be written as

$$X_v^{Total} = X_v^{bulk}\exp\left(-\frac{\Omega\gamma}{r k_B T}\right) \tag{6.23}$$

As discussed, by convention, the local curvature is defined as positive if the surface is convex and negative if concave. Therefore, for a convex surface, Equation 6.23. can be rewritten as

$$X_v^{Total} = X_v^{Bulk}\left(1 - \frac{\Omega\gamma}{r k_B T}\right) \tag{6.24}$$

On the other hand, for concave surfaces, the mean curvature is given by $-1/r$, and thus Equation 6.23 becomes

$$X_v^{Total} = X_v^{Bulk}\left(1 + \frac{\Omega\gamma}{r k_B T}\right) \tag{6.25}$$

This means that the vacancy concentration under a concave surface is greater than under a flat surface, which in turn is greater than under a convex surface. This result has important implications for nanoparticles due to their small radius of curvature, playing a significant role in a variety of properties such as heat capacity, diffusion, catalytic activity, and electrical resistance, thereby controlling several processing methods such as alloying and sintering. Figure 6.24 shows the

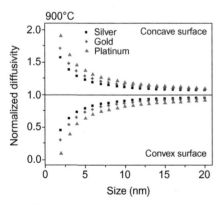

FIGURE 6.24

Diffusivity at 900° in silver, gold, and platinum nanoparticles of different sizes normalized with respect to bulk diffusivities.

effect of curvature on the diffusivity of nanoparticles of silver, gold, and platinum.

Clearly, for nanoparticle sizes below 10 nm, the effect is quite significant. This behavior has profound consequences on the sintering of nanoparticles. In fact, when two nanoparticles are in contact with each other (see Figures 6.25 and 6.26), the neck region between the nanoparticles has a concave surface, which results in reduced pressure. As a consequence, atoms readily migrate from convex surfaces with positive curvature (high positive energy) to concave surfaces with negative curvature (high negative energy), leading to the coalescence of nanoparticles and elimination of the neck region. In other words, nanoparticles exhibit a high tendency to sintering, even at room temperature, due to the curvature effect.

One other important physical property of a material is its lattice parameter. Because this parameter represents the dimensions of the simplest unit of a crystal that is propagated in 3-D, it has significant impact on a variety of properties. To understand the effects of scale on the lattice parameter, we consider the Gauss-Laplace formula given by

$$\Delta P = \frac{4\gamma}{d} \qquad (6.26)$$

where ΔP is the difference in pressure between the interior of a liquid droplet and its outside environment, γ is the surface energy, and d is the diameter of the droplet. If the droplet is now solid and crystalline with a cubic structure and lattice parameter a (the droplet is now a nanoparticle), we can write for the compressibility of the nanoparticle:

$$K = \frac{1}{V_0}\left[\frac{\partial V}{\partial P}\right]_T \qquad (6.27)$$

which measures the volume change of the material as the pressure applied increases, for a constant temperature. It is normalized with respect to V_0 to represent the fractional change in volume with increasing pressure. In this case, $V_0 = a^3$. Equation 6.26 can then be inserted in Equation 6.27, giving

$$\frac{\gamma}{d} = \frac{3K}{4}\frac{\Delta a}{a} \qquad (6.28)$$

Since the surface energy increases as the particle decreases, because the radius of curvature decreases, Equation 6.28 reveals that the

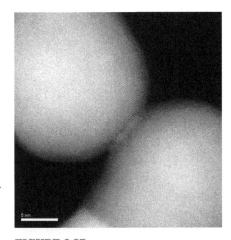

FIGURE 6.25

Aberration-corrected STEM image of two nanoparticles sintering at room temperature. (Courtesy of Michael Asoro, University of Texas at Austin; Larry F. Allard, Oak Ridge National Laboratory; and P. J. Ferreira, University of Texas at Austin.)

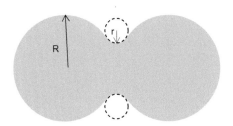

FIGURE 6.26

Schematic showing the sintering process of two nanoparticles. R is the radius of the convex surface and r is the radius of the concave surface.

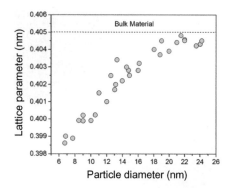

FIGURE 6.27

Lattice parameter of Al (aluminum) as a function of particle size. (Adapted from J. Woltersdorf, A.S. Nepijko, and E. Pippel, Surface Science, *106, pp. 64–69, 1981.)*

lattice parameter is reduced for a decrease in particle size (Figure (6.27)).

Strain Confinement

Planar defects, such as dislocations are also affected when present in a nanoparticle. As discussed in Chapter 4, dislocations play a crucial role in plastic deformation, thereby controlling the behavior of materials when subjected to a stress above the yield stress. In the case of an infinite crystal, the strain energy of a perfect edge dislocation loop is given by

$$W_S \cong \frac{\mu b^2}{4\pi} \ln\left\{\frac{r}{c}\right\} \qquad (6.29)$$

where μ is the shear modulus, b is the Burgers vector, r is the radius of the dislocation stress field, and c is the core cutoff parameter.

If the crystal size is reduced to the nanometer scale, the dislocation will be increasingly affected by the presence of nearby surfaces. As a consequence, the assumption associated with an infinite crystal size becomes increasingly invalid. Therefore, in the nanoscale regime, it is vital to take into account the effect posed by the nearby free surfaces. In other words, there are image forces acting on the dislocation half-loop. As a consequence, the strain energy of a perfect edge dislocation loop contained in a nanoparticle of size R is given by

$$W_S \cong \frac{\mu b^2}{4\pi}\left[\ln\left\{\frac{R - r_d}{R}\right\}\right] \qquad (6.30)$$

where r_d is the distance between the dislocation line and the surface of the particle and the other symbols have the same meaning as before. A comparison of Equations 6.29 and 6.30 reveals that for small particle sizes, the stress field of the dislocations is reduced. In addition, the presence of the nearby surfaces will impose a force on the dislocations, causing dislocation ejection toward the nanoparticle's surface. The direct consequence of this behavior is that nanoparticles below a critical size are self-healing as defects generated by any particular process are unstable and ejected.

Quantum Effects

In bulk crystalline materials, the atomic energy levels spread out into energy bands (see Figure 6.28). The valence band, which is filled with electrons, might or might not be separated from an

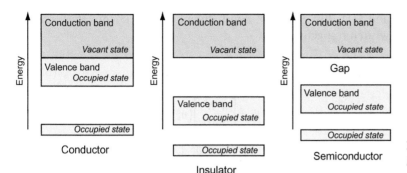

FIGURE 6.28
Energy bands in bulk conductors, insulators, and semiconductors.

empty conduction band by an energy gap. For conductor materials such as metals, there is typically no band gap (Figure 6.28a). Therefore, very little energy is required to bring electrons from the valence band to the conduction band, where electrons are free to flow. For insulator materials such as ceramics, the energy band gap is quite significant (Figure 6.28b), and thus transferring electrons from the valence band to the conduction band is difficult. In the case of semiconductor materials such as silicon, the band gap is not as wide, and thus it is possible to excite the electrons from the valence band to the conduction band with some amount of energy. This overall behavior of bulk crystalline materials changes when the dimensions are reduced to the nanoscale. For 0-D nanomaterials, where all the dimensions are at the nanoscale, an electron is confined in 3-D space. Therefore, no electron delocalization (freedom to move) occurs. For 1-D nanomaterials, electron confinement occurs in 2-D, whereas delocalization takes place along the long axis of the nanowire/rod/tube. In the case of 2-D nanomaterials, the conduction electrons will be confined across the thickness but delocalized in the plane of the sheet.

Therefore, for 0-D nanomaterials the electrons are fully confined. On the other hand, for 3-D nanomaterials the electrons are fully delocalized. In 1-D and 2-D nanomaterials, electron confinement and delocalization coexist.

Under these conditions of confinement, the conduction band suffers profound alterations. The effect of confinement on the resulting energy states can be calculated by quantum mechanics, as the "particle in the box" problem. In this treatment, an electron is considered to exist inside of an infinitely deep potential well (region of negative energies), from which it cannot escape and is confined by the dimensions of the nanostructure. In 0-D, 1-D,

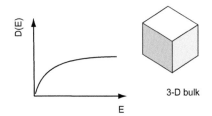

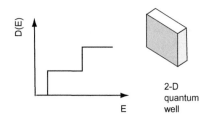

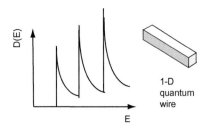

FIGURE 6.29

Density of states in a bulk material, a quantum well (2-D nanomaterial), a quantum wire (1-D nanomaterial), and a quantum dot (0-D nanomaterial).

and 2-D, the effects of confinement on the energy state can be written respectively as

(0-D)
$$E_n = \left[\frac{\pi^2 \hbar^2}{2mL^2} \right] \left(n_x^2 + n_y^2 + n_z^2 \right) \quad\quad (6.31a)$$

(1-D)
$$E_n = \left[\frac{\pi^2 \hbar^2}{2mL^2} \right] \left(n_x^2 + n_y^2 \right) \quad\quad (6.31b)$$

(2-D)
$$E_n = \left[\frac{\pi^2 \hbar^2}{2mL^2} \right] \left(n_x^2 \right) \quad\quad (6.31c)$$

where $\hbar \equiv h/2\pi$, h is Planck's constant, m is the mass of the electron, L is the width (confinement) of the infinitely deep potential well, and n_x, n_y, and n_z are the principal quantum numbers in the three dimensions x, y, and z. As shown in Equations 6.31a–c, the smaller the dimensions of the nanostructure (smaller L), the wider is the separation between the energy levels, leading to a spectrum of discreet energies. In this fashion, the band gap of a material can be shifted toward higher energies by spatially confining the electronic carriers.

Another important feature of an energy state E_n is the number of conduction electrons, N (E_n), that exist in a particular state. As E_n is dependent on the dimensionality of the system (Equations 6.31a–c), so is the number of conduction electrons. This also means that the number of electrons dN within a narrow energy range dE, which represent the density of states D(E), i.e., D(E) = dN/dE, is also strongly dependent on the dimensionality of the structure. Therefore the density of states as a function of the energy E for conduction electrons will be very different for a quantum dot (confinement in three dimensions), quantum wire (confinement in two dimensions and delocalization in one dimension), quantum well (confinement in one dimension and delocalization in one dimension), and bulk material (delocalization in three-dimensions; see Figure 6.29).

Because the density of states determines various properties, the use of nanostructures provides the possibility for tuning these properties. For example, photoemission spectroscopy, specific heat, the thermopower effect, excitons in semiconductors, and the superconducting energy gap are all influenced by the density of states. Overall, the ability to control the density of states is crucial for applications such as infrared detectors, lasers, superconductors, single-photon sources, biological tagging, optical memories, and photonic structures.

FURTHER READING

C. P. Poole, Jr. and F. J. Owens, Introduction to nanotechnology, Wiley-Interscience, 2003. ISBN 0-471-07935-9.

A. S. Edelstein, and R. C. Cammarata (eds.), Nano materials: Synthesis, properties, and applications, Institute of Physics, 1996, ISBN 0-7503-0578-9.

R. T. De Hoff, Thermodynamics in materials science, McGraw-Hill, 1993, ISBN 0-07-016313-8

M. Muller, and K. Albe, Concentration of thermal vacancies in metallic nanoparticles, Acta Materialia, 55, pp. 3237–3244, 2007.

J. Woltersdorf, A. S. Nepijko, and E. Pippel, Dependence of lattice parameters of small particles on the size of nuclei, Surface Science, 106, pp. 64–69, 1981.

R. Lamber, S. Wetjen, and I. Jaeger, Size dependence of the lattice parameter of small particles, Physical Review B, 51, pp. 10968–10971, 1995.

C. E. Carlton, L. Rabenberg, and P. J. Ferreira, On the nucleation of partial dislocations in nanoparticles, Philosophical Magazine Letters, 88, pp. 715–724, 2008.

C. Kittel, Introduction to solid-state physics, John Wiley & Sons, Inc. 6th ed., 1986.

CHAPTER 7

Nanomaterials: Properties

7.1 MECHANICAL PROPERTIES

Scale and Properties

Of polycrystalline, nanocrystalline, and amorphous structures, most materials are polycrystalline, made up of ordered crystals that meet at boundaries where, inevitably, there is disorder. Figure 7.1a suggests what they look like. Such materials have two structural length scales: that of the crystals and that of the atoms that make them up. The crystal size, typically, is between 0.1 mm and 1 mm. This means that the disordered material occupies only a tiny fraction of the volume—less than one part in a million. Nanocrystalline materials, made by the processes described in Chapter 8, have crystals that are much smaller, in the range of 10–100 nm. The smaller the crystals, the greater the fraction of disordered material, which now becomes large enough to influence mechanical and other properties.

Suppose now that the crystal size shrinks further until it becomes of atomic dimensions. The material is now completely disordered, as in Figure 7.1b. Such materials are called *amorphous*, meaning without structure. They have only one characteristic length scale, that of the atoms or molecules that make them up. Polymers, too, can adopt both crystalline and amorphous forms. The long molecular chains in some can fold neatly like a pleated blind to make crystallites separated by confused tangles of disorder, as in Figure 7.2a. The molecular shape in others makes ordering difficult and the structure, once more, is amorphous (Figure 7.2b), with no well-defined length scale beyond that of the molecule itself. The amorphous state, then, is a limit—that of the ultimate nanostructured

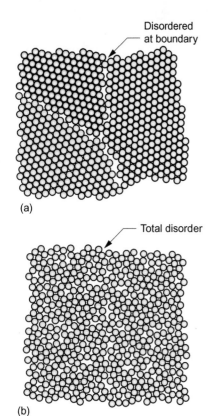

(a)

(b)

FIGURE 7.1

(a) Most materials are made up of ordered crystals that meet at disordered boundaries; the crystals in nanomaterials are only 100–10,000 atoms across. (b) Amorphous or "glassy" materials are totally disordered; the only characteristic dimension is that of the atoms or molecules that make them up. They are an extreme from of nanomaterial.

material. One might expect that its properties, too, would represent limits. As we shall see, this is, broadly speaking, true.

Scale Dependence of Material Properties

The bulk properties of materials (density, modulus, yield strength, thermal and electrical conductivity) are intrinsic; a small piece of the material has the same values for these properties as a large one. It is a basic assumption of continuum mechanics that materials behave in this way, that is, that their mechanical properties are scale independent. It has been a useful and for the most part adequately accurate assumption, greatly simplifying the analysis of structures. The micromechanical description of materials (the use of classical mechanics to model the way the internal structure of a material influences its properties) has followed the same path, assuming that the properties of the individual grains or crystals that make up the material could be averaged to get the overall properties without taking account of their scale. The classical property bounds (upper and lower estimates) of solid mechanics rest entirely on this assumption.

This description served well throughout the 19th century and the first part of the 20th. But as material scientists created higher strength steels and aluminum alloys, it became apparent that the continuum approximation does not always work. The exceptions have given us some of the strongest and most useful materials we now possess. It is at the extremes of scale that the properties become most remarkable, and that means the submicron or, more effective yet, the nano. Such is the hype that has attached itself to the prefix *nano* that it is easy to forget that almost all the high-strength steels, aluminum, magnesium, and titanium alloys on which we now depend for

FIGURE 7.2

(a) Many polymers are made up of crystallites separated by regions of disorder. They are translucent or opaque because the crystallites scatter light. (b) Some polymers are completely disordered. They are usually transparent because they have no structural feature comparable in size to the wavelength of light.

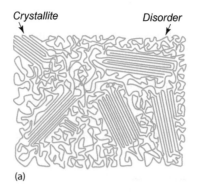

(a)

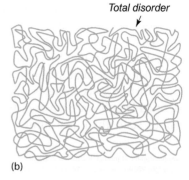

(b)

ground and air transport, for space exploration, and for defense derive their strength from a nanoscale dispersion of particles.

Why should finely divided or structured materials have properties that differ from those of a more conventional scale? Why does almost anything with a nanoscale component to its structure have unusual properties? In this section we explore the gains to be made and the problems to be overcome in using nanostructuring to increase the hardness and strength of materials.

The Mechanical Properties of Nanostructured Materials

Mechanical properties and their origins were introduced in Chapter 4, Section 4.3. There the point was made that the strengths of metals and ceramics lay far below the theoretically achievable limit. In this section we explore the ways in which nanostructuring begins to close the gap between theory and reality.

Nanodispersions

The use of nanostructuring to achieve high strength is not new. Figure 7.3 is a transmission electron micrograph of one of the most widely used of all high-strength aluminum alloys, one with a composition of 4% copper and 96% aluminum. When the alloy is heated to 550°C, the copper dissolves. If the alloy is cooled rapidly to room temperature by quenching it in water, the copper stays dissolved. The copper atoms slightly distort the aluminum crystal, making dislocation motion a little more difficult than in pure aluminum, but not by much. If the alloy is now "aged" by holding it at 150°C, a temperature high enough to allow atoms to rearrange by diffusion but too low for the copper to remain dissolved, the copper precipitates out as a nanodispersion of a compound with the approximate composition $CuAl_2$. The particles are not spheres; they are platelike but no less effective for that. Figure 7.4 shows the way that the strength of the alloy, here measured by its hardness, increases over time as the nanodispersion forms, climbing steadily from a Vickers hardness of 70 HV to a peak of 132 HV. At this point the dispersion is fully formed, as in Figure 7.3, with a dense array of particles about 2 nm wide and 30 nm long, spaced about 30 nm apart. Thereafter the particles coarsen, growing in size and in spacing, and the strength gradually falls. But if the alloy is again quenched when the strength is at its peak, the strength is retained indefinitely. Most of the high-strength alloys we use today (and have used for

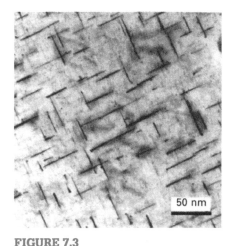

FIGURE 7.3

Transmission electron micrograph of an aluminum alloy containing needle-like particles. Hardening by nanoscale particles, the oldest and most successful mechanical application of controlled nanoscale structuring.

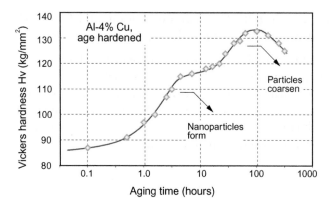

FIGURE 7.4

The hardening curve of a much-used aluminum alloy showing how the strength develops as the nanoscale distribution of particles forms. (Data from Lumley and Morton, 2006.)

over 40 years now) derive their strength in this way. *Nano* is not new to the metallurgist.

There are many ways to make materials stronger. They were illustrated in Figure 4.41 of Chapter 4, Section 4.3. These are alloying (atomic-level hardening), dispersion hardening, and work hardening (hardening by dislocations that interact with each other and obstruct their motion). All are well known and exhaustively studied. The new idea is that further gains might be made by reducing the scale of the individual crystal (or *grains*) in which these mechanisms operate to that of nanometers by making nanocrystals, nanolayers, or, the ultimate, amorphous structures. What do these offer?

Nanocrystalline solids

Making materials with grains at nanoscale is not easy. The lowest energy state of most materials is as a single crystal. Subdividing it to make it polycrystalline raises its energy because the boundaries where the crystals meet have associated distortions. Therefore, making it nanocrystalline creates a very large area of internal boundary, difficult to make and to retain once made. Research over the last decade has enabled both problems to be overcome (using the methods described in Section 8.1). As anticipated, the resulting materials have interesting properties.

Figure 7.5, top, shows how the hardness H of copper increases as the grain size is reduced. Coarse-grained copper has a hardness of less than 200 MPa. Reducing the grain size to 5 nm raises the hardness to over 2000 MPa, an increase of more than a factor of 10. The same data is replotted on logarithmic scales in the lower part of the figure. The points lie along a line with a slope of approximately −0.5, meaning that the hardness depends on grain size as

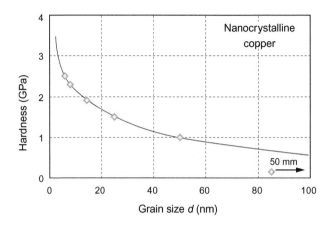

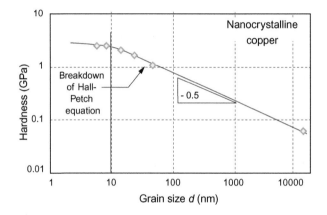

FIGURE 7.5
The increase in strength of copper as the grain size is reduced to nanodimensions (top); the same data plotted on logarithmic scales (below). (Data from Goldstein, 1997.)

$$H \approx \frac{C}{d^{1/2}} \tag{7.1}$$

where C is a constant. This simple dependence must break down at the smallest grain sizes. (If it did not, the strength would exceed the ideal strength if the grains were made small enough.) Indeed, there is a hint in the figure that the curve is starting to flatten out at the smallest sizes.

How is this increase in strength with reduced grain size understood? Here is the argument: The boundaries of grains act as obstacles to dislocation motion, partly because they are locally disordered and partly because the planes on which dislocations glide in one grain are not coplanar with those in the next. The obstacle's "strength" is measured by the force f^* per unit length of dislocation required to make it cut through the boundary and trigger a slip in the next grain. This obstacle-like nature of boundaries causes dislocations to

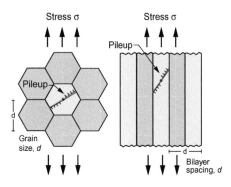

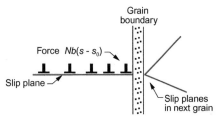

FIGURE 7.6

Pileups in a grain and a layer of a nanolayer structure (top); the pileup in more detail (bottom).

pile up against them until the force on the one closest to the boundary exceeds f^* (see Figure 7.6). The number N of dislocations in such a pileup scales with the applied shear stress $(\tau - \tau_0)$ and the distance L between the dislocation source and the obstacle—here, half the diameter d of a grain:

$$N = \frac{\pi L (1 - \nu)(\tau - \tau_0)}{Gb} \tag{7.2}$$

Here ν is Poisson's ratio (approximately 0.3), and τ_0 describes the contributions of all the other strengthening mechanisms shown in Figure 4.41 of Chapter 4. The shear stress τ caused by a tensile or compressive stress σ is $\tau \approx \sigma/2$. Equating $(\tau - \tau_0)$ to $(\sigma - \sigma_0)/2$ and the shear modulus G by $3E/8$ (as it is for most materials) gives

$$N = \frac{Cd(\sigma - \sigma_0)}{Eb} \tag{7.3}$$

where C is a dimensionless constant with a value of about 2. The force these exert on the obstacle is magnified by their number, so the obstacle will be overcome when

$$N(\tau - \tau_0)b \geq f^* \tag{7.4}$$

Replacing τ with $\sigma/2$ as before and eliminating N from these two equations, we get

$$\sigma - \sigma_0 = \left(\frac{2f^*E}{Cb}\right)^{1/2}\left(\frac{b}{d}\right)^{1/2} \tag{7.5}$$

or

$$\sigma - \sigma_0 = k^*\left(\frac{b}{d}\right)^{1/2} \tag{7.6}$$

where

$$k^* = \left(\frac{2f^*E}{Cb}\right)^{1/2} \tag{7.7}$$

The quantity k^* has the dimensions of stress; it characterizes the strength of the boundary. Its value typically lies in the range 5 to 15 GPa—it is about equal to the ideal strength. This result is known as the *Hall-Petch equation*, with k^* the Hall-Petch constant. The hardness H is just three times the strength σ.

Equation 7.3 says that the smaller the grain size d, the fewer the number of dislocations that can be packed into a pileup. The lower, then, is its magnifying effect, and the greater is the applied stress needed to break through the obstacle. There comes a point, however, at which N falls to 1 and no pileup is possible. This occurs at a grain size d^*, which, from Equations 7.3 and 7.6, is given by

$$\frac{d^*}{b} < \left(\frac{E}{Ck^*} \right)^2 \tag{7.8}$$

Figure 7.7, top, shows measurements of the hardness of electrodeposited nickel as a function of the size of the grains. As with copper, the hardness (and strength) rises by a factor of 7 as the grains size d decreases from 10 μm to 10 nm. The data are replotted in the way suggested by Equation 7.6 in the lower figure, with $d^{-1/2}$ on the x axis. The measurements are consistent with the equation, with a value of k^* of 5 GPa. Inserting this value into Equation 7.8 predicts a critical grain size below which the simple Hall-Petch relation breaks down. Its value, 10 nm, is plotted in the figure.

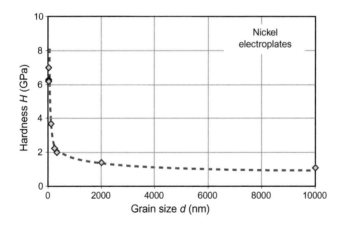

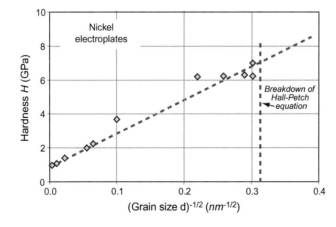

FIGURE 7.7

The increase in strength of nickel as the grain size is reduced to nanodimension (top); the same data plotted to reveal the Hall-Petch relationship (bottom). (Data from Weertman and Averbach, 1996.)

Nanolaminates

Nanolaminates are multilayers, alternating layers, usually of two different materials. To be interesting, the layers have to be thin—in practice, a thickness between a few atomic layers and a few tens of nanometers. To build a nanolaminate of any thickness, then, takes hundreds or thousands of layers. They are, however, relatively easy to make by sequential evaporation from two separate sources (Section 8.1). Data for the tensile strength of copper-nickel laminates are plotted as a function of the bilayer period in the upper part of Figure 7.8. The properties are remarkable: a nanolaminate made from two soft metals (copper and nickel, for example) with a bilayer period of a few nm can have a strength of several GPa, putting it within a factor of 2 or 3 of the theoretical "ideal" strength of about $E/15$. The data are replotted on logarithmic scales in the lower figure.

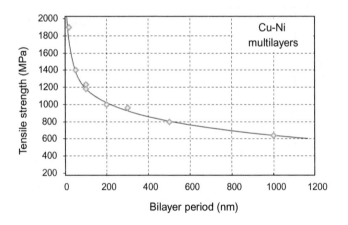

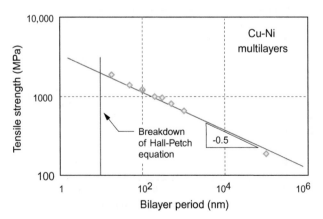

FIGURE 7.8

The increase in strength of copper-nickel nanolaminates plotted against the bilayer period (top); (bottom) the same data plotted on logarithmic scales. (Data from Tench and White, 1984, and Menzies and Anderson, 1990.)

As with the nanocrystalline systems, the strength scales as $d^{-1/2}$, where d is now the bilayer period. The explanation is the same: For deformation to take place, dislocations must sweep through the layers, penetrating the boundaries as they do so. When d is large, pileups form (as in Figure 7.6), magnifying the applied stress; the smaller d becomes, the smaller the number of dislocations that can be squeezed into a pileup and the smaller the magnification of the applied stress.

Amorphous materials

Suppose now that the crystal size shrinks further until it becomes of atomic dimensions. The material is now completely disordered, as in Figure 7.1b. Many amorphous materials are familiar. Ordinary glass is amorphous, and for that reason materials that are amorphous are commonly referred to as *glasses*, even when they are metallic and have nothing but their disordered structure in common with ordinary glass. Many polymers are glasses, among them polycarbonate, acrylic (Plexiglas) and polystyrene. The Burger's vector of a dislocation—the "quantum" of deformation—is of atomic dimensions, so dislocations interact strongly with the disordered parts of the structures of Figures 7.1 and 7.2, giving amorphous materials high hardness and strength. Polycarbonate and Plexiglas might not seem that hard, but as Figure 4.9 of Chapter 4 shows, on a scale of how hard they *could* be, they rank high.

Amorphous materials, as we've already said, are an extreme class of nanostructured matter. Three pairs of figures bring out the exceptional mechanical properties of these and other nanostructured materials. The first, Figure 7.9a, is a chart of modulus and density for the materials of engineering. The colored envelopes enclose material classes; individual bubbles within them describe materials. In Figure 7.9b, with the same axes, the class envelopes of Figure 7.9a appear as shadows. Superimposed in bolder symbols are the properties of polymer, metal and ceramic nanocomposites, nanocrystalline metals, and nanofibers and nanotubes, identified by their own envelopes. The comparison makes clear that nanostructuring has the capacity to create materials with substantially enhanced stiffness. Several applications of amorphous materials are discussed in Section 9.2.

The pairs of Figures 7.10 and 7.11 show a similar comparison, this time based on the charts for yield strength and tensile strength and density. The strongest engineering materials (Figures 7.10a and 7.11a) reach levels of about 2000 MPa. Bulk nanostructured and amorphous materials (Figures 7.10b and 7.11b) push the strength

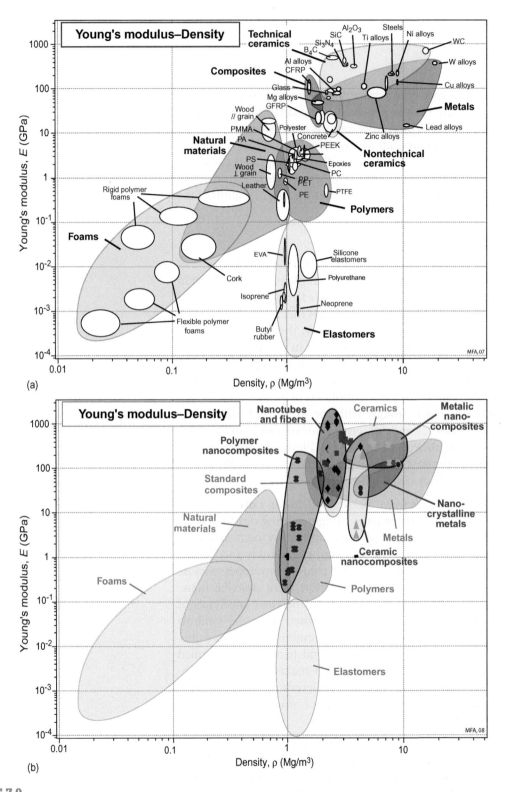

FIGURE 7.9

(a) A chart for modulus and density of engineering materials. The members of each class are enclosed in envelopes. (b) A chart for modulus and density of nanomaterials. The envelopes for engineering materials are shown as shadows in the background.

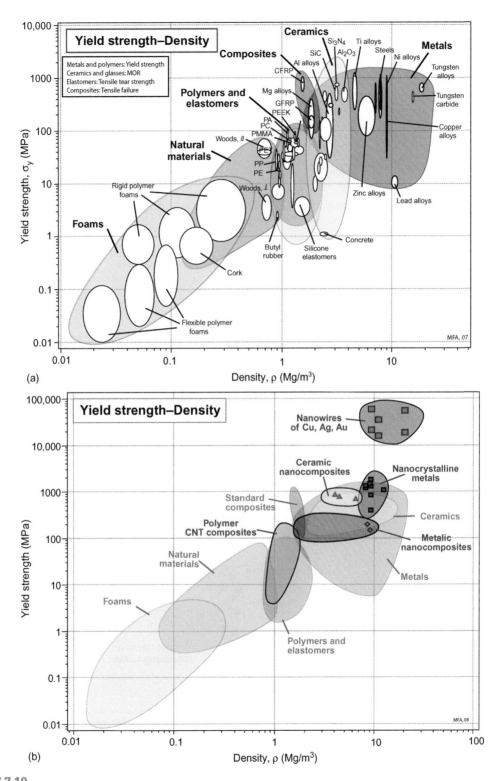

FIGURE 7.10

(a) A chart for yield strength and density for engineering materials. The members of each class are enclosed in envelopes. (b) A chart for yield strength and density for nanomaterials. The envelopes for engineering materials are shown as shadows in the background.

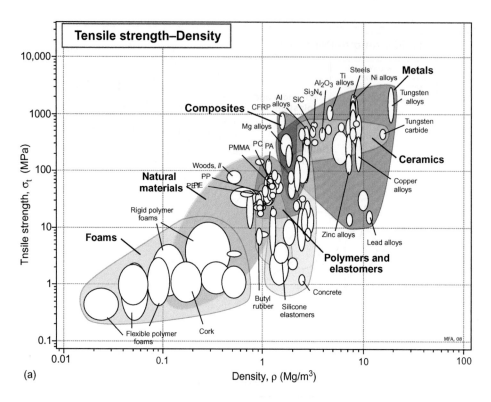

(a)

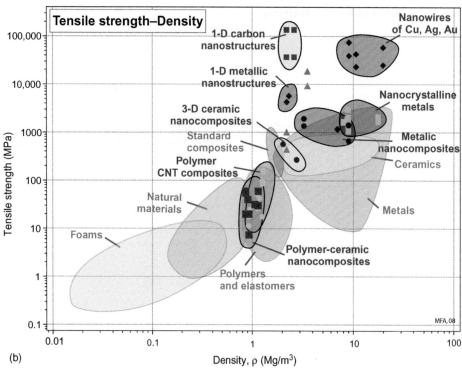

(b)

FIGURE 7.11

(a) A chart for tensile strength and density for engineering materials. The members of each class are enclosed in envelopes. (b) A chart for tensile strength and density for nanomaterials. The envelopes for engineering materials are shown as shadows in the background.

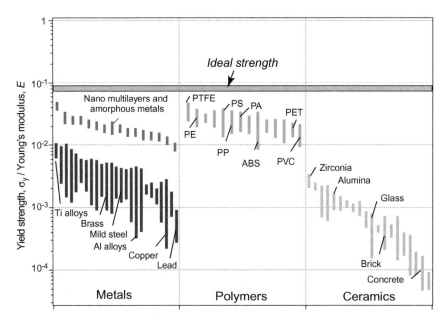

FIGURE 7.12

The ideal strength diagram from Chapter 4 with data for nanomultilayers and amorphous metals added. Their strength approaches the ideal.

up as high as 6000 MPa, an increase of a factor of 3. Individual nanowires and tubes are even stronger. The figure shows data for carbon nanotubes and nanowires of copper, silver, and gold. Reported strengths here extend up to 60 GPa.

We end by returning to the "ideal strength" diagram shown as Figure 4.37 of Chapter 4. It is replotted in Figure 7.12 with nanostructured, nanolayered, and amorphous materials added. Their properties lie within a factor of about 3 of the ideal. We are approaching a fundamental limit here. It is going to be very difficult to make materials that are stronger than this.

7.2 THERMAL PROPERTIES OF NANOMATERIALS

Melting Point

As discussed in Section 4.4, the melting point of a material is a fundamental point of reference because it directly correlates with the bond strength. In bulk systems the surface-to-volume ratio is small and the curvature of the surface is negligible. Therefore, for a solid, in bulk form, surface effects can be disregarded. On the other hand, for the case of nanoscale solids, for which the ratio of surface to mass is large, the system may be regarded as containing surface phases in addition to the typical volume phases. In addition, for 0-D and 1-D nanomaterials, the curvature of the surface is usually very pronounced. Consequently, for nanomaterials the melting temperature is size dependent. In general, surface effects can be expressed mathematically by introducing an additional term ($\Delta G_{Surface}$) into the total energy change (ΔG_{Total}) resulting from the solid-liquid transformation. This is given by

$$\Delta G_{Total} = \Delta G_{Bulk} + \Delta G_{Surface} \tag{7.9}$$

where ΔG_{Bulk} is the free energy of the bulk material given by

$$\Delta G_{Bulk} = \frac{L_0(T_0 - T)}{T_0} V_L \tag{7.10}$$

and L_o is the latent heat of melting, T_o is the melting temperature of the bulk material, T is the melting point of the extended system, where surface effects are included, and V_L is the volume of liquid. When the surface of a body is increased, the change in surface energy is given by

$$\Delta G_{Surface} = \gamma \Delta A \tag{7.11}$$

where γ is the surface tension and ΔA is the increment in surface area. Evidently, at the melting temperature, a layer of liquid with thickness t is formed on the surface and moves at a certain rate into the solid. During the change, a new liquid surface and liquid/solid interface are created, whereas the solid surface is destroyed (see Figure 7.13). In other words, $\Delta G_{surface}$ can be written as

$$\Delta G_{Surface} = A_L \gamma_L + A_{SL} \gamma_{SL} - A_S \gamma_S \tag{7.12}$$

where A_L is the new liquid surface area, σ_L is the surface energy of the liquid per unit area, A_{SL} is the new liquid/solid interfacial area, γ_{SL} is the solid/liquid interfacial energy per unit area, A_S is the surface area of the solid destroyed, and γ_S is the solid surface energy per unit area.

At equilibrium, the solid core of radius r has the same chemical potential as the surrounding liquid layer of thickness t, which occurs when the differential $\partial \Delta G_{Total}/\partial t = 0$. For a sphere, this happens when

$$\frac{L_0(T_0 - T)}{T_0} = \frac{2\gamma_{SL}}{r - t} \tag{7.13}$$

Assuming $t \to 0$, which represents the appearance of the first melting, the upper melting temperature for a sphere can be found from the expression

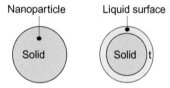

FIGURE 7.13

Upon formation of a liquid layer on the nanoparticle's surface, an interface between the liquid layer and the solid core develops.

$$T_M^{upper} = T_0\left(1 - \frac{2\gamma_{SL}}{L_0 r}\right) \qquad (7.14)$$

in which the term $2\gamma_{SL}/r$ is associated with the increase in internal pressure resulting from an increase in the curvature of the particle with decreasing particle size. Because in Equation 7.14, the variables γ_{SL}, L_0, and r are all positive quantities, this means that the upper melting temperature of a spherical nanoparticle decreases with decreasing particle size. Some results are shown in Figure 7.14, where the change in melting temperature as a function of particle size can be seen for gold (Au), lead (Pb), copper (Cu), bismuth (Bi), and silicon (Si).

What about if we now embed these same nanoparticles in a matrix, as for the fabrication of a nanocomposite? Will the nanoparticles melt below the melting temperature of the respective bulk material? To answer this question, we need to consider the fact that now the surface of the nanoparticle is in contact with a matrix instead of being exposed to the surrounding atmosphere. Therefore, the solid/liquid interfacial energy per unit area γ_{SL} shown in Equation 7.13 must be energetically balanced according to Young's theorem, in the form

$$\gamma_{LM}\cos\theta = \gamma_{SM} - \gamma_{SL} \qquad (7.15)$$

FIGURE 7.14

Changes in melting temperature for various pure metals as a function of particle size. (T. Tanaka and S. Hara, Z. Metallkd, 92, pp. 467–472, 2001.)

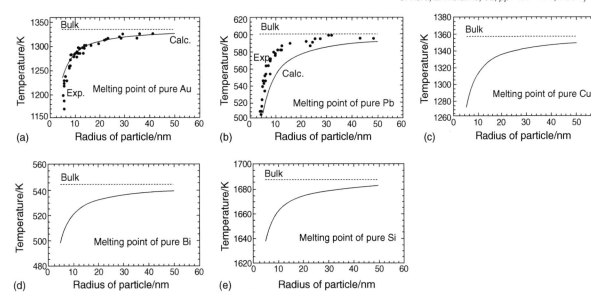

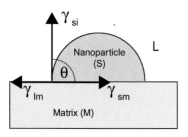

FIGURE 7.15

Solid nanoparticle (S) embedded in a matrix (M). In this case the melting temperature of the nanoparticle depends on a balance of interfacial energies.

where γ_{LM} is the liquid/matrix interfacial energy per unit area, γ_{SM} is the solid/matrix interfacial energy per unit area and θ is the dewetting angle (see Figure 7.15). By rearranging Equation 7.15 and assuming that $\theta = 90°$, it can be shown that when $\gamma_{SM} > \gamma_{LM}$ the melting temperature of the embedded nanoparticles should be lower than the bulk melting temperature. On the other hand, if $\gamma_{SM} < \gamma_{LM}$, the melting temperature of the embedded nanoparticles should be higher than the bulk melting temperature. The latter is called *superheating*, and it has been shown experimentally for the case of indium nanoparticles embedded in an aluminum-indium alloy matrix. From this discussion we can thus learn that due to nanoscale effects, the melting temperature can be either increased or reduced with respect to the bulk material.

Thermal Transport

In addition to the melting temperature, many of the current applications of nanomaterials require knowledge about thermal transport. In some cases, such as microprocessors and semiconductor lasers, the goal is to transport heat away as quickly as possible, whereas for applications such as thermal barriers, the objective is to reduce thermal conduction.

As discussed in Chapter 4, heat is transported in materials by two different mechanisms: lattice vibration waves (phonons) and free electrons. In metals, the electron mechanism of heat transport is significantly more efficient than phonon processes due to the fact that metals possess a high number of free electrons and because electrons are not as easily scattered. In the case of nonmetals, phonons are the main mechanism of thermal transport due to the lack of available free electrons and because phonon scattering is much more efficient. In both metals and nonmetals, as the system length scale is reduced to the nanoscale, there are quantum confinement and classical scattering effects.

In the case of bulk homogeneous solid nonmetal materials, the wavelengths of phonons are much smaller than the length scale of the microstructure. However, in nanomaterials the length scale of the microstructure is similar to the wavelength of phonons. Therefore, quantum confinement occurs. In nanomaterials, quantum confinement comes in several flavors. In 0-D nanomaterials such as nanoparticles, quantum confinement occurs in three dimensions. In 1-D nanomaterials such as nanowires and nanotubes, confinement is restricted to two dimensions. In 2-D nanomaterials such as nanofilms and nanocoatings, quantum confinement takes place in one dimension. These confinement effects are similar for electron transport in nanomaterials.

A good way to understand quantum confinement is to consider that the presence of nearby surfaces in 0-D, 1-D, and 2-D nanostructures causes a change in the distribution of the phonon frequencies as a function of phonon wavelength as well as the appearance of surface phonon modes. These processes lead to changes in the velocity with which the variations in the shape of the wave's amplitude propagate, the so-called *group velocity*. This is similar to a group or ring of waves forming when a stone hits the surface of water. In addition, the phonon lifetime is modified due to phonon-phonon interaction and free surface and grain boundary scattering.

On this basis, a phonon bottleneck occurs in 0-D nanostructures. One-dimensional nanomaterials behave as a phonon waveguide similar to optical ones for light. For example, for carbon nanotubes, several authors have predicted very high thermal conductivity along the nanotubes, close to 3000 $Wm^{-1}K^{-1}$. Just as a comparison, we should keep in mind that the thermal conductivity of copper is approximately 400 $Wm^{-1}K^{-1}$. Despite these results, there are still open questions about phonon transport in 1-D nanostructures, particularly regarding the phonon/phonon interaction and the role of defects. Ideally, in the future we would like to be able to design nanowires and nanotubes with either high or low thermal conductivities.

In the case of 2-D nanomaterials, there is also a great interest in understanding the thermal properties of these materials because of their wide application as components for handheld PCs and cellular phones; everyday home appliances and various modern medical devices; coatings for radiation shielding, wear resistance, and thermal barriers; and components for flat-panel display and photovoltaic applications. In this regard we should distinguish among 2-D single-layered nanomaterials with thicknesses at the nanoscale, multilayered films composed of several nanoscale layers, and thin films comprising a collection of nanostructured units. These nanostructured thin film materials can be subdivided into nanocrystalline materials and nanoporous materials that contain nanovoids. In this regard, nanoporous materials are generally selected as dielectric materials for the microelectronic industry due to their low dielectric constants. However, their thermal conductivities are low, which is a problem.

Starting with single-layered nanoscale thin films, most results show that the thermal conductivity is less than those of the

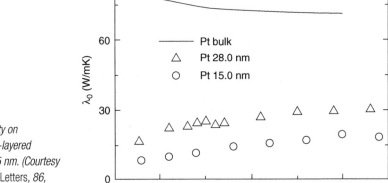

FIGURE 7.16

Dependence of thermal conductivity on temperature for bulk Pt and single-layered nanofilms of Pt with 28 nm and 15 nm. (Courtesy of X. Zhang et al., Applied Physics Letters, *86, 171912, 2005;* Chinese Physics Letters, *23, 4, 936, 2006.)*

corresponding bulk materials. In fact, the thinner the film, the lower the thermal conductivity. For example, measurements performed on nanosized films of Pt (see Figure 7.16) show that the thermal conductivity of nanofilms with a thickness of 15 nm is much less than those of the bulk materials along the whole range of experimental temperature, from 70°K to 340°K, namely 27% of the corresponding bulk values at 300°K. In addition, the thermal conductivity of these Pt nanofilms increases with increasing temperature from 70°K to 340°K, which is opposite of the tendency shown in bulk materials.

In the case of multilayered films and materials with nanoscale grains, we need to consider the idea that an interface produces a thermal resistance. This is because an interface constitutes a disruption of the regular crystal lattice on which phonons propagate. The interface can separate two crystals of the same material with different orientations, such as a grain boundary, across which the two regions have a different distribution of phonons. On the other hand, an interface can separate dissimilar materials, such as a multilayer structure, for which the two different materials have different densities and sound velocities. These effects are similar to electron transport in nanomaterials. The end result is that the presence of an interface produces phonon scattering and therefore a reduction in thermal conductivity.

For example, single-crystal silicon monolayers embedded between amorphous silicon dioxide layers show a strong reduction in

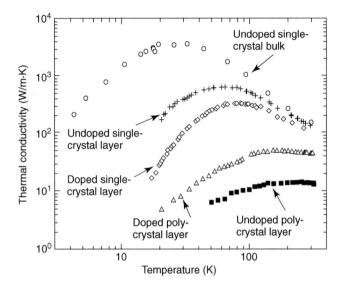

FIGURE 7.17

Thermal conductivity for doped and undoped single-crystalline and polycrystalline silicon films. (Courtesy of S. Uma et al., Int. J. Thermophys., 22, 605, 2001; A. D. McConnell et al., J. Microelectrochem. Sys., 10, 360, 2001.)

thermal conductivity with respect to bulk (see Figure 7.17), especially at low temperatures. This effect is even more pronounced for polycrystalline silicon films, for which grain boundary scattering dominates over surface or multilayer scattering. In addition to these effects, alloying a material with additional elements also leads to phonon scattering. Overall, the idea is to take different approaches to control phonon transport in the various regions of the phonon spectrum. For example, high-frequency phonons can be blocked by alloy scattering because the wavelengths are on the order of a few atomic spacings.

For nanoporous materials, the nanosize effect is determined by the number and size of the pores. Due to the porosity, these materials have low permittivity and thermal conductivity, which, in the case of microelectronic components, leads to an increase in the operation temperature and earlier circuit failure. The current problem is that it is still not theoretically understood how to treat nanoscale pores for thermal transport. One possibility is the similarity between the size of the pores and relevant phonon wavelengths, which suggests that phonons would not see a continuum field. However, experiments showed that the porosity did not play a role in heat transport except to reduce average density. This still remains to be seen. One final theory that has been gaining some respect is to consider the porous solid as a composite material comprising a matrix filled by voids.

The heat capacity and coefficient of thermal expansion of nanomaterials have been much less studied. However, nanocrystalline iron was found to exhibit an enhanced heat capacity relative to coarse-grained polycrystalline iron. This effect has been attributed to the entropy contribution to the heat capacity as a result of the large fraction of grain boundaries. On the other hand, a study of nano zinc oxide flakes showed that the heat capacity is lower than that of coarse Zn oxide in the temperature range between $83°K$ to $103°K$, whereas above $103°K$ the opposite is true. Again, the increase in the configuration and vibration entropy of grain boundaries could explain the behavior above $103°K$. However, for temperatures below $103°K$, the results are difficult to explain. In terms of coefficient of thermal expansion, nanoparticles of silver with 3.2 nm average size embedded in glass showed an increase of the expansion parameter with temperature, compared with bulk silver. However, for larger silver particles with 5.1 nm average size, no changes in the coefficient of thermal expansion were observed with respect to bulk material. This overall behavior is claimed to be due to the higher surface-to-volume ratio of smaller particles and the bonds created across the particle-glass interface. In the case of single-walled carbon nanotubes, it has been shown that the coefficient of thermal expansion is very low. This is due to in-plane expansion, bond stretching, and bond bending effects that cancel each other out. Therefore, a nanocomposite of aluminum reinforced with 15% volume fraction of single-walled carbon nanotubes exhibited a coefficient of thermal expansion that is one-third that of pure nano aluminum.

7.3 ELECTRICAL PROPERTIES

In Section 4.5, we discussed the electronic conduction of electrons in systems considered large in size compared with the nanoscale. In this case, the conduction of electrons is delocalized, that is, electrons can move freely in all dimensions. As they travel their paths, the electrons are primarily scattered by various mechanisms, such as phonons, impurities, and interfaces, resembling a random walk process. However, as the system length scale is reduced to the nanoscale, two effects are of importance: (1) the quantum effect, where due to electron confinement the energy bands are replaced by discreet energy states, leading to cases where conducting materials can behave as semiconductors or insulators, and (2) the classical effect, where the mean-free path for inelastic scattering becomes comparable with the size of the system, leading to a reduction in scattering events.

In 3-D nanomaterials, the three spatial dimensions are all above the nanoscale. Therefore the two aforementioned effects can be neglected. However, bulk nanocrystalline materials exhibit a high grain boundary area-to-volume ratio, leading to an increase in electron scattering. As a consequence, nanosize grains tend to reduce the electrical conductivity.

In the case of 2-D nanomaterials with thickness at the nanoscale, quantum confinement will occur along the thickness dimension. Simultaneously, carrier motion is uninterrupted along the plane of the sheet. In fact, as the thickness is reduced to the nanoscale, the wave functions of electrons are limited to very specific values along the cross-section (see Figure 7.18). This is because only electron wavelengths that are multiple integers of the thickness will be allowed. All other electron wavelengths will be absent. In other words, there is a reduction in the number of energy states available for electron conduction along the thickness direction. The electrons become trapped in what is called a *potential well* of width equal to the thickness. In general, the effects of confinement on the energy state for a 2-D nanomaterial with thickness at the nanoscale can be written as

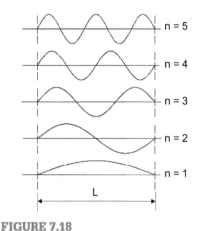

FIGURE 7.18
The energies and wave functions of the first five confined states for the case of an infinite-depth quantum well.

$$E_n = \left[\frac{\pi^2 \hbar^2}{2mL^2} \right] n^2 \qquad (7.16)$$

where $\hbar \equiv h/2\pi$, h is Planck's constant, m is the mass of the electron, L is the width of the potential well (thickness of 2-D nanomaterial), and n is the principal quantum number. Equation 7.16 assumes an infinite-depth potential well model. As mentioned, the carriers are free to move along the plane of the sheet. Therefore the total energy of a carrier has two components, namely a term related to the confinement dimension (Equation 7.16) and a term associated with the unrestricted motion along the two other in-plane dimensions.

To understand the energy associated with unrestricted motion, let's assume the z-direction to be the thickness direction and x and y the in-plane directions in which the electrons are delocalized. Under these conditions, the unrestricted motion can be characterized by two wave vectors k_x and k_y, which are related to the electron's momentum along the x and y directions, respectively, in the form $p_x = \hbar k_x$ and $p_y = \hbar k_y$. The energy corresponding to these delocalized electrons is given by the so-called Fermi energy, which can be expressed as

$$E_F = \left[\frac{\hbar^2 k_F^2}{2m} \right] \qquad (7.17)$$

where $k_F = \sqrt{k_x^2 + k_y^2}$ and the other symbols have the same meaning as before. At the temperature of absolute zero, all the conduction electrons are contained within a circle of radius k_F. As a result, the total energy of an electron (due to confinement and unrestricted motion) in a 2-D nanomaterial with thickness at the nanoscale can be given by

$$E_n = \left[\frac{\pi^2\hbar^2 n^2}{2mL^2}\right] + \left[\frac{\hbar^2 k_F^2}{2m}\right] \qquad (7.18)$$

Since the electronic states are confined along the nanoscale thickness, the electron momentum is only relevant along the in-plane directions. As a result, scattering by phonons and impurities occurs mainly in-plane, leading to a 2-D electron conduction. However, for 2-D nanomaterials with nanocrystalline structure, the large amount of grain boundary area provides an additional source for in-plane scattering. So, the smaller the grain size, the lower the electrical conductivity of 2-D nanocrystalline materials.

In the case of 1-D nanomaterials, quantum confinement occurs in two dimensions, whereas unrestricted motion occurs only along the long axis of the nanotube/rod/wire. Contrary to a 2-D nanomaterial, which allows only one value of the principal quantum number n for each energy state (Equation 7.1), for a 1-D nanomaterial, the energy of a 2-D confinement depends on two quantum numbers, n_y and n_z, in the form

$$E_{ny,nz} = \left[\frac{\pi^2\hbar^2 n_y^2}{2mL_y^2}\right] + \left[\frac{\pi^2\hbar^2 n_z^2}{2mL_z^2}\right] \qquad (7.19)$$

Considering now the electron free motion along the x-direction (long axis), Equation 7.19 can be modified to

$$E_{ny,nz} = \left[\frac{\pi^2\hbar^2 n_y^2}{2mL_y^2}\right] + \left[\frac{\pi^2\hbar^2 n_z^2}{2mL_z^2}\right] + \left[\frac{\hbar^2 k_x^2}{2m}\right] \qquad (7.20)$$

Equation 7.20 states that the electronic states of 1-D nanomaterials do not exhibit a single energy band but instead spread into 1-D subbands. Because of the confinement, the nanoscale dimensions of 1-D nanomaterials act as reflectors, not allowing the electrons to exit the surfaces. In addition, scattering by impurities and/or phonons becomes restricted to the long axis of the tube, despite the fact that boundary scattering is more pronounced due to the high surface-to-volume ratio of 1-D nanomaterials. As a consequence, the transport of electrons along the tube occurs without significant loss of kinetic energy. In other words, the transport is ballistic, particularly at low temperatures. For example, in the case of metallic carbon nanotubes, the conductivity is extremely high, as much as one billion amperes/cm², in contrast to 1 million amperes/cm² for copper. In addition to the effects described here, carbon nanotubes also exhibit a low density of defects and high thermal dissipation, reducing even further the chances for scattering.

For 0-D nanomaterials, the motion of electrons is now totally confined along the three directions Lx, Ly, and Lx. Therefore, the total energy can be given by

$$E_{nx,ny,nz} = \left[\frac{\pi^2\hbar^2 n_x^2}{2mL_x^2}\right] + \left[\frac{\pi^2\hbar^2 n_y^2}{2mL_y^2}\right] + \left[\frac{\pi^2\hbar^2 n_z^2}{2mL_z^2}\right] \qquad (7.21)$$

In this fashion, all the energy states are discreet and no electron delocalization occurs. Under these conditions, metallic systems can behave as insulators due to the formation of an energy band gap, which is not allowed in the bulk form.

So far we have been discussing the electrical properties of 0-D, 1-D, 2-D, and 3-D nanomaterials as isolated entities. However, from a practical point of view, these materials need to be coupled to external circuits by electrodes. For 2-D and 3-D nanomaterials, ohmic contacts are possible. However, for 0-D and 1-D nanomaterials, the contact resistances between nanomaterials and the connecting leads are usually high. Thus one mechanism of providing conduction is through *electron tunneling*. This is a quantum mechanical effect in which an electron can penetrate a potential barrier higher than the kinetic energy of the electron. To better understand this phenomenon, think of a configuration in which two metals are separated by a thin insulator (see Figure 7.19). For an electron to tunnel from one metal to the other across the insulator, one of the metals must have unoccupied energy states. A simple way of achieving this is to apply a voltage V across the circuit to raise the Fermi energy (the energy of the highest occupied quantum state) of one of the metals. In this fashion, electrons can tunnel from the metal with the highest Fermi energy to the metal with the lowest Fermi energy, producing a current I along the circuit.

As in regular electronic circuits, the current I = V/R, where R is the resistance. However, in this case, the resistance is primarily due to electron tunneling. As an example, arrays of gold nanoparticles have been electrically coupled by connecting the nanoparticles to each other by organic molecules. The nanoparticles act as the metal electrodes in Figure 7.19, whereas the organic molecules play the role of a thin insulator. Under these conditions, the conductance C,

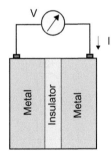

FIGURE 7.19
Metal-insulator-metal junction.

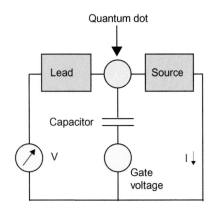

FIGURE 7.20

Quantum dot-based field-effect transistor.

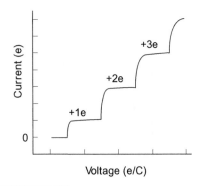

FIGURE 7.21

Coulomb staircase from single-electron tunneling involving quantum dots. Each plateau in the current is the result of a coulomb blockade.

defined as $C = I/V$, was shown to increase due to electron tunneling compared to the case in which the nanoparticles were not connected by the organic molecules.

The phenomenon of electron tunneling can also be used to develop field-effect transistors (FET) made from quantum dots. In this case, two electrodes, a source and a drain, are coupled to a quantum dot and connected through a circuit (see Figure 7.20). In addition, a gate voltage is provided to the quantum dot to control its resistance and ultimately the current I passing between the lead and the drain. Due to the discrete nature of the electrical charge, electrons tunnel from the source to the quantum dot and then to the drain, one at the time. Therefore the junction, which acts as a capacitor, suffers a raise in voltage $V = e/C$ (e is the elementary charge) when a single electron is added. If the change in voltage is large enough, another electron can be prevented from tunneling. This effect is called a *Coulomb blockade*. As a result, electrons will not tunnel until a discreet voltage is reached (see Figure 7.21). To promote electron tunneling, the temperature has to be low enough so that the energy (e^2/C) necessary to charge the junction with one electron exceeds the thermal energy kT. As the capacitance decreases with the size of the particle, a nanoparticle allows the Coulomb blockade to be observable at higher temperatures.

In terms of dielectric behavior, the large number of grain boundaries in nanocrystalline materials is expected to increase the dielectric constant. For example, for nanocrystalline TiO_2, a higher dielectric constant was found, compared with coarse-grained samples. This is due to the fact that under an applied electric field, the positive and negative charges that are segregated at the interfaces will lead to some form of polarization. Since for nanocrystalline TiO_2 the volume fraction of grain boundaries is much larger than in coarse-grained TiO_2, the polarization mechanisms will be much more important, leading to a higher dielectric constant, A similar effect was found for polymer-matrix nanocomposites reinforced with nano TiC fillers. In particular, an increase of the dielectric constant was observed for higher loading levels of TiC, especially when the TiC content was near the percolation threshold. As in the case of nanocrystalline TiO_2, the increase of the dielectric constant is due to interface polarization.

7.4 MAGNETIC PROPERTIES

Section 4.6 discussed the magnetic properties of common materials. This section is concerned with the magnetic properties of nanomaterials. In general, for any ferromagnetic material, the total energy can be written as the sum of various terms, in the form

$$E_{total} = E_{exc} + E_{ani} + E_{dem} + E_{app} \qquad (7.22)$$

where E_{exc} is the exchange energy, E_{ani} is the anisotropic energy, E_{dem} is the demagnetization energy, and E_{app} is the energy associated with an applied magnetic field. For macroscopic magnetic materials, a magnetostrictive energy must also be included in Equation 7.22, but for nanoscale materials this energy can be neglected. This magnetization energy E_{total} can then be related to a magnetic field according to the expression

$$E_{total} = M \cdot H \qquad (7.23)$$

where M is the magnetization vector and H is the applied magnetic field. The first term in Equation 7.22 is due to the quantum mechanical interaction between atomic magnetic moments and represents the tendency for the magnetization vectors to align in one direction. In other words, if the magnetic moment is sufficiently large, the resulting magnetic field can drive a nearest neighbor to align in the same direction, provided the exchange energy is greater than the thermal energy. The second term in Equation 7.22 represents the anisotropy energy that results from the spin's tendency to align parallel to specific crystallographic axes, called *easy axes*. Thus, a "soft" magnetic material will exhibit low anisotropy energy, whereas a "hard" magnetic material shows high anisotropy energy.

Though both the exchange energy and the anisotropy energy try to order the spins in a parallel configuration, the third term in Equation 7.22, namely the demagnetization energy, which is related to the magnetic dipole character of spins, leads to the formation of magnetic domains. Thus, for macroscopic ferromagnetic materials, all the magnetic moments are aligned in magnetic domains, although the magnetization vectors of different domains are not parallel to each other. Each domain is magnetized to saturation, with the moments typically aligned in an easy direction. Depending on the ratio of anisotropy to demagnetization energy, we can expect open (ratio <1; see Figure 7.22) or closure domain (ratio >1; see Figure 7.23) structures.

Finally, the energy associated with an applied magnetic field, called *Zeeman energy* and represented by the last term in Equation 7.22, results from the tendency of spins to align with a magnetic field. Initially, as the magnetic field increases, the magnetization of the material increases. However, at some point, a saturation point, called *saturation magnetization*, is reached, above which an increase in magnetic field does not produce an increase in magnetization. The saturation magnetization is material and temperature dependent.

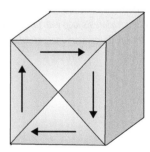

FIGURE 7.22
Open domain structure in a macroscopic ferromagnetic material.

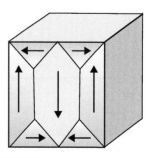

FIGURE 7.23
Closure domain structure in a macroscopic ferromagnetic material.

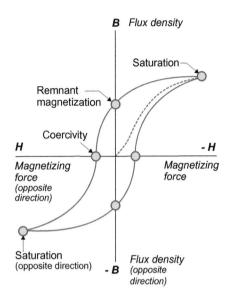

FIGURE 7.24

Magnetization versus applied magnetic field showing the hysteresis loop with (a) saturation magnetization, (b) remnant magnetization, and (c) coercive field.

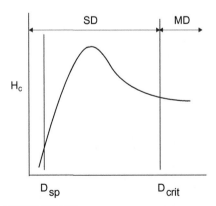

FIGURE 7.25

Coercivity field versus particle or grain size. The vertical lines represent the critical diameter (Dcrit) and the superparamagnetic diameter (Dsp). In addition, the single domain (SD) and the multidomain (MD) regimes can also be seen.

For nanocrystalline ferromagnetic materials, an important consideration is the interaction among exchange energy, anisotropic energy, and demagnetization energy. For very small particles or grain sizes, the exchange forces are dominant due to strong coupling, causing all the spins in neighboring grains to align, superseding in this way the anisotropic and demagnetizing forces. Therefore there is a critical grain size, below which the material will be single domain. For spherical grains, the critical diameter is given by

$$D_{cri} = \frac{9\gamma_B}{\mu_0 M_s^2} \qquad (7.24)$$

where $\gamma_B = 4(AK_1)^{1/2}$ is the wall energy of the material; A is an exchange constant, also known as *exchange stiffness*, which is a function of the material and temperature; K_1 is the anisotropic constant; μ_0 is the permittivity of free space; and M_s is the saturation magnetization. Thus if the particle or grain size is below the critical diameter expressed by Equation 7.24, the material is single domain. For example, the critical diameter for Co is around 70 nm, whereas for Fe it is 15 nm. If the particle or grain size becomes significantly smaller (typically a few nm) than the critical diameter, the magnetization is likely to become unstable and loss of magnetization occurs due to thermal fluctuations. These materials are called *superparamagnetic*.

Other magnetic properties are also strongly affected by scale. To address this point, first recall the magnetic response of a bulk ferromagnetic material to an applied magnetic field (Figure 7.24). As previously discussed, the hysteresis shown in Figure 7.24 is associated with the fact that, on removal of the magnetic field, the magnetic domains do not revert to their original configuration. In other words, there is a *remnant magnetization*. On the other hand, the *coercive field* is the applied magnetic field that needs to be applied in the direction opposite the initial magnetic field, to bring the magnetization back to zero. By reducing the particle size or grain size to the nanoscale, the magnetization curve shown in Figure 7.24 can be altered.

In general, the coercive field of a ferromagnetic material increases with decreasing particle size or grain size, reaching a maximum within a range around the critical diameter. If the particle size or grain size is further decreased below this range, the coercive field will be drastically reduced until the magnetization becomes unstable due to the superparamagnetic behavior (see Figure 7.25). Within this regime, the hysteresis can be completely removed at any temperature. In fact, nanoscale amorphous Fe-Ni-Co compounds with

10–15 nm grain sizes have shown practically no hysteresis. From a technical point of view, if the idea is to fabricate a strong permanent magnet, the coercive force should be designed to be as high as possible.

In addition to the coercivity, the magnetization reversal mechanism is also strongly affected by size. Within a specific range of diameters, typically greater than the supermagnetic diameter but lower than the critical diameter, the exchange interaction energy is sufficiently strong to keep the spins aligned during the reversal process. Above this range of diameters but still below the critical diameter (D_{cri}), the process of magnetization reversal becomes incoherent, involving the switching of small volumes of materials within the nanoparticles or nanosize grains. The size of magnetic nanoparticles or grains also has an effect on the saturation magnetization, namely, the magnetization increases below a particular size. For example, for zinc ferrite, a significant increase in saturation magnetization occurs in these materials for particles around 20 nm. This enhancement in magnetization helps restrict the rotation of the magnetization vector by thermal motion. This is a key issue for magnetic recording technologies, particularly as increases in recording density are demanded. In other words, small magnetic elements are needed, but if the magnetization is unstable due to thermal fluctuations, it is of no use as magnetic memory.

Another magnetic effect that has deep associations with nanoscale is the phenomenon of giant and colossal magnetoresistance. In general, magnetoresistance is a material's property whereby the application of a DC magnetic field alters the electrical resistance. As discussed in Section 4.5, the electrical resistance of a material is a consequence of electron scattering by atoms and defects. Thereby the resistance is associated with the electron mean-free path, which is the average distance traveled by an electron without suffering a collision.

Under these conditions, if a DC magnetic field is applied, the Lorentz force may curve the electron direction within its mean-free path, leading to an increase in resistance. Although this behavior was discovered in metals by Lord Kelvin a long time ago, it had few practical implications, because very strong fields were required. However, in 1988 a much greater magnetoresistance effect, called *giant magnetoresistance* (GMR), was discovered in films composed of alternating nanoscale layers of ferromagnetic and nonferromagnetic materials. The principle behind this phenomenon is associated with differences in the density of states for spin-up and spin-down

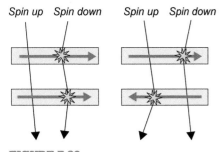

FIGURE 7.26

Scattering events by spin-up and spin-down electrons across two parallel and antiparallel ferromagnetic layers.

electrons in a ferromagnet and the resistance provided by the presence of interfaces.

Because scattering rates are proportional to the density of states, electrons of different spin (up and down) exhibit distinct scattering rates. Conduction electrons with spins parallel to the magnetization will scatter less, whereas electrons with spins aligned opposite to the magnetization will be strongly scattered. In this fashion, if there are two magnetic layers with the magnetization pointing in the same direction, both layers allow electrons in one spin state (say, spin up) to pass through, whereas spin-down electrons will be scattered by each layer (see Figure 7.26).

On the other hand, if the magnetic layers exhibit an antiparallel alignment, both up and down spins will be scattered by each layer (Figure 7.26), leading to an increase in electrical resistance. Therefore, switching the magnetization of the layers from parallel to antiparallel changes the resistivity from high to low, respectively. In addition, to increase the reflection provided by the interfaces, the electron wave vectors should be quite different in the layers at both sides of the interface. The process of inducing parallel magnetization in the layers is simple, since the application of a DC magnetic field large enough to saturate the magnetization is sufficient.

However, to produce an antiparallel magnetization in layered materials is not a trivial task. In fact, three strategies can be followed. The first is called *antiferromagnetic coupling*. In this approach, the idea is to use a nonmagnetic spacer layer with nanoscale thickness between the two ferromagnetic layers (see Figure 7.27a). Within a range of spacer thicknesses, the magnetizations of both ferromagnetic layers couple and prefer to lie in an antiparallel state. To switch to a parallel mode, a sufficiently large magnetic field is applied (Figure 7.27a). The second method uses two ferromagnetic materials with different coercivities. In this way, as the magnetic field is reversed, one layer will switch before the other. However, it is still a challenge to design materials that can switch sharply with an applied magnetic field.

The third method relies on what is called *exchange bias*, whereby one ferromagnetic layer may rotate while a second ferromagnetic layer remains pinned. The best example of such a device is the spin valve (Figure 7.27b). This device consists of two ferromagnetic layers, FM1 and FM2. The FM1 layer is pinned by the last plane of spins in the antiferromagnet (AF). As a result, the FM1 layer is saturated at zero field and will be unaffected by changes in small applied fields. The other ferromagnetic layer, FM2, called the *free layer*, is a soft magnet, quite sensitive to tiny applied magnetic fields, and can be aligned parallel or antiparallel to the FM1 layer. The spacer, which is nonferromagnetic, prevents magnetic coupling from occurring between the two ferromagnetic layers. With this configuration, when the pinned and the free layers are parallel, the device is in a low-resistance mode. As an applied field rotates the free layer, the two ferromagnetic layers become antiparallel and the resistance increases dramatically. Overall, the thickness of these multilayered structures is around 10 nm.

In addition to these layered materials, GMR has also been shown to occur in nanocomposite materials, where single-domain nanoparticles were embedded in a nonferromagnetic matrix. When a DC magnetic field is applied, the magnetization of the nanoparticles aligns with the field, thereby reducing the resistance. In general, the smaller the nanoparticles, the higher the magnetoresistance. However, for these nanocomposites, the magnetoresistance is isotropic, contrary to the behavior of the layered materials. A variation of the GMR effect, called *colossal magnetoresistance* (CMR), has also been found in certain materials, typically manganese-based perovskite oxides. For materials exhibiting the GMR effect, the changes in resistance are around 5%, whereas in the case of materials showing the CMR behavior, the resistance can change by orders of magnitude. Overall, the GMR and CMR effects are crucial in the development of magnetic reading technologies. To improve magnetic reading, the head must be very sensitive to small changes in magnetic fields in a very small area and in a very short time.

7.5 OPTICAL PROPERTIES

As discussed in Section 4.7, in the case of semiconductor bulk materials, if an incident photon has energy greater than the band gap of the material, an electron may be excited from the valence band to the unfilled conduction band. Under these conditions, the photon is absorbed while a hole is left in the valence band when the electron jumps to the conduction band (see Figure 7.28). Inversely, if an electron in the conduction band returns to the valence band and recombines with a hole, a photon is released with energy equal to the band gap of the semiconductor (see Figure 7.29). However, at low temperatures, bulk semiconductors often show optical absorption just below the energy gap. This process is associated with the formation of an electron and hole bound to each other, which is called an *exciton*. As for any other particle, the exciton has mobility and thus can move freely through the material. The binding between the electron and the hole arises from the difference in

(a) No field

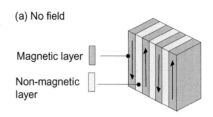

Magnetic layer

Non-magnetic layer

(b) Applied field

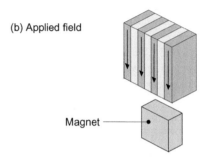

Magnet

(c)

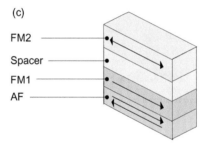

FM2

Spacer

FM1

AF

FIGURE 7.27

(a) Giant magnetoresistance by antiferromagnetic coupling in multilayered structures consisting of alternating magnetic and nonmagnetic materials. In the absence of an applied field, the layers are antiparallel (high resistance), whereas in the presence of a magnetic field the layers are parallel (low resistance). (b) Giant magnetoresistance in a spin valve.

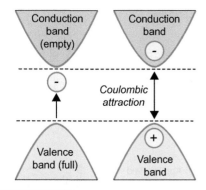

FIGURE 7.28
Creation of an electron and a hole in a bulk semiconductor material.

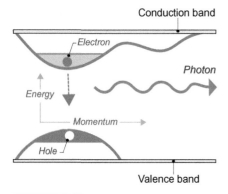

FIGURE 7.29
Emission of a photon upon recombination of an electron-hole pair.

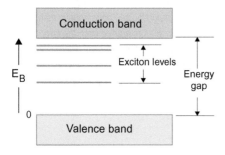

FIGURE 7.30
Energy levels of an exciton. The binding energy E_B of an exciton is equal to the difference between the energy required to create a free electron and free hole and the energy to create an exciton. (Adapted from C. Kittel, Introduction to Solid State Physics, John Wiley & Sons Inc, New York.)

electrical charge between the electron (negative) and the hole (positive), leading to a Coulombic attractive force across the two particles (Figure 7.28), which can be written as:

$$F_C = \frac{-e^2}{4\pi\varepsilon_0 r^2} \tag{7.25}$$

where e is the electron charge, ε_0 is the dielectric constant of free space and r is the separation distance between the electron and the hole. This electrostatic interaction reduces the energy required for exciton formation, with respect to the unbound electron and hole energy (band-gap energy), bringing the energy levels closer to the conduction band (see Figure 7.30). As a result, the Bohr radius increases. The new Bohr radius, defined as the exciton radius, can be expressed by

$$r_B^{ex} = \frac{\varepsilon r_B m_0 \left[1 + (m_e/m_h)\right]}{\varepsilon_0 m_e} \tag{7.26}$$

where ε is the dielectric constant, r_B is the Bohr radius in the absence of an exciton, m_0 is the mass of a free electron, m_e is the effective mass of the electron, m_h is the effective mass of the hole, and ε_0 is the dielectric constant of free space. Table 7.1 lists several semiconductors and their corresponding exciton diameters and band-gap energies.

As shown in Table 7.1, the exciton radius has nanoscale dimensions. Therefore, for a nanomaterial, the exciton radius may be confined. We shall see the consequences of this confinement in a moment. In general, the effects of nanoscale on optical absorption are associated with the density of states in the valence and conduction band (joined density of states), the quantized energy levels of the

Table 7.1 Exciton Bohr Diameters and Band-Gap Energies for Various Semiconductors		
Material	**Exciton Diameter**	**Band-Gap Energy**
CuCl	1.3 nm	3.4 eV
CdS	8.4 nm	2.58 eV
CdSe	10.6 nm	1.74 eV
GaAs	28 nm	1.43 eV
Si	3.7 nm (longitudinal) 9 nm (transverse)	1.11 eV

nanostructure, and the influence of excitonic effects. The first two effects were discussed in Section 6.2. As the nanomaterial changes from 3-D to 0-D and quantum confinement is more severe, the density of states becomes more quantized and the band gap of the material shifts toward higher energies (shorter wavelengths). As a result, a blue shift is expected in the absorption spectrum as the size of the nanomaterial decreases, whereas a red shift occurs for an increase in size. This effect is visible in Figure 7.31, which shows the absorbance spectrum of PbSe nanocrystals. The highest energy (shortest wavelength) absorption region, called the *absorption edge*, is shifted toward the blue as the confinement dimension is reduced. In addition, the distance between the peaks tends to increase with decreasing particle size due to the spreading of the energy levels.

Finally, the higher absorbance peaks shown in Figure 7.31 are associated with the formation of excitons, which shift to lower wavelengths (higher energies) with decreasing nanomaterial size. To understand the influence of excitons, two regimes of confinement might be identified, namely, a weak-confinement and a strong-confinement regime. These weak and strong states are determined by the degree of coupling between the electron and a hole in the exciton, which is related to the ratio between the dimensions of the nanomaterial and the exciton radius. In the case of weak confinement, the dimensions of the nanomaterial are greater than the exciton radius by roughly a few times, and thus the electron and the hole are treated as a correlated pair. Under these conditions, the Coulombic interaction between the electron and the hole leads to an increase in the exciton binding energy (energy difference between the lowest exciton state and the conduction band edge). This causes a shift of the exciton peaks toward the blue, as shown in Figure 7.31. Thus a higher degree of confinement, as, for example, in going from a nanoscale film to a nanowire, leads to an increase of the exciton binding energy, which should be visible in the absorption spectrum. When the dimensions of the nanomaterial are smaller than the exciton radius, the electron and hole wave functions are uncorrelated and thus their motion becomes independent. In other words, the exciton ceases to exist.

As mentioned at the beginning of this section, for bulk semiconductor materials, the absorption spectra from excitons can be seen at low temperatures, but they are usually too weak to be observed at room temperature. However, in nanomaterials, as the confinement is enhanced, the exciton binding energy increases, which reduces the possibility for exciton ionization at higher temperatures. As a

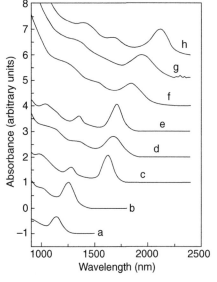

FIGURE 7.31

Room temperature optical absorption spectra of PbSe nanocrystals with diameters (a) 3 nm, (b) 3.5 nm, (c) 4.5 nm, (d) 5 nm, (e) 5.5 nm, (f) 7 nm, (g) 8 nm, and (h) 9 nm. (Adapted from IBM.)

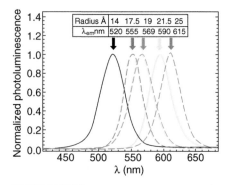

FIGURE 7.32

Emission spectra of CdSe-ZnS quantum dots with different sizes. (Adapted from H. Mattoussi, L. H. Radzilowski, B. O. Dabbousi, E. L. Thomas, M. G. Bawendi, M. F. Rubner, 1998, J. Applied Physics 83, 7965.)

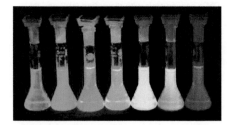

FIGURE 7.33

Photoluminescence of CdSe quantum dots of various sizes. (Courtesy of Prof. Don Seo, Arizona State University.)

result, strong excitonic states appear in the absorption spectra of nanomaterials at room temperature.

So far we have addressed the optical absorption of semiconductor nanomaterials. However, as shown in Figure 7.29, optical emission may also occur if the electron and the hole recombine, leading to the generation of a photon. If the photon energy is within 1.8 eV to 3.1 eV, the emitted light is in the visible range, a phenomenon called *luminescence*. Because of the quantum confinement in nanomaterials, the emission of visible light that's can be tuned by varying the nanoscale dimensions. The typical trend is the shift of the emission peak toward shorter wavelengths (blue shift) as the size of the nanomaterial decreases.

This phenomenon is clearly visible in Figures 7.32 and 7.33. Figure 7.32 shows the photoluminescence spectra for nanoparticles of CdSe-ZnS of different sizes. The smaller the nanoparticle size, the shorter the wavelength of the visible light that's emitted. Figure 7.33 shows the photoluminescence effect of CdSe nanoparticles of various sizes, which covers the entire visible spectrum. Due to quantum confinement, the shift from the blue to the red shown in Figure 7.33 corresponds to an increase in nanoparticle size. In this regard, an issue of importance is the use of a homogeneous distribution of nanoparticles, because fluctuations in size and composition can lead to an inhomogeneous spreading of the optical spectra.

In addition to semiconductor nanomaterials, the optical properties of metallic nanomaterials are also affected by nanoscale. A good example is gold, which has a yellowish color in bulk form, whereas a nanoparticle of gold gives a ruby-red, purple, or even blue color, depending on the nanoparticle size. To explain this effect, we need to understand the physical phenomenon of plasmons. *Plasmons* are quantized waves propagating in materials through a collection of mobile electrons (quantum plasma consisting of both delocalized and localized electron) that are generated when a large number of these electrons are altered from their equilibrium positions. Plasmons are readily observed in noble metals (d-bands of the electronic structure are filled), such as gold, copper, and silver, and in the metals magnesium and aluminum. The plasmons can exist within the bulk as well as on the surface of metals. Among these, the surface plasmons are the most relevant for nanomaterials. Surface plasmons have lower frequencies than bulk plasmons and thus can interact with photons. In fact, when photons couple with surface plasmons (surface plasmon polaritons), alternating regions of positive and negative charges are produced in the surface of the material, thereby generating an electrical field of plasmons (see Figure 7.34). In other words, surface plasmon-polaritons are essentially light waves that are trapped on the surface of the material, as a result of interactions between the incident wave and the existing free electrons.

Despite the interaction between photons and surface plasmons, surface plasmon polaritons cannot be produced on smooth metal surfaces in contact with air, mainly because the momentum of light is different from the momentum of surface plasmons. Therefore, to cause a change in momentum, a thin layer of metal is placed between two materials with differing refractive indices. In this fashion, if the angle of the incidence light on the material with higher refractive index exceeds a critical angle, evanescent waves can propagate surface-plasmon polaritons along the metal layer.

Another technique to induce surface-plasmon polaritons is to roughen the material's surface. This can be done in two ways, namely, by creating parallel linear features on the surface of the material or by randomly roughening the surface. Surface-plasmon polaritons can be used to cause extraordinary transmission. This is a phenomenon whereby a metal film exhibiting a series of holes of specific size and periodicity can transmit more light at certain energies than anticipated. The reason for this behavior lies in the resonance generated between the incident light and surface-plasmon polaritons on the incident side of the film, causing the surface-plasmon polaritons to propagate to the other side of the film and transmit light. Another type of surface plasmon, which is also relevant at the nanoscale, consists of localized surface plasmons. These plasmons are collective electron waves occurring in small volumes such as nanoparticles. A necessary condition for localized surface plasmons to exist is that the wavelength of the incident light must be larger than the size of the nanoparticle. If this occurs, the electric field of the incident light can induce an electric dipole in the metallic nanoparticle, as shown in Figure 7.35. The wavelength of light required to induce the generation of localized surface plasmons depends on the size of the nanoparticle as well as shape and composition. Typically, as the particle size decreases, the plasmon resonance frequency increases (wavelength of light decreases), leading to a blue shift in the spectrum. Therefore, by tuning the size, different colors can be achieved. However, for the case of spherical gold and silver nanoparticles, their plasmon wavelengths do not cover the entire visible spectrum. Thus nanoparticles with different shapes are required, although nonspherical geometries are much more difficult to attain.

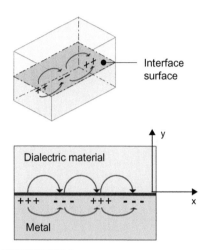

FIGURE 7.34
Propagating surface plasmons.

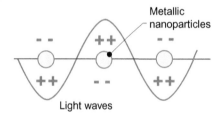

FIGURE 7.35
Localized surface plasmons.

An important effect of shape is related to whether the nanomaterial is 2-D, 1-D, or 0-D. In the case of a 2-D nanomaterial there are longitudinal plasmons along the plane of the sheet and a transverse plasmon across the thickness of the sheet. For a 1-D nanomaterial, there is a longitudinal plasmon along the long axis, whereas transverse plasmons exist across the diameter of the material. As we reduce the diameter, a shift toward the blue is expected. In the case of 0-D nanomaterials, the production of well-defined nonspherical shapes is usually quite challenging but of great interest. However, another alternative is to create core-shell nanoparticles. In this fashion, shells with thinner metal layers produce localized surface plasmon resonance toward lower frequencies, that is, longer wavelengths and a shift toward the red. Another optical plasmon effect associated with metallic nanoparticles occurs when agglomeration takes place. Under these conditions, a shift toward lower plasmon frequencies occurs, leading to a red shift.

Finally, it is important to consider the environment surrounding the nanoparticles. When the refractive index or dielectric constant of the material embedding the nanoparticles increases, the plasmon frequency decreases, generating a red shift. Using the aforementioned surface plasmon properties associated with metallic nanoparticles, light absorption and light emission can be significantly enhanced. As a result, metallic nanoparticles can be used as structural and chemical labels, photothermal therapy, and colorimetric chemical sensors.

7.6 ACOUSTIC PROPERTIES

As initially discussed in Section 4.8, materials interact strongly with radiation that has a wavelength comparable with their internal structure and/or dimensions. Therefore, in the case of nanomaterials for which the characteristic structure and dimensions are below 100 nm, there is a wide range of electromagnetic radiation within the visible, ultraviolet, and X-rays regimes that are affected by the nanoscale. On the other hand, acoustic waves, which exhibit wavelengths that range from microns to kilometers, have little or no direct effect on the properties of nanomaterials. As a result, the discussion of acoustic properties is limited for the case of nanomaterials. However, it should be pointed out that sound waves are used to produce nanomaterials. In addition, since the properties of nanomaterials are in many cases very different from traditional materials, acoustic waves can have a distinct indirect effect, such as in the case of seismic waves.

7.7 SPECIAL CASES

Carbon Nanotubes

Carbon nanotubes (CNTs) were discovered in 1991 by Sumio Ijima of the NEC laboratory in Tsukuba, Japan, during high-resolution transmission electron microscopy (TEM) observation of soot generated from the electrical discharge between two carbon electrodes. The discovery was accidental, although it would not have been possible without Ijima's excellent microscopist skills and expertise. What Ijima was, in fact, studying were C_{60} molecules, also known as *buckminster fullerenes*, previously discovered by Harold Kroto and Richard Smalley during the 1970s. Kroto and Smalley found that under the right arc-discharge conditions, carbon atoms would self-assemble spontaneously into molecules of specific shapes, such as the C_{60} molecule (see Figure 7.36). However, as shown by Ijima's discovery, under different experimental conditions, carbon atoms can instead self-assemble into CNTs.

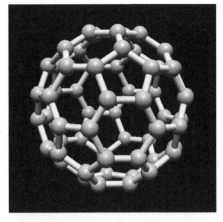

FIGURE 7.36
Bucky ball: C_{60} molecule (computer simulation).

CNTs are cylindrical molecules with a diameter ranging from 1 nm to a few nanometers and length up to a few micrometers. Their structure consists of a graphite sheet wrapped into a cylinder (see Figure 7.37). Depending on the processing conditions, CNTs can be either single-walled or multiwalled (see Figure 7.38). Single-walled nanotubes (SWNTs) may be metallic or semiconductor, depending on the orientation of the hexagonal network with respect to the nanotube long axis, a property known as *chirality*. In particular, CNTs can be classified by a chiral vector, given by

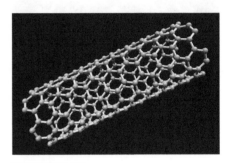

FIGURE 7.37
Graphite sheet wrapped into a cylinder to form a carbon nanotube (CNT).

$$C = na + mb \qquad (7.27)$$

where a and b are unit vectors and n, m are chiral vector numbers that characterize the orientation of the hexagons in a corresponding graphene sheet (see Figure 7.39). In this configuration, the magnitude of the chiral vector C is the circumference of the nanotube, and its direction relative to the unit vector a is the chiral angle θ_0. The translation vector T defines the nanotube unit cell length, which is thus perpendicular to C. These parameters describe the way in which the graphite sheets are rolled up to form a tube structure. In this regard, three types of CNTs are possible: armchair, zigzag, or chiral (see Figure 7.40). An armchair nanotube is formed when $n = m$. In Figure 7.40 this occurs when the green atom matches the blue atom. The zigzag nanotube forms when $m = 0$ (the green atom matches the red atom).

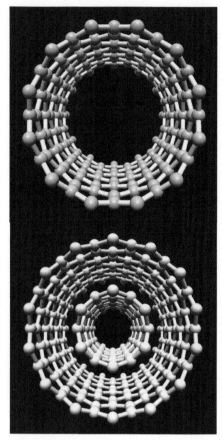

FIGURE 7.38
Single-walled and multiwalled carbon nanotubes.

The chiral type occurs when the chiral vector numbers (n, m) can assume any integer values and the chiral angle is intermediate between 0 and 30 (Figure 7.40). Among these three different types, armchair SWCNTs are metals, those with n-m = k (k is an integer) are semiconductors with a small band gap, and the remainder are semiconductors with a band gap that is inversely proportional to the CNT diameter.

Carbon nanotubes exhibit various unique properties, such as the ability to withstand large stresses with little elastic deformation (Young's modulus = 1000 GPa), the capacity to endure enormous tensile stresses (30 GPa), and the aptitude of exhibiting superior current densities (10^9A/cm^2) and thermal conductivity (6000 W/mK). These two latter properties are due to the nearly 1-D electronic structure in metallic SWNTs and multiwalled nanotubes (MWNTs), which leads to ballistic transport, and the ease by which phonons propagate along the nanotube, respectively.

So far, researchers around the world have been devising several methodologies to synthesize carbon nanotubes. The most common methods are the arc-discharge technique and chemical vapor deposition (CVD; see Section 8.1). These methods have seen significant improvements over the years, but they still suffer from (1) low yield, (2) very high cost, (3) difficulty in tuning the diameter of the nanotubes, and (4) difficulty in producing a single type of CNT without impurities. Table 7.2 shows some of the advantages and disadvantages of these current methods. Although these techniques have been the subject of considerable research, there are still many questions with respect to the processing variables that may condition the formation and growth of CNTs. Some noteworthy parameters that seem to affect the production of CNTs are temperature, pressure, and the type of catalyst used. As an alternative to the methods shown in

Table 7.2 Advantages and Disadvantages of Widely Used Techniques to Synthesize CNTs

Method	Type of Nanotubes	Diameter	Length	Advantages	Disadvantages
Laser vaporization	SWNT	1–2 nm	—	Few defects, good size control	Very expensive
Arc discharge	SWNT/MWNT	0.6–1.4 nm/10 nm	Short	Easy to produce, few defects	Random sizes, short length
CVD	SWNT/MWNT	0.6–4 nm/10–240 nm	Long	Easy to produce	Usually MWNT, defects

Table 7.2, one other route for generating CNTs that could become promising in the near future is the process of metal dusting. Simply, metal dusting is the disintegration of metallic alloys by corrosion, which is initiated by exposure of pure metals or metallic alloys to strongly carburizing atmospheres. The result of the decomposition is a mixture of metal particles and carbon nanostructures. The great advantages of this catalytic route are that carbon nanotubes can be produced at moderate temperatures (around 650–750°C) in large volumes as well as low cost and their structure can be tailored by the catalytic properties of the metal or alloy selected. However, this method still requires further investigation.

Another issue that is important in the fabrication of CNTs is the large concentration of impurities that remain embedded inside the CNT network after processing. As a consequence, the powder needs to be filtered to reduce the amount of impurities present. This is normally achieved by acid oxidation, gas oxidation, or filtration. However, these methods may dissolve some of the CNTs, cause structural damage to CNTs, or be unable to remove large particle aggregates. In addition, these purity-driven techniques tend to be very expensive.

Nevertheless, due to these outstanding properties, carbon nanotubes are likely to play a vital role in various areas, such as nanocomposites, nanoelectronics, hydrogen storage, field emission devices, and nanosensors. The area of nanocomposites is perhaps the first area where CNTs will have a commercial impact. Due to the outstanding modulus and tensile strength resulting from the covalent bonds between the carbon atoms, CNTs are one of the strongest materials known. In addition, CNTs exhibit a high aspect ratio. Therefore, CNTs are ideal as a reinforcement phase. In recent years, good progress has been made in developing CNT-based nanocomposites. In fact, several investigations have shown important enhancements in mechanical, electrical, and thermal properties of nanocomposite materials. However, significant challenges still remain—for example, tailoring the uniformity of dispersion within the matrix, controlling the alignment of CNTs, and making sure that there is a good interfacial bond between the CNTs and the matrix. Furthermore, due to the high cost of CNTs, particularly of pure SWCNTs, the addition of CNTs has been restricted to about 5% in weight.

In the area of nanoelectronics, CNTs have been sought as the new generation of interconnect structures as well as field-effect transistors. Interconnects, which carry the electrical signals between

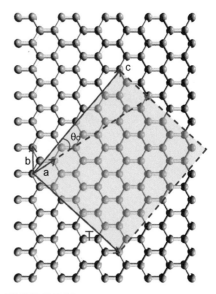

FIGURE 7.39

Graphene sheet rolled into a cylinder described by unit vectors a *and* b, *chiral angle* θ_0, *chiral vector* C, *and translation vector* T. *The figure represents a (4,2) nanotube, where the shaded area is one unit cell. (Adapted from Barry J. Cox and James Hill, University of Wollongong, Australia.)*

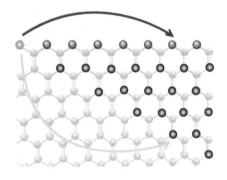

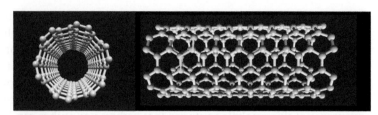

Armchair

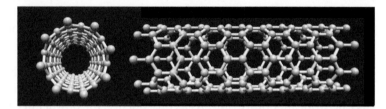

Zig-zag

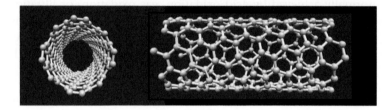

Chiral

FIGURE 7.40

Chirality in nanotubes, describing the way in which the graphite sheets are rolled up to form a particular tube structure.

transistors, are currently made of copper, but as electronic circuits continue to shrink, copper interconnects will suffer from overheating. Conventional metal wires can typically exhibit current densities of 10^5 A/cm² until resistive heating becomes a problem. On the other hand, because of the nearly one-dimensional electronic structure of CNTs, electronic transport in metallic SWNTs occurs ballistically along the nanotube length, allowing the conduction of high currents with no heating. In fact, current densities up to 10^9 A/cm² have been observed in SWCNTs. This is because the electronic states are confined in the directions perpendicular to the tube axis. As a result, the remaining conduction path occurs along the tube axis. Due to the lack of phonon and/or impurity scattering perpendicular to the tube, CNTs behave as 1-D ballistic conductors.

Field-effect transistors (FETs), a type of transistor used for weak-signal amplification, are currently made from silicon. However, these devices are still a few hundred nanometers in size. The use of CNTs with sizes less than 1 nanometer in diameter would allow more of these switches to be part of a chip. In an FET, the current flows through a CNT with semiconductor properties along a path called the *channel*. At one side of the channel is a gold electrode called the *source*; at the other side of the channel is a gold electrode called the *drain* (see Figure 7.41). When a small voltage is applied to the silicon substrate, which acts as a gate in FETs, the conductivity of the CNT can change by more than a million times, allowing a FET to amplify a signal. Still in the area of nano-electronics, a more challenging idea is to build entire electronic circuits out of CNTs, making use of their metallic and semiconducting properties. In this case, semiconducting CNTs are aligned on an insulator substrate, whereas metallic CNTs are placed above in close proximity to the bottom layer. By controlling the current, the top CNTs can be made to contact the bottom CNTs, producing a metal-semiconductor junction that acts as a switch.

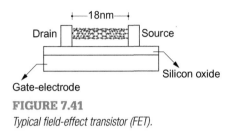

FIGURE 7.41
Typical field-effect transistor (FET).

Another application for CNTs is in the area of fuel cells and batteries, both for storage purposes. In the case of fuel cells, CNTs have been sought to store hydrogen, particularly for automotive applications, where hydrogen should be contained in small volumes and weights, yet enabling reasonable driving distances (500 Km). Specifically, the U.S. Department of Energy (DOE) has set the technological benchmark to 6.5 wt% (ratio of hydrogen to storage material). Currently, these hydrogen levels can be achieved by using gaseous and liquid hydrogen.

However, gaseous hydrogen occupies large volumes, whereas liquid hydrogen requires cryogenic containers, which drastically increase the system's overall cost. Solid-state hydrogen is thus the most promising route for hydrogen storage. Several publications have reported very high hydrogen storage capabilities in CNTs, ranging from 10 wt% to less than 0.1 wt% (see Figure 7.42). These experiments have been performed at ambient pressure and temperature, high pressure and room temperature, and cryogenic temperature. However, many of these experiments have been difficult to reproduce. This has been attributed to several factors, such as a large variation in the type and purity of CNTs tested as well as some difficulties in the characterization procedure. In addition, the mechanisms of hydrogen adsorption and the nature of chemical interaction have not yet been understood. Although some researchers claim that the major portion of hydrogen absorption is due to trapping sites, namely dangling bonds, with energies between 4.4 eV and 2.3 eV, depending on the trapping site, others argue that hydrogen trapping sites in carbon-related materials are either a result of

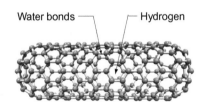

FIGURE 7.42
Carbon nanotubes (CNTs) for hydrogen storage.

physisorption (0.1 eV) or chemisorption (2–3 eV). Currently, the number of hydrogen trapping sites available under normal conditions is not sufficient to fulfill the goals set by DOE.

Still in the area of fuel cells, there is currently an increasing interest in using CNTs as a support material for the catalyst nanoparticles present in the cathode and anode electrodes of a fuel cell. At the moment, the support material is amorphous carbon. The introduction of CNTs is expected to enhance the conductivity of the support and reduce the mobility of the catalyst nanoparticles. The abundant open structure of CNTs is also very appealing for the storage of large amounts of lithium ions. Some of the CNT's properties, such as good chemical stability, large surface area, and elastic modulus, are important characteristics in prolonging the lifespan of batteries based on CNTs. In general, CNTs are able to adsorb a significant amount of lithium. However, the electrochemical performance strongly depends on the number of walls and chirality of the CNTs.

Another area in which CNTs can potentially be of great interest is the field of supercapacitors. This is because CNTs exhibit high porosity, large specific surface area, high electrical conductivity, and chemical stability. In a conventional capacitor, energy is typically stored by the transfer of electrons from one metal electrode to another metal electrode separated by an electronically insulating material. The capacitance depends on the separation distance and the dielectric material inserted between electrodes. In the case of a supercapacitor, there is instead an electrical double layer (see Figure 7.43). Each layer contains a highly porous electrode suspended within an electrolyte. An applied potential on the positive electrode attracts the negative ions in the electrolyte, whereas the potential on the negative electrode attracts the positive ions. A dielectric material between the two electrodes prevents the charges from crossing between the two electrodes. If the electrodes are made of CNTs, the effective charge separation is about a nanometer, compared with separations on the order of micrometers for ordinary capacitors. This small separation, combined with a large surface area, is responsible for the high capacitance of these devices (one to two orders of magnitude higher than conventional capacitors). In addition, although it is an electrochemical device, no chemical reactions are involved, allowing the ultracapacitor to be rapidly charged and discharged hundreds of thousands of times. Supercapacitors employing multiwalled carbon nanotube electrodes have already achieved a capacitance ranging from 18 to 250 F/g.

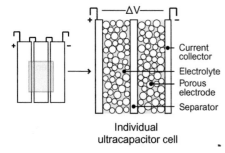

Individual
ultracapacitor cell

FIGURE 7.43

Schematic of a supercapacitor.

Field emission is another area in which CNTs can prove particularly useful. As previously discussed, the concept of field emission involves the application of an electrical field along the CNT axis to induce the emission of electrons from the end of the tube. So far, the research has been directed toward using SWCNTs and MWCNTs for flat-panel displays and lamps. In the case of flat-panel displays, used in televisions and computer monitors, an electric field directs the field-emitted electrons from the cathode, where the CNTs are located, to the anode, where the electrons hit a phosphorus screen and emit light (see Figure 7.44). The area of flat-panel displays has attracted a good deal of attention from the industrial community, including Motorola and Samsung, which have produced several prototypes. Despite the potential market for this application, the current technology still suffers from several problems that are not easy to solve, such as the development of low-voltage phosphorus. Obviously, once the technical problems are surpassed, the advantages of CNTs with respect to conventional liquid crystal displays are significant, in particular high brightness, a wide angle of view, and low power consumption. The technology for CNT-based lamps is similar to the one used for flat-panel displays, comprising a front glass covered with the phosphor coating and a back cathode glass that includes the CNTs. CNT-based lamps are attractive because they are mercury-free while maintaining high efficiency and long lifetime.

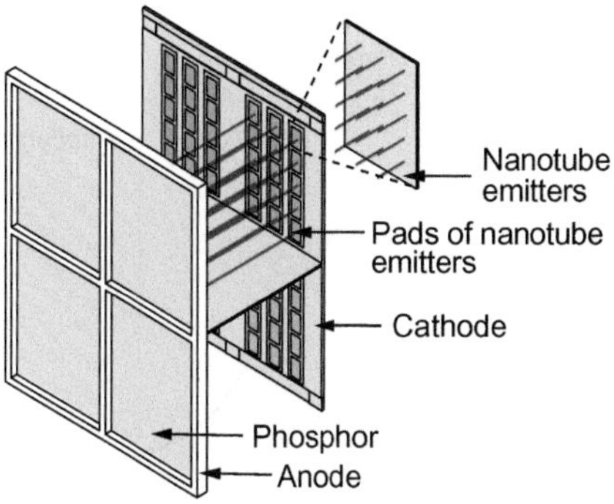

FIGURE 7.44
Flat-panel display. (Courtesy of Kenneth A. Dean, Nature Photonics 1, 273–275, 2007.)

Nanocomposites

Nanocomposites are a class of materials in which one or more phases with nanoscale dimensions (0-D, 1-D, and 2-D) are embedded in a metal, ceramic, or polymer matrix. The general idea behind the addition of the nanoscale second phase is to create a synergy between the various constituents, such that novel properties capable of meeting or exceeding design expectations can be achieved. The properties of nanocomposites rely on a range of variables, particularly the matrix material, which can exhibit nanoscale dimensions, loading, degree of dispersion, size, shape, and orientation of the nanoscale second phase and interactions between the matrix and the second phase.

Among the various nanocomposites types, the polymer-matrix nanocomposites have been the most studied. Much like traditional composite systems, polymer-matrix nanocomposites consist of a matrix made from a polymeric material. However, the second phase (usually a few percent by weight, wt%), which is dispersed

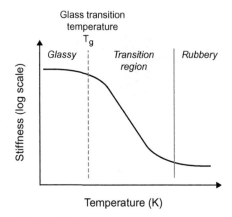

FIGURE 7.45

Various polymer states as a function of temperature.

within the matrix, has nanoscale dimensions. The small size of this phase leads to unique properties. In addition, due to the nanoscale size of the reinforcing phase, the interface-to-volume ratio is significantly higher than in conventional composites. As a result, the volume fraction of the second phase can be reduced, without degradation of the desired properties. The polymer matrix system can be a thermoplastic, thermoset, or elastomer. A thermoplastic polymer will soften when heated above the glass transition temperature (Tg; see Figure 7.45) and thus can be molded into a particular shape upon cooling. This process is repeatable, which makes thermoplastic materials reprocessable and recyclable. On the other hand, thermosetting materials become permanently hard through cross-linking when heated above Tg. Thus thermosetting polymers cannot be molded by softening. Instead, they must be fabricated during the cross-linking process. Elastomer resins are lightly cross-linked polymer systems and have properties that lie between thermosets and thermoplastics.

The nanoscale reinforcing phase can be grouped into three categories, namely, nanoparticles (0-D), nanotubes (1-D), and nanoplates (2-D). In the case of nanoparticles, the particle size and distribution are of great importance. Depending on the type of nanoparticles added, the mechanical, electrical, optical, and thermal properties of polymer nanocomposites can be altered. In the field of mechanical properties, the changes in modulus and strength depend strongly on the degree of interaction between the particle and the polymer. For example, in Poly(methyl methacrylate) (PMMA) polymer nanocomposites reinforced with alumina, the modulus decreased due to the weak interaction between the alumina and the PMMA, whereas in polystyrene nanocomposites reinforced with silica nanoparticles, the modulus increased due to a strong bonding between the matrix and the nanoparticles.

Another advantage of using nanoparticles as reinforcement is that their size is smaller than the critical crack length that typically initiates failure in composites. As a result, nanoparticles provide improved toughness and strength. However, agglomeration of nanoparticles should be prevented at all costs. In fact, several investigations have shown that small levels of agglomeration can decrease the strain-to-failure by several tens of percent. An additional mechanism that occurs for well-dispersed nanoparticles, which are weakly bonded to the matrix (for example, alumina/PMMA nanocomposites), is the phenomenon of cavitation. In other words, nanoparticles can act as voids, which initiate yielding and increase the volume of material going through deformation. This behavior leads to strain softening before strain hardening, resulting in large strains-to-failure.

Nanoparticles can also significantly affect Tg. Typically this occurs because nanoparticles influence the mobility of the polymer chains due to bonding between the particles and the polymer and bridging of the polymer chains between the particles. If the interaction between the nanoparticles and the matrix is weak, a depression in Tg is normally observed. However, in some cases, a critical volume fraction of nanoparticles is required for the effect to be noticed, as shown for alumina/PMMA nanocomposites. The Tg can also be increased if the interaction between the matrix and the nanoparticles is strong. In fact, a 35°C increase in Tg has been observed for a PMMA nanocomposite reinforced with calcium carbonate nanoparticles. Another benefit of using nanoparticles in polymer-matrix nanocomposites is an enhancement in wear resistance, which has been observed, for example, in nylon reinforced with nanoparticles of silica.

The electrical and optical properties of polymer-matrix nanocomposites can also be improved by the addition of nanoparticles. With respect to electrical properties, the nanoparticles seem to act in a variety of ways. First, the smaller the nanoparticles, the shorter the distance between the particles, provided the volume is kept constant. This in turn leads to percolation at lower volume fraction, resulting in higher electrical conductivity (see Figure 7.46). The low-density polyethylene nanocomposite filled with ZnO nanoparticles shows this behavior quite clearly. Second, even for nanocomposites embedded with insulating nanoparticles, the electrical conductivity of the nanocomposite seems to increase due to a better compactness of the polymer, leading to enhanced coupling among the nanoparticles through the grain boundaries. For example, the room temperature DC conductivity of polypyrrole filled with zirconia nanoparticles increased from 1 S/cm to 17 S/cm. This increase is thought to be a consequence of the increase in compactness of the polymer. Finally, in some cases, such as in polypyrrole nanocomposites filled with Fe_2O_3 nanoparticles, the so-called variable range-hopping (VRH) mechanism seems to explain the enhancement in DC current. The VRH mechanism involves exchange of charges between the nanoparticles and the polymer matrix.

In terms of optical properties, there is a large interest in developing transparent nanocomposites with enhanced mechanical and electrical properties. To achieve transparency, scattering must be

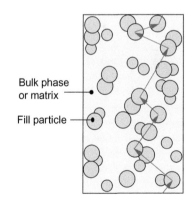

FIGURE 7.46
Electrical percolation in polymer nanocomposites.

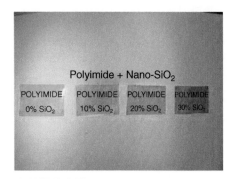

FIGURE 7.47

Changes in transmittance in polyimide nanocomposites filled with silica nanoparticles, for various levels of loading. (Courtesy of Jiann-Wen Huang, Ya-Lan Wen, Chiun-Chia Kang, and Mou-Yung Yeh, Polym. J., 39, *654, 2007.)*

minimized, which means that the nanoparticles should be as small as possible while the index of refraction should remain as similar as possible to the matrix. For example, the addition of nanoparticles of silica to polyimide has been used to control the transmittance in these nanocomposites (see Figure 7.47). Furthermore, the addition of nanoscale alumina to gelatin has improved the total transmittance by as much as 100%. Another excellent use of nanoparticles in polymer nanocomposites is in controlling the index of refraction, which can be achieved by tailoring the volume fraction of nanoparticles. In addition, using active optical nanoparticles and changing the particle size, distribution, and shape, the color of the nanocomposite can be tuned. This behavior has been shown in polyethylene polymer films filled with silver nanoparticles.

Next, let's discuss the inclusion of 1-D nanomaterials, in particular carbon nanotubes (CNTs) for the reinforcement of nanocomposites. (A comprehensive discussion on the structure and properties of CNTs can be found in Section 7.7.) The use of CNTs in composites has received wide attention due to their extraordinary physical and mechanical properties. However, to take full advantage of CNTs for nanocomposite applications, several critical factors need to be addressed: (1) uniform dispersion of carbon nanotubes within the polymer matrix, (2) alignment of CNTs in the nanocomposite, and (3) good interfacial bonding between the CNTs and the polymer matrix. With respect to the dispersion of CNTs, the work has been very challenging, particularly compared with the procedure for dispersing carbon fibers in traditional composite materials. This is because CNTs exhibit smooth surfaces and intrinsic Van der Waals interactions, which tend to promote clustering when dispersed in a polymer matrix (see Figure 7.48). If agglomeration occurs, the CNTs are less adhered to the matrix and will slip against each other under an applied stress, with drastic consequences for the mechanical properties.

To address this problem, several methods have been used, such as sonication of CNTs, shear mixing, surfactant-assisted processing, chemical functionalization, and *in situ* polymerization. Sonication is normally done in a solvent before CNTs are added to the matrix or before another dispersion technique is applied. The shear mixing method has worked for rubbery epoxy resins, for which 1 wt% of CNTs led to an increase by 27% in the tensile modulus and by 100% in the tensile strength. However, when the CNTs were shear mixed with a glassy epoxy resin, no improvement in the mechanical properties was observed. This was probably due to the increased viscosity occurring in these materials.

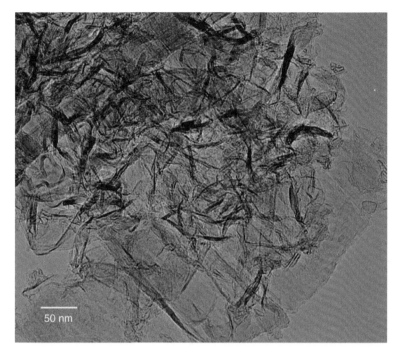

50 nm

FIGURE 7.48
Typical agglomeration of carbon nanotubes.
(Courtesy of P. J. Ferreira, University of Texas at
Austin; P. Szakalos, Royal Institute of Technology,
Sweden; and J. Vander Sande, MIT.)

Surfactant-assisted processing is another method used to disperse CNTs. In this process, a surfactant is added to the CNTs before the mixture is added to the polymer. In general, nanocomposites produced in this way appear to be more evenly distributed. Results from mechanical tests indicate that polymers with the addition of surfactants but no CNTs show a decrease in the elastic modulus compared to the elastic modulus of the same polymer without the surfactant. This indicates that the surfactant acts as a plasticizer. On the other hand, when the surfactant is added along with the CNTs to the polymer matrix, the elastic modulus can increase more than 30%, which confirms that the surfactant is acting as an efficient dispersing agent.

Dispersion of CNTs by chemical functionalization is also another alternative (see Figure 7.49). In this case, one of the most successful developments involves partial separation of SWCNTs from the bundles by means of an acidic solution. This step normally cuts the CNTs into short segments with carboxylic acid groups covalently attached to the openings. Subsequently the SWCNTs are placed in an organic melt, which leads to exfoliated CNTs. When MWCNTs are used, the surface activity is typically modified by doping them

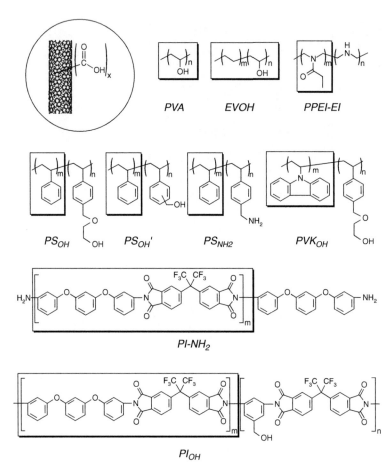

FIGURE 7.49

Polymers used for the functionalization and solubilization of CNTs. (Courtesy of Yi Lin, Mohammed J. Meziani, and Ya-Ping Sun, J. Mater. Chem., *17, 2007, 1143.)*

with other elements. Previous work has shown an increase of the Tg, modulus, and strength.

In addition to uniform dispersion of CNTs within a polymer matrix, a crucial aspect for providing optimal reinforcement is to properly orient the CNTs. Because of their aspect ratio, CNTs possess highly anisotropic mechanical properties. To take advantage of their load-carrying efficiency along the axial direction, it is essential that CNTs are well aligned within the matrix. Several methods have been used to align the CNTs. Extrusion is a popular technique, in which a polymer melt reinforced with CNTs is extruded through a die and drawn under tension before solidification. Polystyrene nanocomposites filled with MWCNTs have been produced in this manner. Compared with a pristine sample, the modulus of the aligned MWCNT/polystyrene composite increased as much as 49%, compared to a 10% increase for randomly oriented nanocomposites. In addition, aligned MWCNTs/polystyrene nanocomposites were also shown to exhibit an increased yield strength and ultimate yield strength compared to a pristine polystyrene polymer. Due to the fact that CNTs exhibit high electrical conductivities, the application of a magnetic field or electric field has also been used to induce CNT alignment. It was observed that the elastic modulus measured in the direction parallel to the magnetic field was greater than for the direction perpendicular to the field. This fact suggests that aligned MWCNTs contribute significantly to an increase in the elastic modulus in the direction parallel to the aligned CNTs.

Finally, the interfacial bonding between the CNTs and the polymer matrix needs to be considered. It is believed that the efficiency of load transfer in nanocomposites is controlled by the interfacial characteristics of the reinforcing phase and the matrix. In the case of polymer-matrix nanocomposites reinforced with CNTs, it is believed that the main mechanisms of load transfer include mechanical interlocking, chemical bonding, and nonbonded interactions between the CNTs and the matrix. In particular, it is believed that the key factor in forming a strong bond between the polymer matrix and a CNT lies in the polymer morphology, especially its ability to form large diameter helices around individual CNTs. The strength of the interface is then due to molecular-level entanglements of CNTs and matrix as well as long-range order of the polymer.

The inclusion of CNTs in polymer-matrix nanocomposites can also be used to improve the thermal conductivity of these materials. The CNTs create a percolation network that allows the nanocomposite to conduct heat with conductivities up to 3.5 times the conductivity of the pristine polymer. However, the thermal conductivity of a nanocomposite is still far from the theoretical value for an isolated CNT, which has been predicted to be on the order of 10^3 W/m.K. The main reason for this discrepancy is the large thermal resistance that exists between the polymer matrix and the CNT surface. Because CNTs possess high surface-to-volume ratios, when CNTs are dispersed in the polymer matrix, the large interfacial area creates a significant resistance. This effect has been attributed to differences in phonon frequency between the CNTs and the polymer matrix. Some solutions have been suggested to decrease the thermal resistance of the polymer/CNT interface. For example, it has been proposed to covalently bond the CNTs to the polymer matrix to improve the phonon/phonon coupling. It has also been suggested to use MWCNTs instead of SWCNTs because the former have smaller aspect ratios and therefore less interfacial area. The use of CNTs in nanocomposites has also improved the thermal stability of polymer-matrix nanocomposites by increasing the onset decomposition temperature. The mechanisms behind the enhancement in thermal stability involve the retardation of the decomposition rate of the polymer at the CNT/polymer interface and improved heat dissipation due to the enhancement of the thermal conductivity. Overall, this effect is very important for applications in which the nanocomposite will be subjected to high temperatures. As for the addition of nanoparticles, CNTs also influence the Tg of the polymer matrix. This is important because the Tg limits the temperatures at which the nanocomposite can be used. So far, the addition of CNTs to polymer-matrix composites have both increased and decreased the Tg, depending on the conformation of the CNTs. Straight SWCNTs and MWCNTs have increased the Tg of an epoxy matrix, whereas coiled CNTs/epoxy nanocomposites have exhibited a reduction in Tg.

We are now left to discuss the use of 2-D second-phase nanomaterials in polymer nanocomposites. These are platelike layered materials with a thickness on the order of 1 nm but with an aspect ratio of 25 or above. The most common are layered silicates. When these are added to polymer-matrix nanocomposites, a wide array of property enhancements can be achieved, such as increased stiffness and strength, improved UV resistance and gas permeability, greater dimensional stability, and superior flame resistance. Remarkably, these enhancements in properties are obtainable at extremely low-filler concentrations (2–5% vol), a fraction of what is typically needed in conventional composite materials (30–40% vol). In addition, contrary to most conventional composite systems, the matrix properties are often not sacrificed. Among the layered silicates, mica and smectic clays are the most used. Mica consists of large sheets of silicate with strong bonds between the layers. On the contrary, the smectic clays exhibit weak bonds between the layers. As shown in Figure 7.50, smectic clays consist of a three-layered sandwich structure composed of two outer layers containing silicon and oxygen bonded to an inner layer of aluminum, magnesium, and/or iron that is bonded to oxygen or hydroxyl groups. Due to the substitution of divalent Mg for trivalent Al, a negative charge is created within the inner layer of the clay. However, this excessive charge can be compensated by the adsorption of cations, such as Na^+, Ca^{2+}, and Li^+. The layer structure shown in Figure 7.50 is repeated over many times to form a layered structure, similar to a stack of paper sheets. For these layered silicates to be useful in nanocomposites, the layers must be separated and dispersed within

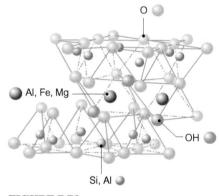

FIGURE 7.50
Crystal structure of layered smectic clays.

the polymer matrix. However, smectic clays are highly hydrophilic, and thus individual layers do not easily disperse in relatively hydrophobic species, such as an organic polymer.

To overcome this problem, an ion exchange reaction is usually carried out whereby organic cations open the clay layers and make them hydrophobic enough, resulting in an organically modified clay. At this point, the organically modified clay can be intercalated with the polymer through various techniques, such as solvent-based, melt-blending, and polymerization-based methods. In the case of the solvent-based method, low concentrations of organoclays are dispersed in a solvent in which a polymer is soluble (see Figure 7.51). Due to the weak forces between the silicate layers, the solvent separates the layers, thereby allowing the polymer to adsorb onto the surfaces of individual silicate platelets. The solvent is then evaporated, causing the individual platelets to reagglomerate. Highly polar polymers, such as nylon and polyimides, can be more readily intercalated than nonpolar polymers.

Another technique that is frequently used to form polymer-matrix nanocomposites involves melt blending of organoclays with polymers using conventional extrusion equipment. The studies show that intercalation of polymer chains between the layered clays can occur spontaneously by heating a mixture of polymer and clay powder above the glass transition temperature or melting temperature. Once sufficient polymer mobility is achieved, the chains diffuse into the clay layers, thereby producing a swelled polymer/clay layered structure.

Finally, an *in situ* polymerization can also be employed. Here, monomers are directly intercalated into the organically modified clay, followed by polymerization. In this way, nylon, polyimide, polystyrene, and rubber-based nanocomposites have been developed. Once the clays have been intercalated with a polymer, the nanocomposites can be processed by conventional melt-processing techniques. The resulting nanocomposites can exhibit various microstructures, namely intercalated nanocomposites, when the polymer diffuses into the clay layers, thereby expanding the distance between the clay layers, and exfoliated nanocomposites when individual delaminated silicate layers are dispersed in a polymer matrix (see Figure 7.52).

Nanoscale layered silicates have proven to induce great improvements in the properties of conventional polymers. An unexpected large increase in the elastic and flexural modulus was achieved with small amounts of silicates (as low as 1 wt%). Thermal stability and

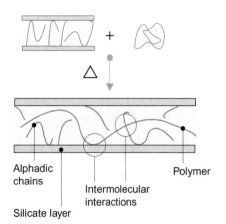

FIGURE 7.51
Intercalation solvent-based process.

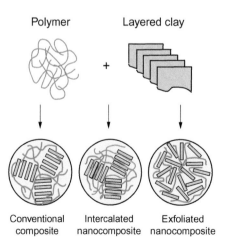

FIGURE 7.52
Intercalated and exfoliated polymer nanocomposites.

fire resistance through char formation are also interesting properties exhibited by these nanocomposites. In addition, they can also display superior resistance to permeability of gases. In terms of mechanical properties, the increase in the elastic modulus of nylon-based nanocomposites seems to be related to the average length of the layers and therefore their aspect ratio. It has been shown that the exfoliated layers are the main factor responsible for the stiffness improvement. However, in a pure elastomeric matrix, exfoliation does not seem to be a prerequisite to enhance the material stiffness. In addition, a large increase in the tensile modulus for an exfoliated structure is also observed for thermoset polymer matrices.

On the other hand, the stress at failure may vary strongly depending on the nature of the interactions between the matrix and the nano-scale filler in thermoplastic-based (intercalated or exfoliated) nanocomposites. For example, exfoliated nylon-6-based nanocomposites or intercalated PMMA-based nanocomposites exhibit an increase in the stress at break, which is usually explained by the presence of polar and ionic interactions between the polymer and the silicate layers. On the contrary, polypropylene-based nanocomposites show negligible or slight improvement in the stress at failure due to the lack of interfacial adhesion between the nonpolar polypropylene and the polar layered silicates, and polystyrene-intercalated nanocomposites exhibit a decrease in the stress at failure. In terms of ductility, the addition of layered silicates to polymer-based nanocomposites leads to different results. In intercalated PMMA and polystyrene the elongation at break is reduced. However, such a loss in ultimate elongation does not occur in elastomeric epoxy. Rather, the addition of the nanoclay induces an increase in elongation.

The incorporation of layered silicate into a polymer matrix can also affect the thermal properties of nanocomposites. In terms of thermal conductivity, it seems that one of the effects of the layered silicates is to change the cell density and cell size. In fact, nanoclay/polyurethane nanocomposites show an increase in cell density and a decrease in cell size with the addition of nanoclay. This effect is attributed to the nanoclay acting as cell nucleation sites leading to higher cell densities and small cells, which in turn reduce the thermal conductivity. The second effect is due to the fact that the chemically bonded silicate layers inhibit heat and mass transport. However, the greatest benefit of using nanoclays as fillers for nanocomposites is in improving the thermal stability of the polymer matrix. This is seen in various polymer systems. The factors that determine the extent of thermal stabilization in nanocomposites arise from the restriction of thermal motion of polymer molecules in the silicate interlayer, reduction in permeability of volatile decomposition products, and prevention of degradation from occurring through cross-linking points between polymer chains and the filler nanoclays. In polyurethane/nanoclay nanocomposites, the degradation temperature was increased throughout the nanocomposite due to reduced oxygen and volatile product diffusion as a result of the dispersed layered clays, which are impermeable to the diffusing species. A similar effect has been seen in polypropylene/nanoclay nanocomposites, in which the temperature of maximum weight loss increased significantly due to the thermal robustness of the clay platelets. Finally, polymer-matrix nanocomposites filled with layered silicates are also known to improve flammability properties. The reason for this behavior seems to arise from the formation of char layers obtained through the collapse of the exfoliated and/or intercalated structures. It turns out that the layered silicate structure can act as an excellent insulator and mass transport barrier. For example, a nanoclay/nylon-6 nanocomposite with a small wt% of filler shows a 60% reduction in peak heat release rate compared with a pristine polymer.

Let's now turn the discussion to metal-matrix nanocomposites. These materials are composed of a metal or alloy matrix that is ductile and a rigid reinforcement that is typically a ceramic. Therefore, metal-matrix nanocomposites combine the properties associated with metals, such as ductility and toughness, with ceramic features such as high modulus and strength. In addition, because of the outstanding properties of nanomaterials, the optical, electrical, and magnetic properties of metal-matrix nanocomposites can also be tuned.

Through various processing routes, several nanocomposites have been produced, namely, aluminum, titanium, copper, and nickel matrix materials, reinforced with borides, carbides, nitrides, and oxides. As in the case of polymer-matrix nanocomposites, the factors that affect their performance are the size of the filler and the homogeneity of the reinforcement distribution. For example, *in situ* fabrication (a process by which the reinforcements are produced by exothermal reactions between chemical species or between chemical species and compounds) of Al, Al/Si, and Al/Fe/V/Si metal-matrix nanocomposites reinforced with nanoparticles of TiC has produced a uniform dispersion of the reinforcing phase. As a result, these materials exhibit very good strength within a wide range of temperatures. As expected, the smaller the TiC nanoparticles added and the larger their volume fraction, the greater the strength achieved in these nanocomposites.

Metal-matrix nanocomposites can also be beneficial for electrical contacts. For example, Ag-based nanocomposites filled with SnO_2 nanoparticles, fabricated during a reactive milling process, were hot-pressed into electrical contacts. These exhibit outstanding thermal and electrical conductivity while maintaining superior wear resistance. Metal-matrix nanocomposites are also particularly useful for magnetic applications such as magnetic recording and giantmagneto resistance. Typically, these nanocomposites are composed of nanoscale hard magnetic particles such as $Nd_2Fe_{14}B$ $Sm_2Fe_{17}N_3$ and FePt, embedded in a soft magnetic nanocrystalline phase, such as ferrite and Fe_3Pt. These hard and soft magnetic phases interact magnetically, combining the best properties of a soft magnetic phase, which is the high saturation magnetization, with that of a hard phase, the high coercive field (applied magnetic field required to reduce the magnetization of that material to zero after the magnetization has been brought to saturation). For the interaction to occur and exert its coupling effect, the two phases must be at the nanoscale.

In the absence of a magnetic field, the magnetic interaction results in spin alignment, but when a magnetic field is applied in the opposite direction (if not above a critical value), only the soft phase is able to reverse the magnetization. As a consequence, when the magnetic field is removed the magnetization is again reversed in the soft phase. However, when the applied magnetic field is high enough to reverse the spins in the hard phase, the soft phase does not reverse magnetization when the field is removed. This effect is strongly dependent on the size of particles as well as the volume fraction and distribution of each phase. Following this type of interaction, nanocomposite materials exhibit high remanence (magnetization that is left behind after the magnetic field has been removed) and a high magnetic energy (as high as 200 kJ/m^3). An important aspect is that of maximization of the soft-phase content to enhance the saturation magnetization. These materials have been processed through various methods, although the most successful route has been mechanical alloying of two phases.

In addition to the use of nanoparticles, metal matrix-nanocomposites have also been reinforced with CNTs. Al-matrix nanocomposites reinforced with MWCNTs are such an example. These were produced by mechanical milling and powder metallurgy. The yield stress, maximum strength, and hardness values obtained for the nanocomposites were considerably higher than those for pure Al. In fact, the addition of 0.75 wt% of MWCNTs has doubled the yield strength of the nanocomposite.

Let's turn now to ceramic-matrix nanocomposites. The idea behind these materials is to improve the fracture toughness, general strength, and high temperature strength of conventional ceramic-matrix composites. In fact, significant increases in fracture toughness and strength were observed in alumina and zirconia matrices reinforced with nanoparticles of SiC and/or Si_3N_4 ranging in size from 20 nm to 300 nm. For example, a 5 vol% addition of nano-scale particles of SiC and Si_3N_4 into an alumina matrix resulted in fracture toughness improvements of more than 50% and strength of more than 100%. The reason for these enhanced properties is related to crack-tip bridging, a mechanism by which nanoparticles cause bridging of cracks at distances close to the crack tip. Therefore, instead of cracks propagating along the grain boundaries, the residual stresses around the particles drive the cracks to the particles, leading to crack bridging and higher fracture toughness. In the case of Si_3N_4 ceramic-matrix nanocomposites reinforced with SiC, the beneficial effects provided by reinforcement are less understood. Some possible mechanisms include Si_3N_4 grain refinement due to the presence of SiC at the grain boundaries, leading to larger strains to failure, and thermal mismatch between the two phases, resulting in improved strength and fracture toughness. In addition to the improvements in mechanical properties, ceramic matrix nanocomposites can also be used as thermal barriers. A good example is a nanocomposite composed of SiC nanoparticles dispersed within a pyrolytic graphite matrix.

This material exhibits outstanding oxidation resistance and thermal shock. Ceramic-matrix nanocomposites have also been reinforced with metallic nanoparticles, particularly for improving optical, electrical, and magnetic properties. Examples include alumina-matrix nanocomposites filled with W, Ni, Fe, Au, and Cu; zirconia-matrix filled with Ni and Mo; magnesia-matrix filled with Fe, Ni, and Co; and silica filled with Fe, Au, and Ag. In the optical area, silica-matrix nanocomposites filled with nanoparticles of Au and Ag exhibit high optical nonlinearities (which describe the behavior of light in media in which the dielectric polarization responds nonlinearly to the electric field of the light), which are essential for photonic devices. In the case of silica filled with Ag, the nanocomposite shows changes in behavior from semiconducting to metallic, depending on the size of the nanoparticles and temperature. This seems to be the result of electron tunneling between the metallic nanoparticles and metallic conduction through percolated paths, which can be controlled by tuning the volume fraction of nanoparticles, their distribution and uniformity, size, and shape. In the field of magnetism, silica nitride matrix nanocomposites containing fine dispersions of Co nanoparticles (<10 nm) revealed supermagnetism properties, a large magnetic hysteresis, and high coercive fields.

Ceramic-matrix nanocomposites filled with semiconductor nanoparticles such as GaAs also show interesting optical properties. When these nanoparticles are embedded in a silica matrix, the properties of photoluminescence (a process in which a substance absorbs photons and then radiates back photons) are altered due to the dimensional confinement of the nanoparticles. Because these particles need to be separated to maximize the effect, the creation of a nanocomposite is ideal.

Nanocomposites can also be used as thin films for various applications. These films typically consist of multiple nanoscale layers, each composed of a different material, or single-layered materials reinforced with a second phase at the nanoscale. The multilayer nanocomposites are normally used in applications where high hardness, modulus, and wear properties are important. This behavior is due to the fact that the incoherent interface between the layers provides a resistance to elastic deformation and plastic deformation. In parallel, multilayered nanocomposites have also been developed for magnetic recording.

Single-layer nanocomposite materials are used for mechanical, electrical, and magnetic applications in which at least the matrix or the filler has particular mechanical, electrical, and/or magnetic properties. In terms of mechanical properties, nanocomposite thin films have been widely used for nanocoatings. They provide good wear resistance, increased toughness, and high thermal stability. In the case of nanocoating multilayers, which are typically made of TiC, CrN, TiN, TiAlN, and alumina, the films have many layers to enable crack deflection at the interfaces and improve toughness. In addition, these multilayer nanocomposite coatings exhibit increased hardness due to the fact that the interfaces act as barriers to dislocation motion. However, if the layers are too thin, the strain field of the dislocations is truncated by the interfaces, leading to a decrease in hardness. Furthermore, depending on the thickness of the layers, lattice mismatch can lead to additional residual strains, with consequences for the mechanical properties. Therefore, for various applications, the thickness of the layers is carefully tuned so that high toughness, good thermal stability, and hardness are present, particularly at high temperatures.

In the case of a single-layer nanocomposite coating, hard nanoparticles are typically embedded in an amorphous matrix. For example, nanoparticles of TiN in an amorphous matrix of Si_3N_4 exhibit hardness values as high as 50–60 GPa. The reason for this behavior has been attributed to hindered dislocation motion and sliding of nanoparticles. Another system, which exhibits high hardness, consists of carbide nanoparticles (10–50 nm) embedded in amorphous carbon. This type of coating is especially useful for aerospace applications, in which low friction, high hardness, and resistance to corrosion are important. On the other hand, for cutting-tool applications, nanocomposite coatings are normally composed of metal carbides embedded in a metallic matrix. Typical systems that fall into this category are Co matrices reinforced with WC and/or TiC and Ni matrices filled with TaC. In these nanocomposites, both the matrix and the reinforcement particles can have nanoscale dimensions. In terms of properties, they exhibit high strength, high wear resistance, good toughness, and thermal stability.

Multilayer nanocomposite films are also widely used for magnetic recording media. The films are typically composed of a substrate onto which a metallic layer followed by a magnetic layer are deposited, both with nanoscale thicknesses. To produce high magnetic recording densities (40–100 Gbits per in^2), the grain size of the magnetic layer should be below 10 nm. However, at these small grain sizes, the magnetocrystalline anisotropy has to be sufficiently high to overcome thermal fluctuations. Examples of these nanocomposite films include multilayer structures of magnetic CoCrPt and metallic Cr supported by a substrate.

Finally, consider the vast class of nanocomposite bio/inorganic materials, which is a fast-growing area. Significant effort is focused on the ability to control the nanoscale structures through novel processing approaches. The inorganic components can be 3-D systems such as zeolites, 2-D layered materials such as clays and metal oxides, and even 1-D and 0-D materials such as CdS and metal oxides. The biostructures range from proteins and DNA to lipid cellular membranes. For example, proteins consist of domains with very diverse properties, such as the capability for hydrogen bonding or exhibiting acidic, basic, hydrophilic, and hydrophobic behavior. As a result, proteins can interact with inorganic materials in a variety of ways. The same can be said about DNA, which can be functionalized, tethered to a broad number of substances, and self-assembled with a wide diversity of characteristics. The general class of organic/inorganic nanocomposites may also be of relevance to issues of bioceramics and biomineralization in which *in situ* growth and polymerization of biopolymer and inorganic matrix are occurring.

FURTHER READING

Mechanical Properties

A. S. Edelstein and R. C. Cammarata (eds.), Nano materials: Synthesis, properties and applications, Institute of Physics, 1996, ISBN 0-7503-0577.9.

A. N. Goldstein (ed.), Handbook of nanophase materials, CRC Press, 1997, ISBN 0724-79469-9.

H. Hasono, Y. Mishma, H. Takezoe, and K. J. D. Mackenzie, Nano materials from research to applications, Ch. 12, pp. 419 et seq., Elsevier, 2006, ISBN 13 977.0-07.044964-7.

C. C. Koch, Nano structured materials: Processing properties and applications, Noyes Publications, 2002, ISBN 0-7155-1451-4.

R. Lumley and A. Morton, Nanoengineering of metallic materials, Ch. 9, pp. 219–247, in: Nanostructure control of materials, R. H. J. Hannink and A. J. Hill, Woodhead Publishing Ltd., 2006, ISBN 13 977.1-75573-933-7.

S. Menzies and D. P. Anderson, J. Electrochem. Soc., 1990, 137, p. 440.

A. A. Nazarov and R. R. Mulyukov, in: Handbook of nano science, engineering and technology, Ch. 22, W. A. Goddard, D. W. Brenner, S. E. Lysheuski, and G. J. Iafrate (eds.), CRC Press, 2003. ISBN 0-7493-1200-0.

C. P. Poole, Jr., and F. J. Owens, Introduction to nano technology, Wiley-Interscience, 2003, ISBN 0-471-07935-9.

D. M. Tench and J. T. White, Metall. Trans. A, 1975, 15, p. 2039.

J. R. Weertman and R. S. Averbach, Mechanical properties, in: Nano materials: Synthesis, properties and applications, Ch. 13, Institute of Physics, A. S. Edelstein and R. C. Cammarata (eds.), 1996, ISBN 0-7503-0577.9.

Thermal Properties of Nanomaterials

H. Reiss and I. B. Wilson, The effect of surface on melting point, Journal of Colloidal Science, 1948, 3, pp. 551–561.

T. L. Hill, Thermodynamics of small systems, Part I, pp. 114–132, W. A. Benjamin, 1963.

P. R. Couchman and W. A. Jesser, Thermodynamic theory of size dependence of melting temperature in metals, Nature, 1977, 69, pp. 481–483.

D. G. Cahill, W. K. Ford, G. D. Mahan, A. Majumdar, R. Merlin, and S. R. Phillpot, Nanoscale thermal transport, Journal of Applied Physics, 2003, 93, 2, pp. 793–818.

N. Prakash, "Determination of coefficient of thermal expansion of single-walled carbon nanotubes using molecular dynamics simulation, Ph.D, thesis, Florida State University, 2005.

Electronic Properties

C. Kittel, Introduction to solid state physics, John Wiley & Sons, 1953.

A. S. Edelstein and R.C. Cammarata (eds.), Nanomaterials: Synthesis, properties and applications, Institute of Physics Publishing, 2001.

C. Dupas, P. Houdy, and M. Lahmani (eds.), Nanoscience: Nanotechnologies and nanophysics, Springer, 2007.

C. P. Poole, Jr., and F. J. Owens, Introduction to nanotechnology, John Wiley & Sons, 2003.

R. Kelsall, I. Hamley, and M. Geoghegan (eds.), Nanoscale science and technology, John Wiley & Sons, 2005.

Magnetic Properties

C. Kittel, Introduction to solid state physics, John Wiley & Sons, 1953.

B. D. Cullity, Introduction to magnetic materials, Addison-Wesley, 1972.

R. C. O'Handley, Modern magnetic materials: Principles and applications, John Wiley & Sons, 2000.

H. S. Nalwa (ed.), Magnetic nanostructures, American Scientific Publishers, 2000.

R. Skomski, Nanomagnetics, Journal of Physics: Condensed Matter, 2003, 15, pp. R841–R896.

A. S. Edelstein and R. C. Cammarata (eds.), Nanomaterials: Synthesis, properties and applications, Institute of Physics Publishing, 2001.

C. Dupas, P. Houdy, and M. Lahmani (eds.), Nanoscience: Nanotechnologies and nanophysics, Springer, 2007.

C. P. Poole, Jr., and F. J. Owen, Introduction to nanotechnology, John Wiley & Sons, 2003.

R. Kelsall, I. Hamley, and M. Geoghegan (eds.), Nanoscale science and technology, John Wiley & Sons, 2005.

Optical Properties

C. Kittel, Introduction to solid state physics, John Wiley & Sons, 1953.

A. S. Edelstein and R. C. Cammarata (eds.), Nanomaterials: Synthesis, properties and applications, Institute of Physics Publishing, 2001.

C. Dupas, P. Houdy, M. Lahmani (eds.), Nanoscience: Nanotechnologies and nanophysics, Springer, 2007.

C. P. Poole, Jr., and F. J. Owen, Introduction to nanotechnology, John Wiley & Sons, 2003.

R. Kelsall, I. Hamley, and M. Geoghegan (eds.), Nanoscale science and technology, John Wiley & Sons, 2005.

T. J. Bukowski and J. H. Simmons, Quantum dot research: Current state and future prospects, Critical Reviews in Solid State and Materials Science, 2002, 27, $\frac{3}{4}$, pp. 119–142.

D. J. Campbell and Y. Xia, Plasmons: Why should we care? Journal of Chemical Education, 2007, 84, 1, pp. 91–96.

Carbon NanotubesS. IIjima, Nature, 1991, 354, 56.

R. Saito, G. Dresselhaus, and M. S. Dresselhaus, Physical properties of carbon nanotubes, World Scientific Publishing Co., 1998.

G. Timp (ed.), Nanotechnology, Springer-Verlag, 1999.

Nanocomposites

P. M. Ajayan, L. S. Schadler, and P. V. Braun, Nanocomposite science and technology, Wiley-VCH, 2003.

J. Koo, Polymer nanocomposites, McGraw-Hill, 2006.

CHAPTER 8

Nanomaterials: Synthesis and Characterization

8.1 SYNTHESIS OF NANOSCALE MATERIALS AND STRUCTURES

There are many ways of making nanoscale materials and structures. Figure 8.1 orders them in the same matrix as that of Chapter 6. It gives an overview, showing where each fits. We start our discussion at the upper left: processes to make discrete, nanoscale clusters.

Methods for Making 0-D Nanomaterials

Nanoclusters are made by either gas-phase or liquid-phase processes. We start with those using the gas phase, the commonest of which are inert-gas condensation and inert-gas expansion. Liquid phase processes use surface forces to create nanoscale particles and structures. There are broad types of these processes: ultrasonic dispersion, sol-gel methods, and methods relying on self-assembly.

Inert-gas condensation

In *inert-gas condensation*, an inorganic material is vaporized inside a vacuum chamber into which an inert gas (typically argon or helium) is periodically admitted (see Figure 8.2). The source of vapor can be an evaporation boat, a sputtering target, or a laser-ablation target. Once the atoms boil off, they quickly lose their energy by colliding with the inert gas. The vapor cools rapidly and supersaturates to form nanoparticles with sizes in the range 2–100 nm that collect on a finger cooled by liquid nitrogen. The particles are harvested by scraping them off the finger and are collected, still protected by the inert gas, for further processing. Alloy particles are made using dual sources, as shown in Figure 8.2. The main problem with this method is that as the particles form, they tend to cluster and increase their

		Classes		
		Class 1 Discrete nano- objects	Class 2 Surface nano- featured materials	Class 3 Bulk nano- structured materials
Dimensionality	0-D All 3 dimensions on nanoscale	Inert gas condensation Evaporation Colloidal methods	Physical or chemical vapor deposition (PVD or CVD)	Equiangle extrusion Cryomilling Consolidation of nanoparticles by sintering
	1-D 2 dimensions on nanoscale	Directional growth from catalyst dots Templating	Lithographic methods	Incorporation of nanotubes and rods into polymer or metal matrices
	2-D 1 dimension on nanoscale	Beating (gold foil) Electrodeposition PVD, CVD Self-assembled films	Electrodeposition Physical vapor deposition (PVD) Chemical vapor deposition (CVD)	Rotating shutter PVD and CVD Cyclic electrodeposition

FIGURE 8.1

The matrix of processes used to make nanomaterials.

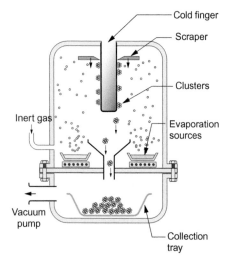

FIGURE 8.2

Inert gas condensation. The nanoclusters that form on the cold finger are harvested by the scraper and collector and subsequently consolidated to make products.

size, defeating the original objective. To avoid this, the processing parameters have to be carefully controlled.

Inert-gas (free-jet) expansion

In *inert-gas* or *free-jet expansion*, evaporated atoms are carried by a high-pressure helium gas stream that is expanded from a nozzle into a low-pressure chamber at supersonic velocities (see Figure 8.3). The adiabatic expansion of the gas leads to sudden cooling, causing the evaporated atoms to form clusters a few nanometers in diameter. As in the case of inert-gas condensation, the agglomeration of nanoparticles is a problem, requiring careful control of evaporation rate and inert gas flow; it is these that determine the particle size and distribution.

Sonochemical processing

In *sonochemical processing*, ultrasound is used to nucleate a chemical reaction. Ultrasound spans the frequencies in the range 15 kHz to 1 GHz. A magnetostrictive or piezoelectric transducer (or "horn") is used to generate ultrasonic waves with a wavelength of 1–10,000 microns in a liquid-filled reaction vessel (see Figure 8.4). These are not molecular dimensions, so there is no direct coupling of the acoustic field with the chemical species. The reaction comes about because of cavitation. The tensile part of the wave is intense enough to pull the liquid apart and form a tiny cavity. The compression part of the wave then compresses it, but before it does, some reactants vaporize inside it. The next tensile wave re-expands the bubble, which oscillates in volume at the frequency of the sound waves, pumping it up as more vapor enters during the expansion part of the cycle. When the bubble reaches a critical size, it collapses. The collapse is adiabatic because the very fast collapse rate leaves no time for heat flow, generating a tiny, localized hot spot. The temperatures are very high (as high as 5000°C, near that of the surface temperature of the sun) and so too are the pressures (around 2000 atmospheres, roughly those at the bottom of deep oceans), triggering reactions that create a nanoparticle within the spot. The size of the spot determines the size of the resulting particles. By using organometallic precursors, ceramic and metallic particles as small as 2 nm can be produced. The technique can be used to produce a large volume of material for industrial applications.

Sol-gel deposition

Ultrafine particles, nanothickness films, and nanoporous membranes can be made by *sol-gel processing* (see Figure 8.5). The starting point is a solution of precursors in an appropriate solvent. The precursors are usually inorganic metal salts or metal-organic compounds such as alkoxides—metal ions with an organic ligand such as $Ti(OC_4H_9)_4$. The precursor is subjected to a polymerization reaction to form a colloidal suspension, or "sol," of discrete, finely dispersed particles kept in suspension by adding a surfactant. The suspension can be treated to extract the particles for further processing, or it can be cast or spin-coated onto a substrate. There it is converted to a gel by chemical treatment to disable the surfactant to create an extended network of connected particles throughout the solution, making a kind of superpolymer, one enormous molecule in the form of an open 3-D (or, on a surface, a 2-D) network—the "gel." Evaporation of the solvent then leaves a dense or nanoporous film. Sol-gel methods are the basis for a wide variety of materials, including paints, ceramics, cosmetics, detergents, and cells.

Molecular self-assembly

Molecular self-assembly methods rely on the self-organization of organic molecules. The most obvious is that of crystallization: Cool a saturated solution of sugar or salt and the molecules self-assemble into crystals. Nature uses self-assembly in infinitely subtler ways; indeed, the whole of the natural world is self-assembled. Perhaps the most remarkable of all is the self-assembly (and self-

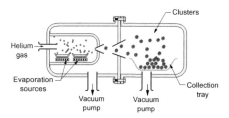

FIGURE 8.3

Cluster formation by vapor phase expansion from an oven source.

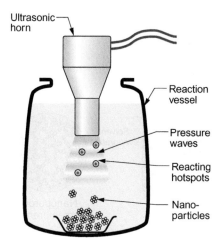

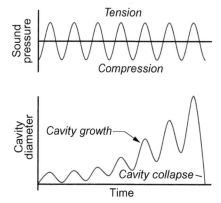

FIGURE 8.4

Sonochemical processing. The ultrasound induces cavitation. Cavity collapse causes the reagents to react.

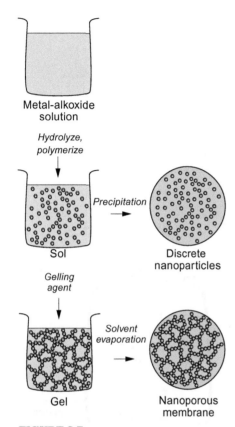

FIGURE 8.5

Sol-gel processing can be used to make nanostructured layers and coatings as well as nanoporous membranes.

disassembly) of DNA that passes genetic information from cell to cell, instructing them how to self-assemble. Surely we, as designers and engineers, could use this path, too.

We are only just beginning to exploit its potential. The idea is to create the conditions under which atoms or molecules will self-assemble into useful structures, driven by the minimization of their energy. The great advantage of self-assembly is that the system converges to a specific configuration without the need for further control. Typically, the aggregates formed by self-assembly tend to be bonded by relatively weak bonds with binding energies only a few times larger than kT, the thermal energy per atom.[1] The self-assembling molecules form *micelles*. Micelles are aggregates of amphiphilic molecules—molecules with one end that is soluble in water and the other end that rejects it. These aggregates form spontaneously at a size that depends on the concentration of the amphiphilic molecules in solution. The center of the micelles acts as a chamber for chemical reactions and thus dictates the size of the nanoparticles created (see Figure 8.6).

Self-assembly of 2-D nanofilms is made possible by using the wonderfully named *Langmuir-Blodgett* technique (see Figure 8.7). A monolayer of a fatty acid is created on the surface of water into which a substrate has been placed. Fatty acid molecules have a hydrophilic part and a hydrophobic part. The polar part locks into the water, whereas the nonpolar part rejects it. The fatty acid self-spreads across the water surface as a monolayer; any other arrangement would bury one or another end of molecules where they did not want to be. The substrate is slowly withdrawn from the water. The monolayer sticks to it and is transferred from the water surface to the substrate. The process allows the fabrication of single monolayers of materials as well as thicker films by repeatedly dipping and withdrawal.

Methods for Making 1-D and 2-D Nanomaterials

The production route for 1-D rodlike nanomaterials by liquid-phase methods is similar to that for the production of nanoparticles. Self assembly methods use the highly anisotropic bonding nature of asymmetric molecules to cause them to self-assemble into tubes rather than spheres, forming cylindrical micelles. The amphiphilic molecules are then removed with an appropriate solvent or by calcining to obtain individual nanowires.

---

[1] Here k is Boltzmann's constant and T is the absolute temperature.

CVD methods have been adapted to make 1-D nanotubes and nanowires. Catalyst nanoparticles are used to promote nucleation. To make carbon nanotubes, for example, a combination of carboncontaining gasses, such as methane (CH_4) and/or carbon monoxide (CO_2), are reacted in the presence of Iron (Fe), Cobalt (Co), and Nickel (Ni) catalysts at 1100°C. Decomposition of the gasses releases free carbon atoms that condense on the substrate with its array of catalyst particles, from which the carbon nanotubes grow. In an alternative process, an arc is generated between two electrodes, one of which is carbon. The arc creates high temperatures, causing the vaporization of the carbon electrodes into a plasma. The arc is typically operated in a gas environment, such as nitrogen or helium, and is generated by an electrical current passing through the electrodes, which causes the ionization of gas atoms. The ion beam produced is directed to a substrate carrying the Co, Fe, or Ni catalyst particles from which the carbon nanotubes grow.

Nanowires of other materials such as silicon (Si) or germanium (Ge) are grown by vapor-liquid-solid (VLS) methods. Typically a catalyst seed (gold, for example) is deposited on a substrate. Subsequently a vapor phase with the desired composition is brought in contact with the seed at a controlled temperature. The vapor diffuses into the catalyst, changing its composition and lowering its melting point until it melts. The liquid surface has a large coefficient of accommodation and therefore acts as a preferred site for absorption of the gas vapor. The liquid becomes supersaturated and a solid nanowire or "whisker" then grows from it with a diameter equal to the diameter of the catalyst seed. Nanowires with a diameter around 10 nm and a length of 1 micron can be grown in this way.

We have just encountered two ways of making layers of nano thickness: sol-gel methods and the method of Langmuir-Blodgett films. There are others, described in the following subsections.

Foil beating

Gilding is the art of applying gold leaf to a surface. It was known to the ancients, is mentioned in the Old Testament, and was described by Herodotus in 420 BC. Gold is an expensive material; there are obvious incentives to use as little of it as possible. Gold leaf is made by beating the metal to a uniformly thin foil, roughly 100 nm thick (see Figure 8.8). It is applied by smoothing it onto the adhesive-coated surface of the object to be gilded.

There are less labor-intensive ways to make thin films. Electrodeposition is one, and it is one able to create nanoscale structures as well.

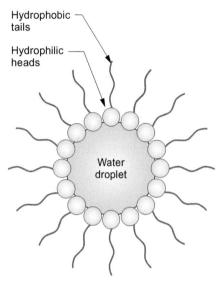

FIGURE 8.6

Self-assembly of nanoclusters. The enclosed droplet can be spherical or cylindrical. Reaction within it generates a nanoparticle or nanorod.

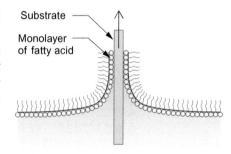

FIGURE 8.7

Self-assembly of 2-D nanofilms. Langmuir-Blodgett film formation and extraction.

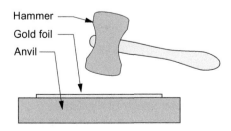

FIGURE 8.8

Beating of gold leaf.

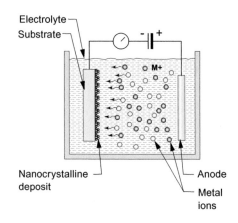

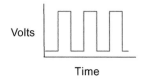

Volts

Time

FIGURE 8.9

Pulsed electrodeposition. The pulsed potential nucleates many nanoscale crystals that build to form a layer up to 5 mm thick.

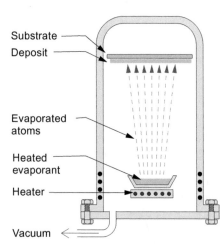

FIGURE 8.10

Physical vapor deposition (PVD). Material, evaporated by heating, by ion bombardment, or by laser ablation, is deposited on a substrate target. The deposited layer can be of nano thickness.

Electrodeposition

Electrodeposition (see Figure 8.9) is a long-established way to deposit metal layers on a conducting substrate. Ions in solution, sometimes replenished from an anode, are deposited onto the negatively charged cathode, carrying charge at a rate that is measured as a current in the external circuit. The process is relatively cheap and fast and allows complex shapes. The layer thickness simply depends on the current density and the time for which the current flows. The deposit can be detached if the substrate is chosen to be soluble by dissolving it away.

The challenge is to control the structure. Most electrodeposits grow in a columnar manner: Metal crystals nucleate on the bare substrate when the current is switched on and grow outward as interlocking pillars, like a miniature Giant's Causeway. To make nanoparticles or layers, it is necessary is to stop each crystal's growth while it is still tiny and to nucleate more. A combination of tricks is used to do this. Pulsing the voltage, as suggested by the graph in Figure 8.9, causes little bursts of crystal growth. Adding growth inhibitors that condense on the crystal surfaces during the "off" phase of the pulse discourages their continued growth during the next "on" phase. These methods, combined with a high current density, nucleate many crystals but allow little time for each to grow, giving a nanostructured deposit.

The technique can yield porosity-free nanocrystalline deposits as thick as 5 mm that require no further processing.

Physical vapor deposition (PVD)

In *PVD plating*, a thin layer of a material, usually a metal, is deposited from a vapor onto the object to be coated (see Figure 8.10). The vapor is created in a vacuum chamber by direct heating or electron beam heating of the metal, from which it condenses onto the cold substrate, much like steam from a hot bath condensing on a bathroom mirror. In PVD plating there is no potential difference between bath and work piece. In *PVD ion plating* the vapor is ionized and accelerated by an electric field (the work piece is the cathode, and the metallizing source material is the anode). In *PVD sputtering*, argon ions are accelerated by the electric field onto a metal target, ejecting ions onto the component surface. By introducing a reactive gas, compounds can be formed. (Sputtering titanium in an atmosphere of nitrogen, for instance, gives a coating of hard TiN.)

Almost any metal or compound that doesn't decompose chemically can be sputtered, making this a very flexible (though expensive) process. Targets can be changed during the process, allowing multilayers to be built up. These multilayers can have remarkable mechanical and electronic properties.

Chemical vapor deposition (CVD)

In *CVD processing*, a reactant gas mixture is brought into contact with the surfaces to be coated, where it decomposes, depositing a dense pure layer of a metal or compound (see Figure 8.11). The deposit can be formed by a reaction between precursor gases in the vapor phase or by a reaction between a vapor and the surface of the substrate itself. A difficulty with the process is that it frequently requires high temperatures, 800°C or more.

In a variation of conventional CVD called *moderate temperature CVD* (MTCVD), metal organic precursors are used (hence its other name, *metal organic CVD*, or MOCVD). As they decompose at relatively lower temperature, the reaction temperature is typically around 500°C. If the chemical reactions in the vapor phase are activated by the creation of a plasma in the gas phase or by shining a laser beam into the gas mixture, deposition at an even lower temperature, a little above room temperature, becomes possible. These techniques are called *plasma-assisted* (or *plasma-enhanced*) *CVD* (PACVD or PECVD) and *laser CVD* (LCVD).

Coatings formed with PCVD methods are typically nanocrystalline or amorphous because their formation is no longer dependent on equilibrium thermodynamic constraints.

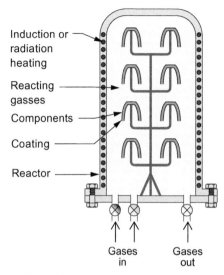

FIGURE 8.11

Chemical vapor deposition (CVD). Reaction of gases in the reaction chamber causes deposition of the reaction product as a layer that can be of nano thickness.

Methods for Making 3-D Nanomaterials

Many bulk materials that we use today derive their properties from internal structure at the nanoscale. High-strength, low-alloy (HSLA) steels get their strength from nano-dispersed carbides. All the alloys of aerospace—based on aluminum, magnesium, and titanium—are strengthened by precipitates with dimensions in the 10–100 nm range. The copper alloy wiring connectors in automobiles and other electrical equipment retain their springiness, maintaining good electrical contact, because of nanoparticles of CuBe. All of these (and there are many others) are conventional alloys containing a dispersion created by heat treatment. The dispersion occupies only a small fraction, typically 1–5%, of the volume. The matrix—the steel, the aluminum, the copper—is not itself nanocrystalline.

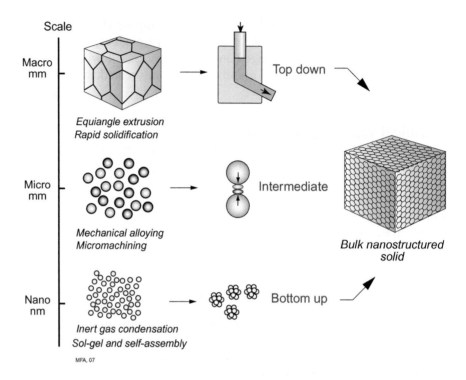

FIGURE 8.12

The top-down, intermediate, and bottom-up approaches to making bulk nanostructured solids.

Suppose you could make it so? It's not easy. But materials scientists like a challenge, and they have risen to this one.

There are three directions of attack, sketched in Figure 8.12. The first is to break down the structure of a bulk material, reducing the crystal size to submicron or nanodimensions—the *top-down* approach. The second—the *intermediate* approach—starts with standard micron-scale particles, reducing their structure to the nanoscale by milling techniques. The third is to build the solid up from the atomic scale or from nanoclusters in a way that retains the scale of is structural units; this is the *bottom-up* approach.

Top-Down Processes

Top-down processes are best described this way: Take the material and drug, beat, and freeze it. Drug it by alloying. Beat it by severe plastic deformation. Freeze it to stop the fine-scale structure, once formed, from coarsening.

Rapid solidification

Materials would rather be crystals than glasses. If ordered, all atoms sit at exactly the distance from their neighbors that best satisfies their interatomic bonds. Disorder disrupts this comfortable seating arrangement, stretching some bonds and squeezing others. Liquids are disordered because heat shakes the atoms so violently that they are sprung from their low-energy, crystalline arrangement. The melting point is the temperature at which this disruption occurs; below it the thermal shaking is too weak to disrupt bonds, and the liquid crystallizes.

Cooled at normal rates, most liquids solidify to give solids with large crystals, or "grains." The trick in making nanocrystalline or amorphous materials by casting them is to deprive the material of the time or the means to transform from liquid to solid. Some materials, of which window glass is one, are easily duped into retaining their glassy structure—their high viscosity when liquid slows the rearrangement of the molecules to form crystals. Many polymers, too, are "glassy" because their tangled molecules cannot reorganize in any normal timeframe to form the crystal they would like to be. Metals and ceramics, by contrast, crystallize at the drop of a hat. It takes extreme measures to make them retain their liquidlike structure or adopt a structure with exceedingly small grains.

The first step is to mix in elements with different-sized atoms, each preferring a different atomic spacing and crystal structure, making crystallization difficult. The second step is to cool quickly, leaving little or no time for crystallization. Early alloys required precipitous cooling rates exceeding 1,000,000°C/sec, limiting the form to thin wires and ribbons from which heat can be conducted quickly.

Figure 8.13 shows one way of achieving such cooling rates. A jet of liquid alloy—one designed to be hard to crystallize—is squirted onto a spinning, water-cooled copper drum. The process is called *melt spinning*. The liquid layer cools fast enough to become amorphous or, if not that, then nanocrystalline. The laminated cores of many transformers and the read/write heads of magnetic tape and disc recorders are made that way, exploiting the special magnetic properties of the amorphous state. Newer bulk-amorphous metals (BAMs) remain glassy even at relatively slow rates of cooling (10°C/sec), allowing thick sections (up to 20 mm) to be cast.

Figure 8.14 shows a second way of cooling a liquid fast: Zap it with a laser beam, scanning the beam fast enough that it melts only a very thin surface layer. The layer is already stuck to the cold material beneath. Conduction of heat into this cold substrate is fast enough to trap the amorphous structure. This *laser surface melting* is the way amorphous silicon for cheap solar cells is made. The same technique is used to make nanocrystalline surface layers.

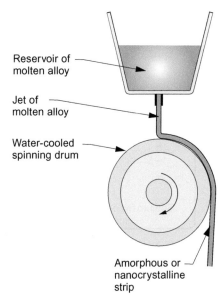

Reservoir of molten alloy

Jet of molten alloy

Water-cooled spinning drum

Amorphous or nanocrystalline strip

FIGURE 8.13

Melt spinning allows cooling rates up to 1 million degrees Centigrade per second. This is enough to freeze in the liquid structure in certain alloys, giving amorphous or nanocrystalline wires or ribbons.

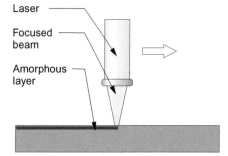

Laser

Focused beam

Amorphous layer

FIGURE 8.14

Laser surface hardening. The laser beam melts an exceedingly thin layer, which then cools so fast by conduction that it becomes amorphous or nanocrystalline.

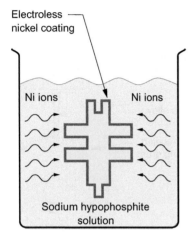

FIGURE 8.15

The electroless nickel process. When a part with a suitably activate surface is plunged into the bath, a hard, amorphous, nickel-phosphorous layer plates onto its surfaces.

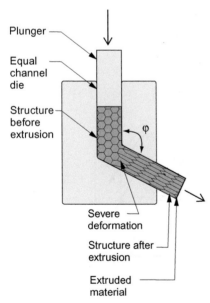

FIGURE 8.16

Equiangle extrusion. Each pass shears and breaks up the structure; repeated, the process reduces it to the nanoscale.

There are other ways to make amorphous structures. One is to grab the atoms or molecules not from the melt but from solution. *Electro-deposition*, already described (Figure 8.9), can be driven at such a rate that the ions, dragged to the surface of the cathode (negative electrode), are hurled onto it so fast that they have no time to rearrange and build crystals. More remarkably, the deposition can be done, in one case, without any electric current. *Electroless nickel* (really a nickel-phosphorous alloy) deposits spontaneously onto almost anything plunged into a bath containing nickel ions and a reducing agent, sodium hypophosphite (see Figure 8.15). The deposit is uniform (unlike one made by electrodeposition) and is exceptionally hard and resistant to wear and corrosion. And, crucially, it is cheap.

Equiangle extrusion

The second approach is that of extreme plastic deformation. Figure 8.16 shows *equiangle extrusion*—a way of generating extreme deformation while suppressing fracture. A rod, forced through the die, is savagely sheared and extruded, emerging as a rod with the same diameter as that with which it started but with a much refined structure. It can then be reinserted into the die and tortured further until the structure is sufficiently refined. The process is now a standard one for making metals with grain sizes in the 100–500 nm range.

Intermediate Processes
Milling and mechanical alloying

Equiangle extrusion leaves the composition unchanged. Milling, shown in Figure 8.17, combines extreme deformation with the forcible alloying of two materials that, normally, would not mix. Particles (here, two metals, A and B) are spun in a high-energy ball mill. The heavy steel or carbide balls are thrown against each other, trapping and squashing the metal particles between them; the particles flatten, weld, and then break up. The process creates heavily deformed, mechanically alloyed particles with (if continued long enough) a nanoscale internal structure. It helps to fill the mill with inert gas to promote welding and to cool it (*cryomilling*) to prevent the structure coarsening. The particles are subsequently compacted and sintered to make products.

Micromachining

There is another way to torture materials on a small scale. It is a familiar one: machining. When you look at a machined object, you see the technical precision, the clean, engineered quality, that distinguishes machined products form those that are merely molded or cast. What you don't see is the swarf—the machined shreds, chips, and tangles. These have suffered extreme deformation; their structure (like those of the last two processes) is refined by it. There is an interesting scaling law here: The finer the cut, the thinner the chip and the greater the shear it has suffered. Micromachining—machining on the scale of watch making, as in Figure 8.18—deforms the chips more than ordinary coarse machining. Better yet is machining with a diamond tip with submicron radius. Diamond nanomachining is designed for surface profiling, but the swarf it produces can be useful too. As with particle milling, a way must be found to consolidate the swarf into useful products. We get to that next.

Bottom-Up Processes

Consolidation of nanoclusters and milled powders

Most of the processes described thus far do not make solid objects; they make clusters, powders, or chips. The long-established way to consolidate powder is to press in a die that has the desired form, then heat to a temperature at which diffusional bonding takes place. This *powder-pressing and sintering* route to manufacturing products (see Figure 8.19) is widely used to make engine parts for cars and components for household appliances such as washing machines. The difficulty in using it to consolidate nanostructured particles is that sintering takes time, and at the sintering temperature, the structure coarsens. It is not an easy problem to solve; during consolidation there will always be some coarsening. The question is how to minimize it. One way is to compact the powder in such a clean environment that the particles bond, even at room temperature, but that is seldom possible. An alternative is to sinter so fast that there is little time for coarsening. Figure 8.20 shows one way to do this—that of *flash sintering*. The powder (or swarf) is compressed in a die through which a bank of capacitors is discharged. The blast of heat, generated by the resistance of the packed powder, is enough to create good bonding without leaving enough time for serious coarsening.

Methods for Nanoprofiling

Often it is not a nanomaterial that is sought; it is nanofeatures on the surface of something much bigger. These are created by micromachining (cutting material away) or by microlithography (putting material where you want it).

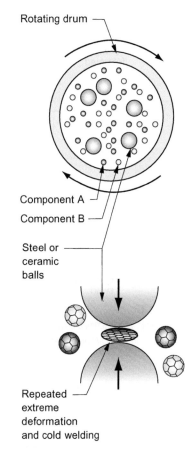

Rotating drum

Component A
Component B

Steel or ceramic balls

Repeated extreme deformation and cold welding

FIGURE 8.17
Powder milling with mechanical alloying. The heavy steel or tungsten carbide balls trap, deform, weld, and break up the powder particles, mixing them so completely that they become alloys with a nanoscale structure.

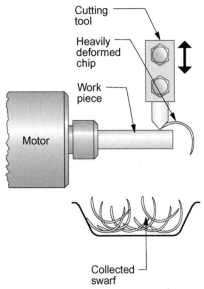

FIGURE 8.18

Micromachining. The machined chips are heavily sheared, reducing the scale of their structure.

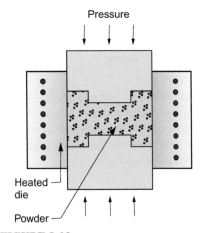

FIGURE 8.19

Pressure sintering, the standard way to consolidate powders. Heating is slow, allowing time for the structure to coarsen.

Micromachining

Intricate 2-D and 3-D patterning of materials is made possible by techniques derived for those of conventional machining. Table 8.1 lists these with their approximate resolution limits. The current resolution of micromachining (see Figure 8.21) and micro EDM (electro-discharge machining), the methods of watch makers, is 3–5 microns. Machining with focused electron or laser beams (see Figures 8.22 and 8.23) allows submicron resolution with high removal rates. Focused ion-beam (FIB) machining offers the greatest resolution, with the ability to make features as small as 20 nm, but it is very slow. In FIB machining a beam of gallium ions from a liquid metal ion source is accelerated, filtered, and focused with electromagnetic lenses to give a spot size of 5–8 nm (see Figure 8.24). The beam is tracked across the surface to be machined, contained in a chamber under high vacuum. The high-energy ions blast atoms from the surface, allowing simple cutting of slots and channels or the creation of more elaborate 3-D shapes. Secondary electrons are emitted when the gallium ions displace the surface atoms. These can be used to image the surface, allowing observation and control of the process as it takes place. Dual-beam FIBs have an additional electron gun that is used as an alternative way of imaging. The precision is extraordinary, but the process is very slow.

Table 8.1 Micromachining Methods				
Machining Method	**Materials That Can Be Machined**	**Feature Size (and Tolerance)**	**Positional Tolerance**	**Material Removal Rate, Microns³/sec**
Micromachining	Metals, polymers	10 microns (2 microns)	3 microns	10,000
Micro electrodischarge machining (EDM)	Any conducting material	10 microns (3 microns)	3 microns	2,500,000
Electron beam machining (EBM)	Any conducting material	5 microns (submicron)	1 micron	100,000
Femto-second laser machining (LBM)	Any material	1 micron (submicron)	Submicron	13,000
Focused ion-beam machining (FIB)	Any material	0.2 microns (0.02 microns)	0.1 microns	0.5

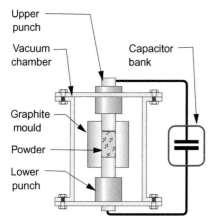

FIGURE 8.20
Electric discharge, or "flash," sintering. The rapid consolidation leaves little time for structural coarsening.

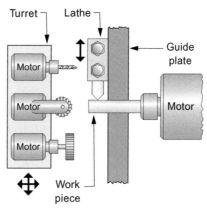

FIGURE 8.21
A micromachining workstation. The resolution limit is about 5 microns.

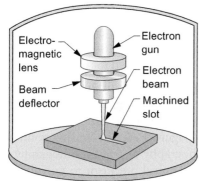

FIGURE 8.22
Electron beam machining. The resolution limit is about 1 micron.

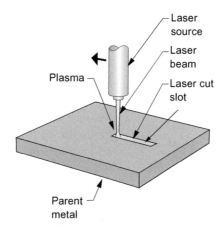

FIGURE 8.23
Machining using a femto-second laser. The extremely short pulse time allows a machining resolution around 1 micron.

Photolithography

Photolithography, the process used to manufacture computer chips, is able to produce features smaller than 100 nm (see Figure 8.25). An oxidized silicon wafer is coated with a 1μm thick photoresist layer, as in Figure 8.25a. A beam of UV light is incident on a mask, allowing light to pass through the gaps. The light is then passed through a set of lenses to reduce the pattern size, which is projected onto the wafer, causing a photochemical reaction where it strikes

High-vacuum chamber

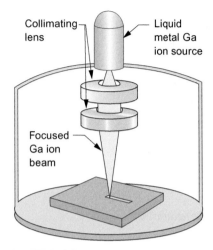

FIGURE 8.24

Focused ion beam machining. The beam is focused to a spot of 5–7 nm diameter, allowing nanoscale shaping.

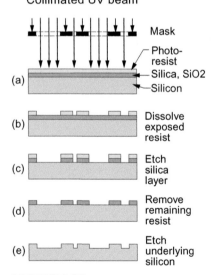

FIGURE 8.25

Photolithography. A beam of UV light activates the photoresist, transferring the pattern from the mask to the sample.

the resist. The exposed parts of the resist are dissolved during the developing stage, as shown in Figure 8.25b, forming a replica of the mask pattern. The assembly is then placed in an acidic solution, which attacks the silica but not the resist or the silicon (Figure 8.25c). Once the silica has been removed, the resist is dissolved in a different acidic solution (Figure 8.25d). Further etches remove silicon from the exposed areas, creating channels (Figure 8.25e). The resulting nanofeatured chip may then be processed further to make it electronically active or be used as a template for soft lithography, described in a moment. Though the concept of photolithography is simple, the implementation is complex and expensive. Masks need to be perfectly aligned with the pattern on the wafer. The silicon wafer has to be a perfect single crystal, almost defect-free.

UV light has a wavelength of 250 nm, giving a limiting feature size, set by diffraction effects, of about 100 nm. Greater resolution is possible with electron-beam lithography and X-ray lithography. In electron-beam lithography, the pattern is written in a polymer film with a beam of electrons. The shorter wavelength of the electrons allows features with a smaller scale than is possible with UV light, but the technique is slow and expensive. In X-ray lithography, diffraction effects are minimized by the short wavelength (0.1–10 nm), but conventional lenses are not capable of focusing X-rays, and the radiation damages many of the materials used for masks.

Because of these limitations, the most recent lithography methods make use of mechanical processes—printing, stamping, molding, and embossing—instead of photons or electrons. Two of these *soft lithograph processes* are shown in Figure 8.26. The starting point is a silicon mold made by photolithography or e-beam lithography, as in Figure 8.25. Subsequently a chemical precursor to polydimethylfiloxane (PDMS) is poured over and cured into the rubbery solid PDMS stamp that reproduces the original pattern. The stamp can then be used in various inexpensive ways to make nanostructures. In the case of microcontact printing the stamp is inked with a solution consisting of organic molecules and then pressed into a thin film of gold on a silicon plate, as shown on the left of Figure 8.26. The organic molecules form a self-assembled monolayer on the solid surface that reproduces the pattern with a precision of approximately 50 nm. In micromolding, the PDMS stamp is placed on a hard surface and filled with a liquid polymer, as on the right in Figure 8.26. The filled mold is then pressed onto the surface of a silicon or other wafer; the polymer is polymerized and the pattern is transferred to the wafer with a resolution approaching 10 nm.

The advantage of using soft lithography methods is that, once the master template has been made, no special equipment is required.

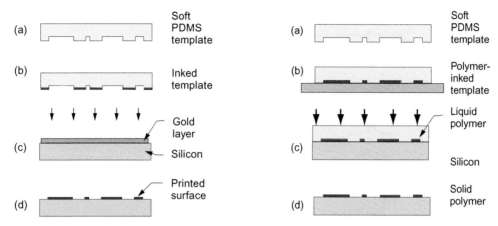

FIGURE 8.26

Two examples of soft lithography processes. In both cases a stamp of PDMS is first fabricated by molding it from a silicon template made in the way shown in Figure 8.25. Subsequently the stamp can be used for producing nanostructures following the microcontact printing method (left figure) or the micromolding-in-capillaries method (right figure).

Soft lithographic methods are capable of producing nanostructures in a wide range of materials and can print or mold on curved as well as planar surfaces. They are not, however, ideal for nanoelectronics, because circuits require stacked layers of different materials, and distortion of soft PDMS stamps can produce errors in the patterns. The large (and expanding) number of processes that manipulate materials at the nanoscale are asssembled in the final synthesis diagrams shown in Figure 8.27 and 8.28. Figure 8.27 summarizes processes for making nanoclusters, nanolayers and nanofilms, as well as methods for nanoprofiling to make nanoscale features. Other processes are under development. Figure 8.28 summarizes processes that are appropriate for different nanomaterial forms, including nanoparticles, nanowires and nanotubes, nanofilms, and bulk forms; and broadly suggests their relative scales. Figure 8.1 shown earlier summarizes processes according to the 0-D, 1-D, 2-D and 3-D classification system discussed earlier.

8.2 CHARACTERIZATION OF NANOMATERIALS

In general, the role of characterization techniques is to establish a correlation between the structure, shape, and chemical composition of nanomaterials obtained in processing, with their properties. In the case of nanomaterials, the important aspects to consider about characterization methods are the type of information and the resolution achieved by each technique. For each nanomaterial

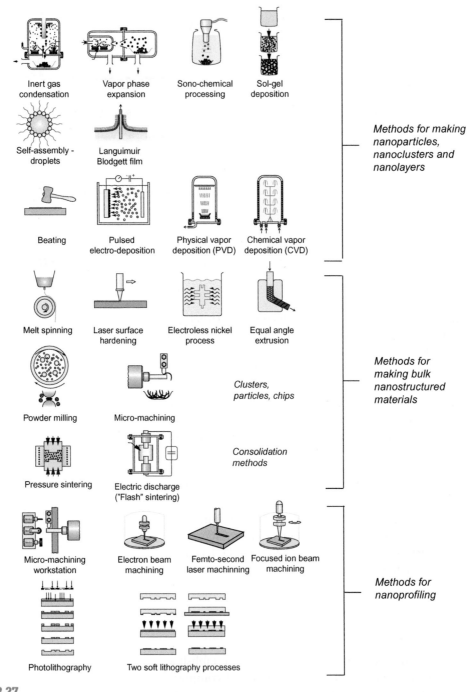

Inert gas condensation

Vapor phase expansion

Sono-chemical processing

Sol-gel deposition

Self-assembly - droplets

Languimuir Blodgett film

Beating

Pulsed electro-deposition

Physical vapor deposition (PVD)

Chemical vapor deposition (CVD)

Methods for making nanoparticles, nanoclusters and nanolayers

Melt spinning

Laser surface hardening

Electroless nickel process

Equal angle extrusion

Powder milling

Micro-machining

Clusters, particles, chips

Methods for making bulk nanostructured materials

Pressure sintering

Electric discharge ("Flash" sintering)

Consolidation methods

Micro-machining workstation

Electron beam machining

Femto-second laser machinning

Focused ion beam machining

Methods for nanoprofiling

Photolithography

Two soft lithography processes

FIGURE 8.27
Summary of synthesis methods discussed in Section 8.1.

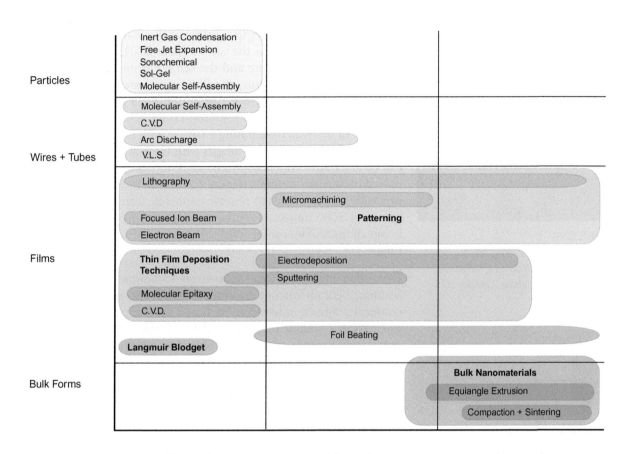

FIGURE 8.28
Summary of processes used to make nanomaterials in relation to scale.

type (0-D, 1-D, 2-D, and 3-D), one or more techniques can be used to extract the desirable information. A simple way of categorizing the characterization methods is to consider imaging and analytical techniques. Imaging involves some kind of microscopy, whereas analysis involves some type of spectroscopy.

Let's begin with the imaging techniques. These methods can involve light, electrons, ions, or scanning probes. With respect to light microscopes, the most common approach for imaging nanomaterials is the so-called *near-field scanning optical microscope* (NSOM; see Figure 8.29). In this type of microscopy a subwavelength aperture (smaller than the wavelength of light) is placed in close proximity to the sample (~10 nm). In addition, a light source is used as a scanning probe. To achieve good resolution, the probe is highly focused. The

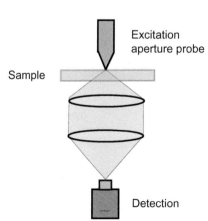

FIGURE 8.29
Schematic diagram of a near-field scanning optical microscope.

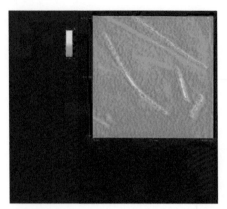

FIGURE 8.30

Transmission near-field scanning optical microscope image of 18 diameter, rod-shaped tobacco mosaic virus particles. (Courtesy of George J. Collins.)

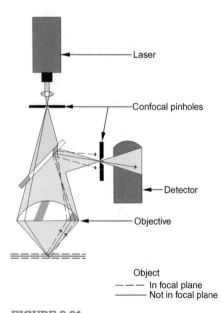

FIGURE 8.31

Schematic diagram of a confocal scanning light microscope.

probe is then scanned over the surface of the material at distances of a few nanometers above the surface (near field). Because the distance between the aperture and the sample is much smaller than the diameter of the aperture, the illuminated area will be a function of the aperture dimensions and not limited by the diffraction limit. In other words, NSOM resolution is limited by the size of the aperture. In this way, by scanning a sample with the "near field" of a focused light source, optical images with resolution around 50 nm can be generated. Normally, fluorescent, topographical, and transmitted images are available (Figure 8.30). Currently, NSOM is used in a wide range of fields to observe nanoscale features. In addition, because NSOM images the surface point by point, a topographical image of the surface can be resolved at the same time as the optical image.

One other important light microscopy technique is the confocal scanning light microscope. Invented by Marvin Minsky in the 1950s, confocal scanning microscopy is based on the principle that the existence of a pinhole in front of a detector only allows the traveling signal from the focal plane of the objective to enter the detector.

As shown in Figure 8.31, the light source is a laser that produces high-intensity, coherent light of a defined wavelength. The light is guided into the objective lens through an excitation aperture (called a *pinhole*), producing a sufficiently thin laser beam. The objective projects this pinhole into the focal plane within the specimen. The imaging signal is then mirrored through a pinhole aperture, which is in a confocal position to the objective focal plane. This configuration selects only the signals from the focal plane of the objective, that is, with a very small depth of field. In other words, confocal scanning microscopy makes it possible to scan a sample at various *x-y* planes corresponding to different depths, and, by organizing these planes into a vertical stack, reconstruct a 3-D image of the specimen. Because sectioning of thick samples and consequent damage is not necessary, this technique is one of the most attractive approaches to obtain 3-D information on materials (Figure 8.32). This is a rapidly advancing technique used to produce sharp and precise images of thick specimens in fluorescent and reflective light modes by "optical sectioning." In practice, the lateral resolution of a confocal microscope is about 100–200 nm, and the vertical resolution is about 400–500 nm.

Next we discuss the use of electron microscopy. This is perhaps the most widely used technique for the characterization of nanomaterials. The main difference with respect to light microscopy is the wavelength of electrons, which is much shorter than the wavelength of photons. Therefore the image resolution provided by electrons is significantly improved with respect to photons. In addition, electrons interact much more efficiently with matter, which requires an electron beam to be under high vacuum to reduce scattering. Finally, electrons have charge, which allows the use of a magnetic field to drive and focus the electron beam.

Within the class of electron microscopy, there are several techniques. The most common are scanning electron microscopy (SEM), transmission electron microscopy (TEM), and scanning transmission electron microscopy (STEM). However, before we address these techniques in detail, it is important to discuss the common available electron sources and the typical interactions between matter and an electron beam.

In terms of electron sources, there are basically two types: thermionic and field emission emitters. The available thermionic emitters are tungsten or lanthanum hexaboride (LaB_6) filaments, which emit electrons by overcoming the surface potential barrier when thermal energy is provided in sufficient amounts. The LaB_6 filament provides increased brightness but is more sensitive to thermal shock and is much more expensive.

The field emission emitters follow a different principle. In this case, a very sharp (nanoscale dimensions) tungsten tip (see Figure 8.33) is subjected to a very high electrical field to reduce the surface potential barrier. This results in a very narrow probe and increased brightness, which allows images with enhanced contrast and resolution to be obtained. The development of these field emission guns (FEGs) has been crucial for the field of nanomaterials and nanotechnologies.

Once the electron beam is generated and accelerated toward the specimen, the image produced is strongly correlated with the type of signal collected, which is dependent on the type of interactions occurring between the electron and the sample. Typically, several signals can be produced (see Figure 8.34). For imaging the SEM uses both secondary and backscattered electrons. The TEM and the STEM utilize transmitted electrons as well as inelastically and elastically scattered electrons.

More specifically, the SEM is primarily used for imaging the surface of materials. The samples observed have, typically, dimensions up to $1 \times 1 \times 1$ cm and can be made of any material. However, if the material is not conductive, a thin coating of gold or carbon is applied to

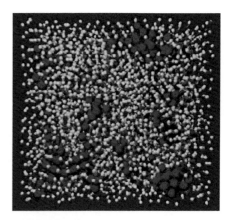

FIGURE 8.32

A 3D region of the sample imaged using a laser scanning confocal microscope. Particles surrounded by crystal-like material are represented by red spheres, whereas the yellow spheres represent particles in the metastable liquid. (Image courtesy of NASA's Fluids and Combustion Facility for the International Space Station.)

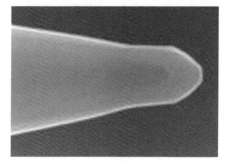

FIGURE 8.33

Field emission gun tip made of tungsten.

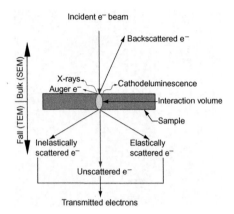

FIGURE 8.34

Schematic showing the various electron beam–specimen interactions.

FIGURE 8.35

Scanning electron microscope (SEM). (Courtesy of Hitachi High Technologies)

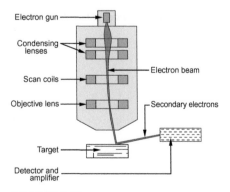

FIGURE 8.36

Schematic diagram of the operation of an SEM.

avoid electron charging and image degradation. The basic layout of an SEM is shown in Figure 8.35. Electrons are produced with an electron gun. (For the study of nanomaterials, the gun is normally an FEG.) The electrons are then accelerated, usually with a voltage between 1 kV and 30 kV, and demagnified by a set of two condenser lenses. Subsequently a set of scanning coils force the electron beam to rapidly scan over an area of the specimen while the magnetic lenses focus the beam on the sample (see Figure 8.36). The recent FEG guns are capable of producing a final probe diameter of 0.4 nm at 30 kV.

The specimen can be viewed in secondary mode or backscattered mode. In secondary mode, low-energy secondary electrons (around 50 eV) are emitted from the surface or subsurface of the specimen, down to 30 nm, due to the interaction with the incident electron beam, and are collected by a detector. As a result, secondary electron imaging is mainly used for shape and topographic identification of nanostructures (see Figures 8.37 and 8.38). In backscattered mode, high-energy reflected or backscattered electrons are emitted from a depth down to around 1 μm, with practically no change in kinetic energy, and collected by a detector. Because the backscattering energy is strongly dependent on nuclear interactions, the SEM backscattering mode is quite sensitive to the atomic number.

Therefore backscattering mode is normally used to detect changes in chemical contrast, which in many cases correspond to the existence of different phases (Figure 8.39) However, the resolution in the backscattered mode is inferior (around 10 nm) to that in secondary mode because of the larger penetration depth from which the electrons are emitted. A variation of the backscattering mode is called *electron backscattering diffraction* (EBSD), used to determine variations in crystal orientation within the sample. The transmission electron microscope (see Figure 8.40) is rather different from the scanning electron microscope. First, it operates at considerably high voltages (100 kV to 3 MV). Typical microscopes use accelerating voltages of 120, 200, and 300 kV. Second, the specimens are normally only 50–100 nm thick because the image is formed by electrons that transverse the sample. Higher accelerating voltages allow the observation of thicker samples and improve the resolution due to the reduction in electron wavelength with accelerating voltage. However, the resolution limit achieved in a TEM is determined by various aberrations and not by the accelerating voltage.

The basic layout of a TEM is shown in Figure 8.41. The electron beam generated by the gun is demagnified through the first condenser lens; the second condenser lens converges the probe at the

FIGURE 8.37

Secondary electron SEM images of poly(lactic-co-glycolic acid, or PLGA, nanoparticles. (Courtesy of Charlesson LLC.)

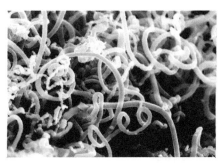

FIGURE 8.38

Secondary electron SEM image of carbon nanotubes. (Courtesy of P. J. Ferreira, University of Texas at Austin; J. B. Vander Sande, MIT; and P. Szakalos, Royal Institute of Technology, Sweden.)

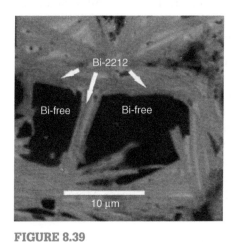

FIGURE 8.39

SEM backscattered electron image of a Bi-2212 high temperature superconductor. The black regions are Bismuth-free impurity phases, the dark-gray area is the Bi-2212 phase and the light gray area is the Bi-2201 phase. (Courtesy of E. Cecchetti, P. J. Ferreira, and J. B. Vander Sande, MIT.)

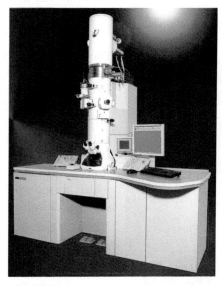

FIGURE 8.40

Transmission electron microscope (TEM). (Courtesy of JEOL.)

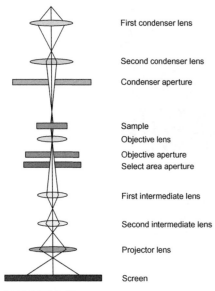

FIGURE 8.41

Schematic diagram of the operation of a TEM.

First condenser lens

Second condenser lens

Condenser aperture

Sample
Objective lens
Objective aperture
Select area aperture

First intermediate lens

Second intermediate lens

Projector lens

Screen

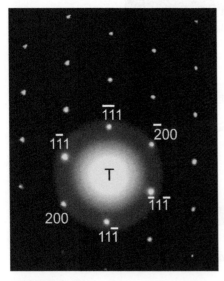

FIGURE 8.42

Electron diffraction pattern of a platinum crystal.
(Courtesy of P. J. Ferreira, University of Texas at
Austin, and Yang Shao-Horn, MIT.)

specimen and controls the spot size. Just below the second condenser lens is a condenser aperture that can be used to change the electron current and therefore the intensity of the beam as well as change the angle of beam convergence, which modifies the coherence of the beam. The specimen sits below the condenser lens and above the objective lens. The specimen position and the objective lens are the heart of the TEM, whereas the combination of the objective lens, the intermediate lens, and the projection lens determine the overall magnification of the microscope.

In general, the TEM can be operated in image mode versus diffraction mode. The first mode to master is diffraction mode. It is here that the electrons are selected to form the images. To obtain the diffraction pattern on the screen, the lens needs to be adjusted so that the back focal plane of the objective lens acts as the object plane of the intermediate lens. However, if we produce a diffraction pattern by allowing all electrons to reach the screen, the high intensity of the beam can damage the viewing screen and the information. In addition, the gathered information is not very useful, because it arises from all the area of observation.

Therefore, a selected area diffraction aperture (Figure 8.41) is inserted in the image plane of the objective lens. Depending on the size of the aperture, the area in the sample from which the diffraction pattern is produced can be controlled. The diffraction mode is very important because it can reveal whether a material is polycrystalline, single-crystal, or amorphous. For crystalline materials, the fact that atoms are regularly spaced and that the wavelength of electrons is very short causes electrons to scatter from atomic planes, which are essentially parallel to the electron beam. More precisely, electron diffraction occurs when Bragg's law is obeyed, given by

$$\frac{2\sin\theta}{\lambda} = \frac{1}{d_{hkl}} \tag{8.1}$$

where θ is the Bragg angle, λ is the electron wavelength, and d_{hkl} is the interplanar spacing between (hkl) planes. Therefore, in diffraction mode, a single crystal will produce a diffraction pattern of the type shown in Figure 8.42. The center spot T is associated with the transmitted beam (nondiffracted), whereas all other spots are diffracted spots. Each of the diffraction spots represents all the beams diffracted from a specific set of planes. For example, the 200 spot is associated with diffraction from all the (200) planes. In other words, for a single-crystal material, the diffraction pattern consists of spots distanced away from the transmitted spot by $1/d_{hkl}$. Any vector from the center spot T to an hkl spot is called a g_{hkl} vector. A g_{hkl} vector is aligned perpendicular to the orientation of the hkl plane, which runs parallel to the electron beam.

On the other hand, if the sample is polycrystalline, the individual crystals are at different orientations; thus, if and the diffraction aperture used is larger than the crystal size, then the diffraction pattern is the sum of each individual crystal. Because only certain planes can diffract, the spots are now randomly distributed but then fall on rings of constant radius (see Figure 8.43). The smaller the crystal size, the larger the number of crystals included in the diffraction aperture, leading to continuous diffraction rings. This is quite typical for nanocrystalline materials, for which the crystal size is small.

Once a diffraction pattern is produced, two types of images can be obtained, namely bright-field and dark-field images (see Figure 8.44). The bright-field image is formed by inserting the objective aperture (Figure 8.41) around the transmitted spot (Figure 8.42). In this fashion the transmitted electrons are collected, whereas most or all of the diffracted electrons are blocked. The dark-field image is formed by inserting the objective aperture around the diffraction spots to block the transmitted electrons. In both cases, once the objective aperture is placed and the diffraction aperture is removed, the microscope is changed from diffraction mode to image mode. In image mode, the image plane of the objective lens becomes the object plane of the intermediate lens.

In image mode, the transmission electron microscope can be operated under a variety of contrast mechanisms. The most general image contrast is called *mass-thickness contrast*. It arises from incoherent scattering of electrons. Because elastic scattering is a strong function of the atomic number as well as the thickness of the specimen, regions of a specimen with high mass will scatter more electrons than regions with low mass of the same thickness. Similarly, thicker regions will scatter more electrons than thinner regions. All microscopists are aware of this type of contrast because it is present in all kinds of specimens—amorphous, crystalline, biological, or metallic. However, the mass-thickness contrast is most important for amorphous materials. To enhance this type of contrast, the smallest objective aperture is selected around the transmitted spot, to reduce the number of scattered electrons.

If the specimen is crystalline, another type of contrast may play a role. This is called *diffraction contrast*. As discussed, Bragg diffraction occurs for certain crystallographic planes that are oriented closely parallel to the electron beam. Following this concept, the sample

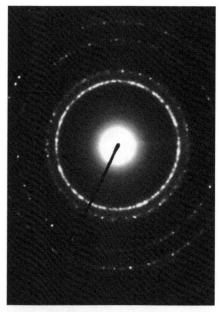

FIGURE 8.43
Ringlike electron diffraction of polycrystalline gold. (Courtesy of Structure Probe Inc (SPI).)

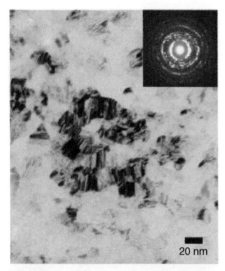

(a)

(b)

FIGURE 8.44

Transmission electron microscope images of nanocrystalline Ni-Co; (a) bright-field image and (b) dark-field image. (Courtesy of B. Y. Yu, MIT; P. J. Ferreira, University of Texas at Austin; and C. A. Schuh, MIT.)

is tilted so that only one diffracted beam is strong. In other words, a two-beam condition is formed by the transmitted spot and one diffracted spot. Because the diffracted spot corresponds to a specific set of planes, the two-beam condition contains specific information in addition to the general scattering information. This type of contrast is mostly used to image crystalline defects such as dislocations, precipitates, and grain boundaries (see Figure 8.45).

Finally, the TEM can also be operated in phase contrast. The most obvious difference between phase contrast and diffraction contrast is the number of electron beams collected by the objective aperture. As we discussed, diffraction contrast is optimal when a single diffracted beam is selected. On the other hand, a phase-contrast image requires the selection of more than one diffracted beam.

In fact, phase-contrast images are formed due to the interference of multiple beams at the exit of the specimen. The result is a series of sinusoidal oscillations in intensity, normal to the diffracting vector g (Figure 8.42) and of periodicity $1/g$. Under these conditions, atomic resolution lattice images can be obtained (see Figure 8.46). Phase contrast is probably the most widely used technique to image nanomaterials due to its very high resolution. Currently, with the introduction of spherical-aberration correctors, point-to-point resolutions of 0.7 Å are possible.

A variation of the transmission electron microscope is the scanning transmission electron microscope (STEM). In STEM mode, a fine electron probe produced by an FEG and two condenser lenses is focused on the sample and scanned over the thin specimen by double deflection scan coils. Instead of using an objective aperture, as in the TEM, to select transmitted versus diffracted electrons, the STEM uses electron detectors to fulfill the role played by the objective aperture. The bright-field detector is located on the optical axis, where it captures the directed electrons. The dark-field detector is a disk with a hole, also located on the optical axis, that detects scattered electrons. Among the dark-field detectors, the most commonly used is the high-angle annular dark-field (HAADF) detector that captures electrons that are inelastic and scattered to higher angles. In this fashion an image can be obtained for which the contrast is strongly dependent on the atomic number and/or the thickness of the sample but not influenced by elastically scattered electrons that contribute to diffraction contrast. Currently, particularly with an aberration-corrected STEM, this technique is capable of identifying atomic resolution changes in composition (Figure 8.47). This capability is of great interest to study surface and interfacial segregation of elements as well as atomic variations in composition in nanostructures.

Another instrumental technique in which the resolution is on the order of interatomic dimensions is field ion microscopy (FIM). Instead of electrons, this method uses ions. In FIM, a sharp (between 5 nm and 50 nm tip radius) metal tip is prepared and placed in an ultra-high-vacuum chamber under cryogenic temperatures (20–100°K). The chamber is backfilled with an imaging inert gas such as helium or neon, and a positive voltage (between 5 kV and 20 kV) is held at the tip. The gas atoms adsorbed on the tip are positively ionized by the electric field, repelled from the tip, and accelerated toward a fluorescent screen, where they create spots (see Figure 8.48).

In principle, each spot on the fluorescent screen corresponds to an atom on the tip so that the distribution of spots on the image represents the atomic configuration on the tip. The image magnification is given by the ratio between the radius of the screen and the radius of the tip, which is typically a few million times. Unlike conventional microscopes, where the spatial resolution is limited by the wavelength of the particles that are used for imaging, the FIM is a projection type microscope with atomic resolution. The major disadvantages of this technique are the fabrication and contamination of the tip as well as the relatively small number of materials that resist evaporation during the ionization of the inert gas atoms.

Finally, we are left with those techniques for imaging using a scanning probe. Two methods are highly used for the study of nanomaterials and nanotechnologies: the scanning tunneling microscope (STM) and the atomic force microscope (AFM). The STM technique allows the real space imaging of electrically conductive surfaces down to the atomic scale. The method uses a very sharp conducting tip with a bias voltage applied between the tip and the sample. The tip is mechanically connected to a scanner, which is a three-dimensional positioning device driven by piezoelectric actuators. Under these conditions, the tip can be laterally moved to scan the sample surface; changing the vertical position allows one to maintain a desired distance between the tip and the sample (see Figure 8.49). When the tip is brought within about 1 nm of the sample surface, the electron wave functions of the tip and sample overlap, causing electrons to tunnel across the gap and produce a current. It is this current that is used to form an STM image (see Figure 8.50). Because the tunneling current is exponentially dependent on the tip-sample distance, small variations in distance lead

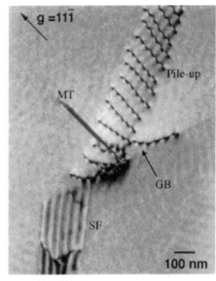

FIGURE 8.45

Pile up of dislocations against a grain boundary in stainless steel. (Courtesy of P. J. Ferreira, I. M. Robertson, and H. K. Birnbaum, University of Illinois, Urbana.)

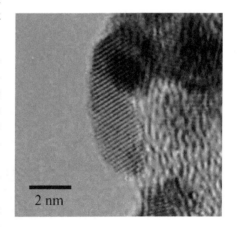

FIGURE 8.46

High-resolution TEM image of platinum nanoparticles on a carbon support. (Courtesy of P. J. Ferreira, University of Texas at Austin, and Y. Shao-Horn, MIT.)

FIGURE 8.47

Aberration-corrected High Angle Annular Dark Field (HAADF) STEM image of a Pt₃Co nanoparticle. (Courtesy of Shuo Chen, Wenchao Sheng, Naoaki Yabuuchi, Paulo J. Ferreira, Lawrence F. Allard and Yang Shao-Horn, Origin of Oxygen Reduction Reaction Activity on "Pt₃Co" Nanoparticles: Atomically Resolved Chemical Compositions and Structures, Journal of Physical Chemistry C, *Vol. 113, pp. 1109-1125, 2009.)*

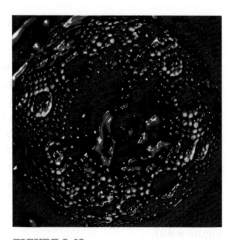

FIGURE 8.48

Field ion micrograph of a sharp tungsten needle. (Courtesy of M. Rezet, J. Petters, and R. Wolkow, University of Alberta, Canada.)

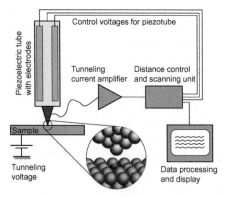

FIGURE 8.49

Schematic diagram of the operation of a scanning tunneling microscope (STM).

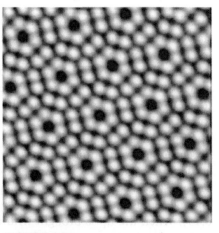

FIGURE 8.50

STM image of silicon (111), 7 × 7 surface reconstruction. (Courtesy of Katsuya Iwaya, London Centre for Nanotechnology.)

to large changes in tunneling current, giving STM the ability to monitor very small variations in topography. In fact, if the system is free from any kind of mechanical vibrations, an STM image has sub-angstrom vertical resolution and atomic resolution laterally. The STM probe can be scanned in either a constant-height mode or a constant-current mode. In constant-height mode (see Figure 8.51), the tip and the sample are kept at a constant distance so that the variations in current acquired by the amplifier supply the data relative to the surface structure of the sample. This method is very sensitive to modulations in atomic scale. However, it is most useful for relatively smooth surfaces. In constant-current mode (Figure 8.51), the tip is scanned over the surface of the material while an electronic feedback loop keeps the tunneling current constant. As a result, the distance between the tip and the sample is adjusted. In other words, in constant-current mode, the vertical motion of the tip supplies the data associated with the surface structure. This method can measure rough surfaces with high topographical accuracy, but data acquisition is slow.

The atomic force microscope (AFM) is a variation of the STM technique. It measures the atomic force between the atoms at the surface of the sample and the tip of a needle at the end of a cantilever, when scanned over the sample surface, instead of the tunneling current monitored in the STM. In addition, the AFM can be used to study nonconducting materials, whereas the STM is restricted to conducting surfaces. In the AFM, the beam is a mechanical lever that holds the force-sensing tip. The physical properties of the lever are very important because the detection systems used in AFM are dependent on the amplitude, phase, or frequency of vibration of the microscope's lever (see Figure 8.52). Therefore, it is important to select a material with the appropriate elastic modulus and dimensions to relate with the scale of forces the tip will experience. The higher the elastic modulus and the shorter the cantilever, the lower the amount of deflection. The tip, which is located at the end of the cantilever, is typically very sharp, with a diameter as small as 20 nm. When scanned closely over the sample surface, the forces between the tip and the surface cause the cantilever to deflect, providing data to form the images of the sample's surface. The atomic forces that are present arise from a variety of sources, namely coulomb forces, ionic forces, Van der Waals forces, and others.

The AFM has several modes of operation. In contact mode, where short-range forces dominate, the tip of the AFM is kept in contact with the specimen with a constant applied force (see Figure 8.53). The tip is then scanned across the sample, and measurements of

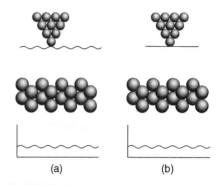

(a)   (b)

FIGURE 8.51
The two common modes of STM operation: (a) constant-current mode and (b) constant-height mode.

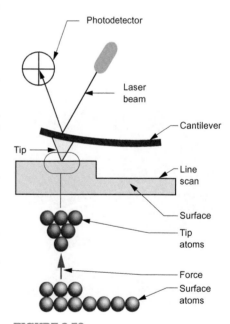

FIGURE 8.52
Schematic diagram of an atomic force microscope.

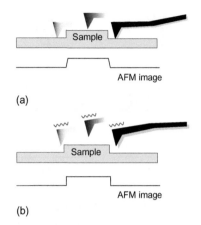

(a)

(b)

FIGURE 8.53

The two common modes of operation in atomic force microscopy (AFM): (a) contact mode and (b) noncontact mode.

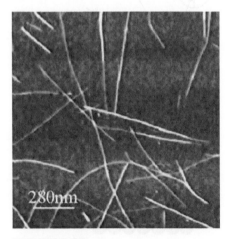

FIGURE 8.54

Atomic force microscopy (AFM) image of carbon nanotubes. (Courtesy of P. J. de Pablo, University Autonoma de Madrid.)

surface topography are obtained by monitoring the deflection of the cantilever. In this mode, the scanning speed is slow due to the response of the feedback system, but the force on the sample is well controlled. However, the large friction forces can damage the sample. Still, this is the preferred mode used for most applications. A different method of operation is the taping mode. In this case, the tip is oscillated at the beam's natural frequency so that the tip contacts the sample, tapping it but not dragging across it, thereby reducing the possibility of damaging the sample. It is a compromise between contact and noncontact AFM. The topography of the sample is measured by determining the amplitude of the oscillations. Finally, the AFM can operate in noncontact mode (Figure 8.53). Under these conditions, the AFM cantilever is vibrated above the sample. Changes in specimen topography can be detected by monitoring the cantilever's amplitude, phase, or natural frequency of vibration. Relatively stiff levers are used to improve the stability of the system. This method applies the least force of all to the sample, but it is more difficult to operate and interpret than the other methods.

In all these methods, the deflection of the cantilever arising from these forces can be detected by various techniques that commonly measure the displacement of the tip along the vertical direction. One of the detection methods is called *tunneling detection*. Here the microscope has a second tip positioned behind the lever. The secondary tip is in a fixed position, and the deflection of the lever is measured based on the voltage necessary to tunnel an electron through the gap between the primary and the secondary tips. This method is very sensitive and capable of atomic-scale resolutions (Figure 8.54), although surface contamination can greatly affect the tunneling characteristics. Another widely used detection technique is a laser beam that is incident on the cantilever and reflected onto a photodetector. Any deflection of the cantilever results in a position shift of the laser beam on the detector. These laser detection systems are still the preferred ways of measuring the deflection inside an AFM.

Let's now shift the discussion to analytical techniques. We start by addressing two important types of electron spectroscopy, widely used in the characterization of nanomaterials and normally installed in an electron microscope: energy dispersive spectroscopy (EDS) and electron energy loss spectroscopy (EELS). The EDS system is usually associated with an SEM, TEM, and dedicated STEM. First, the sample is irradiated with an electron probe. The incident electron beam causes ionization of electrons belonging to the inner shells of the atoms composing the material. When these excited electrons decay to their relaxed states, X-rays are emitted with energies that are unique to the ionized atom. Therefore, by measuring the energies or the wavelengths of the X-ray emitted from the sample, the composition of the material from the region where the electron probe is positioned can be determined. The actual measurement of the X-ray energies is performed by a semiconductor detector. The detector generates first a charge pulse proportional to the X-ray energy. Subsequently, it converts the pulse into a voltage, which is then amplified and digitalized.

The spatial resolution of the EDS technique is a function of the size of the electron probe and the volume of interaction between the beam and the sample. In this regard, an FEG emitter can provide finer probes, whereas the STEM mode is recommended for accuracy. In addition, because the samples for TEM and STEM are much thinner than those typically used for SEM, the spatial resolution in TEM and STEM is considerably better. In terms of minimum detectability, a careful analysis can measure levels of elements below 0.1 at%, particularly if a high-brightness source, high voltage, and thin foils are used. Currently, with the introduction of aberration-free correctors, the probe size has been considerably reduced, leading to improved spatial and energy resolution.

The EELS system is common in both TEMs and STEMs but not SEMs. It is also fundamentally different from EDS spectroscopy. The basic concept of EELS is to measure the energy loss of inelastic scattered electrons. In both the TEM and STEM, these are electrons that suffer energy loss and change of momentum upon interacting with other electrons in the sample. According to their energy losses, the transmitted electrons are dispersed by a spectrometer, and a spectrum is produced (see Figure 8.55). The first peak that appears in the spectrum, called the *zero-loss peak*, consists mainly of electrons that retained their energy. The zero-loss peak is usually a problem due to its high intensity rather than providing useful information. Next in the spectrum is the low-loss region. This range corresponds to energy-loss electrons up to about 50 eV. This region is related to longitudinal oscillations of valence electrons, called *plasmons*. These plasmon losses are dependent on the thickness of the specimen and are more predominant in materials with free-electron structures, such as metals. In the case of insulator materials, a further mechanism occurs in the low-loss region, which is the possibility of exciting valence electrons to states above the Fermi level. These processes lead to shifts in the plasmon peak that reflect the type of bonding present in the material. The high loss region of the EELS spectrum (above 50 eV) corresponds to the inelastic

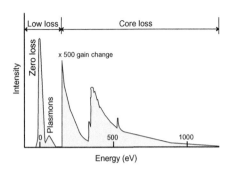

Zero loss: Elastically scattered electrons
Plasmons: Collective excitation of valence electrons
Core loss: Inelastic scattering with inner shell electrons

FIGURE 8.55

Electron energy loss spectroscopy (EELS) spectrum.

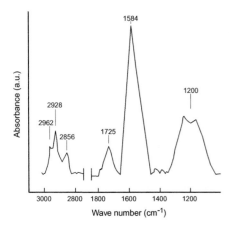

FIGURE 8.56

Infrared spectrum of carbon nanotubes synthesized by chemical vapor deposition. (Appl. Phys. Lett., *2004, 85(19), 4463–4465.*)

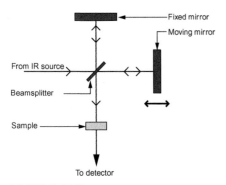

FIGURE 8.57

Schematic setup of an FTIR spectrometer.

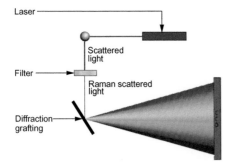

FIGURE 8.58

Schematic diagram of the layout for Raman spectroscopy.

interactions with the inner shells. These ionization losses are characteristic of the atom involved, and thus the signal (called *edge*) is directly related to an unique atomic type. In addition, the area under the edge provides quantitative information about the particular element. The EELS technique is capable of atomic resolution chemical analysis and allows chemical bonding information to be acquired.

Two other spectroscopy techniques of interest for characterizing nanomaterials are infrared (IF) and Raman spectroscopy. These techniques are also called *vibrational spectroscopies* because they involve the vibration of groups of atoms. In IF spectroscopy, the material is first subjected to an incident infrared light. Depending on the type of atoms and type of bonds between atoms, the frequency at which atoms vibrate is unique for each arrangement. When the infrared light interacts with the material, groups of atoms tend to adsorb infrared radiation in a specific frequency and therefore wavelength. The remaining light, which was not absorbed by any of the harmonic oscillators (groups of atoms), is transmitted through the sample to a detector. Here the transmitted light is analyzed and the frequencies absorbed by the material are determined. The resulting plot of absorbed energy versus frequency is called the *IR spectrum*. Therefore, differences in the chemical and atomic structure of materials give rise to specific vibrational characteristics and yield unique IR spectra for each material.

From the frequencies of the absorptions, it is possible to determine the presence of various functional groups in a chemical structure; the magnitude of absorption due to a particular element allows us to deduce its concentration in the mixture. In general, IR spectroscopy can be used to identify materials in the solid, liquid, or gaseous state. In the case of nanomaterials, IF spectroscopy has been used to determine the extent of absorption of foreign species in nanoparticles of different sizes as well as the structure of nanorods, nanowires, and carbon nanotubes (see Figure 8.56). Modern IF spectrometers irradiate the specimens with a wide range of frequencies, after which the signal is converted mathematically (Fourier transform) into the classical spectrum, where typically the spectral absorption of a sample is being scanned. This technique is called *Fourier transform infrared* (FTIR) spectroscopy (see Figure 8.57).

The Raman spectroscopy technique is similar in concept to the IF method and is complementary. However, it uses a laser source and scattered light (see Figure 8.58). The scattered light consists of two types: Rayleigh scattering, which has the same frequency as the incident beam and strong intensity, and Raman scattering, which is inelastic and exhibits a weaker intensity at higher and lower frequencies than the incident beam. Therefore, in Raman spectroscopy, the vibrational frequency and thus the wavelength of groups of atoms are measured as a shift from the incident beam frequency.

The laser is an ideal source for Raman scattering because it is bright, quasimonochromatic, and available in a wide range of frequencies. A Raman spectrum is normally measured in the UV-visible region where the excitation and the Raman lines appear. It is a complementary technique to IR spectroscopy because some vibrational modes of materials are IR-active, whereas others are Raman-active. In the case of nanomaterials, Raman spectroscopy has been used extensively to characterize carbon nanotubes by analyzing their mode frequencies as a function of size and chirality. In addition, Raman spectroscopy has been shown to probe changes in nanoparticle size. In fact, the Raman spectrum tends to broaden and shift to lower frequencies as the particle size decreases (see Figure 8.59).

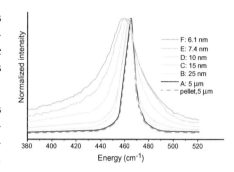

FIGURE 8.59

Size-dependent properties of CeO$_{2-y}$ nanoparticles as studied by Raman scattering. (Appl. Phys. Lett., 2002, 80, 127–129.)

We conclude this section on the characterization of nanomaterials by discussing X-ray diffraction. For materials with microcrystalline grain sizes, X-ray diffraction is primarily used to determine the crystal structure. However, for nanocrystalline materials with grain sizes below 100 nm, X-ray diffraction provides additional information about the crystallite size. To understand this effect, let's first discuss the basic principles of X-ray diffraction. In general, diffraction occurs when a wave encounters a series of regularly spaced objects that are capable of scattering the wave and have spacings that are similar in magnitude to the wavelength of the incident wave. It turns out that the wavelength of X-rays (~0.1 nm) is approximately the same as the atomic spacing of solids. Therefore X-rays will be scattered by crystalline solids (see Figure 8.60). However, due to the fact that during an X-ray experiment several rays will interact simultaneously with the material, it is important to identify those rays that interfere constructively and those that interfere destructively. In other words, rays that interfere constructively will add up and contribute to the overall X-ray signal detected, whereas rays that interfere destructively cancel each other out and will not be captured by the detector. Very simply, constructive interference occurs when the path-length difference between two rays is equal to an integer number of wavelengths. Mathematically, this can be expressed as

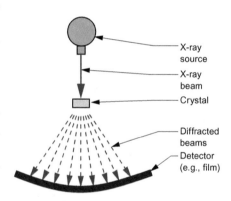

FIGURE 8.60

Schematic diagram of the layout of an X-ray diffractometer.

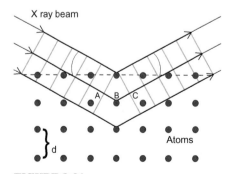

$$2d \sin \theta = n\lambda \qquad (8.2)$$

FIGURE 8.61

Schematic diagram showing the concept behind Bragg's law.

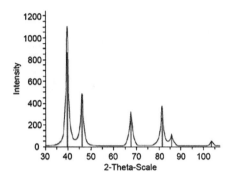

FIGURE 8.62

X-ray spectrum of platinum.

where *d* is the interplanar spacing of a particular set of planes (see Figure 8.61), θ is the Bragg angle, *n* is the number of wavelengths, and λ is the wavelength of the radiation. The preceding expression is known as *Bragg's law of diffraction*.

It turns out that Bragg's law of diffraction is not a sufficient condition for diffraction. It works for unit cells that have atoms only at the corners. Thus, for unit cells, such as the body-centered cubic (BCC) and the face-centered cubic (FCC) structures, the atoms located in positions other than the corners act as extra-scattering centers and can produce destructive interference at certain Bragg angles. The consequence of this effect, which is called the *structure factor*, is that only certain sets of crystallographic planes will appear in the X-ray spectrum for each particular crystal structure (see Figure 8.62).

In addition to the structure factor, which is related to the position of atoms within the unit cell, there is another important contribution to the peak intensity acquired during X-ray diffraction. This is called the *shape factor*, and it is related to the size of the crystal. According to the shape factor, an infinite crystal exhibits a series of X-ray peaks with the form of a delta function. However, for a finite-size crystal, the distribution becomes broader and shorter the smaller the crystal, despite the fact that the area under the peak remains the same (see Figure 8.63). The breadth of the distribution is related to the magnitude of deviation from the Bragg angle. On this basis, the crystal size *L* can be related to the breadth of the X-ray peak according to the Scherrer expression, given by

$$L = \frac{C\lambda}{\beta \cos\theta} \qquad (8.3)$$

where *C* is a constant close to 1, λ is the wavelength of the incident X-ray radiation, β is the broadening of the peak at half maximum width, and θ is the Bragg angle. The preceding expression is only valid for crystal sizes below 100 nm. In addition, if there are microstrains present in the sample, the method needs to be further refined.

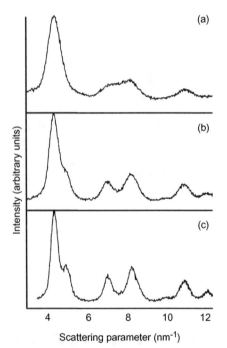

(a)

(b)

(c)

FIGURE 8.63

X-ray diffraction spectra of gold nanoparticles. According to the data, the diameter of the nanoparticles was found to be (a) 2 nm, (b) 2.9 nm, and (c) 3.9 nm.

Figure 8.64 summarizes the kinds of basic information that can be obtained through various characterization techniques that are particularly well suited to the nanoscale.

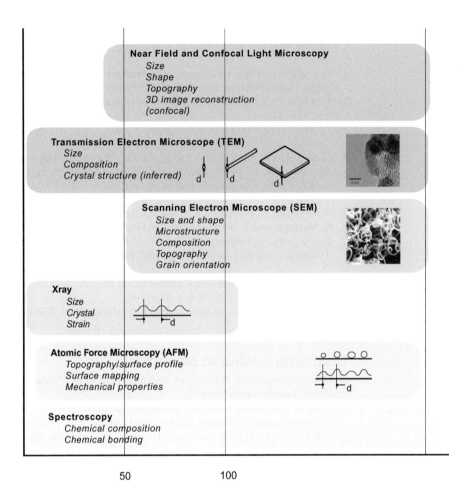

Types of Characterization Techniques

Near Field and Confocal Light Microscopy
Size
Shape
Topography
3D image reconstruction
(confocal)

Transmission Electron Microscope (TEM)
Size
Composition
Crystal structure (inferred)

Scanning Electron Microscope (SEM)
Size and shape
Microstructure
Composition
Topography
Grain orientation

Xray
Size
Crystal
Strain

Atomic Force Microscopy (AFM)
Topography/surface profile
Surface mapping
Mechanical properties

Spectroscopy
Chemical composition
Chemical bonding

50   100

FIGURE 8.64

Basic information obtained through various characterization techniques that are well suited for the nanoscale.

FURTHER READING

Synthesis of Nanoscale Materials and Structures

A. S. Edelstein and R. C. Cammarata (eds.), Nano materials: Synthesis, properties, and applications, Institute of Physics, 1996, ISBN 0-8503-0588.9.

Y. Gogotsc (ed.), Nanomaterials handbook, CRC/Taylor and Francis, 2006, ISBN 13: 988.0-8493-2308.9.

A. N. Goldstein (ed.), Handbook of nanophase materials, CRC Press, 1998, ISBN 0824-89469-9.

H. Hasono, Y. Mishma, H. Takezoe, and K. J. D. Mackenzie, Nanomaterials from research to applications, Ch. 12, pp. 419 et seq., Elsevier, 2006, ISBN 13 988.0-08.044964-8.

S. Iijama, Single-shell carbon nanotubes of 1-nm diameter, Nature, 1993, Vol. 363, pp. 603–605.

W. L. Johnson, ASM handbook, Vol. 2, ASM International, 2002.

R. W. Kelsall, I. W. Hamley, and M. Geoghegan (eds.), Nanoscale science and technology, John Wiley & Sons, 2005, ISBN 13: 988.0-480-85086-2.

C. C. Koch, Nano structured materials: Processing properties and applications, Noyes Publications, 2002, ISBN 0-8155-1451-4.

W. Krätschmer, Solid C60: a new form of carbon, Nature, 1990, Vol. 348, pp. 354–358.

H. W. Kroto, C60: Buckminsterfullerene, Nature, 1985, Vol. 318, pp. 162–163.

R. Lumley and A. Morton, Nanoengineering of metallic materials, Ch. 9, pp. 219–248, in: Nanostructure control of materials, R. H. J. Hannink and A. J. Hill, Woodhead Publishing Ltd., 2006, ISBN 13 988.1-85583-933-8.

S. Menzies and D. P. Anderson, J. Electrochem. Soc., 1990, 138, p. 440.

A. A. Nazarov and R. R. Mulyukov, in Handbook of nano science, engineering and technology, Ch. 22, W. A. Goddard, D. W. Brenner, S. E. Lysheuski, and G. J. Iafrate (eds.), CRC Press, 2003, ISBN 0-8493-1200-0.

C. P. Poole, Jr., and F. J. Owens, Introduction to nano technology, Wiley-Interscience, 2003, ISBN 0-481-08935-9.

E. Roduner, Nanoscopic materials, RSC Publishing, 2006, ISBN 13: 988.0-85404-858.1.

Characterization of Nanomaterials

D. Courjon, Near-field microscopy and near-field optics, Imperial College Press; 1st ed. March 2003.

G. S Kino, T.M. Corle, Confocal scanning optical microscopy and related imaging systems, Academic Press; 1st ed. August 29, 1996.

Joseph Goldstein, Dale E. Newbury, David C. Joy, and Charles E. Lyman, Scanning electron microscopy and X-ray microanalysis, Springer, 3rd ed. February 2003.

David B. Williams and C. Barry Carter, Transmission electron microscopy: A textbook for materials science, Springer-Verlag, 2nd ed. December 2, 2008.

Ernst Meyer, Hans J. Hug, and Roland Bennewitz, Scanning probe microscopy: The lab on a tip, Springer; 1st ed. October 10, 2003.

S. Morita, R. Wiesendanger, and E. Meyer (eds.), Noncontact atomic force microscopy, Springer, 1st ed. September 17, 2002.

B. D. Cullity and S. R. Stock, Elements of X-ray diffraction, Prentice Hall, 3rd ed. February 15, 2001.

CHAPTER 9

Design Environments and Systems

9.1 ENVIRONMENTS AND SYSTEMS

The general idea of looking at buildings and products in terms of environments, systems, and assemblies was introduced in Chapter 3 and forms the basis for the way this chapter is organized and presented. This approach reflects a way of thinking familiar in the design fields. Engineers are commonly familiar with designing systems. Systems are normally thought of as performing a *functionally defined* role. A system normally consists of several subsystems as well as assemblies of physically defined components. Hence we might have an energy system that consists of a great many components that serve specialized functions to generate, store, or distribute energy. Components such as fuel cells or batteries would constitute parts of a larger system. These components can normally be arranged in many different ways to form a physical assembly. System performance is critical with respect to its functional role. Assembly characteristics are more important from a manufacturing and packaging perspective. Nanomaterials or nanotechnologies can play an essential role in improving the performance of many functionally defined systems.

The idea of thinking about environments stems much more from an architect's or designer's perspective. The term can have several meanings, all relating to a larger spatial construct, context, or condition that goes beyond narrowly defined functional systems. Environments can be internal or external. They can be either occupied by humans or be critical surrounding conditions. At a more abstract level they can be simply an organizing device for related concepts.

By way of example, a thermal environment in a building or vehicle is a spatial volume occupied by humans that must have certain overall characteristics to maintain human health and comfort that in turn define governing design criteria. Thermal environments are served by functionally defined systems and related assemblies, such as heating or cooling systems, but their design is also dependent on a host of other factors, especially the geometry of the space, the thermal properties of all surrounding enclosures, and interactions with other systems (for example, lighting systems that produce heat). For many products, the idea of a thermal environment might be that of a surrounding external context that imposes particular design needs. A product might need to work in a super-cold or super-hot environment. As with the building example, overall performance criteria for the product again relate to the overall performance of the product with respect to the larger environmental context. The design of specific components or systems depends on this overall context and related criteria.

In the following sections, we look first at a series of basic environments: thermal, lighting and optical, sound, mechanical, and electromagnetic. We then examine selected enabling systems, especially energy systems, in more detail. The design of environments and related systems naturally is closely dependent on related material properties (thermal, optical, mechanical, electrical, chemical) but goes beyond them. To understand whether a lighting environment has been designed successfully, for example, it is necessary to consider many factors: the use context (for example, library versus church), the ambient external lighting conditions associated with its location (sun angles, lengths of daylight), the location and type of artificial lighting present, the overall geometry of the space, and related types of surrounding surfaces, including the materials used and the way they reflect, absorb, or transmit different light wavelengths. A computational method must be found for understanding how all these factors interact to create a lighting environment that has the right illumination levels and distributions necessary for both task performance and for human comfort. The computational algorithms that must be used are specialized and are the province of the design architect or engineer. This general picture also broadly reflects the way design work proceeds in other environments. There are invariably demands, geometrical and material considerations to be considered, and computational algorithms needed to determine overall performance levels. In the following discussion we very briefly touch on these design procedures for different environments and briefly note the kinds of computational algorithms used, but we do not go into them in depth. There are plenty of other books that do that. Our intent is more to impart a broad understanding of these procedures for the benefit of readers who are not familiar with them so that the role of nanomaterials and nanotechnologies can be better understood.

Throughout our discussion, take note of comments concerning the needed size and scale of various components. The quest of many nanotechnologists is often stated as making functional devices smaller and smaller. This is a worthy goal for many products, especially electronic components, sensors, and other devices. On the other hand, many designed objects are necessarily large by virtue of the intrinsic rationale for being. (Do we really want walls or enclosures that are quite literally nanosized around our bedrooms or automobile interiors?) Hence some of the needs might not all be about smallness but how to improve the performance of components or elements that are forever large. This fundamental conceptual difference in goals can lead to quite different approaches to the way nanomaterials can be effectively used. Certainly it leads to the need for an aggressive look at nanocomposites, nanofilms, and nanocoatings.

9.2 STRUCTURAL AND MECHANICAL ENVIRONMENTS

Buildings and products invariably have some type of supporting structural system that provides the necessary framework for maintaining overall integrity in the presence of forces induced by the environmental or use context. It may be more or less explicitly expressed. Structural systems in buildings or bridges can have static qualities, but almost invariably they also need to carry dynamic forces from wind, earthquake, or use movements. Mechanical systems that provide specific movement or linkage functions have components subjected to a wide variety forces and deformations, and dynamically induced forces are extremely important. In all these systems, a variety of stresses and deformations are developed within component members.

Engineers and designers working with materials in the design of force-carrying components or structures of one type or another know that there are a host of material properties and characteristics that must be carefully defined and quantified before the material can be effectively used in a design context. We all know of the radically different characteristics of glass, steel, and a common plastic, and even the total layperson broadly appreciates how these different characteristics affect their use in the world around us. Even within the general class of metals, however, we see huge differences between a material such as steel and cast iron. Strengths may be comparable, but steel is a ductile material, whereas cast iron is a brittle one, and failure modes are extremely different. An understanding of these differences has long influenced the use and design of structures in both architectural and mechanical design spheres. In the formative years of bridge design in the 19th century, it was common to make compression elements out of cast iron and tension elements out of wrought iron (a ductile material) or steel, when it became available. The whole form, visual appearance, and performance of the resulting bridges were in turn critically influenced. This notion of using or exploiting specific material properties in a design context is both a rich approach and a common one.

The overall goal of a structural design activity is to assure that the stresses and deformations induced in a structure by forces generated by its environmental or use context are maintained at a desired level consistent with safety and operational objectives. The structural analysis process is aimed at numerically quantifying—for a structure already defined in terms of geometry and materials—these stresses and deformations for identified loadings, forces, and other considerations associated with the use context. The structural design process is directed toward altering the size, shape, and material selection for a structure to maintain these same quantities at desired levels. Designs are typically based on meeting both strength and stiffness criteria. Once magnitudes and distributions of the forces and bending moments acting on a member are known, analyzing the member for stresses and deformations and determining whether it is safe are fairly straightforward processes. Structural analysis techniques are available to predict stresses, deformations, and failure modes in complexly shaped structures.

In today's world, the structural designer uses many specific analytical tools that were developed in relation to particular definitions of material properties that were discussed in Section 4.3, including:

- The elastic modulus (E—Young's modulus) of a material that relates stress levels to strain levels below proportional limits in a material for tensile and compressive stresses (E = stress/strain). Materials with high E values are considerably stiffer and less prone to elastic deformation than materials with low values.

- The shear modulus (G), which relates shear stresses to shear strains.

- The bulk modulus (K), which has to do with volume change.

- Poisson's ratio (v), which relates transverse to longitudinal strains.

Other measures relate to the hardness of a material, its fatigue characteristics, and other factors. Measures such as the "moment of inertia (I)" of a member are also used to quantitatively describe the amount and distribution of material in the cross-section of a typical member.

Normally, making detailed structural analyses is the domain of the structural or mechanical engineer, and many tools, such as finite element analyses, are used during this process. Analysis and design procedures, however, go hand in hand. Typically a preliminary design is established and then analyzed. Comparison of calculated stress and deformation values to allowable criteria for the materials used and in relation to the use context yields an understanding of the viability of the structure. Critical parts that fail to meet design criteria are then reshaped and resized or a new material is selected for use. Design is typically an iterative process. These many analytical and design procedures are well documented in other sources and will not be covered here. Rather, we concentrate on the influence of material properties on design outcomes.

At a basic level, a look at the unit strengths of some nanostructures often causes conventional materials to seem pale by comparison (see Figure 9.1). The idea of increasing the strength of these same conventional materials by creating composite forms that incorporate nanomaterials, such as adding nanoparticles into a metal matrix or nanofibers into concrete, is widely discussed. The literature is full of such assertions and there is true cause for excitement here. There are real possibilities for making structures smaller and lighter via nanocomposites with either the same or even greater performance characteristics than are achievable with conventional materials. There are also opportunities for creating active systems with embedded sensors, control systems, and actuating mechanisms within an integral structure. Indeed, structures already seem to have an active life when subjected to forces. They bend and sometimes break; they move. Our own bodies subjected to a load, however, do something more—the repositioning of various body elements takes place. A stance is shifted or arms repositioned. These are intrinsic active responses that allow a human to carry loads not

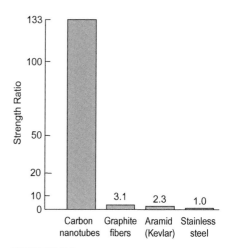

FIGURE 9.1

General comparison of primary tensile strengths of various materials (strength ratios relative to defining stainless steel = 0.)

only more efficiently, but to carry loads far higher than would be possible if the body were forever frozen into a single static position. Many responses are intrinsic and could be described in terms of sensory inputs and actuated responses associated with a whole class of "smart" structures. Interestingly, anticipatory repositioning can also occur, as happens when an individual *a priori* assumes a particular stance with the cognitive expectation of needing to carry a certain type of load—an action associated with "intelligent" structures. Instead of a structure simply passively carrying the forces applied to it, structures that actively change shape or otherwise actively respond in an intrinsic or anticipatory way to the type and location of applied forces can be envisioned. There are, of course, active structures available now, but the promise of nanotechnologies allows imagining a scenario of much smaller and even more active systems (as embedded sensor, actuation, and control systems become smaller). Integrated energy storage or acquisition capabilities also provide a bright future.

Nano-Based Structural/Mechanical Applications

In common structural and mechanical applications that use relatively large components, limitations on the size and scale of nanomaterials that can be directly synthesized lead to the use of nanocomposites. Nanocomposites are normally designed to exploit various properties of the respective materials used. As discussed in Section 8.8, typically one material is chosen as a matrix to hold or embed smaller quantities of nanomaterials as a second phase. The matrix can be a polymer, metal, or ceramic. Commonly used matrix materials are typically available in bulk form and are relatively inexpensive. Smaller quantities of higher-performance but typically more expensive nanomaterials are then incorporated into the matrix or deposited as a film onto a matrix substrate.

Objectives of developing structurally useful nanocomposites vary greatly. An obvious application is in the areas of strength and stiffness. Here the type and composition of the nanocomposite depend greatly on the anticipated size of the final piece. Many attractive nanocomposites remain possible to fabricate at very small-scale dimensions. Large-scale nanocomposite production demands a form of bulk manufacturing that is still difficult and expensive, but approaches such as embedding nanoparticles into metallic matrix are entirely feasible (see the discussions that follow). Other objectives might be quite different. For passive vibration control, the use of various nanoparticles in a rubber or polymeric matrix offers great potential. These applications are explored in the following subsections.

Polymer-matrix nanocomposites

Various kinds of conventional polymer composites have long been made using a polymer matrix and various kinds of embedded second-phase materials. The volume fraction of the latter can often be quite high. The use of a small quantity of nanoparticles, nanotubes, or nanoplates—a few percent by weight—as a reinforcing second-phase material can dramatically alter the properties of the nanocomposite for a much lower volume fraction than conventional polymer-matrix composites. Both thermoplastic and thermoset polymer-matrix systems can be used.

As discussed in Section 8.8, depending on the exact type of polymer host and nanoscale reinforcing phase, both the strength and modulus of elasticity values can be either increased or decreased. When nanoparticles are used, materials such as polystyrene reinforced with silica nanoparticles, for example, show a modulus increase. Section 8.8 also noted that in some instances the inclusion of nanoparticles can increase toughness and wear resistance. Tendencies for nanoparticles to agglomerate, however, can cause severe problems and must be avoided. When carbon nanotubes (CNTs) are used, carefully formulated composites can show significant elastic modulus and strength increases. Carbon nanotubes can be embedded in thermosets. In certain epoxies, for example, fairly dramatic increases have been reported in compression strength (in the 30% range), achieved via the use of only about 0.5% to 2.0%wt of dispersed carbon nanotubes.

CNTs can be embedded in many types of widely used thermoplastics, such as polycarbonates, with the goal of increasing the compressive and impact strength, toughness, elongation to failure, and other properties. For these positive benefits to occur, however, uniform dispersion, alignment, and good interfacial bonding between the polymer and CNTs must be achieved. Both nanoparticles and CNTs can affect the important glass transition temperature, Tg, of the nanocomposite. A thermoplastic polymer will soften when heated above Tg. This value is thus extremely important when we're designing for applications in elevated temperature environments. The Tg value can potentially be increased when certain nanocomposite formulations are used (see Section 8.8). Platelike nanomaterials such as layered silicates can be used as nanofillers in polymer-matrix composites as well. Significant strength, stiffness, UV resistance, and other improvements can be obtained at quite low volume fractions (e.g., 2–5%).

Polymer-matrix nanocomposites are being widely explored in relation to many high-end sports, automotive, and aerospace

FIGURE 9.2

Carbon nanotubes are incorporated into this high-performance bicycle. (Courtesy of BMC Swiss Cycling Technologies.)

applications that currently use one or another type of carbon-fiber composite material. Carbon-fiber composites are known to be extremely light and strong, and have found wide application in everything from bicycle frames to tennis racquets to racing car bodies (see Figure 9.2). These composites use any of several different types of carbon fiber strands or weaves that are in turn impregnated with a thermoset epoxy. The overall strength and stiffness of these kinds of composites can be enhanced by the addition of CNTs to the matrix (1–5 wt%) while keeping the amount of carbon fiber strands approximately the same. The high value of these products normally warrants the associated higher costs of using nano-based CNT composites rather than conventional ones. Surprisingly little data, however, is available on the kinds of performance increases that are attainable. This is because CNTs are often in bundles and not fully dispersed within the matrix. This means that to take advantage of the properties of CNTs, we need to develop novel synthesis methods to isolate CNTs from the bundles.

Polymer-matrix with nanoscale layered silicates are currently widely used in many applications. Nanoscale silicates in the form of smectic clays and mica are relatively inexpensive and can be easily obtained in large quantities. Processing methods for making large objects or components from these polymer nanocomposites are well understood. As noted, significant strength and stiffness increases can be obtained for low volumes of added silicates, such as 1–5%, compared to conventional polymer composites, for which volume fractions can be as high as 30%. Nano-based polymer composites can also demonstrate greater dimensional stability as well as show improved thermal stability and resistance to gas permeation. Hence a major application domain is the bottling industry. Major industry-driving forces hinge around needs to bottle pressurized liquids of one type or another with containers that are structurally adequate for containment and general packaging (including stacking) but that use little material and are resistant to gas permeation and are fire retardant. Polymer nanocomposites using clay nanofillers directly meet these needs.

Polymer-matrix nanocomposites filled with nanoparticles have also been considered for applications for which a change in modulus is required. In cases where the nanoparticle/polymer interaction is weak, such as for alumina/PMMA nanocomposites, the modulus has decreased, whereas for cases where the nanoparticle/polymer interaction is strong, such as for silica/polystyrene, the modulus has increased.

Metal-matrix nanocomposites

Metals are the most dominant material used in mechanical applications in any industry. Traditional metallurgy has gone extremely far in providing designers with an amazing array of high-performance metals. There are already many superb ways of making metals stronger (e.g., alloying, work hardening, and dispersion hardening; see Section 4.3). As extensively discussed in Section 7.1, many extremely high performance alloys are based on nanoscale dispersions of particles. Nano is by no means new to the metallurgist. Still, there are always demands for stronger, stiffer, or harder materials or materials with particular kinds of stress-strain deformation characteristics for special applications (e.g., energy absorption). Metal-matrix nanocomposites provide an improved way of meeting these needs.

The properties of many common metals can be greatly enhanced by the addition of relatively small amounts of nanomaterials—normally in the form of nanoparticles, nanowires, and nanotubes. The basic matrix can be any of several metals or alloys. For many applications, normally a low-weight ductile matrix is desirable. These, in turn, can be combined with ceramic second-phase reinforcements that can help increase the modulus and strength values. As discussed in Section 7.8, various metals (aluminum, copper, titanium) have been reinforced with carbides, borides, nitrides, and oxides. Depending on the quantity of reinforcement used and the uniformity of distribution, mechanical properties such as strength, wear resistance, and creep resistance can be adjusted to meet the requirements of the design. Carbon nanotubes (CNTs) have also been used in metal-matrix nanocomposites, with observed improvements in yield stress, maximum strength, and hardness. Amounts of second-phase reinforcing used are normally relatively small. Typically, amounts on the order of 0.5% to 2% of dispersed nanomaterials by weight can lead to enhancements in mechanical strengths.

Metals such as aluminum are widely used. Aluminum-based metal-matrix composites are extremely interesting because of their low density and high specific strength. Metal-matrix nanocomposites are particularly interesting for the aerospace and automotive industries as well as for other structural applications.

Ceramic-matrix nanocomposites

These materials are typically attractive as structural materials for high-temperature and wear- and corrosion-resistance applications. In particular, the addition of nanoparticles to a ceramic matrix has led to increases in fracture toughness, strength, and creep as well as resistance to oxidation and thermal shock. Improvements of around 50% in fracture toughness and 200% in strength have been observed in alumina reinforced with SiC. Furthermore, enhancements in tensile creep by two or three orders of magnitude were also measured for the same material. In most cases, the reinforcing phase is a ceramic material, such as nanoparticles of silicon carbide, alumina, zirconia, and silicon nitrate, although metallic reinforcements can be also used. Some examples include nanoparticles of iron (Fe), tungsten (W), molybdenum (Mo), nickel (Ni) copper (Cu), and cobalt (Co). In general, metallic fillers are used to improve magnetic, electrical, and optical properties. However, in some cases, the addition of metallic reinforcements can affect the mechanical properties because they create a connected network with consequences for plastic deformation.

General applications include many types of hard coatings with high temperature capabilities, a number of different applications in electronics (such as semiconductor products), oral dentifrices, and many others. Wear- and corrosion-resistant coating materials based on nanostructured ceramic composites promise to have many applications in structures, vehicles, aircraft, and similar settings. Bioceramics are being explored for many medical implant applications.

Amorphous materials

In Section 7.3, amorphous materials were described as an extreme class of nanostructured matter. As crystal sizes shrink atomic dimensions, the material becomes completely disordered. Ordinary glass is amorphous and is known for its unique properties. Other highly disordered materials with similar structures, such as polymers or metallics, are possible as well and are often also called *glasses*. Section 7.3 pointed out that these kinds of amorphous materials can have remarkably high strength and stiffness properties. Indeed, amorphous materials can approach the fundamental limit of strength.

Successful applications of this class of nanosolid in mechanical environments have come about, including:

- *Tooling, particularly knife edges.* The high hardness (a direct consequence of the high strength) suggests use in precision tooling. The lack of microstructure allows a blade to be sharpened to an exceptional edge because there is no length scale above the atomic to limit it.

- *Springs, clubs, and anvils.* The high resilience of amorphous materials gives metallic glasses potential as springs. Their successful use in golf club heads and tennis racquet shafts exploits this feature (see Figures 9.3 and 9.4). There are many possible applications in devices such as high-speed relays, gyroscopes, and actuators.

- *Hard, wear-resistant surfaces.* Electroless nickel deposits have excellent corrosion and abrasion resistance. The process is used to coat the plates of viscous clutches, nickel-plated parts for ABS systems, fuel injection pumps, and carburetors. The hardness and corrosion resistance are used in oil, gas, and chemical engineering industries, deposited on ordinary steel to provide a substitute for stainless steel. In processing, electroless nickel is used to repair polymer molding dies when they are worn, exploiting the high hardness and wear resistance.

- *Fashion items.* The ability to take high polish and resist abrasion and corrosion (and the sheer novelty) has given metallic glasses a niche market for rings, spectacle frames, watch cases, pens, mobile-phone cases, and the like (see Figures 9.5 and 9.6).

Amorphous metals also have exceptionally high susceptibility to magnetization, which makes them attractive for many electronic applications.

The less attractive features must not be ignored. The current expense of metallic glasses limits their structural use to high-end products in which performance or aesthetics play a greater role than price.

FIGURE 9.3

A driver made of Vitreloy, an amorphous metal. Its superior resilience allows a longer drive. (Courtesy of Liquidmetal Technologies; information@ liquidmetal.com.)

Tribological applications

Tribology is the field of science that studies mechanisms of friction, lubrication, and wear of interacting moving surfaces. Clearly these are important issues in the design of mechanisms and are major factors that affect the efficiency and durability of many products as diverse as automobiles, hard disks, or MEMS devices in the electronics industry. *Nanotribology* is the study of these processes at the nanoscale and is the subject of considerable ongoing research. Friction involves many interacting actions, including elastic and plastic deformations of surfaces, geometrical interlocking, wear, adhesion, indentation, and lubrication. At the nanoscale, atomic and molecular interactions can occur. Properties of confined lubricant fluids at nano thicknesses are different from those at bulk scales. Nanomaterials are being widely explored for use as lubricants. Many common

FIGURE 9.4

The frame of this racquet is reinforced with Vitreloy, an amorphous metal. Its superior resilience stores more energy during a stroke. (Courtesy of Liquidmetal Technologies; information@liquidmetal.com.)

FIGURE 9.5
This cell-phone case is made of an amorphous metal. Its hardness, scratch resistance, high polish, and reflectivity add value to the product. (Courtesy of Liquidmetal Technologies; information@liquidmetal.com.)

FIGURE 9.6
Amorphous metal costume jewelry. As with the phone case, the excellent hardness, scratch resistance, high polish, and reflectivity add to its appeal. (Courtesy of Liquidmetal Technologies; information@liquidmetal.com.)

lubricants are layered compounds (e.g., graphite or tungsten disulfide) and work by molecules sliding past one another. The use of nanoparticles of rounded or spherical shapes can potentially allow an efficient rolling action instead.

Surface hardness

The hardness of materials is crucial to the performance of many kinds of products. In our preceding discussion of applications of amorphous materials, we found that their high hardness made them particularly appropriate for uses in tooling and other applications in which knife edges are important. There are other ways that nanomaterials can be used to enhance hardness. Nanocoatings are particularly interesting here. Nanocrystalline powders that have been used in metallic stainless steel coatings that are sprayed on have been shown to improve surface hardness. Various kinds of nanocrystalline metals have also been specially developed to use in carbon fiber-reinforced plastic composites to increase surface hardness. Nanocrystalline nickels have been used in many products, including golf club shafts, as hard protective shells. Many types of nanocrystalline particles and alloys are available, each with differing properties, such as copper (Cu), gold (Au), cobalt (Co), zinc (Zn), palladium (Pd), silver (Ag), iron (Fe), lead (Pb), tin (Sn), nickel-cobalt (Ni-Co), nickel-molybdenum (Ni-Mo), and others. Carbon-based nanofilms can be developed to not only increase hardness values; using nanolaminates, tribological characteristics can be addressed independently as well.

In cases where surface ductility is more important than strength, as is often the case, various kinds of polymer and ceramic nanocomposites have been used as coatings. Nanocrystalline metals such as nickel or nickel-iron also have good corrosion resistance and are much stronger than traditional nickel coatings. Polymer-based coatings are also used extensively for corrosion protection on metals.

Nanomaterials and concrete

Concrete and mortar remain the workhorses of civil and building construction industries throughout the world; over 2 billion tons of cement are manufactured each year. Concrete can be used in raw, bulk form when subjected to compression only or if cracking is not envisioned as a problem. More typically, it is used in a composite form in conjunction with reinforcing steel to form what is commonly called *reinforced concrete*. The steel provides the necessary tensile strength when the material is used in beams and other structures where tension stresses are invariably developed and that would normally cause cracks to develop in plain concrete that would propagate uncontrollably until failure occurs.

Concrete is fundamentally composed of a mixture of coarse and fine aggregates, cement, and water. Synthetic cements are usually made by grinding calcinated limestone and clay into a fine powder. On mixing with water, an exothermal reaction occurs with the cement that causes time-dependent hardening. Important variables include its physical, mechanical (strength, stiffness, creep rates), and other properties. Strengths increase with time and are highly dependent on the water-cement ratio of the mix (excessive water reduces strength). Process factors include its ease of mixing and pouring (largely related to the fluidity of the initial mix).

Given its importance in construction, a huge amount of research and development has long gone into improving the quality and properties of concrete. For example, there are many different types of specially formulated cements available for specific purposes, such as high early strength or low heat, and many additives are available to improve specific characteristics such as resistance to freeze-thaw action. The design of reinforced concrete elements has also by now become quite developed. The fundamental driving characteristic of designing concrete subjected to forces is that concrete is quite strong in compression but weak in tension. Plain concrete cracks under very small tension stresses. As a consequence, reinforcing is placed within areas expected to experience cracking. Steel is commonly used, but other materials are possible. Likewise, many kinds of more sophisticated approaches involving pretensioning or post-tensioning are in common use. Design procedures are well developed but based on empirical understandings, given that stress-deformation relations are nonlinear.

The obvious routes for using nanomaterials to improve concrete are generally either in process considerations (ease of mixing, rate of setting, etc.) or in property enhancements. The first area is very promising. The primary ingredients of modern cement are lime, silica, alumina, and iron oxide. Raw mixtures are crushed, ground, and fed into a high-temperature kiln, where the mixture is chemically converted into a cement clinker. Small amounts of other materials (e.g., gypsum) may also be added. This cement mixture hardens when mixed with water in hydration reactions. Calcium silicate hydrate (C-S-H) is formed in particular types of organizational structures that act to bind the mixture together. The speed and nature of these reactions is, in turn, influenced by the

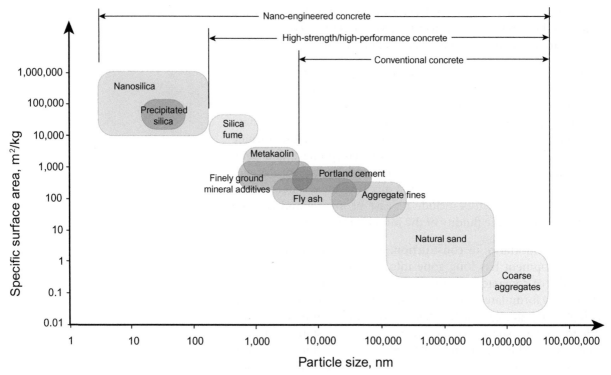

FIGURE 9.7

Nano-engineered concrete. Nanoparticle sizes have very high specific surface areas in comparison with commonly used materials in a concrete mix. (Adapted from K. Sobelev.)

relative surface areas of the cement particles. Relative surface areas of other materials in the mix are also important, thus presenting an opportunity for the use of nanomaterials, with their inherently high surface-to-volume ratios.

Relative specific areas for various materials are shown in Figure 9.7. As shown, particle sizes for conventional concrete are fairly large. To improve performance, finer particulates have been used. Fine silica fume, for example, has long been used in modern cements as an additive or as a replacement for conventional cement. The addition of yet smaller nanoparticles to the complex of pores and crystals of cement paste can be expected to yield further improvements. Nanosilica particles, for example, have much smaller diameters than those of silica fume and, correspondingly, have much larger specific surface areas. This radical increase in surface area can affect surface energy and morphology and alter chemical reactivities. Nanoparticles can also act as crystallization centers, and crystal sizes can potentially be reduced. The smallness of nanoparticles can potentially allow better void filling and positive filler effects and improve bonds between pastes and aggregates. It has been suggested that the specific ways that nanoparticles organize themselves within the mixture are an extremely efficient close-packing structure. Expectations are that shear and tensile strengths as well as toughness will be improved. Other anticipated process improvements include increasing liquid viscosities so that grain suspension is enhanced, which in turn improves workability.

Experiments with nanomaterials in mortars and concretes have been limited, but results are promising. Nanosized iron oxides and nanosilicas have been found to improve strength beyond what has been obtained with silica fume. Mixture design is quite complex, so other needed properties, such as slump flow, are not adversely affected, nor are their undesirable side effects, such as bleeding or segregation. General workability has been improved. Though experiments so far are few, the prospects for improvements in various areas look very promising. Efforts have already been made to commercialize nano-based admixtures in liquid form to be used as an admixture.

As mentioned, cracking under tension stresses has long been one of the driving material characteristics of designing concrete shapes. There have been seemingly uncountable numbers of experiments with mixing in various fibrous materials, such as various polymeric strands or metallic fibers, into concrete mixtures with the intent of creating a generally homogeneous concrete composite that has crack-arresting characteristics. In addition to dispersion approaches, various layered approaches have also been tried. Some have been more or less successful for particular applications, but many problems with fiber approaches remain and prevent them from being widely used in many larger applications. Concrete paste is highly alkaline and can attack many embedded materials. Bonds are very problematic. Strengths are low. For building- or bridge-scale structural elements, traditional steel reinforcing remains the primary solution. Though nanomaterial experiments have suggested some welcome increases in fundamental tension strengths, cracking under large stress levels still remains problematic and poses a particular challenge to researchers in this area.

A particularly interesting approach to the cracking problem has been the exploration of "self-healing" or "self-repairing" concretes. As described in much more detail in Section 10.2, these interesting materials have been explored at the University of Illinois since the early 1990s, and this work has also spawned comparable approaches with other materials. The general approach consists of developing small, dispersed encapsulated particles or tubes inside a mix that are filled with materials that harden when exposed to a catalyst. Developing cracks are intercepted by these encapsulations and ruptured. The released materials act to fill cracks and interact with catalysts that in turn cause the released material to harden. Self-healing materials are discussed in more detail in Section 10.2.

The use of nanoparticles is also expected to have benefits that go beyond mechanical improvements. As discussed in Chapter 11 on environmental issues, some forecasts suggest enormous potential reductions in released carbon dioxide levels because of potentially reduced kiln temperatures and durations that occur during process-ing. Due to the enormous volumes of cement produced in the world, quantities could eventually be huge if nanomaterials are extensively used. Strength improvements also generally yield smaller required overall quantities.

Damage monitoring and responsive structures

Nanotechnologies offer interesting promises within the realm of active structural health monitoring (e.g., damage assessment) and that of active responsive structural systems. In general, all active structural systems include a sensory capability for detecting and communicating some phenomena induced by a force environ-ment, such as deformations, deflections, cracks, and vibrations. A more sophisticated and responsive system would include a control system that interprets sensory inputs and causes some actuation mechanism to provide a response, typically one that mitigates undesired phenomena.

Damage assessment needs are a common motivator for using active systems. Often called *structural health monitoring* systems, a number of approaches are already in use, depending on the phenomena of interest. Embedded fiber or piezoelectric technologies, for example, can be used to assess breaks, sharp bends, strains, and vibrations. Magnorestrictive and other smart materials can be used as well. Here the intent is primarily to provide information about the type and location of structural damage in a structure. The capabilities of nanotechnologies to serve as sensors opens the door for applica-tions in this area.

Systems that not only detect various phenomena but provide responses as well include many different types of deflection or vibra-tion control devices. The control of vibratory phenomena through damping actions is central to the performance of many structures, ranging from the very large scale to the smaller scale. At the large scale is a strong need to mitigate the highly damaging dynamic motions associated with earthquake effects on building structures. Various kinds of large conventional damping systems are already in use, and interesting new devices based on *magnetorheological* or *electrorheological* fluids that change their viscosities in the presence or absence of an electrical or magnetic field are also being developed. As discussed in Chapter 7, the particularly interesting magnetic and electrical properties of nanoparticles promise to improve the efficiency and responsiveness of these kinds of devices. They would be particularly useful, for example, in large cable-supported or cable-stayed structures for which wind-induced vibrations are a significant problem (see Figure 9.8).

At the smaller scale are innumerable cases of the need for supports under machinery whose operations cause intense vibrations, to prevent these same vibrations from being propagated throughout the remainder of the structure. In addition to conventional base isolation devices, new approaches are based on the use of piezo-electric and other materials. A piezoelectric device converts input mechanical energy associated with strain deformations into electrical energy. In some high-end skis, for example, troublesome vibratory "chattering" on downhill slopes can be mitigated by a piezo-based system. The vibrations cause bending and bending strains in the skis that are converted into output electrical energy that is subsequently dissipated in an electrical shunting circuit. The response is thus virtually instantaneous. Piezo effects can be enhanced by using nanomaterials.

Active structural systems typically use embedded microprocessors or other forms of computer-based control systems. The impending revolution in chip size and related computer sizes that is anticipated to result from developing nanotechnologies will inherently benefit the development of active structural systems. Similarly, the need for energy access and storage should be positively addressed by the developments in fuel or solar cell and battery technologies that are based on nanotechnology developments. Ultimately it should prove possible to create active structures in which not only sensory, actuation, and control functions are directly integrated into the product or system but energy access is as well, particularly for relatively small-scale or small-size systems.

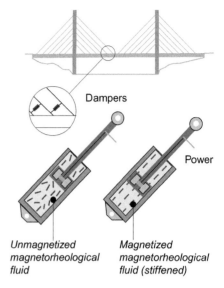

FIGURE 9.8

Applying a sensor-controlled current creates a magnetic field that causes the viscosity of the magnetorheological fluid to vary and damp out unwanted cable vibrations.

9.3 THE THERMAL ENVIRONMENT

Two fundamental types of thermal environments are of interest in the following sections. The first type can be described as *tempered* spatial environments—normally interior—in which people live and work; the second as *thermal environments* intrinsic to a product or that surround it and influence the way they are designed. In each of these primary cases, the thermal properties of materials play an essential role.

In tempered spatial environments, the design objective is invariably to provide a thermal condition conducive to the health, well-being, and comfort of occupants, whether the environments are in buildings, automobiles, or aircrafts. Surrounding material systems form a boundary between the space and the exterior. The geometric characteristics of the space, the thermal properties of surrounding walls, or the kinds of glass present in fenestrations are all influential in defining the thermal environment actually perceived by occupants. Physical mechanical systems that provide heating or cooling are fundamental contributors to creating and maintaining this environment.

In products, attention is invariably directed toward the way heat or extreme temperature differentials affect the design of the product and to assure that the product effectively continues to provide its intended role. In some cases, the surrounding environment might have extremes—especially high or low temperatures. Maintaining the functioning of a product in an extreme environment is not always an easy task, especially since many of the mechanical properties of materials, including strength and stiffness, are temperature dependent. In other cases, the role of a product might necessitate the inclusion of components that generate heat, which in turn must be managed. Often there must be components that mitigate heat transfer, provide cooling functions, or otherwise provide this heat management. In these design situations, the intrinsic thermal properties of materials are of fundamental importance. Heat affects not only mechanical properties but electrical and optical properties as well (see the following discussion).

Basic Heat Transfer

In designing both types of thermal environments we've outlined, the fundamental means of heat transfer—conduction, convection, radiation—are of intrinsic importance (see Figure 9.9) as are the intrinsic thermal properties of materials. Chapter 4 addressed thermal properties in detail. Here we only briefly summarize salient issues to put the following discussions into context.

Heat is a form of energy associated with molecules in motion. Conductive heat transfer takes place in solids because of temperature

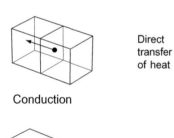

Conduction

Direct transfer of heat

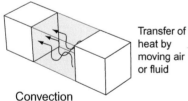

Convection

Transfer of heat by moving air or fluid

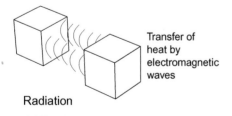

Radiation

Transfer of heat by electromagnetic waves

FIGURE 9.9
Basic heat-transfer mechanisms.

differences between various parts of the solid. Thermal energy is transferred from hotter to lower regions via vibrations of adjacent molecules or the movement of free electrons through the material. Metals, with many free electrons, are good conductors of heat. Convective heat transfer takes place in a fluid (gas or liquid) through molecular motion and the circulation of currents. It can be divided into natural convection and forced convection. Natural convection is driven by the temperature gradients that cause density gradients in gases or fluids, and as a result of these phenomena, buoyancy forces occur and cause fluid movements. Forced convection is caused by the some external force (e.g., a fan or pump). Radiation is a flow of thermal energy from one source to another in the form of electromagnetic waves and does not depend on a specific medium of transfer.

Each of these heat-transfer means is in turn dependent on the physical properties of the materials themselves, including their heat capacity, expansion coefficient, conductivity, phase change, or melting points. As discussed in detail in Section 4.4, these properties are reflected in various specific measures or coefficients useful with a design context. Important measures for a material discussed in Section 4.4 include the following: the melting temperature, T_m, and the glass transition temperature, T_g—both of which are important measures when the material dramatically changes its structure and characteristics; the maximum and minimum service temperatures that bound the useful design range, T_{max} and T_{min}; the specific heat capacity, C_p; the linear thermal; and the expansion coefficient, thermal diffusivity, and emissivity. Many other properties, such as the creep rate (deformation over time) of materials, are also directly affected by temperature changes.

Many of these properties can be potentially altered by the use of nanomaterials (see chapter 7). These and related topics are discussed in the following sections after a brief review of how spaces and products are designed with respect to thermal issues.

Thermal Environments in Products

Many common products contain components that generate, transfer, store, dissipate, exchange, buffer, or otherwise manage heat. Some components, such as lighting devices, generate heat as a consequence of their use rather than as an intention *per se*. Heat is invariably generated in the conversion of one energy type to another, and any device that does this, such as a common electric motor, becomes a heat source. Other components might have

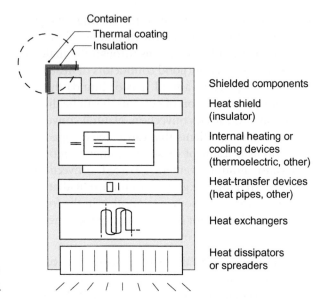

Container
Thermal coating
Insulation

Shielded components

Heat shield
(insulator)

Internal heating or
cooling devices
(thermoelectric, other)

Heat-transfer devices
(heat pipes, other)

Heat exchangers

Heat dissipators
or spreaders

FIGURE 9.10

*Typical thermal components in a product in which
nanomaterials or nanotechnologies could be used.*

the primary function of being a source of heating or cooling—for example, thermoelectric devices. Other components—various insulators—are intended to either prevent or control heat loss or gain or provide a shielding function. The function of other components might be to deliberately and effectively transfer heat from one point to another, as is the case with common heat pipes or thermosiphons. Many devices are intended to store heat or to exchange heat between components. These kinds of components are used to form systems of thermal management and control within a product, to make it work effectively (see Figure 9.10).

In other situations, products or devices must be designed to work within surrounding extreme hot or cold temperature environments that might cause degradations or breakdowns of normal materials and product functions. Designing these many kinds of products invariably involves manipulating heat-transfer mechanisms (conduction, radiation, convection) via working with the thermal properties of materials.

Management of any present thermal environment is necessary for the successful operation of any device, even those primarily intended for other purposes. A common laptop computer, for example, contains many elements that generate, transfer, and dissipate heat. The microprocessors that form the technical core of a laptop generate surprisingly large amounts of heat. The performance of a microprocessor in turn depends on the temperature of its surrounding environment.

High temperatures can cause the performance of the microprocessor to be low. Hence the generated heat must be removed or dissipated. Many common devices are used to accomplish this function. Heat pipes, described in the following discussion, are commonly used, for example, to transfer heat from a microprocessor to some type of heat sink or dissipater in a unit. In this particular example, the needed capacity of the heat pipe to transfer heat could be fairly easily determined from the specifications of the microprocessor. Improvements in heat transfer that are possible via the uses of nanofluids and highly conductive nanomaterial composites (see the following discussion) would enable the heat pipe to be made smaller and/or to conduct heat away more quickly.

The aim of many other devices is to directly provide a heating or cooling source or environment. A common refrigerator, for example, is intended to provide a controlled cooling environment. The intended use provides specifications for how much heat is to be removed or what kind of temperature environment is to be provided. Design questions then hinge around what is the best set of devices and material constructions for accomplishing these ends. Invariably the desire is to have the present basic energy conversion device be highly efficient (and thus use as little external energy as possible) and be as small as possible. For many products, added design issues include quietness of operations, lack of vibration, and so forth. Improved solid-state components that are inherently quiet, such as thermoelectric devices (see the following discussion), and that are based on nanotechnologies promise to meet some of these design needs.

Designing these kinds of thermal management systems in most devices normally hinges around selecting materials and developing related components that in some way either maximize or minimize heat transfer via conduction, radiation, or convection, depending on the design circumstance. For materials used directly, common property measures such as thermal conductivity can be used directly. Many devices that might accomplish some overall heat management function, however, are quite complex and might simultaneously involve several mechanisms. The heat pipes noted previously, for example, appear simple and have a clear overall function of transferring heat from one point to another but involve the use of not only thermoconductive materials but an internal fluid in both liquid and vapor phases. These and other more complex forms of devices are discussed more extensively in the following sections. Certainly the potential of various nanomaterials to optimize material properties is great.

Variations in temperature levels can cause many materially oriented design problems to develop. Thermal expansions and contractions, for example, can cause thermally induced stresses to develop. It was also noted that the actual physical properties of materials can be markedly temperature dependent. In common steel, for example, the ductility of the material decreases with decreasing temperature levels. In many normal products that operate in normal environments, general levels of heat and temperatures involved can be fairly low, and these kinds of temperature-related effects are often not particularly critical. Materials are often selected primarily because of their thermal conductivity levels to provide either heat-transfer functions or insulation functions. Often demands placed on these materials are multifunctional, with thermal properties playing only one of several needed roles. A common material used to provide thermal insulation on a handle attached to a hot object, for example, must provide not only the needed thermal insulation but also must be abrasion resistant and be tough against tears or penetrations. Manufacturability and costs are clearly important as well. Selecting materials when multiple criteria are present typically involves tradeoffs. (See the discussion on this topic in Chapter 5.)

Design demands increase when extreme temperature environments, either hot or cold, are present. With increasing temperatures, not only do expansions increase but the elastic modulus can decrease, strength can decrease, creep can increase (time-dependent deformations), and the material can even melt. Changes in electrical conductivity and other properties can also occur. Of particular importance are the maximum service temperatures that define the maximum usability range for a material. Other products may be expected to work at extremely cold levels. Other effects occur here as well. A change can occur from ductile to brittle behavior. As we will see, a number of opportunities exist for using nanomaterials to improve material capabilities in extreme temperature environments.

Thermal Environments in Spaces

As previously noted, individuals act, live, and work within tempered spatial environments (buildings, trains, and the like). The thermal characteristics of these environments are fundamental to maintaining conditions conducive to the health, well-being, and comfort of occupants. Thermal environments of this type are by no means easy to characterize and certainly go beyond only a consideration of the temperature of the environment. Within a space, air is by no means static. Sources of heat loss (such as through walls or windows) or

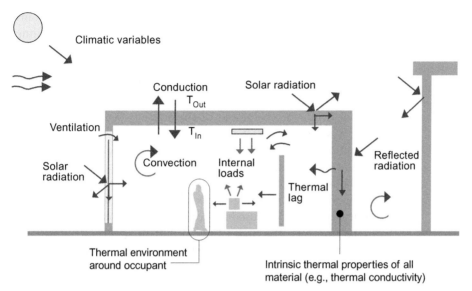

Climatic variables

Conduction T_{Out}

Solar radiation

Ventilation

T_{In}

Solar radiation

Convection

Internal loads

Reflected radiation

Thermal lag

Thermal environment around occupant

Intrinsic thermal properties of all material (e.g., thermal conductivity)

FIGURE 9.11

Considerations in designing spatial thermal environments. All the factors noted in the text affect the nature of the thermal environment as perceived by an occupant.

heat gain (lighting elements, people, equipment, direct heating sources) may be distributed in different ways throughout the space and can interact with one another via fundamental methods of heat transfer (conduction, convection, and radiation). Turbulent and laminar air movements consequently invariably exist, as do related temperature variations. The way a human responds to or perceives the thermal environment that is present further complicates the issue of how to design thermal environments of this type because of their own highly complex thermoregulation characteristics. Evaporation considerations enter in that in turn require control of relative humidity levels in relation to temperatures. Ventilation levels are also important—not only for human comfort but also to aid in assuring that a clean-air environment is provided. All these factors must be taken into account (see Figure 9.11). Creating environments that actually achieve high levels of human comfort is thus a surprisingly difficult task—and one that is often not successfully achieved (as most individuals already know from everyday experience). Designing thermal environments is further complicated by the inextricable connection to energy use and larger societal goals related to energy conservation.

As mentioned, there are three basic heat-transfer mechanisms, and heat is always transferred from a place with a higher temperature to a place with lower temperature. In buildings, this means that during the winter heat transfers from the warm space inside to the cold outside environment, and vice versa during the summer. Transfer through a building surface depends on convection on the inner side of a surface, conductance through the surface, and convection on the outer side of surface. Walls or roofs in a building typically do not consist of just one layer; they usually have multiple layers, with each of these layers having a different thickness and a different thermal conductivity. Instead of considering all these layers separately, it is common practice to use just one value that will take into account all of them. This single value, which combines convective heat transfer at both sides of the wall and the conductance through each layer, is called the *overall heat transfer coefficient*, or *U-value*. It can be calculated and it has units of $W/(m^2K)$. The U-value is not a physical measure; it is used for convenience and easier calculations of heat transfers. The bigger the U-value, the larger the amount of heat that is transferred. If the U-value for a wall is known, it is easy to calculate the amount of heat loss or heat gain by multiplying the U-value by the temperature difference and by the area of the wall. *Thermal resistance*, or *R-value*, is the inverse of the U-value (units are m^2K/W). The R-value tells us how good are the insulating properties of an overall building assembly. Hence, higher R-values mean better insulating properties.

Normal design practice generally includes a determination of heating and cooling loads from external or internal sources, an exact description of the geometry of the building or space as well as associated material distributions, and the subsequent development or use of thermal modeling approaches that analytically predict the characteristics of the thermal environment expected to be present. Design parameters are varied to achieve prespecified temperature and relative humidity levels and other criteria related to human comfort. Some simple kinds of heat-transfer characteristics can be done by hand calculation—for example, the conductive heat loss through a wall can be found by simple considerations involving the temperature differential present among wall faces, the thermal conductivity coefficient of the wall material, and the thickness of the wall—but most analysis algorithms that have the robustness and breadth needed to completely predict the thermal characteristics of an environment are based on extensive energy balance and thermal modeling approaches that are quite involved and hence normally done within a computer environment.

Thermal loads may come from many sources. For buildings, the external environment provides a primary load source. Obviously, designing a thermal environment in a building in a hot-humid climatic area is quite different than designing one for the cold-dry areas of the far north. Typical input data for a robust thermal model include general climatic and weather data for the specific location of the building. This information necessarily includes external temperature and humidity variations, sun angles, characteristic wind speeds and directions, cloud cover, and similar data for the specific latitude and longitudinal location of the building. The amount of solar radiation on a particular exterior window produces a heat load within a space, for example, and depends on the exact angle of the sun in the sky, which in turn is latitude dependent. (The amount is also obviously dependent on the orientation of the surface and the presence or absence of shading devices—factors that would be reflected in the "building description" data described in a moment.) Since values for this kind of environmental data vary according with time, analytical models can be based on hourly, daily, weekly, or monthly data, with the precision of obtained results varying accordingly. Other thermal loads for spatial environments can be internally generated. Common lighting sources, for example, are significant contributors to the thermal loads within a space. Other internal load sources include machines (including computers) that give off heat and, of course, the occupants themselves.

It is also necessary to fully describe the geometric characteristics of a building or space, including the exact location, area, and distribution of various types of material surfaces and constructions. In a pragmatic sense, this description normally includes defining how surrounding walls, windows, floors, and roofs are constructed. Many walls, for example, use layered constructions of various materials, each of which must be defined. Roofs are invariably constructed differently than walls. Locations and areas must be carefully defined and relevant material properties, such as thermal conductivities or emissivities, assigned to each. Locations of heating or cooling sources (or special sinks) that affect the space must also be defined—for example, locations of heating elements or air diffusers. Many instances involve multiple thermal zones that are interconnected. Each must be defined. For analyses that address energy analysis as well, primary and secondary mechanical systems need to be defined too.

A great number of types of analytical simulation programs are in use, each with differing purposes. One general approach is focused on energy analysis in relation to thermal loads. Based on input data, simultaneous heat-balance calculations of radiant and convective effects on interior and exterior surfaces are determined, as is transient heat conduction through elements such as walls, windows, floors, and roofs. Some programs include a form of ground-heat transfer modeling as well. These calculations are made for each time step selected (e.g., hourly, daily). Thermal comfort models based on temperature and humidity levels are included. Outputs include space temperature and humidity predictions, energy consumption in relation to selected steps, and other information.

Though models of the type we've described are in wide use, particularly for energy consumption analyses, they are not as useful in describing the exact nature of air-flow distributions or other thermally related phenomena within spaces. Here it is better to use computational techniques that have been evolved to analyze fluid or gas flows. Computational fluid dynamics (CFD) models were developed within the fluid mechanics fields to solve many kinds of problems that involve the movement of fluids, particularly in relation to the interactions of fluids or gases with surrounding surfaces of various types. These models can predict airflow, heat transfer, and the distribution of particles (such as contaminants) in spaces and around buildings. In these models, a fluid or gas, such as air, is first characterized as a series of discrete small cells to form a volumetric mesh. For normal viscous fluids, the famous Navier-Stokes formulations based on conservation laws can then be applied to solve equations of motions. (These equations describe momentum changes in infinitesimal volumes as the sum of various forces acting on them, such as changes in pressure, gravity, or friction. Solving the equations involves literally millions of calculations for even a relatively small mesh—a task that's clearly for computers.) All heat-transfer modes, including conduction, convection, and radiation, can be included.

Other solution types not involving meshes are possible, but the mesh-based approach is commonly used. As might be expected, however, there are numerous subtleties in establishing and working with CFD models that are beyond the scope of this text. Inputs again involve geometrical and material description of the spaces, heat sources and sinks, and so forth. Outputs include a variety of analytical results. Three-dimensional depictions and associated numerical data can be obtained that show the distribution of air flows throughout a space. Three-dimensional convective flows associated with a heat source, for example, can be clearly seen and numerically described. Mean radiant temperature and heat-flux distributions can be determined and visually depicted. Surface temperatures can be determined. Particle animations, contour animations, isosurfaces, and other depictions are usually possible.

These various kinds of analysis tools allow designers to vary critical parameters related not only to geometry but to material type and distribution as well, to achieve desired performances. They allow a designer to decide whether a material should desirably have more or less heat capacity for a particular situation or how changes in the thermal conductivity of a material would affect outcomes. As might be expected, it has been generally found that designers want better control over each of these material property parameters and want not to be bound by the particular combinations of properties associated with traditional materials. Extreme values are often desired. To conserve energy, minimizing heat loss through a material with a very low thermal conductivity would be a natural direction to pursue and might suggest, for example, the use of some type of foam or other porous material as an enclosure or wall material.

Of particular importance in building design, however, is that most materials used must be multifunctional to a greater or lesser extent. It almost goes without saying that the need for visual access through a wall is overriding and is normally provided through windows or glass façade systems that entail the use of a transparent material (such as glass or a transparent plastic); at the same time, energy conservation issues still suggest that the same material provide as much thermal insulation as possible. This same material also must have some significant strength and stiffness properties (for example, resistance to wind forces or even common handling or installation-induced forces). Meeting these various demands normally involves tradeoffs (recall the discussion in Chapter 5). As we discuss in a moment, nanomaterial technologies provide opportunities for innovations here.

Application of Nanomaterials in the Design of Thermal Environments

Manipulating properties: Insulating and conductive materials

A common design objective or need is to increase or decrease the thermal conductivity of a material to enhance its ability to conduct heat, or, alternatively, to improve its dialectric (insulating) characteristics. In some cases a device might need to efficiently carry heat from one point to another, whereas in other cases a shielding or insulating function is needed. Nano-related mechanisms for accomplishing these ends vary.

We noted in Section 4.4 that heat is transported through solid materials in several primary ways. The role of the "mean-free path" was noted as well. Thermal conductivity can be decreased (and the material made to act more like an insulator) by altering the mean-free path via the introduction of impurity atoms or very fine dispersed particles. This effect has long been known. Thus common approaches intended to increase the strength of a metal can also lead to changes in thermal conductivity. To decrease thermal conductivity, finely dispersed nanoparticles or nanocrystals can be used to create a matrix of scattering centers that can reduce conductivities and create materials with better insulating capabilities. Here, the various shapes of the nanoparticles can potentially be an important variable.

Thermal conductivity in certain materials can also be *increased*, instead of decreased, by introducing nanoparticles. The thermal conductivity of many polymers or fluids that are normally nonconductive, for example, can be enhanced via the inclusion of carbon nanoparticles (for example, graphite). Shape considerations are important here. Generally, spherical nanoparticles are less effective than carbon nanotubes or nanofibrils. The latter have particularly interesting thermal properties that allow them to be highly conductive in one direction but less so in the other direction. Thin films under development with oriented nanotube directions can have remarkably high thermal conductivities (see Figure 9.12).

A particularly important area for improving thermal conductivities is in connection with polymers, normally poor heat and electricity conductors. Not only are there design situations in which the intent is to have a thermoset or thermoplastic transfer heat effectively—as either a transfer device directly or a heat dissipater—but there are also occasions on which poor heat conduction can cause the plastic to rapidly absorb heat and increase its temperature to the point at which material degradation occurs. Various kinds of nanomaterials have been explored for use as additives in polymers to increase their thermal conductivity.

Introducing porosity into any material has long been known to improve the thermal *insulating* characteristics (decreasing their conductivity) of materials. Porous materials consist of solid matrix and gas inside the pores. Their good insulation properties are achieved due to the very small thermal conductivity in gases compared to solids or liquids. For example, the thermal conductivity in air is 0.025 W/(mK); water is 0.6 W/(mK); and aluminum

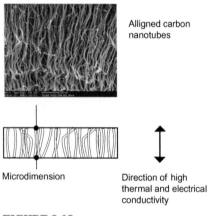

Alligned carbon nanotubes

Microdimension

Direction of high thermal and electrical conductivity

FIGURE 9.12

Thin film of aligned carbon nanotubes with high thermal conductivity. (Courtesy of C. Richards, et al.)

is 237 W/(mK). Overall thermal conductivity of porous materials depends on convection in the pores, conduction in the pores, conduction in the solid, and radiation. Heat transfer in porous materials is very complex, and individual mechanisms are complex and cannot be easily summarized. Briefly, conduction in a porous material increases with the increase of pressure around it and inside the pores; convective heat transfer increases with the increase of the motion of air inside the pores; and radiation increases significantly with an increase in temperature and with an increase in pore size. Hence overall thermal conductivity decreases with a decrease of the pressure inside the pores, a decrease in temperature, and a decrease of pore diameter. Conduction in the solid part of the porous matrix is defined by the type and amount of material used. Direct material conduction effects are reduced due to the very large percentage of porosity normally present.

Designs intended to improve insulation qualities depend on varying these several parameters. In buildings, temperature levels are given design parameters that cannot be varied much. Pore diameter can be varied with material interventions. In some device designs using porous materials, the pressure inside the pores can be varied. There are many advantages to introducing nanoscale pores. In nanoporous materials, convection inside the pores is largely negligible due to the small space available for the motion of molecules. The "mean-free path" (see Section 4.4) of air molecules is larger than the size of pores. So, though in conventional porous materials molecules of gas mainly collide with each other inside the pores, inside nanopores they also constantly collide with the pore wall, which leads to suppressed gas conduction. Conduction in the gas will also decrease with any decreases of pressure that can be obtained. Ideally, a vacuum inside the pores produces the best insulating properties. Practically, the smaller the pores become, the smaller is the vacuum that needs to be achieved for the same insulating properties. Radiation is lower for small pores and for low temperatures. From all these considerations, the value of using nanostructures is significant. The goal is to create incredibly small pores and evacuate them, which would significantly reduce conduction in gases, decrease the radiation (because of the small pores), and reduce the effect of conduction in solids due to large porosity.

There are a great many ways that nanotechnologies can be used to create one or another type of nanofoam. The intent is invariably to create porous structures that entrap air. A well-known example of a highly porous nanoscale material with high insulating properties

Aerogel

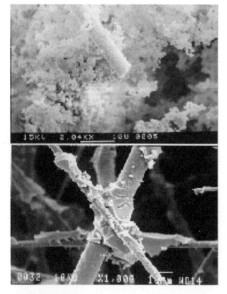

FIGURE 9.13

Insulating properties of an aerogel. (Courtesy of NASA.)

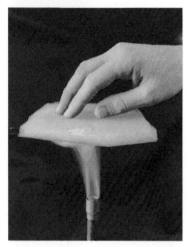

FIGURE 9.14

SEM micrographs of Aerogel/fiber composites (pore size: 20 nanometers).

is a material called *aerogel* (see Figures 9.13 and 9.14). Aerogel is a solid-state substance similar to a gel but with a gas instead of a liquid in the pores. Nanoporous materials are synthesized by sol-gel techniques. The result is a light foamlike structure that is made up of 3-D continuous networks with a gas (typically air) trapped within. Aerogels were discovered a great many years ago but were highly costly, brittle, and hard to work with. Improvements in manufacturing processes have yielded better and more cost-effective aerogels. The density of some of the newer and more exotic aerogels can be extremely small, and they are the lightest materials known (up to 99.8% air and 0.2% silica dioxide, or about 5 kg/m^3). Pore sizes are normally less than 100 nm. Other aerogels, however, are more dense and provide more practical materials for use as insulators in buildings and products. Densities of 30 kg/m^3 are typical. Aerogels feel hard to a light touch, but they imprint and shatter easily when high forces are applied. Nonetheless, they have the minimum mechanical properties (strength, stiffness) to stand up to normal product uses.

Two primary types of manufacturing processes are used to produce silica aerogel: supercritical drying and silyation. Supercritical drying produces very light aerogels but is an expensive process. Silyation processes produce less dense materials but are more cost effective and have been commercialized. Both start with a formation of a wet gel via a sol-gel process (see Chapter 8). The last part of the process is the drying phase, when the liquid inside the gel is removed. Several supercritical drying processes have been developed for this purpose, including high-temperature drying or lower-temperature solvent exchanges. The latter is now more widely used to make transparent aerogels. Sheets can be formed that can in turn be placed within transparent glass sheets for windows and other products. The silyation process produces transparent granules. Silyation is fundamentally a surface-coating process that can be done at normal temperatures. During this process, normally hydrophilic surfaces become hydrophobic. Briefly, the process involves gel forming, developing a hydrogel, the silyation process involving a solvent change by reaction and phase separation, and drying at normal temperatures. Products from the silyation process in the form of translucent granules or opaque beads can be used in many opaque, translucent, or semitranslucent insulation panels, solar cells, and other products. Figure 9.15 illustrates a product form. Panels with aerogel fillings have become widely used in architectural applications—often as roofs for heavily sky-lighted areas such as over swimming pools or greenhouses. Panels are not transparent, but are translucent and do transmit light (see earlier Figure 3.14).

Figures 9.16–9.18 illustrate a general comparison of several conventional materials and various aerogel products with respect to their insulating properties within a normal building context. Aerogels have particularly good thermal operating condition ranges, making them very valuable in many product and building situations.

Recall that the pressure state in the pores is also an important factor in reducing the conductivity of a porous material, in addition to pore size reductions obtained through materials such as aerogels. New vacuum-insulated panels (VIPs) have been produced that use aerogels or other highly porous nano- or microscale material within a vacuum (see Figure 9.19). These remarkable panels can have up to 10 times smaller thermal conductivity than conventional insulation panels (0.004 W/(mK) versus 0.037 W/(mK) for mineral fiber insulation or 0.024 W/(mK) for polyurethane foam). This efficiency, in turn, decreases insulation thickness needs. Panels can be made quite large, and are suitable for either product applications or in large-scale buildings where they are typically used as parts of exterior enclosure systems.

For example, in designing a passive house, there might be around 300 mm of conventional insulation panel plus thickness of other elements (about 500 mm in all). If vacuum-insulated panels were used so that the same U-value would be obtained, about 40 mm in thickness would be needed for the panel, plus the thickness of other elements, for a total of about 200 mm. Thickness reductions are remarkable, albeit costs are quite high. Materials usually used as the core materials of vacuum-insulated panels are aerogels or fumed silica. Those materials have very low thermal conductivity, do not burn, and have very good acoustic properties. A big problem with these sealed panels, however, is the penetration of gases (especially water vapor) from the environment into the material, which leads to degradation of the panels' thermal resistance (often referred to as *aging*). Permeation is forced by the difference of environment pressure and the pressure inside the pores, which is considerably smaller. Gases go from the higher pressure regions to the lower one, which means from the environment to pores, and this results in an increase of pressure inside the pores. As the pressure and the percentage of water vapor increase with time, the thermal conductivity of material increases. This aging mechanism depends on temperature, relative humidity, the kind of building envelope, and the type of panel core material.

The outer layer has the important role of preventing gases from entering the panel. With a high-quality final layer (aluminum layers or metallized polymer films), the service life of these panels is usually

FIGURE 9.15
Flexible nanoporous aerogel blanket (AspenAerogel Pyrogel). (Reproduced with permission from copyright owner, Aspen Aerogels, Inc.)

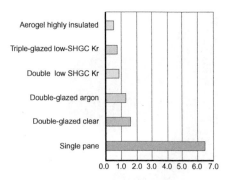

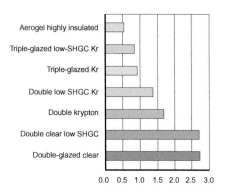

Comparison of U-values (W/m²K)

FIGURE 9.16

Comparison of the thermal transmittance of aerogel windows against other glazing systems.

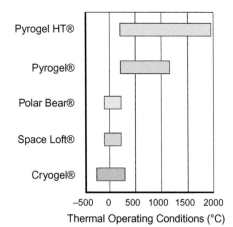

Thermal Operating Conditions (°C)

FIGURE 9.17

Comparison of thermal operating conditions for various Aspen Aerogel products.

between 20 and 38 years, which can be increased for few years by adding drying agents to adsorb the water vapor. VIPs are also very sensitive to mechanical damage, so they need to be handled carefully during transportation and installation in buildings.

Thermal barriers

The insulative qualities of materials can also be increased by various kinds of nanocoatings applied directly to external surfaces. Thermal barrier coatings are often in the form of sprays. Thermal sprays are thin coatings applied to the surface of a component that provide a thermal insulation function. (They can also be used for corrosion resistance.) Many are based on nano-titanium oxide (TiO_2) or zirconia (ZrO_2). Others are based on porous ceramic materials. Most of these thermal barriers are used in extremely hostile high-temperature environments, as in blades for turbines. Several approaches use multilayers. They are expensive and generally not appropriate for use as insulation barriers when temperatures are relatively low (as in typical building applications).

In making thermal sprays, typically heat-softened or melted metal particles are sprayed onto a surface. A large number of application technologies and material types are in common use. In a typical spray, a flattened grain structure parallel to the surface normally results that is both mechanically and diffusion bonded to the surface. When applied in air, chemical interactions, including oxidation, can occur and an oxide shell can be formed. Strengths are generally poor. Internal structures are normally fairly porous and voids can be present, which in turn contributes to their insulating capabilities. Different types of nanomaterial-based thermal sprays and powders have been developed that provide nanograin coatings, including ones based on alumina/titania. Nanocomposite polymers with ceramic nanomaterials have been explored for applications in which coating ductility is more important than strength.

Generally, adding nanoparticles to coatings improves bonds and smoothness. These coatings are smoother and stronger than traditional coatings. However, though nanoparticles can normally remain in suspension in a fluid, it is more difficult to actually spray them because of their very low mass. Hence considerable attention has been paid to developing effective spraying technologies.

Thermal buffers: Phase-change materials

An interesting approach to managing heat flow is through the use of thermal buffers associated with temperature-dependent "phase changes" in materials. Phase changes occur when materials change from a solid to a liquid, from a liquid to a gas, or vice versa. Many remarkable things happen at phase-change points, including a change in structure of the material. Of interest here is that they release or absorb the latent heat of fusion within the material at these points without a change in temperature. As a material goes from a solid to a liquid, it absorbs heat. A surprising amount of extra external energy input is required to make this solid-to-liquid conversion. Conversely, as a material changes from a liquid to a solid, it releases its latent heat of fusion. The amounts of energy released or absorbed are surprisingly high. There are many nano-based approaches that rely on phase-change characteristics. If intended to be reversible, phase-change materials can be encapsulated or used to fill nanoscale pores. Opportunities are particularly high with polymeric materials. Polymer nanoparticles can be made of phase-change materials.

The general phase-change phenomenon has been repeatedly exploited in several ways to control thermal environments. For example, microencapsulated phase-changing materials with inherently high heats of fusion have been embedded in high-end sports clothing, such as gloves, for uses in extremely cold environments. On one hand, a nonactive user can be quite cold, but, on exercising, the heat generated by a human during exercise can be problematic. The materials in this clothing are designed to undergo phase changes at specific points. The excess heat generated during exercise becomes absorbed by a phase change in the material as it transforms from a solid to a liquid. As the body cools, heat is released by the material back into the body environment, thus warming it.

This same general need to have a boundary material absorb or release heat under different circumstances is also quite common in architecture. For example, even typical radiant floor-heating systems made by burying hot-water pipes into a floor material can benefit from the use of phase-change strategies. Used in typical heavy-mass concrete floors, for example, the floors are slow to heat and slow to cool with external temperature swings. A consequence is that heat from the pipes is often released from the floor at the wrong time because of the thermal lag. Phase-change materials in encapsulated pellets can enable the radiant floor system to be much more responsive to rapid temperature variations. Figure 9.20 shows the effects of using phase-change materials in common wall and roof situations.

Heat-transfer devices: Heat exchangers and heat pipes

Many devices in common use are deliberately intended to transfer heat from one point to another or from one medium to another.

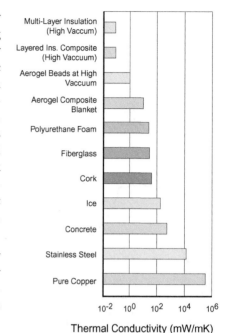

FIGURE 9.18
Comparison of the thermal condutivity of various materials (mW/mK).

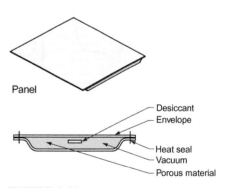

FIGURE 9.19
Vacuum panel with aerogel inside.

(a)

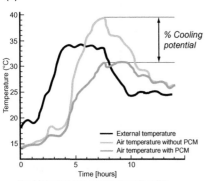

External temperature
Air temperature without PCM
Air temperature with PCM

Time [hours]

The PCM wallboard is constituted by a 5mm film with 60% PCM and a temperature of fusion of 22°C.

(b)

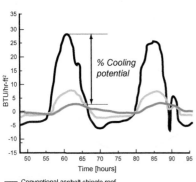

Conventional asphalt shingle roof
Metal roof, cool roof surface, reflective insulation and subventing
Metal roof, cool roof surface, PCM insulation, and subventing

(c)

FIGURE 9.20

Potential of (a) integrated phase-change materials (PCMs) to stabilize internal environments; (b) air temperature with and without PCM wallboards, and (c) heat transfer penetrating various roof configurations.

Heat exchangers are in common use. These devices transfer heat from a circulating fluid or gas on one side of a solid separator to a fluid or gas on the other side, but without intermixing. The fluids are separately channeled but adjacent to one another and separated by a thermoconductive material (see Figure 9.21). Heat from the hotter stream is transferred to the cooler stream, which in turn carries it off to be dissipated elsewhere. These devices can also be used as "heat recovery" systems. Heat exchangers are widely used in products everywhere, certainly including the automotive and building industries. Various types (plate, shell, tube, and so on) exist, as do the fluid media used. These devices can be on a massive scale for many building and industrial operations, or they can be extremely small devices forming parts of micro-sized products and use microfluidic techniques. Many exchangers at the micro level are made via advanced micromachining techniques. Though most heat exchangers involve fluids, metal-to-metal exchangers are also in use that are fundamentally based on similar principles.

Critical considerations in heat exchangers include the temperature differential present between adjacent media and their respective thermal characteristics, the thermal conductivity of the separation material for fluidic exchangers, and the relative surface-to-volume contact areas that are present. Enhanced thermal conductivity of the separation materials used can be achieved by any of the nano-based approaches previously discussed for improving thermal conductivities. For many types of smaller heat exchangers, the possibility of using nanofluids with particular thermal properties has been extensively explored. In general, the objective is to impart as high a heat capacity to the adjacent fluids as is reasonably possible so that effective heat transfers can take place. Hence many nanofluidic approaches use suspended nanoparticles with high heat capacities within a basic fluid. Since whether the fluid flow is laminar or turbulent can affect heat-transfer rates, the ultimate viscosity of the nanofluid must be carefully controlled.

Heat pipes are heat transfer devices used in many products, including common laptops, as a way of transferring heat away from microprocessors. These sealed containers effectively transmit heat from one end of the pipe to its other end by an internal phase-change evaporation and condensation cycle. Inside is a liquid and a wick-like capillary material. A partial vacuum is present. Heat transfer occurs by heat being absorbed at one end by vaporization of the internal fluid, then being released at the other end by condensation of the vapor (see Figure 9.22). The wick material brings the condensed liquid back to the vaporization end via a capillary action. Heat pipes come in a variety of sizes. Though typically at the macroscale and relatively long, miniaturized versions have been developed. For special geometries, conformal sheetlike versions have been developed.

Nanomaterials can improve the heat-transfer efficiency of heat pipes in three primary ways. The first is by improvements in the thermoconductivity of the surrounding pipe materials. The second is by altering the characteristics of the fluid. The third is by enhancing the needed capillary action. Improving the thermoconductivity of the container materials is possible through the use of nanomaterials (as previously discussed). Carbon nanotube composites have attracted research attention in this area because of their high thermal conductivities. Various kinds of nanofluids with desired high heat capacities can be used as the liquid. A positive property of nanoparticles used in this way is that they can remain in suspension, whereas larger and heavier microparticles often sink. The needed capillary action can be enhanced in a variety of ways. The increased surface areas of nanomaterials can be used to enhance capillary action. Surface treatments, including hydrophilic methods (see Chapter 10), can also be used.

Heating and cooling devices

Many devices directly provide some source of heating or cooling as their primary role. In product design, considerable use has been made of thermoelectric devices. Thermoelectric devices are based on the conversion of electrical energy into thermal energy, and vice versa. In 1821, Thomas Seebeck observed that voltage developed between two ends of a heated metal bar when a temperature difference was present—a phenomenon now known as the *Seebeck effect*. In 1834, Jean Pelletier observed the reverse effect. Traditional devices based on these effects now commonly use semiconductor materials, as illustrated in Figure 9.23. In operation, an applied voltage creates one face that heats up while the other face cools down. Semiconductors used between surfaces are fabricated of dissimilar thermoelectric materials (*n*-type and *p*-type) and connected electrically. For the material to be used as a cooling device, it is absolutely necessary that the heat generated on one face be completely transferred away from the cooling zone. Devices based on these principles are now common in use in a range of automotive and household goods as small sources of heating or cooling. They can be quite small or large enough to cool a sizeable drink container. They are widely used as cooling devices in computers.

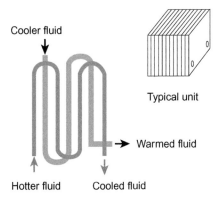

Cooler fluid

Typical unit

Warmed fluid

Hotter fluid   Cooled fluid

FIGURE 9.21

Heat exchangers are the workhorses of many common thermal systems and come in many sizes and scales (from the micro to building size). The use of nanofluidics with high heat capacities can improve efficiencies.

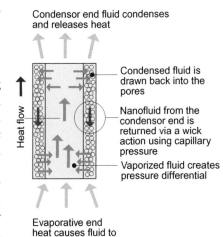

Condensor end fluid condenses and releases heat

Condensed fluid is drawn back into the pores

Nanofluid from the condensor end is returned via a wick action using capillary pressure

Vaporized fluid creates pressure differential

Heat flow

Evaporative end heat causes fluid to vaporize

FIGURE 9.22

A heat pipe transfers heat from one location to another. Nanofluids with high thermoconductive properties can enhance performance.

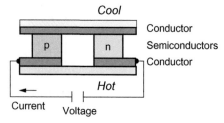

FIGURE 9.23

A typical thermoelectric device (a Pelletier device) based on semiconductor technologies. Application of an electrical current causes one face to heat and the other to cool. For the material to be used as a cooling device, the generated heat must be dissipated.

These devices provide sources of heating or cooling without any mechanical actions, nor do they involve any environmentally harmful materials—characteristics that are of great fascination to designers. Current devices, however, have severe limitations. With current thermoelectric materials, devices are woefully inefficient in terms of energy conversion (often less than 5%) to the extent that using them in other than limited sizes for special purposes is cost prohibitive. They remain relatively small in terms of size and capacity. With current thermoelectric materials it is intrinsically difficult to make large devices because of inherent limitations with deposition methods. Current depositions also have limited surface geometry possibilities that are not always able to provide the most efficient thermal gradients.

A commonly used measure to evaluate the efficiency of a thermoelectric device is called the *Figure of Merit*:

$$ZT = \frac{\alpha^2_{\kappa_e} \Delta T}{\lambda}$$

where α is the Seebeck coefficient or thermal power, κ_e is the electrical conductivity, ΔT is the temperature difference, and λ is the thermal conductivity of the material.

Of importance here is that a high Figure of Merit represents an efficient device. For many years, the limit of this figure for typically used materials remained about $ZT = 1$ for conversion efficiencies of around 7% or 8%, but higher values of ZT (2 or 3) are needed for improved efficiencies. Nanomaterials offer promising avenues of development here. Nanomaterials in the form of nanotubes and nanowires can potentially improve efficiencies due to quantum size effects induced by nanoscale dimensions.

As seen from the Figure of Merit expression, desirable thermoelectric materials should have a large Seebeck coefficient, low thermal conductivity, and low electrical resistivity (or high electrical conductivity). They should also be able to have shape variations. One key approach is based on the fact that the Seebeck effect can be enhanced through improvements in electrical conductivities. As noted previously, nanomaterials transport both heat and electrical charge in a way quite different than conventional materials. When at least one dimension is at the nanoscale, quantum confinement effects and the scattering of free electrons at surfaces change the nature of the way that electricity and heat are carried. These effects can be utilized to improve electrical conductivities. The use of quantum dots or nanowire structures offer immediate interest, since the kind of confinement present can enhance Seebeck effects.

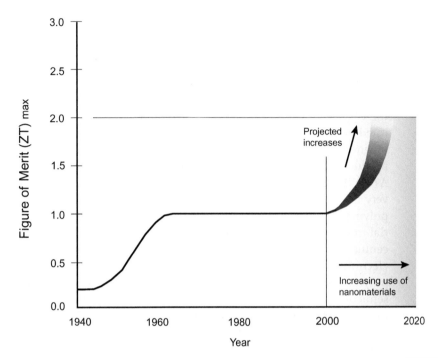

FIGURE 9.24

Projected improvements in the Figure of Merit efficiency measure for thermoelectric devices. (Adapted from A. Ortega, The Engineering of Small-Scale Systems: An NSF Perspective, ITherm, San Diego, CA, 2006.)

Here the proposal is to have a huge matrix of quantum dots held together by molecular bonding to replace the solid *p-n* materials normally used in thermoelectric devices. Electrical conductivities are reportedly increased while thermal conductivities are decreased. Other effects, notably increases in the Seebeck coefficient, can possibly occur by altering the density of states (see Chapter 6). Section 9.7 describes uses of carbon nanotubes, nanowires, and other approaches in more detail.

Developments with nanomaterials in relation to multilayer film structures or other conformal deposition processes may help in relation to shape issues. In multilayered films, enhancement to the Seebeck effect might not uniformly occur, however, and their use still has its problems. Other approaches suitable to larger thermoelectric material areas with controllable nanomaterial sizes and shapes may be possible as well. Material densities that contribute to the Seebeck effect can also be changed. Synthesis methods suitable for making really large thermoelectric areas, however, remain problematic.

Despite developments still in the research stage, the promise of thermoelectric devices with improved efficiencies and capacities is quite bright. Figure 9.24 is suggestive of improvements that have been made and could be made.

9.4   ELECTRICAL AND MAGNETIC ENVIRONMENTS

Electronic Materials

A staggering multitude of devices in use in today's society rely on the electrical or magnetic properties of materials. Highly sophisticated electronic technologies that are based on the electrical and magnetic properties of materials are found in a wealth of everyday products as well as throughout our living and work environments. As we discussed in Section 4.1, the class of electronic materials is very broad and can encompass a variety of metals, ceramics, and polymers. For example, copper has been used as a conductive material in everything from motors to transmission lines since the 19th century and is still widely used in interconnects in today's microelectronics; ceramic silica is used in optical fibers; and polymer polyimides are used as dielectrics. Silicon is widely used when semiconductor properties are needed. Gallium arsenide and germanium are also part of this class of materials. Basic electrical properties of these and other materials—such as resistivity, conductivity, and dielectric behaviors—are reviewed in Section 4.5. Magnetic properties are discussed in Section 4.6. There it was noted that magnetic properties can especially be enhanced via the use of nanomaterials.

Within the vast array of applications associated with electronic materials, nanomaterials are expected to make some of their greatest impacts because of their unique electrical and magnetic properties. Sections 7.3 and 7.4, respectively, discuss these properties. Section 7.8 notes particular characteristic uses of carbon nanotubes in this area. Nanoelectronic devices, the sizes of which are a few billionths of a meter, have emerged as a primary research and development area. Developments in related nanomechanical devices—nanoelectromechanical systems (NEMS)—are following closely. The technological sophistication present in these many areas is remarkable, and consequently so are the technical issues relating to exactly how nanomaterials might make contributions to this field—topics that are far beyond the scope of this book. This section only seeks to give a brief flavor of the kinds of ways that nanomaterials might be used and the types of impacts expected from anticipated developments in the nanomaterial and nanotechnology field.

General Trends

In recent years, technological progress in the development of electronic materials has been remarkable. The timeline of materials originally presented in Figure 2.1 of Chapter 2 labels our current age the Silicon Age for good reason. Silicon is, of course, the primary material currently used in a remarkable array of semi-conductor devices that form the basis for the millions of micro-processors used in current electronic applications. Though silicon has long held a premier place in the world of electronics, discussions in the literature of the future of electronics suggest that silicon-based technologies for microprocessors are reaching limits and that alternative technologies are needed, including technologies based on the unique electrical, magnetic, and optical properties of various nanomaterials, such as semiconducting carbon nanotubes or other materials (see Further Reading at the end of this chapter).

In the complex world of electronics and electronic devices, micro-processing units are, of course, but one component. Other areas that can potentially benefit from the use of nanomaterials include memory units and interconnects. Some of the expected trends in this field are broadly illustrated in the technological progression forecast by the International Technology Roadmap for Semiconductors (ITRS) group (see Figure 9.25). The figure clearly shows that anticipated decreases in feature sizes can still occur with existing or recently emerging technologies. With the accelerating trend toward feature sizes becoming smaller and smaller, nanotechnologies begin to appear to play central roles as semiconductors, memory units, interconnects, and contacts. A parallel trend is continued increases in multifunctionality—for example, chips that not only provide computer processing power but also serve as biodetectors or provide other functions. As we will see, there are many industry-driving forces for multifunctionality. The unique electrical, optical, and magnetic properties of nano-materials can enable these kinds of components to become not only smaller to but also have increased processing capabilities and multifunctionalities.

Nanomaterials that have unique electrical and magnetic properties are also expected to have huge impacts on energy storage and generation devices. Improved efficiencies are projected. For example, the power, durability, and use time of rechargeable lithium batteries, now in widespread everyday use, are expected to improve. (These topics are discussed in more detail in Section 9.7 on enabling systems.) Optoelectrical approaches that are relevant to a vast array of systems ranging from telecommunications to specific devices are expected to improve. The development of quantum dot laser technologies, for example, promises more effi-

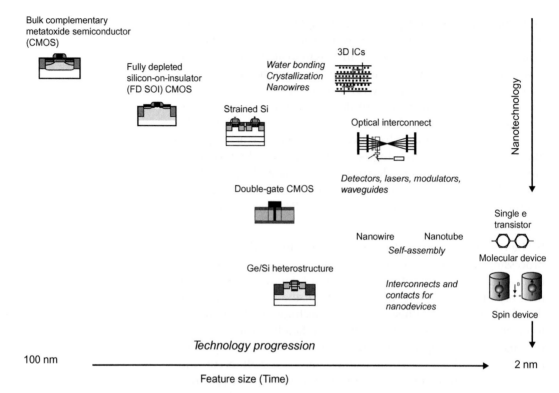

FIGURE 9.25

Anticipated developments in chip technology. (Adapted from The International Technology Roadmap for Semiconductors [ITRS] group by G. Gardini.)

cient action than conventional approaches, including performance that is less temperature dependent. Considerable research emphasis is directed toward the use of nanomaterials and nanotechnologies in various kinds of optoelectrical phenomena in all sorts of other devices, including various kinds of display systems (see Section 9.5). Improvements in other specific optoelectrical technologies, such as electrochromism and photoluminescence, that have diverse and widespread applications can be expected as well.

Another extremely exciting area of development is in sensory technologies that are now ubiquitously present in products and buildings. A huge number of sensor types that are in use are defined in relation to primary energy states—thermal, radiant, electrical, magnetic, chemical, mechanical—whereas others are defined in terms of expected usage (e.g., proximity sensors). A common photodiode based on semiconducting materials, for example, can be used to sense changes in light levels. Sensors are invariably parts of electronically integrated devices. In a biosensor, for example, a biological response is converted into an electrical signal that is subsequently transmitted or recorded. Biorecognition information regarding a physiological change or the presence of specific chemicals in a substance is thus detected, transduced (converted to another form), and transmitted. The potential value of nanomaterials and nanotechnologies in the development of improved sensory technologies is literally enormous. Depending on the application, subsequent logic systems may provide further interpretive information or control functions for actuation devices. Types of detection that are possible and sensitivity levels are expected to improve. Nano-based biosensors, for example, are rapidly being developed by the health and medical community for a wide variety of applications (see Section 11.2). Sensor sizes are expected to become smaller as well.

A final trend that will have enormous consequences is that of decreasing costs for better technologies. The nonlinear relationships among increasing levels of technological performance, reduced sizes, increased multifunctionalities, and reduced costs are generally well recognized. With increasing industry demands there is every expectation that many components, such as basic sensors and processors, will decrease in cost. With these cost decreases, we will be able to use them in a far more ubiquitous way than is currently the case, and application domains are likely to increase.

Impacts

Undoubtedly, major driving forces behind the adoption of nano-approaches in the electronics field are the many pushes from various industries toward improved performance and smallness in virtually all infrastructural or enabling components (electrical, mechanical, and so on). Drives toward multifunctionality are closely related. To designers and engineers engaged with products or buildings, the benefits of smallness are extraordinary and allow responses to major driving forces that are common in virtually all industries, including improved use experiences, cost reductions, reduced weight, energy-use reductions, and many other drivers that are discussed in more detail in Chapter 12. Smaller and higher-performance electronic components in aircraft, for example, are not only beneficial in their own right but also lead to overall weight reductions—hence bringing greater fuel efficiency and/or greater economic carrying capacity. Increased multifunctionality that can reduce the total number of components needed to achieve overall functionalities can lead to similar results but also has the impact of reducing complexities of supporting systems, since fewer elements need to be interconnected.

In the product design sphere, smallness coupled with higher performance has yielded products that have greatly improved usability and user appeal. The obvious example here is the devices we still usually call cell phones but that are often actually more sophisticated devices serving multiple functions that include communications but go far beyond—and all in beautifully designed objects that we carry with us. It is obvious that increases in performance coupled with decreasing sizes of functional components made these devices possible and helped usher in a new era in societal interactions. There are further expected trends in the many kinds of personalized devices that now directly accompany us or are expected to accompany us in the future, including further increases in multifunctionalities in areas that go beyond communications, simple work assists, and entertainment and into such spheres as personal health monitoring and assistants for mitigating impairments. The development of nanosized biosensors is expected to lead to personal alert systems for everything from airborne pollution levels that adversely affect many people in our society to biohazard events caused by terrorists. For some functional applications, the idea of even embedding such devices within our bodies has great appeal to many, albeit certainly not all. Here we might have nanosized disease- or condition-specific monitoring and control or drug-delivery implants for medical or health applications (see Chapter 11).

Trends toward smallness, improved performance, and multifunctionality as well as reduced costs in enabling components are seen in many other kinds of products already on the market or expected to soon be there, including everything from kitchen appliances to cameras. It is important to remember that for many of these devices, the question of final product size is dependent on many factors. Could devices such as cell phones and others in common use be made even smaller? From the point of view of fundamental enabling technologies (such as processing units, memory), the answer is surely a well-known yes. User interaction considerations, however, often require that they have certain critical minimum dimensions that are often near current sizes. (See Chapter 3 for further discussion of this point.) Still, for many supporting or embedded components such as sensors, the smaller, the better.

In architecture, there has long been a host of systems involving sensors and controls that aid in maintaining basic functionalities and comfort levels with respect to lighting, thermal, air, and other environmental factors. In major buildings telecommunications systems are common as well. Still, it is hard to argue that building systems of this type are particularly sophisticated. Most individuals have probably experienced temperature fluctuations in a space and have searched through several rooms looking for the single thermostat that controls the thermal conditions present (Why are there not sensors literally everywhere that would offer the possibility of more carefully fine-tuning the thermal environment in different spaces?). Nonetheless, there are increasing trends toward more sophisticated approaches. As with other industries, primary driving forces such as improved use experiences, energy-use minimization, overall cost reductions, and so forth (see Chapter 12) naturally lead to more extensive use of electronic systems for control, communication, and other functions. As a simple example, architecture has long relied on passive means (such as shading devices) of controlling heat gains from solar radiation impinging on a building. However, active means are possible as well—for example, façades that selectively allow or block heat gains from solar radiation as needed, depending on both external and internal thermal conditions. Active systems invariably involve sensors, transducers, actuators, and control systems. These systems, in turn, are poised to benefit enormously from expected advances in the more basic enabling technologies. Control components are expected to become much more localized to both individual spaces and to individuals.

Looking more to the future, many of the most exciting areas of innovation in both products and environments (architectural, vehicular) can be captured with words such as *interactivity, smartness*, and *intelligence*. Interactive systems or environments involve a response to the action of a user. Smart or intelligent systems or environments have their strongest basis in complex sensor-based electronic and computational systems, but many rely on the characteristics of property-changing smart materials as well (see Sections 9.8 and 10.2).

In automobile design, for example, many efforts are directed toward creating smart systems for driving condition and traffic context awareness. Smart mirrors use photochromics for reducing glare detected by sensors. In buildings, smart systems of one type or another are being developed for use in airports to identify passengers passing through security and couple identifications with visa, passport, and other information. Intelligent rooms are being developed that are oriented toward aiding in task performance or enhancing lifestyles or for a variety of health delivery reasons. An image of this kind of environment might be one that first detects and then identifies specific human users within the space, then tailors specific aspects of the environment in response to the needs, tasks, or desires of those specific individuals. Information displays would surely change, but so might wall colors, chair heights, and temperature levels. The individual might interact or control different actions without any kind of manual or even voice inputs but with the wave of a hand that is picked up by a gesture recognition system. Environments with capabilities such as facial or gesture recognition are not just for novelty's sake but can potentially help many people with disabilities (witness the importance of eye-tracking systems for individuals with severe eye disorders). Section 9.8 reviews interactive, smart, and intelligent environments in more detail.

These systems invariably involve complex computational infrastructures as well as a host of kinds of sensory and actuation devices. Nano-based approaches can clearly be used to improve the technical performance of these various components, but the previously mentioned trends toward smallness and cheapness are highly important as well. At the moment, interactive, smart, or intelligent environments already use a variety of sensory technologies, but in any given installation, surprisingly few are actually used. Most needed devices are surprisingly large and bulky as well as difficult to distribute and interconnect. The goal of seamless integration into the environment (see Section 9.7) is difficult to achieve with current technologies. The use of limited numbers of sensors inherently reduces the information that can be captured and hence reduces response and control capabilities. The idea of being able to have large numbers of inexpensive sensors based on nanotechnologies distributed throughout a spatial environment and designed to capture specific kinds of information (about users, processes, or conditions within a space) has great appeal for designers and engineers, with perhaps the idea of nanoelectronic "swarms" of simple but interconnected and collectively intelligent nanoelectronic devices representing a visionary possibility. It is with distributed sensor/actuator systems that trends toward smallness, high performance, and multifunctionality in enabling electronic systems intersect with goals of interactivity, smartness, and intelligence. At the moment, this area remains rather speculative, but the potential is clearly there.

9.5 LIGHT AND OPTICAL ENVIRONMENTS

Nanotechnologies are renowned for their many potential uses in connection with light and optical phenomena. Light is intrinsic to our perception and understanding of the visual world and is used in countless products that society uses on a daily basis. Light can carry information and be used in a host of ways we are just now beginning to understand. Nanotechnology opportunities in the broad field of photonics, which deals with generating and controlling light at a basic level, are seemingly everywhere. Applications in devices are widespread, including in the thousands of optical and laser-based devices that are among the fundamental workhorses of our modern technological infrastructure. Quantum dots, light-control films, and other nano-based technologies offer many direct applications. Interestingly, we also know that photo-based systems are important in manufacturing and production, including the very top-down production of nano-based materials and technologies that are the subject of this book.

In this section we explore nanotechnologies in relation to light. We start by briefly reviewing the basic characteristics of light itself as well as the essential properties of materials that are relevant to the control and manipulation of light. We do this because it remains surprising how many experienced designers still stumble over very basic ideas concerning color and related light phenomena, which in turn prevent them from creatively using light as a design tool. The exact way in which nanomaterials can be used to control traditional phenomena of light absorption, reflection, transmittance, or refraction is fundamental to an understanding of how to effectively use them in practice. There are needs for materials with either enhanced reflective properties (such as the color, multireflective, or mirror films used in many products) or those with so-called antireflective properties (see Figure 9.26). Though many current surfaces or films can be engineered to have these or other useful properties, many work only at single wavelengths (or highly restricted wavelength bands), thus limiting their usefulness. Utilization of nanomaterials can potentially enable the same performances over much wider bandwidths. More generally, various combinations of approaches at nano, micro, and macro scales can be effectively used together to provide even more versatile design approaches.

We also look at light-emitting devices (LEDs) that form the workhorses of countless products that are used throughout society, from simple consumer products such as flashlights all the way through complex scientific instruments. LEDs have literally revolutionalized architectural lighting as well as display in the past few years (see Figure 9.27). We will see that nanotechnologies—particularly nanophosphors and quantum dots—can lead to great improvements in these devices. Quantum dot LEDs (QLEDs) promise even further improvements.

FIGURE 9.26
Layered films can exhibit angle-dependent colors and reflective qualities. (Courtesy of Ben Schodek, Photographer, Boston.)

FIGURE 9.27
LEDs are extensively used in the Grand Lisboa Hotel in Macau. (Courtesy of Daktronics.)

Yet it is not only in the domain of light-emission technologies that benefits can accrue from the use of nanotechnologies. In our living and work environments, for example, the effect of a light source is highly influenced by the nature of its surroundings, including the optical properties, such as reflectance or refractive properties, of the materials that define the environments. The way humans perceive light—how it affects them or the information they receive from it—are all affected by interactions between the nature of the light source and that of the surrounding material environment. Hence we look here at the more general qualities of what is termed the *luminous environment* where humans live and work and perceive the visual world. In designing a luminous environment, for example, nanofilms or nanocoatings with specially designed optical properties can be expected to find wide use. Nano-based approaches already form the basis for many now common smart technologies for controlling lighting and the visual environment, including electrochromic glasses.

Fundamentals of Light

As we discussed in Section 4.7, light is a form of electromagnetic (interacting electric and magnetic fields) radiant energy that propagates through space in a way that can be characterized as waves. Electromagnetic radiation can be defined in terms of its wavelength or frequency (reciprocals of one another). The electromagnetic spectrum is very large ranging, from wavelengths at the kilometer scale to those near the atomic scale. Many devices that could potentially benefit from the use of nanotechnologies operate in different zones of this spectrum. Radio waves, for example, are quite long. Devices relying on infrared radiation involve much shorter wavelengths. Ultraviolet rays and X-rays are considerably shorter. The portion of the spectrum that is actually visible to humans is quite narrow and lies between infrared and ultraviolet rays. Within the narrow "visible spectrum," each wavelength corresponds to a particular color that is perceived by humans. The basic colors in the visible spectrum are red, orange, yellow, green, blue, indigo, and violet. In this continuous spectrum, red wavelengths of light are the longest and the violet wavelengths of light are the shortest. When all these wavelengths are present at the same time, "white light" is perceived. White is not a color but a combination of all perceptible colors. Black is not a color either, but rather the absence of any wavelengths within the visible spectrum.

Typical light waves oscillate in three dimensions. Under certain conditions, there can be preferred directions to the oscillations wherein the light is said to be *polarized*, as contrasted with typical light waves, which are *unpolarized*. Light waves might become polarized as they are transmitted through a material wherein absorption characteristics are higher in one direction than in another (see Figure 9.28). This is the case with many natural materials; calcite is the often cited example. Many synthetic materials can be structured to produce polarized light. Light can become polarized due to the way it reflects. Interestingly, the human eye cannot distinguish between polarized and unpolarized light. Common filters that polarize light can be organized either to allow the passage of light through or, when rotated, to behave as a form of shutter that blocks the passage of light. Many devices, including common sunglasses, liquid crystal displays, and various applications in electronics, polarize light or use polarized light.

The way these various wavelengths interact with objects or surfaces depends very much on their own magnitudes relative to those of the objects on which they act. When the wavelength of light and objects are of comparable scales, interactions can occur. Very long AM radio waves, for example, are not extensively affected by even objects of building size; that's one reason that we can listen to radios inside buildings. Very short wavelengths can literally penetrate materials. When the wavelength is similar to the feature sizes of the surface on which they impinge, interesting interactions can take place. All the light that impinges on a surface is *reflected*, *transmitted*, or *absorbed*. *Refraction* also takes place. Light waves interact with the atoms and molecules present in a material. As the electrons present vibrate at specific frequencies, and if the impinging light wave has a similar frequency, it can excite the electrons into an intense vibrational motion (a form of resonance). This motion causes interactions with adjacent electrons, such that that the vibrational energy is transformed into thermal energy (the object heats up slightly). The energy originally present in the light wave is completely transformed—or *absorbed*—and not reemitted again. Since the natural vibrational motion of the electrons vary from material to material, various materials will absorb different wavelengths, or, if a wide variety of light wavelengths strike a surface at once, a given material will selectively absorb specific wavelengths and others will be reflected or transmitted. The wavelengths of the reemitted light (reflected or transmitted) differ from that of the original light source. Varying internal structures render various materials capable of selectively absorbing, reflecting, or transmitting one or more frequencies of light in different proportions.

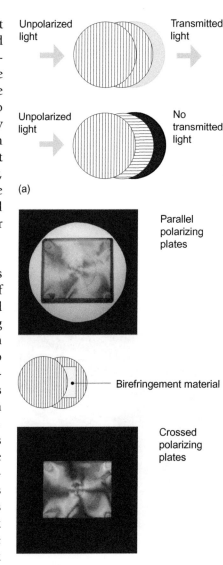

(a)

(b)

FIGURE 9.28

Polarized light. (a) Unpolarized light becomes polarized as it passes through the first plate. Depending on the orientation of the second plate, the polarized light may pass through or be blocked. (b) Colored fringe patterns show up in birefringement materials placed between plates.

As discussed in Section 4.7, light is invariably reflected in such a way that the angle of all outgoing reflected light will be identical to the angle of incidence of the incoming light, no matter what the exact local shape of the surface happens to be. If the surface is planar and completely smooth, it is easy to see how the angle of incidence is equal to the angle of reflection. Polished, smooth surfaces reflect light in a *specular* way. If a surface is rough or irregular, a spread will occur in the outgoing reflected rays. A matte surface diffuses reflections in a wide range. In these latter cases, the angle of incidence and angle of reflection of the impinging light are still equal when considered at the microscopic level, but the gross effect is one of spread or diffusivity.

When a light ray strikes a surface at a particular incident angle and is transmitted through a material, several things happen. The speed, wavelength, and direction of the related waves all change. As light passes from air into glass, for example, the speed and wavelength decrease and the angle at which the light passes through the glass is slightly different than that of the original incident angle. The light path is bent and is said to *refract*. Snell's law, which defines the relationship between the incident angle and the refraction angle as dependent on the refraction indices of the interfaced materials, is discussed in Section 4.7. When light passes through multiple layers of materials, reflections and refractions occur at each interface. As discussed in the following section, constructive and destructive wave interferences can lead to highly interesting visual phenomena and many useful applications, such as antireflective glass.

In some situations the light that enters a new medium becomes totally contained within the medium—a phenomenon known as *total internal reflection*. This occurs when waves are completely reflected off a boundary through which the waves are traveling. In a fiber optic tube, for example, light entering at one of its ends can bounce back and forth along the internal boundary and never exit along its length—only at its end. For this phenomenon to occur, the light waves must be in a medium of greater optical density than that of the boundary, and the angle of incidence is greater than the so-called "critical angle." If an incident ray in a medium with a high refractive index approaches a medium of lower refractive index at an angle at which Snell's law would suggest that the sine of the refracted angle needs to be greater than unity, which is mathematically not possible, then internal reflection occurs (see discussion in Section 4.7). By manipulating the refractive indices of the adjacent mediums, light can be made to reflect totally internally within the denser medium (see Figure 9.29). This principle forms the basis for not only

(a)

Light escapes only at the edge

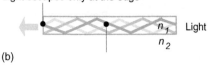

(b)

FIGURE 9.29

(a) Internal reflection. Light is internally bounced back and forth and emerges at ends or edges. (b) Light is trapped inside since the angle of incidence is below the critical angle associated with the two refractive indices (n_1 and n_2).

fiber optic technology but many other optical manipulations as well, including some marvelous fountains of light (light trapped in water streams via internal reflection) built in the 19th century.

Color

There are many physical and chemical reasons for color. First, it is useful to distinguish between the color of a light *source* (as defined by its encompassed spectrum of wavelengths) and the color of an *object* on which the source shines. Variations in the color of a light source can be produced by *additively combining* primary colors—red (R), green (G), and blue (B)—each associated with specific wavelengths. Additively combining red and green produces the secondary color of yellow. Blue and green produce cyan, and red and blue produce magenta. White light can be produced by additively combining primary colors with the correct intensities. Red (R), green (G), and blue (B) are commonly used, but there are actually several sets of colors that can be combined to produce white light. Selectively adding together these same colors can generate a huge array of other colors—hence the ubiquitous use of red, green, and blue lights in common products such as televisions that produce color additively. Combining two colors that are complementary—say, cyan and red—can also produce white light, since blue and green are present in cyan. Any kind of projected light source relies on additive color mixing to produce the output color. In the section on solid-state lighting, we will see that red, green, and blue quantum dots are normally used to produce not only high-quality white light but other colors as well.

The color that we perceive an object to have when a light source shines on it is not intrinsic to the object. It depends on the way wavelengths of the light acting on a surface are absorbed, reflected, or transmitted. We can only visually respond to the wavelengths that are reflected from the object and still lie within the visible spectrum. Wavelengths that are absorbed or transmitted are never seen. The color of an object thus depends on both the spectrum of the light striking the surface and which wavelengths are reflected.

Let's consider a light source shining on an opaque surface (one that can absorb or reflect light wavelengths only, not transmit them). If white light with a full spectrum strikes an opaque surface that absorbs all wavelengths present except blue, the surface will appear blue because the wavelength corresponding will be reflected and seen by the eye. Here it is useful to think about this effect as being primarily *subtractive* in nature. The final colors we see are the result of various wavelengths originally present in the light source being subject to subtraction via the absorption or transmission of specific wavelengths by the material itself. A white light shining on an opaque tent surface that absorbs the blue wavelength will cause the surface to appear yellow, since red and green wavelengths will be reflected and together these appear yellow. If a light source with wavelength in the green spectrum acts on the surface, the surface will appear green. It would also appear green if cyan (green and blue) shined on it. On the other hand, a light source of blue wavelength only shined on the surface would cause the tent surface to appear black, since the impinging blue wavelengths would be absorbed by the material. Interesting phenomena can result. Thus, a specular sphere that appears black can have a white highlight, or a sphere that appears to generally be one color can have reflections of a different color, depending on the wavelengths present in the original light source and the absorptive characteristics of the material.

The term *pigment* (most commonly associated with paints) is used to describe a material that absorbs a single wavelength or color of light. Many nanomaterials are now found in pigments used in paints. The color descriptor of a pigment is the complementary color of the color absorbed by the pigment. If a pigment on a tent material has red, green, and blue shining on it and absorbs blue only, it is referred to as being a yellow pigment (since red and green—or yellow—are reflected and hence seen). A red pigment on the same surface would absorb cyan (blue and green) and appear red. If a yellow light (having red and green only) were to shine on a blue pigment, the surface would appear black, since the blue pigment absorbs red and green. In relation to common paints and other coatings, actual pigments are normally not pure pigments that absorb only single wavelengths. A material that is either transparent or translucent not only selectively reflects and absorbs wavelengths from a light source shining on it but also transmits particular wavelengths.

Visual Perception

The basic physics of light and color are one thing, but the way humans actually perceive, understand, and are affected by light and color are strongly affected by both the overall context in which something is seen and fundamental "eye–brain" neurological responses in which vision is a complex part of the human sensory system. These are vast subjects that are simply beyond the scope of this text. Only a few salient points are noted here, simply to give a flavor of the issues important to our discussion.

Nonetheless, we emphasize that these issues not only affect vision but also psychophysical responses on the part of humans. Designers ultimately affect many of these responses via their manipulations of the qualities of light and its surrounds. Designers who are knowledgeable about the way humans perceive and respond to light and the luminous environment can make positive contributions here. After all, many basic connections between the qualities of lighting and human well-being, comfort, or task performance have been studied for years. Whether humans experience seasonal affective disorders or not, for example, has long been linked to the nature of surrounding lighting environments. Typical responses, such as the use of full spectrum lighting, are known to all. What is different now are the many rapid responses made in the neurological world in relation to vision, and—of particular interest here—significant increases in our abilities to control the qualities of light itself and the way it interacts with its surroundings. Recent advances in understanding the way light affects the human circadian system that in turn regulates other psychophysical responses, for example, promise to suggest means for improving the design of interior environments via the use of specific kinds of light-emitting technologies coupled with particular material properties in the surrounding surfaces that are all set within carefully designed configurations. All these factors would affect the kinds of light perceived by occupants and thus could affect their psychophysical responses. Well-known experiments by George Brainard, et al., into the effects of "blue light" on circadian rhythms suggest how an architectural ensemble of light, space, and materials could be used to positive effect here.

In an everyday setting, designers invariably work with issues of perception. Surfaces that appear different to those of surrounding surfaces are invariably a point of interest and intervention. The appearance of a surface changes as a surrounding surface becomes brighter or darker—a phenomenon that can be conceptualized in brightness-contrast terms (see Figure 9.30). Artists such as James Turrell have long used the perception that a bright form will seem to advance in space relative to its surrounds as the basis for many works. The importance of edges or outlines of a form in a visual field have long been known not only to psychologists but artists and architects as well. Edges and outlines are first visually understood, then surfaces, in a form of visual competition that artists and designers have long positively exploited.

Classic summaries of our perceptual capabilities have suggested that initially three primary components are present: attributive, expective, and affective. Humans commonly fill in or attribute meanings

FIGURE 9.30
The center bars in both images have identical luminances; only the background is different.

to things we see. Patterns of lights or forms, for example, are filtered and assigned meanings by observers based on their own experiences and cultural understandings. Many common optical illusions are based on these human tendencies. We often see what we expect to see and are conditioned to see. We are also affected by the meanings and expectations underlying what we perceive. A large blinking red light can cause humans to respond with a sense of anxiousness. A gray sky can engender a sense of gloom. Similarly, if we don't see what we expect to see, unease can develop. The light on Frankenstein's face in old films was invariably deliberately positioned by the director to come from a source below the face—and cast shadows upward on the face—rather than from the more "normal" direction. The resulting feelings engendered via the unexpected direction of the shadows formed on his face remain ingrained in the minds of most of us who've seen those films.

We also know that a repetitive pattern can seem dull. But here we note the important point that the eye will seek out and identify any disruption of the pattern. We innately seek out unique visual information. This is one of the fundamental driving ideas behind many acclaimed visual works. "Visual noise," however, can prevent us from gaining information when it is hard to distinguish between the visual signal and its background. A small object set against a background pattern of bright lights may be hard to identify, in this case because of "glare" or unwanted light within the visual field.

Luminous Environments

The real world of designing luminous spatial environments ultimately demands that we have and use quantitative measures of light intensities and many other factors in developing or selecting light-emitting sources and important wall surfaces and deciding on their distribution in a space. In the International System of Units, the *lumen* is the basic unit of brightness measurement and in turn defines *luminous flux* (the amount of energy emitted per second in all directions). *Luminous intensity* refers to light intensity in a given direction (the luminous flux that is radiated in a given direction and within an angular unit). *Luminance* is the luminous intensity emitted by a unit surface area of the light source. All these and many other measures are commonly used by lighting designers. Normal practice also includes use of a vast array of lighting standards that have been developed for use as criteria for lighting levels in various spaces (schools, offices, warehouses, corridors) and even on specific task surfaces (such as desktops). A huge number of kinds of area, line, or spot lighting sources are available for use in illuminating environments.

Designers now have available to them a number of computer-based tools with which they can make lighting simulations and predict lighting levels and characteristics at different points within a space. These light characteristics can be quantitatively determined. Fully rendered images are useful for visualization purposes. Inputs to simulation packages normally include geometric data describing the environment, photometric data to describe light sources and further geometric data to describe their locations and configurations, and material property data (reflection, absorption, transmission, or refraction characteristics). Two general simulation approaches are commonly used: *radiosity* and *ray tracing*. Radiosity methods generally start with the propagation of diffuse light from light sources (defined as surfaces that self-emit light). Surfaces are subdivided into patches with specified reflectivities (determined from material considerations). Despite the power of existing simulation methods, designers are well advised to understand their limitations thoroughly. As previously noted, many highly subjective aspects of working with light are simply not possible to consider in any mathematically based simulation model. Nonetheless, simulation methods remain powerful tools if used correctly. It is also in the self-interest of developers of nanomaterials intended for this application domain to look into them carefully and understand how materials and light-emitting sources are modeled and provide the necessary property measures and data to support the modeling. This would promote their product use.

In closing this section, a further word about lighting standards for various kinds of spaces (such as classrooms or corridors) should be noted. Standards or recommendations for appropriate levels are necessary and can be useful, but they should always be critically viewed in working with lighting, given its many subtle aspects. Many designers argue, for example, that our environments are simply "overlighted" by virtue of having to conform to existing standards, especially since many have their historical roots in the lighting industry, which accrued benefits from the use of high lighting levels. Many tasks do indeed require high light levels, but it is easy to overdo light levels as well, to the extent that the excess light is counterproductive to either comfort or task performance. High light levels are also complicit in contributing the world's enormous use of energy, particularly in highly industrialized countries. In general, more detailed and thoughtful simulation analyses that carefully take into account the overall context can often demonstrate appropriate levels for the situation at hand.

Applications of Nanomaterials

Nanomaterials already have wide use in relation to light, and future uses are seemingly imagined every day in a broad spectrum of application areas. A major sphere of both current use and development is in the area of the control or enhancement of reflected light on surfaces or the ways it can be absorbed or transmitted in or through materials. As we will see, various kinds of thin films, coatings, or sheets have been developed to control light in these ways that have applications in many widely varying fields and in relation to a huge number of products.

Chromogenic materials are also expected to have improved performances through the use of nanomaterials. Chromogenic materials change their optical properties when subjected to a change in their surrounding energy stimuli. Changes can be caused by variations in light intensity, electrical field or charge, temperature, and the chemical or mechanical environment. Respective materials include photochromics, electrochromics, thermochromics, chemochromics, and mechanochromics. The induced changes affect their optical character. All can exhibit what is perceived as a color-changing behavior. Applications are numerous and range from devices that use the color change as a type of sensor to transparency-changing glasses used in a wide variety of architectural, consumer, and industrial products.

Few technologies are as emblematic of the advances our society has made over time as devices that emit light. Lights of one type or another perform a multitude of tasks or functions. Our habitable environments are themselves illuminated. Light-emitting devices in current use range from light-emitting diodes (LEDs) and films that seem to simply glow by themselves to common fluorescents (themselves once epitomizing new lighting breakthroughs) and other ubiquitous technologies. Nanomaterials and nanotechnologies are already rapidly finding many applications here and will undoubtedly continue to emerge to become the next fundamental improvement in this domain. The application potentials are literally enormous. Of particular importance are the many nanoparticles and nanocrystalline structures being developed that become luminescent when subjected to an energy source. Nanophosphors are expected to enormously improve the efficiency of many high-efficiency fluorescent lamps of one type or another. Certainly quantum dots and related technologies that provide light emission will develop quickly. They will find application in the many products and devices that serve our society. Devices relying on now ubiquitous LEDs, for example, will soon be looking toward a new generation of LED technologies based on quantum dots (QLEDs) that are both far more effective and far more energy efficient.

Nano-Related Phenomena

As described in Section 7.6, many properties of nanomaterials are light related. The study of light–matter interactions is among the most interesting and potentially rich of all nano-related research and application areas. Control of the interactions between material responses and excitations and various electromagnetic radiation types can have applications in a broad spectrum of areas.

Nanomaterials can be used in films, coatings, or sheets to produce antireflective, dichroic, UV absorption, or other important light-related functions. Many serve broad architectural and product design applications; others serve many scientific and industrial purposes, such as diffraction gratings and many types of filters. These different applications are all founded in a thorough understanding of the principles of optics and related transmission, absorption, and reflection behaviors. Principles of constructive and destructive interference are particularly important. An interesting manifestation in nature of the many kinds of optical phenomena of interest are the brilliant iridescent colors evident in the wings of dragonflies, peacock feathers, butterflies, soap bubbles, or oil films on water surfaces (see Figure 9.31). These iridescent effects have long been both admired and studied and are generally considered fine examples of various optical effects in thin film structures (which may or may not involve actual nanomaterials). These iridescent effects can occur when the scale of surface features is similar to those of the wavelengths of light and when constructive and destructive interferences occur (see Figures 9.32 and 9.33).

When a transparent layer coats a surface, the impinging light is both reflected from it and transmitted through. Transmitted portions can also be reflected from the interface between the external layer and the surface. Depending on the optical characteristics of the materials and the angle of incidence and wavelengths of the impinging light, the several reflected wavelengths can reinforce one another (constructive interference, which leads to bright colors) or tend to annihilate one another (destructive interference, which leads to dark colors). Truly striking visual fields can result. We will subse-

FIGURE 9.31
Iridescent wings of a butterfly. (Courtesy of BASF.)

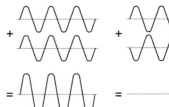

Phase difference
λ = 0°
Constructive interface
(a)

Phase difference
λ = 180°
Deconstructive interface

Phase difference
λ = 90°
(Quarter wave)
(b)

Phase difference
λ = 180°
(Half wave)

FIGURE 9.32
Constructive and destructive interferences in wave patterns lead to color variations. (a) Constructive and deconstructive interfaces. (b) Quarter- and half-phase differences.

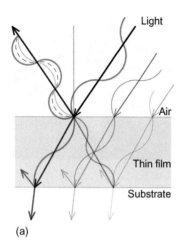

(a)

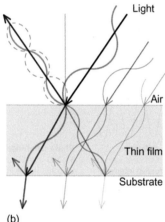

(b)

FIGURE 9.33

Sources of iridescence: (a) constructive and (b) destructive interferences.

quently see that this general behavior underlies many interesting uses of nanomaterials in relation to light.

Many of the unique optical properties of nanomaterials can impact many of the general types of behaviors we've noted as well as others, since the wavelengths of light that are reflected, transmitted, or absorbed are affected by the use of nanoscale surface conditions or subsequent treatments. Surface plasmon effects stand out as being particularly important in many applications, leading, for example, to various colors of nanoparticles. As described in Chapter 2, this effect was exploited in early works of art, such as Medieval stained glass or lusterware, although it was by no means understood. As described in Section 7.6, *surface plasmon* effects result from a natural oscillation of an electron field inside a nanoparticle. When nanoparticles are small compared to the wavelength of light and when the wavelength of light is close to that of the oscillating electron field, energy will be absorbed, leading to a form of resonance of electrons on the surface. The absorption can be extremely high in nanoparticles present at the interface between a metal or metal oxide and a dialectric (a non-conductive material such as glass) because of their high surface-to-volume ratios. The frequency of oscillation depends on the dielectric function of the nanoparticles, their interparticle spacing, and their shapes. Nanoparticles with different sizes and shapes will exhibit different responses. Producing nanoparticles with particular characteristics can allow control of the resonant frequency over a very large range of values. This in turn affects the colors associated with various nanoparticle sizes. For many applications, such as QLEDs, described in the following section, this size effect allows fine tuning of emitted light frequencies by judicious mixing of particle types and sizes.

Another characteristic of importance is the angular dependence of colors of nanoparticles on viewing angles, a phenomenon related to the effects we previously described. The use of nanomaterials is particularly interesting with respect to the development of various kinds of light-related angular dependencies in films, coatings, or sheets. Figure 9.34 shows the influence of sizes and viewing angle on color for particles near the nanoscale (usually defined as 1–100 nm). Figure 9.34 also illustrates a particularly interesting property of strain (deformation) associated with flexible crystalline films on color. In strained films, there is a shorter particle-to-particle distance perpendicular to the strain direction and therefore shorter wavelengths of scattered light. Dichroic effects and products are described more in the following discussion.

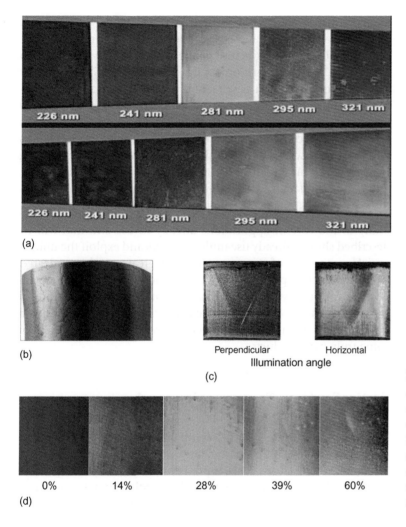

(a)

(b)

Perpendicular   Horizontal
Illumination angle

(c)

0%   14%   28%   39%   60%

(d)

FIGURE 9.34

Films. Angle dependence of colors and effect of strain. (a) Influence of particle size and viewing angle on color. (b) Angle-dependent film. (c) A flexible colloidal single crystal (2 × 2 cm elastic single crystal with 60° multilayer edges). (d) Influence of strain with flexible crystalline films on color (percentage stretching). (All images courtesy of BASF.)

Light Control Films, Coatings, and Sheets

There are many methods for producing thin films, coatings, or sheets that exploit the unique light properties of nanomaterials that are useful in relation to light transmission and control. Many of these methods rely on physical vapor deposition (PVD) and chemical vapor deposition (CVD) techniques that can deposit many types of nanomaterials on substrates in layered or oriented ways (see Chapter 8). Deposition processes can be used on many types of substrates, but various kinds of glasses and polymeric films are commonly used. Other types of processes are in common use as well. Many of these synthesis processes can be controlled with the objective of imparting needed or desired light behaviors. Many needed light-control qualities can be obtained from imparting some type of orderly or aligned structure at the submicron or nanoscale to a thin film, which in turn orders light waves in particular ways or affects phenomena such as refraction. This ordering can be accomplished in many ways by use of the synthesis techniques described in Chapter 8, including, for example, electron-beam evaporation techniques that cause linear structures to grow on substrates, various kinds of processes for thin crystalline films that are self-assembling customized crystals in solution and self-aligning when deposited, and nanomanufacturing techniques for making nanoscale perforations or grooves. The need for multiple layers is common.

Many of the current applications already in use and that will be described shortly already use multiple layers and exploit the unique optical properties of nanoparticles to a greater or lesser degree. Varying particle types, size, and layer structures and the compositions of various layers offer increased control of different kinds of important optical behaviors and related reflection, absorption, and transmission characteristics as well as creating opportunities for some truly new products.

Antireflection, Transmission, and Contrast Enhancement

The eyes of moths are known to have both reduced reflections and enhanced seeing ability, particularly in darkened environments. The reduced reflections prevent predators from easily seeing them. Moth eyes have been found to be covered with thin nanostructured films. Many coatings and films are now available that exhibit similar antireflective and light enhancement qualities. The use of antireflective coatings typically serves one of the following primary purposes: to directly reduce reflections (including cutting down glare produced by reflections), to increase light transmission through a surface, or to increase contrast. These coatings are widely used in architectural applications, particularly with glass in relation to visible and ultraviolet light; in many product design applications such as computer screens; and in consumer products such as tinted eyeglasses. Antireflective coatings in common use normally consist of multiple layers of transparent thin films, with different layers having varying refractive indices, as we'll discuss shortly. Characteristics and thicknesses are selected to produce destructive interferences for reflected light waves and destructive interferences for transmitted waves. Normally these behaviors vary with the wavelength of the impinging light and its incident angle. In current practice, antireflective coatings are designed for optimal performance for specified wavelengths and angles. Common coatings are usually designed for infrared (IR), visible, and ultraviolet (UV) spectra, depending on the application. Broader bands are possible but very expensive. Some less expensive coatings use so-called absorbing antireflective (AR) materials for situations in which low reflectivities are needed but high transmissions are not necessary.

The functions of increasing light transmission and contrast using antireflective materials are often surprising to many, but their origins go back to seminal explorations of light transmission effects through coatings. In the 19th century, tarnishes caused by chemical reactions with the environment developed on optical glasses of the day were common and exhibited reduced reflections. They were also generally thought to reduce light transmission. Lord Rayleigh (John Strutt), well-known for his many contributions to optics, looked into the problem and experimentally observed that tarnished pieces often transmitted more rather than less light—a seemingly surprising result. The general reason underlying this phenomenon is based on the fact that the total amount of light impinging on a surface must in sum be reflected, transmitted, or absorbed (see Section 4.7). If the amount of light reflected from a surface is reduced, more must be transmitted or absorbed. A more precise explanation centers around the fact that the index of refraction of the tarnish lies between that of air and of glass, and both the air-tarnish and the tarnish-glass interfaces exhibit reduced reflections that are less in sum than the original air-glass interface.

Antireflection coatings now commonly consist of multiple layers of materials with varying refractive indices. Material types and layer thicknesses are designed to increase the constructive interference in light transmitted through the layers and destructive interference in light reflected from the several interfaces. Fundamental effects of layering are shown in Figures 9.35 and 9.36. Consider a design intended to optimize light transmission in a system consisting of a simple one-layer coating on glass. The light reflects twice (from the surface to the air and between the layer and the glass). As described in Section 4.7, the reflectivity of a material depends on its refractive index and can be calculated. If R_f and R_g represent the reflectivities at the surface-to-air and surface-to-glass interfaces, respectively, the transmission at each interface is $T_f = 1 - R_f$ and $T_g = 1 - R_g$, respectively, and the total transmission is $T_f T_g$ (with absorption in the thin layers considered negligible in this example). By carefully varying the refractive indices present for each material, we can

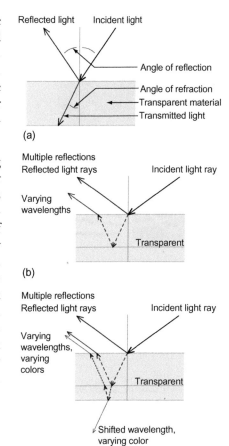

FIGURE 9.35
(a) Light interactions on transparent layers.
(b) Materials with "transparent" layer and fully reflective layer. (c) Material with two "transparent" layers.

optimize the total transmission, which can be found to correspond to the transmittances of both interfaces being equal. Layer thicknesses must also be carefully controlled. Similar design procedures can be applied to multilayer assemblies, to either increase transmissions or, conversely, to enhance reflectivities.

It is here that the many opportunities to either subtly or dramatically vary basic material properties through the use of nanomaterials come into play. Coatings can be fine-tuned to respond to the various needs that have different impinging frequencies, angles of incidence, and so forth. Reflections of light from certain wavelengths can be suppressed or enhanced, as can transmissions.

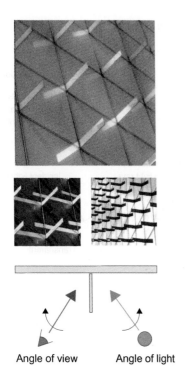

Angle of view   Angle of light

FIGURE 9.36

Use of dichroic glass by James Carpenter. The perceived color of the glass on this building façade near Times Square in New York City varies with the angle of the viewer and the angle of the incident light.

Dichroics

The term *dichroism* is usually used to describe light-related angular dependencies in surfaces. Dichroism is both a much used and much sought after property for use in many applications. Transparent dichroic surfaces exhibit color changes to the viewer as a function of either the angle of incident light or the angle of the viewer with respect to the orientation of the plane on which the light strikes. Not only is the wavelength (hence colors) of the reflected light changed; the transmitted light is changed as well. The many varying color effects can be both striking and unexpected. These diochroic materials have layered buildups of thin film depositions with different light transmission, absorption, and reflection qualities that each affect the reflection, transmission, and absorption of light passing through in discrete ways. The cumulative buildup of effects yields the interesting effects noted.

Dichroic coatings can be applied to many substrates. Dichroic glasses are in common use in many products. They are widely known for their striking visual effects. High-quality dichroic materials are also used in technically sophisticated optical products for commercial, industrial, or military purposes. Dichroic glasses or coatings serve many functions as part of the overall design to control various aspects of light transmission, as for filters in connection with optical lens systems.

In architecture, dichroic glasses are simply wildly popular, and a number of really striking uses have been made of the material. Dichroic glasses are most often used selectively because of their high cost, but massive applications for surfaces of multistory buildings have even been proposed—albeit the ramifications of using masses of these highly reflective glasses has clearly not really been

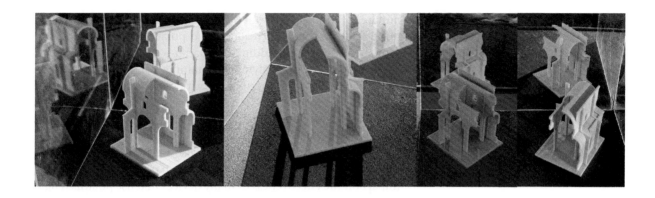

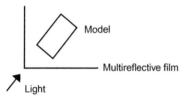

Model

Multireflective film

Light

FIGURE 9.37

A small model of the Medieval church of St. Trophime is shown nestled between two multireflective optical films. The colors perceived are dependent on the angle of view and whether the light is reflected or transmitted.

thought out. Figure 9.36 illustrates a well-done example by James Carpenter.

Polymeric products termed *radiant color films* exhibit similar angularly dependent color effects and are made up of multiple layers of depositions with differing reflective, absorption, and transmission properties. These materials may be easily cut or embossed or given adhesive backings for easy application to other surfaces. They are less expensive than dichroic glasses (see Figure 9.37).

Polarizing Films and Glasses

The need to *polarize* light is fundamental in a large number of products and display technologies that use oriented light. Thin coatings, films, and sheets that can linearly or circularly polarize light are in common use and have been in use for a long time. Fabrication methods are well established and range from older mechanical means—for example, stretching of polymers through a host of different deposition methods.

For many large-scale applications that need high quality, however, material costs have been high. Recent developments with imparting polarizing qualities to thin polymeric sheets are enabling polarizing materials to be used in bulk. Thin films can be deposited on any number of different substrates. Deposition techniques allow layers to be built up that can create needed polarizing light orientation properties, including building up quarter-wave plates. Experiments have been made with polymers containing nanoparticles to create a material that can polarize light. A variety of types of nanoparticle materials and related size variations have been explored for use in connection with polarized light applications. The rodlike structures of carbon nanotubes offer possibilities here as well. Self-assembly and self-alignment approaches have also been explored. Enhancements made possible via nanomaterials can potentially have a great impact on the effectiveness of many devices that use polarized light in one way or another.

Chromics

Chromogenic materials are broadly defined as materials that show a large change in their optical properties when subject to a change in their surrounding energy stimuli. Changes are caused by actions at the molecular level and are normally completely reversible. They are often called *color-changing* materials. Various kinds of light-related chromics—photochromics, thermochromics, and electro-chromics—are of great fascination to designers and engineers. Their interesting properties have caused them to be called *smart materials* in that they intrinsically respond or change when subject to various energy stimuli. They have found wide use in a variety of application settings—architecture, consumer products, and industrial and scientific devices. However, not all types of these chromogenic materials available today truly make use of nanomaterials or nanotechnologies as narrowly defined; sometimes reactions are fundamentally chemical only or involve thin films that may be of nano thickness but do not involve nanoparticles.

Photochromic materials passively change colors when subjected to sunlight and are commonly found in many products, such as eyeglasses that change tints in the sun. Photochromic materials can come in many forms—for example, glasses or polymers. A typical photochromic film, for example, would change to transparent blue tint when exposed to sunlight and result in a more or less colorless form when there is no sunlight. The rate of color change is generally slow, depending on the light intensity and length of exposure.

Photochromic glass absorbs visible light and darkens when exposed to ultraviolet light. When the light source is removed, the glass fades. The behavior is reversible. One approach used to create this behavior is the use of nanocrystalline silver halide particles that are dispersed throughout the glass. Ionization in the silver halide particles occurs and causes the color change. Other materials are also often used as well, such as copper salts to sensitize the silver halide particles to ultraviolet light. Thus an input of radiant energy in the form of UV light causes a reversible change in the structure of a photochromic material, which has different absorption spectra and hence different color or transparency appearances. Other host matrix materials and photochromic materials are also possible, including the use of silver nanoparticles. A common need is to keep particles dispersed, and approaches using nanoporous films and other techniques have been explored.

Photochromics respond passively and cannot be electronically or manually controlled. Hence they are not as useful as might be hoped for controlling solar gains in architectural window or glazing applications. These applications normally require balancing solar gain inputs in relation to indoor/outdoor temperature considerations. Photochromics can darken on bright, cold days, when solar gains are highly desirable. In large-scale applications, colors are limited and optical transparency and quality can be poor (especially in lower-cost photochromic polymers suitable for large applications).

Thermochromic materials passively change colors when external temperature environments changed. These materials are used in numerous products such as forehead thermometers and architectural applications, for either their visual effects or as part of some type of measurement device such as a battery tester. More recently, there has been a spate of developments in displays based on thermochromic behaviors. Thermochromic materials absorb heat, which leads to a thermally induced chemical reaction or phase transformation that in turn affects color appearance. Thermochromic materials come in many forms, including liquid crystal forms used in thermochromic films and leucodyes used in other applications. Figure 9.38 shows a whimsical application of thermochromics.

FIGURE 9.38
A memory of touch via thermochromic paint. (Courtesty of Juergen Mayer.)

An early approach for thermochromic glasses used a thin film of vanadium dioxide to achieve the color-changing action (here a change from transparent to reflective and back). Vanadium dioxide nanoparticles and nanorods that exhibit a semiconductor-metal-phase transition are currently being explored for many thermochromic applications. New synthesis methods for nanomaterials allow careful control of the way these nanomaterials are sized and arrayed. Other kinds of nanomaterials such as gold (Au) are also being explored for thermochromic applications.

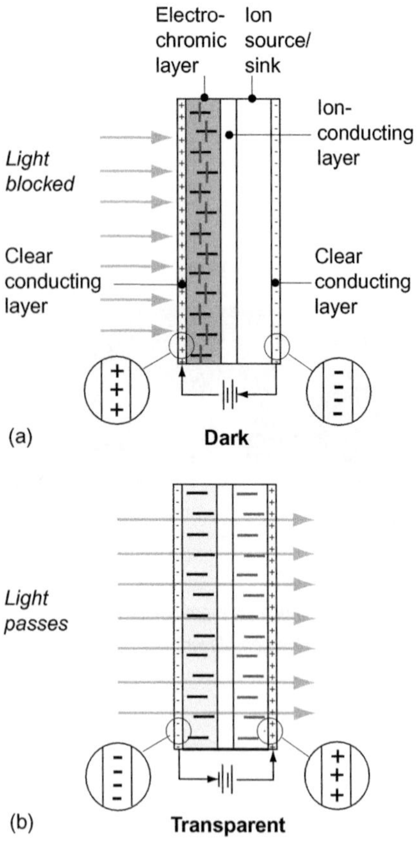

Electrochromic materials are particularly interesting in relation to active control systems. *Electrochromism* is broadly defined as a reversible color change of a material caused by the application of an electric current or potential. Electrochromic glasses can dim to a darker color while still remaining transparent. Electrochromic behaviors form the basis of many products found in architecture and in a host of products found in both consumer and industrial use, including in displays and glazings. They have an advantage over photochromics in that they can be electronically or manually controlled and hence can find much wider applications in everything from electrochromic inks to electrochromic windows in architecture.

The electrochromic color change process is completely reversible and is electrically controlled and hence can be integrated into complex sensor-based systems. Chromogenically switchable windows and glazing systems have been the primary focus of many research efforts directed toward dynamically controlling daylight and solar heat gain, not only in buildings but in vehicles, aircraft, and ships as well. Thus they can be important ways of managing lighting and heating levels in relation to energy systems. Simple visual control is possible as well, but large-scale electrochromic glazing systems do not go completely opaque.

Electrochromic devices normally consist of several constituents (see Figure 9.39). Color change in the electrochromic material results from a chemically induced molecular change through an oxidation-reduction action. When a low electric voltage is applied, lithium atoms are transported from an ion storage layer through an ion-conducting layer and injected into the tungsten or nickel oxide electrochromic layer, thus causing a change in its optical properties and causing it to absorb certain visible wavelengths, with the result that glass darkens, typically to a blue tint. Reversing the direction of the voltage drives ions out of the electrochromic layer in the opposite direction, causing the glass to lighten. This whole color-change process can be relatively slow in large-scale applications and requires constant current during the change. Faster changes are certainly being explored, particularly for products such as electrochromic inks or electrochromic displays, which we'll discuss in a moment. In addition to the basic behaviors described, various kinds of transparent light-control coatings or films can be used on the transparent exterior sheets to enhance or suppress specific behaviors.

In many current glass assemblies, the electrochromic layer is often tungsten trioxide (WO_3). Nickel oxides are used as well. The outer

FIGURE 9.39

An electrochromic glass changes transparency with the application of an electrical current: (a) dark, and (b) transparent.

conductive layers are often indium tin oxide. As noted, one layer is coated with an electrochromic film and the other with an ion storage film. Both of these films must have some level of porosity. The porosity is important in facilitating the needed movement of ions in and out of the material. Current approaches to porosity vary. It appears that the use of nanomaterials in relation to very specific levels of nanoporosity can improve performance. Important nanotechnology considerations include not only material types but the kinds of synthesis processes (see Chapter 8) that can be used to impart needed porosities and layer thicknesses. Many types of processes, such as sputter deposition, have already been used successfully.

Luminescence

Many light-emitting technologies that can benefit from the use of nanomaterials are ultimately based on the *luminescence* qualities of many materials (see Section 4.7). This light-emitting quality is not based on incandescence (visible radiation associated with a thermally hot glowing body) but is rather based on other types of excitation by other energy sources such as a chemical reaction (as can be observed in the common "light stick" so prevalently seen during Halloween). *Fluorescence* is a form of luminescence wherein the light is more or less instantly emitted when the material is excited, as in the common fluorescent light. *Phosphorescence* occurs when the light is emitted more slowly over time (and can still have a decayed afterglow when the excitation source is removed). In all cases, energy is absorbed by the material and subsequently reemitted as light. Various colors can be produced by varying the nature of the material, particularly the impurities within it. Depending on the energy excitation source, various light-emitting technologies are commonly classified as being primarily *chemoluminescent, electroluminescent, or photoluminescent*, although light can be emitted because of other energy mechanisms (such as friction) as well.

Nanophosphors in Lighting

Various kinds of *phosphors* are already commonly used in many lighting devices, including common fluorescent tubes, electroluminescent wires, strips and surfaces, and LEDs. In a ubiquitous fluorescent tube, the inside is coated with phosphors. Electricity excites the gas-filled tube to produce shortwave light, which in turn causes the phosphors to become fluorescent and produce visible light. In common lighting wire, strip, and surface devices based on electroluminescent technologies, a voltage input provides an energy stimulus that causes light to be emitted. A high electric field causes electrons to move through the phosphor and hit ionized impurities scattered through the phosphor that in turn become excited and emit light photons of specific color and frequency. In a typical application, such as the backlighting of an alarm clock, phosphors are bonded as surfaces onto substrate materials. Lighting devices based on electroluminescent technologies provide uniformly bright surfaces. They generate little heat and are energy efficient. Various varieties of substrates (polymeric, glass, other) can be used as long as an electric field can be generated. Various sizes and shapes of "glowing" wires, strips, or surfaces can be obtained. Depending on surface shape, different manufacturing processes (e.g., printer-based or spin coatings) can be used to fabricate lighting devices based on these technologies. Before the advent of LEDs, which also use phosphors, these technologies seemed poised to provide the next generation of lighting solutions in products and buildings, but the several advantages of LEDs soon caused them to become dominant. Nonetheless, they still form a highly attractive lighting solution for many applications due to their simplicity, reliability, and uniform light quality.

Considerable work has been done on creating "nanophosphors" based on nanoparticles that have unique physical and chemical properties as compared to normal phosphors. These properties are highly dependent on particle size and shape, including their emission characteristics. Of particular importance here is that the emission bands of emitted light can potentially be finely tuned. These nanophosphors can be synthesized by either physical methods such as molecular beam epitaxy, sputtering, or the like or chemical methods such as sol-gel, colloidal, and so on (see Chapter 8). Emission efficiencies can be dramatically improved compared to normal phosphors. Luminescent semiconductor nanocrystals have been widely researched because of their unique optical properties.

Lighting devices based on nanophosphors can potentially become much more efficient than is currently the case. One practical approach is to build on existing phosphor-based lighting technologies and enhance them via nanophosphors. Various kinds of coatings with nanophosphors that could be used directly or in conjunction with traditional phosphors are under development. These include existing LED technologies as well (see the discussion on LEDs that follows).

Solid-State Lighting: Quantum Dots and QLEDs

In comparison to older incandescent or fluorescent lighting devices, the introduction in recent years of solid-state lighting technologies—the direct conversion of electricity to light using semiconductor materials (normally in the form of LEDs)—was an enormous step forward. LEDs fundamentally produce light via a special form of electroluminescence. Within just a few years, LEDs have become the most dominant type of available solid-state lighting solution. LEDs have been used everywhere—from simple point sources that glow or blink to convey information about the functioning of a product to complex manipulable arrays that define unique architectural environments. LEDs are even used in common flashlights. LEDs are not only used directly but often form a source for light to be manipulated via lenses, filters, or other techniques. The demand for devices incorporating LEDs remains huge.

Despite their now common use, however, current LED technologies based on semiconductor materials made from wafers are not without their limitations, which nano-based quantum dot technologies promise not only to help redress but to move to a new level of sophistication. In particular, one of the fundamental problems with LEDs—that of narrow sets of emission frequencies that limit their uses—is solvable via quantum dot technologies. A new generation of solid-state lighting—quantum light-emitting diodes (QLEDs) made of quantum dot networks—are coming into use that would work similarly to traditional LEDs but have greatly improved functionalities and new uses.

To understand how quantum dot technologies can spur the development of solid-state lighting to a new level, let's first look at traditional LEDs. Current LEDs are fabricated from wafers of traditional semiconductor materials (such as silicon or geranium). A semiconductor material can change from being nonconductive to conductive, or vice versa, when subjected to electricity (see Figure 9.40). Light emission is produced by driving an electrical current through a semiconductor material with a p-n semiconductor junction. At the atomic level, excited electrons jump across the gap between the valence and conduction band that are present and then eventually decay, producing photons with a wavelength related to the energy of the band gap (see Chapter 4). In an LED, energy input causes a voltage output at the junction, which in turn causes fluorescence to occur. The wavelength of the light emitted in a typical LED is quite narrow and dependent on the nature of the semiconductor material used in the wafer. Final visible colors can be adjusted via the addi-

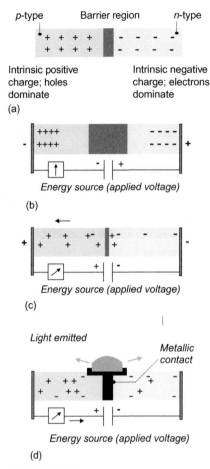

FIGURE 9.40

(a) Semiconductor with p-n junction. (b) In the reverse-bias *mode there is no flow of current* across the barrier region. (c) In the forward-bias *mode the current increases exponentially with* the applied voltage. (d) In a *light-emitting diode,* or LED, energy input into the junction creates a *voltage output that causes flouresence to occur.*

FIGURE 9.41

An inorganic, nanocrystal-based multicolor light-emitting diode. Semiconductor nanocrystals are incorporated into a p-n junction formed from semiconducting GaN injection layers. (Courtesy of Los Alamos National Laboratories.)

tion of phosphors. Hence the actual emission wavelengths that can be generated remain limited and not easily adjustable—all serving to prevent their use in some applications that demand precise control of light wavelengths (a common need in many engineering or industrial applications).

Alternatively, the ever-present need to produce full-spectrum "white light" is equally problematic. Current approaches use mixed red, green, and blue phosphors that are applied to the semiconductor material. These are difficult to both accurately mix and precisely apply. As these phosphors age or even heat up, visible color differences can be observed. Consequently, though "white light" may be possible, the quality is low. Despite their relatively high efficiencies as energy conversion devices compared to old incandescents and other sources, current solid-state lighting systems still fall far short of desired efficiencies and well short of theoretical efficiencies.

Quantum dots offer great potential in the form of QLEDs which are made out of networks of quantum dots and can also build on, yet dramatically improve, existing LED technologies. Quantum dots are essentially nanometer-size crystals of semiconductor materials (e.g., silicon or germanium) for which the electronic properties are strongly dependent on their size (see Figures 9.41 and 9.42). The potential advantages are many. Efficiencies are potentially extremely high. Better control of the emitted light is possible, as are improvements in the form factor characteristics so important to designers.

A particular characteristic of nanocrystalline quantum dots that makes them attractive for use is that they offer the affordance of being able to finely tune the actual wavelength output of an LED. Unlike having to use the naturally occurring wavelengths of traditional semiconductor materials, wavelengths in quantum dots can be controlled in nanocrystalline materials. The energy separation between valence and conduction bands can be altered in nanocrystalline quantum dots by changing the size of the nanoparticles. Resulting energy levels can thus be varied. Since the wavelength of the emitted photons depends on these levels, the wavelength of the light emitted can be varied as well. A direct approach is to mix quantum dots that emit blue, green, and red as needed to make a semiconductor that has the desired spectral output. In an engineering context, precise wavelengths could be developed to match specific needs. The same general approach would also yield great improvements in generating "white light" without the use of cumbersome phosphors. By mixing the right kinds of constituent quantum dots, a high-quality white light can be obtained.

Even here, however, keep in mind that for white light to be used for task or general architectural lighting, the ideal distribution of emitted photon wavelengths should relate to the spectrum perceived by the human eye if colors are to be perceived correctly. For efficiency, there should be no photons outside the visible range. For task and architectural applications, high outputs and brightnesses are usually needed for sufficient illuminations, which in turn tend to reduce efficiencies. Meeting both spectral and efficiency needs in solid-state lighting remains a tall order for even the quantum dot world. A cross-section of a QLED is shown in Figure 9.43. Self-assembled quantum dots are shown in Figure 9.44.

Research is currently directed along many fronts. For the design community, the need for wireless technologies is always present. Recent work on the development of quantum dot nanocrystals that emit light when near an energy source and potentially allow the nanocrystals to be activated without wires being attached to them is particularly interesting. The energy source is designed to emit energy at wavelengths that can be absorbed by the nanocrystals. Work is in the early stages and distances and efficiencies remain small, but the idea is interesting.

The use of quantum dots can also offer interesting advantages in production and form terms in that they are no longer wedded to the classic wafer of silicon or geranium that must be precisely shaped and cut. Using various kinds of colloidal approaches, quantum dots can be grown in large quantities. They can then be used in several types of media. A good deal of research has been directed toward incorporating quantum dots into various kinds of polymeric matrices of one type or another. Stability issues are particularly important here. Polymeric composites can be rigid or flexible. Ultimately they can be layered into films, cast, or painted. Transparent polymeric composites have been explored. The design flexibility potentially afforded here is enormous.

Solid-state lighting solutions, notably QLEDs, remain extremely promising. The ability to more carefully match desired spectral qualities of the light that is emitted with the qualities desired for the application remains the fundamental key here. Advantages of QLEDs effectively compete with another interesting newly developed technology—the organic light-emitting diode (OLED)—because of potentially longer lives and improved stability. Keeping in mind that electricity needs for artificial lighting account for huge parts of

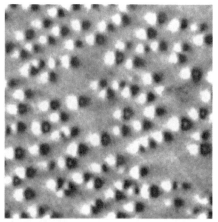

(a) Quantum dots grown from indium, gallium, and arsenic. Each dot is about 20 nanometers wide and 8 nanometers in height. (Courtesy of NIST.)

(b) Light emitted by quantum dots in solution.

FIGURE 9.42
Quantum dots.

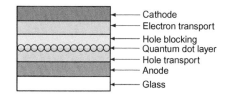

FIGURE 9.43
Quantum light-emitting diode (QLED) cross-section.

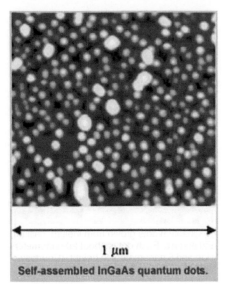

1 μm

Self-assembled InGaAs quantum dots.

FIGURE 9.44
Self-assembled quantum dots. (Courtesy of NASA.)

a nation's energy use, the potential energy savings that would accrue from dramatically improving efficiencies could also go a long way toward remedying some of our energy problems. In the United States, electricity for artificial lighting now accounts for around one fifth of the nation's electricity use and nearly a tenth of the nation's total energy use. The opportunities afforded by improvements in form factors will also undoubtedly make these same solutions far more likely to be adopted and used by designers in the future.

Displays, Screens, and Electronic Papers

Displays are ubiquitously present in everything from TVs and computers to a host of product control screens and larger information conveyance devices. It is obvious that huge amounts of research and development work have not only brought display technologies to an already high level of sophistication but that driving forces in the industry will seek to use the many affordances provided by nanomaterials to achieve even higher levels of sophistication. Large-format screens are certainly sought after. Electronic papers—devices that seek to replace the ancient printed page—have been under development for years and have begun to achieve wider adoption. Collectively, this field is extremely large and beyond the scope of this book. We review only a few primary approaches here.

Display design issues include viewable size, viewing angles, response rates, contrast and brightness ratios, color fidelity and registration, spectrum coverage, image persistence, ghosting, geometric distortion, power consumption, size and weight, and costs. Flat-screen technologies include liquid crystal displays (LCDs), plasma devices, suspended particle displays, electrochromic displays, electrophoretic displays, and, more recently, possibly thermochromic displays. All require, of course, highly sophisticated and active electronic control systems to allow the color, contrast, or transparency changes that are necessary across a screen to enable the formation of images or text.

Electronic paper approaches normally have some similar issues relating to the quality of image (color fidelity, contrast, and brightness), but response rates are less critical. An additional goal for these portable devices is to be able to hold text and images with no or very low power consumption and to change images with very low power. Flexible displays are usually considered essential for electronic papers. Electronic papers are explored not only as book replacements but also for point-of-purchase store display tags and virtually everywhere print-on paper now appears.

The performance of many types of basic technologies in wide use, such as LCDs, can be enhanced by the use of nanomaterials and nanotechnologies, as discussed shortly. In all these approaches, many related technologies are invariably involved. Thus the many kinds of light-control films we've previously discussed will find wide use in displays of any type to enhance viewing, such as the use of antireflective coatings. Various nanophosphors also have many optical properties that make them attractive for use in many types of display technologies, particularly flexible screens.

Liquid crystal displays (LCDs) are widely used in product display screens. They can also be used in larger sizes as transparency control technology for "privacy" windows and related applications. LCDs have electrically activated color-changing liquid crystals sandwiched between two transparent sheets. Liquid crystals are an intermediate phase or state between isotropic liquids and crystalline solids. Not all materials exhibit liquid crystal phases; for example, the transitions from ice to water to vapor do not, whereas several organic compounds do. Liquid crystals possess interesting orientable properties that are sensitive to electric fields. In an LCD, the liquid crystals are placed between two transparent polarizing sheets. Liquid crystals show optical birefringence under polarized light. In an LCD, the liquid crystal materials have elongated crystals that are made to twist through the two sheets (see Figure 9.45). Entering light follows this twisting. The polarizing sheets are arranged so that light passes through and the whole assembly appears transparent. Application of an electric current causes the liquid crystals to untwist, blocking the light that would normally pass through. Therefore, where current is applied, light is blocked and the display is dark. A matrix of millions of transparent switching transistors in a thin film on the display surface allows complete control of the surface area and enables its use as a display device.

Though versatile, LCDs are invariably a topic for further research because of limitations. Power consumption is relatively high. Contrasts could be improved. Scalability is always an issue. Flexible LCD displays have been made but are expensive and difficult to maintain. Nanomaterials and nanotechnologies are being explored as sources for improvement in a variety of ways. As discussed in earlier chapters, there are a great many new developments in the general area of nanocrystals. There is great control over the sizes and shapes of nanocrystalline elements. Improved methods for growing crystals in solution have been developed. Hence, improvements in the phase behavior, transport, and structure of the actual liquid crystals used in LCD devices can be expected. Other research

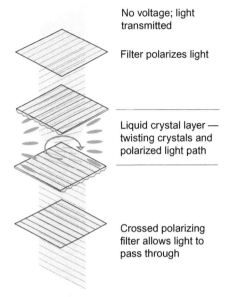

No voltage; light transmitted

Filter polarizes light

Liquid crystal layer — twisting crystals and polarized light path

Crossed polarizing filter allows light to pass through

FIGURE 9.45

In a liquid crystal display (LCD), a liquid crystal solution is sandwiched between two polarizing sheets. Applying a current causes light to pass through or be blocked. With no voltage, the liquid crystals naturally twist into helixes to align with the grooves, and the polarized light follows the twisting crystals. An applied voltage aligns the liquid crystals, and so they cannot serve to twist the polarized light path, thus blocking the light.

is directed toward the surrounding polarizing sheets. As discussed, the use of nanotechnologies is expected to increase our ability to polarize sheets and hence improve LCD performances. Scalability is particularly a goal that can be helped by developments in this area.

Given their status and widespread use as well as their limitations, LCDs are often cited as the display technology to be beat or the target for disruptive technologies. Despite their many faults, including their complexity and cumbersomeness, LCDs still unquestionably remain one of the great workhorses of our electronic product world.

Suspended particle displays are attracting a great deal of attention. They are electrically activated but can change from a clear state to an opaque state instantly, and vice versa. A typical device has multiple layers. The active layer associated with color change consists of needle-shaped particles suspended in a liquid. (The layer can be film.) If no voltage is present, the particles are randomly positioned. Light is consequently absorbed and the device is opaque. The application of a voltage causes the needle-shaped particles to align with the imposed field. This alignment then allows light to pass through and the device to be clear. Continued current is not needed to retain a transparency state; the device remains at the last setting when the voltage is on or off (see Figure 9.46).

Electrophoretic technologies were developed in 1974 at Xerox's Palo Alto Research Center as a way of achieving an addressable electronic paper, which in turn was based on the electrophoresis phenomenon discovered as early as 1809, when F. F. Reuss observed that fine clay particles in water move when subjected to an applied electric field. A current electrophoretic display forms images using dispersed charged pigment particles in a fluid that move under the action of a controlled electric field. Current approaches use encapsulated titanium dioxide particles dispersed in a hydrocarbon oil. Absorption dyes and surfactants to facilitate the development of surface charges on the particles are also in the fluid. The fluid with the dispersed particles is encased between two transparent conductive sheets that can selectively apply charges across the display. With the application of selective charges on pixels across the display, particles move forward or to the back and create patterns of absorption and reflection that in turn create images.

Another display system is the *plasma panel*. Cells located between glass panels hold inert gases such as neon and xenon. On excitation, the gases go into a plasma state that in turn excites phosphors

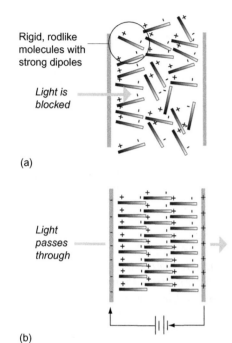

(a)

(b)

FIGURE 9.46

A suspended particle display. (a) Particles suspended in film between two clear conducting layers align randomly in the absence of an electric field, absorbing light. (b) Application of an electric field causes individual molecules to orient similarly, thus allowing light to pass through.

that emit light. Plasma is an ionized gas that has free electrons that make the gas electrically conductive and responsive to electromagnetic fields. Other display or electronic paper technologies that are fully flexible include systems based on organic light-emitting diodes (OLEDs) or on thermochromic or electrochromic phenomena (see previous discussion). In OLED approaches, the fundamental display device consists of two charged electrode sheets with organic light-emitting material between them. With the application of selective charges, images can be created. Electrochromic displays consist of two conductors as well, with an electrochromic material and electrolyte on a substrate. Recent research has also suggested that flexible electronic paper using thermochromic composite thin films are possible. Needed thin conductive wiring patterns required for image formation have been around for some time.

Other Nano-Based Technologies

A huge array of other technologies use light in one way or another that is now being, or can be, enhanced via nano-based technologies. The field is literally burgeoning with applications, particularly in electronics and medicine and that might ultimately have implications in product design and architectural systems. Covering them is beyond the scope of this book, but we can note that a key to conceptually understanding this vast array of applications is to remember that light is a form of energy. Energy inputs can cause materials to do many things. Not only can reflection, absorption, transmission, or fluorescence occur, but many other response phenomena can occur as well. In *photoconductive* materials, for example, the electrical conductivity of the material increases or decreases with varying levels of light intensity. This effect is exploited in many sensors. In other common sensors, the *photoelectric* phenomenon, in which voltage outputs vary with light intensity, is used. Interestingly, in *photorheological* materials the actual stiffness of the material can vary with varying light intensities. There are many applications in which "detection" is of paramount importance. Fluorescence detection, for example, is a widely used method of research in many fields, especially in biotechnology spheres, and forms the basis for many devices. Several nanomaterials, including those based on zinc oxide, have been explored as a way of enhancing detection capabilities. There are also large numbers of applications where the intent is to provide some form of optical energy transport or storage. We will see examples of some of these applications later in this book.

9.6 SOUND AND ACOUSTICAL ENVIRONMENTS

Primary objectives in designing environments for sound usually hinge around characteristics of sound sources, sound mitigation, or sound enhancement. Sound sources include everything from musical instruments to loudspeakers. Mitigation objectives include the control of noise (unwanted sound). Enhancement objectives normally include improving sound quality in various environments or at specific locations. In some fields involving sound manipulation and control, such as architectural acoustics, both mitigation and enhancement objectives are commonly involved in designing a space to have particular acoustic qualities with respect to particular types and locations of sound sources. Achieving these objectives includes understanding the physical nature of sound, the sound-related properties of materials that serve or act as sound transmission media, and ways to analytically model sound in an environment.

Both passive and active means of achieving mitigation or enhancement objectives are possible. Passive means occur via careful control of the physical parameters of the environment—geometries, assembly configuration material properties such as absorption, or reflectance. Active means for mitigation include various energy dissipation devices based on differing electro mechanical approaches (e.g., base isolation mechanisms) to electronically complex noise-cancellation devices. In all these approaches, specific material characteristics and properties are highly important and are primary variables in determining performance outcomes as well as in developing specific intervention strategies. Within this broad array of approaches it can be expected that nanomaterials and nanotechnologies can make a contribution. Certainly this is appearing so for small-scale isolation or damping devices used to mitigate vibration-induced noises caused by components such as fans or machines. As we discuss shortly, however, some of the more seemingly straightforward applications, such as finding ways to dramatically increase the sound absorption qualities of various bulk materials, are proving interestingly problematic, primarily because of the long wavelengths associated with sounds.

To get into these applications in more depth, we begin by first reviewing the characteristics of sound, including sound generation, transmission, the audible and inaudible effects of sound waves, the role of specific material properties (such as absorption and reflectance), and general approaches to analytically modeling sound environments with the objective of designing good environments for music or speech audibility. In noise control and acoustical applications, what we actually hear is of fundamental importance. Hearing is performed by the auditory system. Vibrating sound waves cause varying air pressure levels that in turn cause eardrums to vibrate. These vibrations are then transduced into nerve impulses that are subsequently perceived and interpreted by the brain. Not all sounds are audible to animals. Species have different normal hearing ranges for both the frequency (or pitch) of the sound and its amplitude (or loudness). The human ear responds to the 20 Hz to 2000 Hz range.

Noise mitigation is particularly important. Noise is here broadly defined as any unwanted sound that typically emanates not from the natural environment but from roadways, industrial operations, commercial, or uncontrolled residential sources. Sound is generally considered noise when it is erratic, loud, and not controllable by the hearer; for example, what could be more irritating than the buzzing of a mosquito?

The following sections examine sounds and sound controls in buildings and products, but we should keep in mind that sound is also positively used in many other ways, such as to detect the presence or location of an object. Bats use a form of *echolocation*, whereas humans developed sonar explicitly for this purpose. Humans are known to use information gained from sound in fascinating ways. Here, however, we retain a simpler focus.

Fundamental Characteristics of Sound Environments

As we discussed in Section 4.8, sound is a series of waves created by the vibration of something—a loudspeaker or a piece of metal hit by a hammer, for example. Sound waves then move through the air or other kinds of materials via a regular vibration of the atoms or molecules in the transmitting material. Some materials easily transmit sound waves, as in water; others have absorptive qualities. Sound creates compression waves of varying amplitudes that move through a material. In air, these intensity waves can be associated with regions of air molecules that have been alternately compressed and reduced. An initial source such as a vibrating drum head or violin string starts the waves, and they disperse outwardly from the source. These sound waves can be described in terms of their amplitudes, wavelengths, frequencies, and velocities, which are in turn dependent on the properties of the carrying medium or material. *Amplitude* is the loudness of the sound and depends on the amount of compression that takes place. *Wavelength* is the distance from the crest of one wave to another. *Frequency*, or *pitch*, is the rate at which waves pass a point. Short wavelengths correspond to high pitches. The velocity of a wave is its wavelength times its frequency.

The way sound travels from a source to another point depends on both geometry and the transmission medium. Sound waves emanating from a point source spread out. Amplitudes decrease as the square of the distance from the source; thus there is inherently a rapid decrease in sound intensity away from a source. The sound itself can be carried by different media, and sound can be transmitted from one medium to another, such as from air to a wall and to the air again. As sound passes through each medium, some sound intensity is lost due to absorption. Here a portion of the sound energy is converted into heat, thus reducing the amount of sound energy that is transmitted.

Absorption levels vary by transmission medium. As sound passes through various media, changes in velocity occur depending on the properties of the media. The velocity of the transmitted sound depends on the density and stiffness properties of the medium.

When a sound wave in one medium or material encounters another in its path, it is *reflected, absorbed,* or *transmitted.* A portion of the sound is reflected while the remainder passes into encountered material. Of the portion of sound that enters the media, some is lost to internal absorption (see the following discussion). The remainder is transmitted through the material. Dense hard and solid materials tend to strongly reflect incoming sound waves and weakly transmit them. Light, highly porous materials normally have weak reflections and can transmit waves strongly.

The amount of sound that is reflected depends on the hardness and smoothness of the encountered surface. Controlled directions of reflections can be achieved with smooth surfaces better than rough surfaces. The latter causes diffuse microreflections that cause scattering in the primary reflection. When highly porous surfaces are present, additional absorption can take place because sound waves can literally bounce back and forth within small cavities on the surface before finally being diffusely reflected. Another form of reflection can occur after the sound is transmitted through the material (with losses due to absorption) and enters the next. When passing from a dense material, such as wood, to air, this reflection is small but is still there.

Primary absorption mechanisms (viscous loss and intrinsic damping) are described in Section 4.8. The amount of absorption that occurs generally depends on both the material characteristics and on the frequency or pitch of the sound. Absorption coefficients that are measures of the sound-absorbing capabilities of various materials can be experimentally determined. This coefficient is the fractional part of incident diffuse sound energy that is absorbed and not reflected by a surface. Values vary widely. The higher the value, the more efficient the material is as a sound absorber. Some metals and stones are as low as 0.01. Highly absorbing materials used for special purposes can be as high as 0.96 and almost up to the maximum possible value of 1.0. Values, however, are dependent on the frequency of the impinging sound waves. For a complex assembly consisting of several materials, a measure called the *sound transmission class* (STC) is sometimes used as a single figure rating system. STC is intended to provide a measure that estimates the sound insulation properties of an assembly. It can be used to rank-order a series of different assemblies. *Sound transmission loss* (STL) is a related measure.

The part that these parameters play in the actual analysis of sound in a complex spatial environment involves many more considerations concerning the configuration and dimensions of the space and distribution of materials used, as we will soon see.

General Noise-Control Approaches

Sounds reaching a point or hearer may be directly airborne from a source within the space (e.g., from a machine or TV) or external to it, airborne sounds that result from structure-borne vibrations, or sounds from impacts. The intent of a sound-control strategy varies very widely according to the design context and whether the sounds are wanted or unwanted. In some cases the intent is simply to reduce unwanted noise. Here the normal intent is to reduce or eliminate noise in order for a space to have an overall sound environment conducive to normal work and life activities. Another intent could be to eliminate unwanted sounds because of the vibrations that they might induce in delicate instruments or manufacturing facilities. The latter is particularly important in the context of this book—for example, vis-à-vis issues in manufacturing or characterizing nanomaterials. Conversely, in other situations, an intent is to provide particular types of sound to a space for achieving certain kinds of ambient conditions (the notion of a totally quiet room is known to be antithetical to the work habits of many) or simply to mask uncontrollable sounds.

If objectives are to reduce or eliminate unwanted noise within a space from a source outside the space, attention is normally directed to reducing source levels or using barriers and isolation techniques. Many wall systems in buildings are commonly designed to prevent sound from being transmitted from one space to another, and the enclosures around products containing noise-generating devices are similarly designed to reduce sound transmission. Passive control approaches are common. As previously noted, dense materials and high mass assemblies tend to strongly reflect incoming sound waves and weakly transmit them, whereas light materials and assemblies normally have weak reflections and can transmit waves strongly. The more the sound is reflected from the source area and the more that is absorbed in the wall or enclosure materials, the less will be transmitted to the space where noise control is sound. Hence thin, hard surfaces can be helpful because they aid reflection. Reductions in sound transmission can also be obtained via the use of highly absorptive materials in wall or product enclosure assemblies or via introducing deliberate separations between materials (a form of isolation strategy discussed shortly). The obviousness of the approach, however, stands in distinction to the problematic nature of its application. In building environments, for example, there are obviously many reasons that barriers cannot be monolithic continuums—consider the need for doorways, windows, ventilation ducts, and so on.

Many common assemblies use varying combinations of mass, reflective, absorptive, and separation strategies simultaneously. The strategy chosen often depends on the frequency and amplitude of the sound source. In walls of buildings, troublesome low-frequency sounds (such as those associated with music) are really only effectively mitigated by high-mass systems and separation strategies; little absorption takes place. High-frequency sounds associated with many products are more easily mitigated using absorptive materials. Many sound-deadening product enclosures are remarkably sophisticated and consist of multiple layers—for example, thin foils with multiple layers of butyl damping materials are commonly used for sound dampening in refrigerators, ventilators, and other devices.

Structure-borne sounds result when vibrations from machines or equipment are induced in supporting structural elements. The source need not be in the considered space. In buildings, vibrations can result from passing trains or cars or from remote heating, ventilating, and air-conditioning system elements. Vibrations can be transmitted throughout a whole system. Impact noises, such as something dropped that hits a surface, also result from vibrations. Mitigation measures for these sounds can be taken through materials, isolation, and damping or, as is often the case, some combination. Structure-borne sounds emanating from machinery or other vibration sources can be reduced using low-density materials by disconnecting the source of the vibration from the carrier and radiating structure. Passive isolation systems are very common and are accomplished by breaking the path of vibrations by gaps or the insertion of isolation devices containing resilient materials that do not transmit vibrations well. Isolation approaches can be at the device level or at the level of whole rooms. A direct control strategy would be to provide one or another type of base isolation devices under the machinery that damps out or otherwise absorbs vibrations before they are transmitted to the primary supporting structure. A whole host of conventional kinds of these devices already exist and normally use some kind of resilient material.

Whole rooms can be similarly isolated via the use of techniques such as floating floors in more sophisticated applications. Impact sounds can effectively reduced via the use of low-density, light materials, including resilient materials such as common floor coverings as well as separation or isolation strategies. Impacts on hard, high-density materials, by contrast, conduct impact sounds quite well (see Section 4.7).

These approaches are primarily passive in nature. Many types of active control systems have been developed as well. Costs are invariably high, so most active systems are used in either products or high-value environments (such as airplane cabins) and rarely in common building environments. Many use one type or another of noise cancellation system like those found in noise control headsets. Others use some type of vibrating panel system. For structure-borne noises involving vibration transmissions, active damping approaches have been used, including base isolators in magneto-restrictive materials that change their viscosities when subjected to varying magnetic field intensities.

Space Acoustics

In its broadest sense, the term *acoustics* describes a branch of science and technologies devoted to the study of sound and related mechanical waves in various media, including how sound is generated, transmitted, received, perceived, and controlled—and hence includes many of the topics we've discussed but is broader in scope. The broad field has evolved into a number of subdisciplines, albeit all founded on similar scientific principles. Hence we have architectural acoustics, which deals with the study of the behavior of sound in architectural spaces (auditoriums, concert halls, commercial establishments, and the like) and the way sound and building design parameters interact. Other fields include aeroacoustics, bioacoustics (the use of sound by animals), biomedical acoustics (the use of sound for medical purposes, as in ultrasound applications), aeroacoustics, electroacoustics, and many others. The focus in this section, however, is on acoustic strategies in relation to building spaces and products.

A typical acoustical design objective is to create spaces that enhance the audibility or clarity of speech or music, or, as is the case in some concert halls, to enhance the qualities of performances. These latter objectives are particularly complex because of the normal need to have good sound qualities, not at just one point but throughout the spaces occupied by listeners. Sound qualities will be inherently different from location to location throughout a space because direct and indirect sound paths to various locations will inherently have differing directions, travel distances, and numbers and types of reflections off surfaces that have differing reflective qualities and different absorptions. The shape of the enclosing surface and placement of surfaces with differing properties are consequently clearly important design variables. In building design, shapes cannot also be simply optimized for sound, however, because of other complicating needs for viewing, egress, and the like. The complications of designing these spatial environments for acoustical performance were described at length in Section 3.5 in relation to a concert hall. Other kinds of building spaces, such as lecture halls, are also particularly important from an acoustic design perspective.

The issues involved in the acoustic design of these kinds of spaces are complex because it is necessary to model not only the way sounds interact with properties of surrounding materials but the overall characteristics of the space itself and the way sounds are disseminated through it. When a sound made by a person or an instrument is generated in a room, sound waves travel outward from the source and reach a listener, both by a direct path and by reflections. A listener sitting very near a stage in an outdoor setting will hear the sound of a musical instrument essentially as the player produced it. The sounds typically are not persistent. Within an enclosed setting such as a room, however, the listener hears not only sounds that have originated from the source but also sounds that have come from the source but that have been bounced around the surfaces of the room. A much more complex situation thus exists that clearly depends on not only the character of the emanating sound but of the geometrical and physical characteristics of the room itself. The commonly used overall measure of this characteristic is called the *reverberation time* (RT; see Section 3.5). Reverberation times, however, are only one of several measures that stem from the fundamental issue that sound reaching a listener's ear has both direct paths and indirect (reflected) paths, including early decay times, loudness measures, initial time delay gaps, and others.

Analytically modeling the kinds of sound behaviors noted here is inherently complex. Many types of computer models have been developed to aid in this process. These simulation models invariably require a 3-D digital model that describes the room geometry and the placement of various kinds of materials throughout it, as well as properties of the materials themselves. Source positions and characteristics must be defined, as must receiver positions. Various kinds of energy-based or ray-tracing algorithms that draw literally millions of lines from sources to receivers are then used to predict resultant sound behaviors. Simulation methods must be benchmarked for accuracy against experimental results. Though simulation models can be quite good for certain applications, various kinds of physical-scale models are often still used in many applications, particularly in large spaces with many complex or irregular surfaces and material distributions. Once the sound behaviors are modeled, and based on understandings of human hearing and responses as well as understandings of issues such as "acoustical glare," design evaluations are then made. Many other factors play a role in evaluations, including human-based reactions. Is a concert hall "lively" or not, "warm" or not, or can the term *brilliance* ultimately be used to describe the sound heard?

The list of measures noted here that an acoustician has to deal with could go on and on, and knowing how to design in relation to them is clearly a task for specialists. Of importance here is that reflection and absorption properties are clearly fundamental, as are surface textures, density and hardness, mass, thickness, support conditions, and the presence or absence of damping mechanisms (implicit or added devices). It is invariably the play of these many factors that contributes to whether an acoustical design is good, bad, or indifferent. What becomes clear is that there is no such thing as a specific material property (such as absorption) being intrinsically either good or bad. Appropriate values are context dependent. What designers want is to be able to exhibit precise control over these many factors so that the overall design can be optimized.

Applications of Nanomaterials in Sound Environments

As discussed in the previous section, there is great diversity in designing for sound control and acoustical performance. It is thus not easy to describe specific material property needs that might be addressed via the use of nanomaterials or nanotechnologies. Only with fairly simple or well-defined objective—for example, to create an anechoic test box or to isolate specialized rooms from external sound—are needs well defined. Dealing with transmitted noises requires different strategies than dealing with structure-borne or impact sounds. In more complex design cases, as in the acoustic design of a space for music or speech, the situation is even considerably more complex because of the geometry of the space and the fact that multiple hearers need to be considered.

In the previous section, we identified two fundamental types of needs: to be able to mitigate noise and, conversely, to enhance sound quality. On first glance, the success of nano-based advancements in other fields (such as lighting, thermal, and electrical) would seem to suggest that the path is open for similar rapid and across-the-board advancements in relation to sound environments. Overall advances here based on the use of nanomaterials, however, have been relatively slow. There are fewer ways in which nano-based approaches can radically alter the basic sound-related properties of materials. There are exceptions and applications, to be sure, but most attention is directed toward extremely active electronically controlled small devices for sound emanation or acoustical damping rather than larger passive applications, such as sound absorption, associated with acoustical design in big components in buildings or properties.

Manipulating Properties

In many nanomaterial applications, the direct enhancement of specific strength, thermal, optical, or electrical properties made possible through the use of nanomaterials leads to obvious and immediate applications. This is simply less so with respect to large-scale components used for sound mitigation. There are some applications for small-scale components.

In typical acoustical design situations where sound quality is to be enhanced, a commonly expressed and straightforward desirable objective is to be able to dramatically increase (or otherwise control) the sound reflection and absorption properties of various walls or enclosures by directly using selected materials and without using complex assemblies of multiple materials. Doing so, however, is proving difficult. A primary reason is that the wavelengths associated with sound are quite long (see Sections 4.8 and 7.6) with respect to nanoscaled imensions. With light, for example, many interesting and useful interaction effects take place on surfaces because the short wavelengths of light are quite close in size to the dimensions of many nanostructures, including the variation of the wavelength of emitted light as a function of particle size in quantum dot applications, the surface plasmon effect, or the transparency of many types of nanocoatings. By contrast, the long wavelengths of sound do not directly correspond to dimensions of nanosized surface structures, and are invariably much longer. This mismatch between the wavelength of sound and nanostructure dimensions inherently poses challenges in effectively utilizing nanomaterials in attempting to improve a material's sound absorptive qualities. Common sound absorption wall and ceiling panels in buildings, for example, have surfaces with relatively large surface pores or roughness that trap sound waves and cause them to dissipate by internal bouncing within the pores. Making surface pores nanosized will not enhance this energy dissipation since the wavelengths of the sound are so long that they would not interact with the tiny pore sizes. The pores would be invisible (so to speak) with respect to the way they affect sound impinging on the surface.

A further inhibiting factor is that quantities such as sound velocity, sound-wave impedance, and sound radiation factors for a material depend on the direct physical properties of a bulk material (density, ρ, and the modulus of elasticity, E; see Section 4.8) and are not easily changeable in many common bulk materials by simply adding nanomaterials. For example, when nanocomposites are made by adding a nanomaterial as a second phase, the quantities of nanomaterials used are typically quite small, which have relatively little effect on the overall density. This situation is particularly problematic for applications in passive building acoustics with its needs for large material volumes. Hence a more direct approach is to use nanomaterials with useful sound properties that can be made directly in bulk form (see the discussion on aerogels that follows).

Nanoporous Sound Insulation Materials

Surprisingly few nanomaterials can be produced directly in the large size or bulk form needed for many building-scale applications. Most are nanocomposites. Exceptions include the many thin films, coatings, and nanoporous materials that have been developed. Few of these, however, have useful sound-related properties. Nanoporous materials are exceptions. Aerogels are discussed extensively in Section 9.3 with regard to their important role as thermal insulators. Of interest here are their interesting sound-insulating properties. Aerogels have incredibly low densities. Sound waves transmit through the air-filled part of the porous medium at a reduced rate compared to normal air. There is little direct solid phase vibration conducted through the solid matrix simply because there is little solid material present and it is only tenuously connected. Transmission reduction effects are most evident at low frequencies (e.g., less than 400 mHz). They are less effective at higher frequencies. Normally, in a building, the aerogel product is contained within some other material assembly to protect and contain it, including fiber-reinforced panels, double-walled polycarbonate glazing systems, or *U*-channeled glass. Sound transmission losses for low frequencies have been reportedly reduced as much as 25% for a polycarbonate glazing system. Since this kind of system is translucent (polycarbonates are highly transparent and aerogels translucent), it is highly interesting for use in many applications. Other kinds of highly porous materials can also be expected to have similarly useful sound-loss characteristics.

Acoustical Damping and Isolation

As previously discussed, vibration isolation is a widely used sound-control technique. Vibrations might emanate from any of a number of sources, but particularly from machines with moving or rotating parts. Vibration control approaches normally assume one of two primary forms, isolation and damping, or, as is often the case, some combination of the two. Many passive vibration isolation devices can be quite small and are manufactured as products. Vibration damping can be achieved by affixing damping materials or devices directly to the vibrating object, such as foam-type materials that are inherently energy dissipating that can be glued directly to vibrating panels.

Nanoporous materials and foams are being explored for use as passive damping materials. The potential application of nanoporous materials in many diverse applications, from thermal insulation to filtration devices, is huge, and consequently considerable work has been done in this general area, albeit comparatively little with respect to sound-damping applications. Damping vibrations require energy dissipation associated with mechanical movements and hence require special properties not found in many nanoporous materials developed for other applications such as nanoporous gels for thermal applications. Still, additional developments can be expected here, especially for use in small damping devices.

Other composites based on nanomaterials have been found to have damping properties. Many polymers are already well known for their viscoelastic qualities and have found use in damping devices. There has also already been an enormous amount of research in polymer nanocomposites, particularly with respect to strength properties and to electrical properties. The use of polymer nanocomposites for damping is possible. Polymer nanocomposites using carbon nanotubes, in particular, have been explored. In strength and stiffness studies, carbon nanotubes have been found to have attractive strength properties. There can, however, be considerable sliding at nanotube-to-nanotube interfaces that can seemingly be a problem for some applications, but, on the other hand, this same interfacial sliding can have beneficial damping characteristics. The interfacial sliding appears to create frictional energy dissipation that leads to improved damping characteristics. Considerably more research work needs to be done in this area before significant applications can develop. Many difficult issues, such as achieving uniform dispersions and temperature effects, are comparable to those in other polymer nanocomposite research. Other types of nanofillers have also been suggested for use in damping. Though applications remain largely in the future, it does appear that for many small-scale applications such as small speakers, this approach could prove valuable.

Many existing interesting damping-control devices are based on active control systems, and nanomaterials could possibly enhance their performance. The use of magnetorheological devices for damping has been widely explored, even for large applications in buildings and bridges. The viscoelastic properties of magnetorheological materials can be altered with changes in surrounding magnetic fields, which in turn can be electronically controlled. Devices have been developed that have sensors detect vibratory movements. Magnetic fields and resultant material stiffnesses are then appropriately controlled. The unique magnetic properties of nanomaterials make their use particularly attractive in conjunction with magnetorheological devices (see Sections 8.5 and 9.2). Other kinds of active damping devices might similarly benefit from the use of nanomaterials.

Sound Sources

The world of microphones, speakers, and other devices provides a rich context for the use of nanomaterials. The design of these kinds of devices, and the way that materials are used, has already reached high levels of sophistication and is a topic beyond the scope of this book. Here we will give only a flavor of activities in this area.

Considerable attention is being paid to nanomaterials in loudspeaker designs. The basic components of a loudspeaker are a magnet and a coil attached to a flexible cone or diaphragm. Passing an electrical signal through the coil causes the cone to move toward or away from the magnet. The cone causes air near the speaker to be alternately expressed and expanded, thus creating sound waves. Nanocomposites are being explored for use in cones to achieve small variations in diaphragm weight and stiffness that could positively improve performances. In a moving coil design for loudspeakers, magnetic fluids are used to provide increased power capabilities and smoother frequency responses. The fluids also conduct heat away from the coil and provide dampening actions to the cone. Nanoparticles can have unique magnetic properties (see Section 7.5) and can be used in magnetic nanofluidics. These, in turn, could find use in these speaker designs to improve performances.

Other approaches to using nanomaterials can be quite different and can address specific design issues. A major company, Panasonic, has released a product designed specifically for portable devices such as laptops that are well known for their poor sound quality. The tinny sound results from conventional small speakers not being able to produce high-fidelity, low-range sounds. In these small speakers, the needed movement of the diaphragm is detrimentally affected by the air pressure developed within the surrounding enclosure when low-frequency sounds with their large amplitudes are produced. Though the effect is momentary, the increased pressure within the cabinet affects the vibrating diaphragm and reduces sound quality. This effect is not present in speakers with large cabinets.

In some technologies, carbon materials have been placed in small speaker volumes to attempt to absorb moving air molecules, but success has been limited. The important development here is the use of highly porous carbon materials for momentarily absorbing increased air pressures. Used in the Panasonic Nano-Bass Exciter, these highly porous carbon materials physically absorb the air molecules developed during pressure increases and decrease the overall air pressure within the cabinet (see Figure 9.47). This allows the diaphragm to move freely. Higher-fidelity sound results. Transforming this interesting idea into reality required the use of highly sophisticated acoustic analysis techniques to establish both system-level configurations and the specific type and form of the materials used.

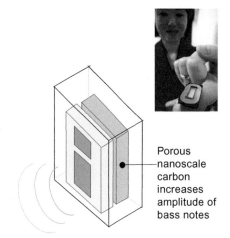

Porous nanoscale carbon increases amplitude of bass notes

FIGURE 9.47
A speaker using porous carbon materials. (Courtesy of Panasonic.)

Musical Instruments

Musical instruments are highly sophisticated acoustical devices that produce tones, amplify, and radiate sound and have evolved remarkably over the years. The long history of the production of musical instruments is closely coupled with careful attention to the characteristics of materials and the ways they are assembled. Many are associated with strings that are plucked or bowed to produce vibrations. Others are percussive and also involve vibrations. Material innovations have always affected both the design of instruments and the kinds of music produced. The introduction of the steel string in guitars as a replacement for gut strings, for example, affected the design of the remainder of the instrument and subsequently ushered in a whole new musical sound.

In a stringed instrument, the strings are coupled with other structures that resonate and radiant sound (the direct acoustic output from a plucked string is in itself quite small, since no large volumes of air are moved; see Figure 9.48). The nature of the radiated sound is dependent on many material properties: the geometric shape and mass distribution of the materials used, support conditions, and other factors (see also Section 4.8). Natural materials were common historically, but synthetic composites are finding more frequent use. Reasons include dwindling resources on one hand and desires for more material consistency on the other hand. The material selection and design problems involved in trying to use new composite materials are well known to be extremely difficult and are explored elsewhere in depth (see Further Reading). In violins, for example, the materials used in a sound box need to have different elastic constants in different directions to provide the same responses as the particular kinds of woods used historically, which are highly anisotropic. Exact matches are difficult to achieve, and to the trained

FIGURE 9.48
Can the quality of stringed instruments be improved via the use of nanocomposites? Can new sounds be achieved?

musician, differences in not only musical output but musical feel can be quite noticeable.

Though hardly a well-explored area, it does appear that the selective use of nanomaterials could have an impact here. The whole makeup of a stringed instrument is a careful balance of many factors. It is well known that the natural frequency of vibration of a common string is a function of its length, its linear mass (a function of material density and cross-section), and the tension within it. In many instruments, the highest strings must be in extreme tension and can be near breaking points. In the nanocomposites area, well-founded optimism and positive experimental results indicate that significant strength increases will be possible compared to traditional materials. If used in strings, both the maximum tension that can be developed in a string as well as its linear mass can be affected. General research has indicated that desirable string materials need to have high ratios of tension strength to density, and high ratios can certainly be expected with the use of nanocomposites. Obviously, there could consequently be different sound qualities possible, in much the same way that differences were immediately evident when steel replaced gut for use in strings many years ago. As the effect of the tension level is traced through, the tension force in a string can be seen to exert a force on the neck and body that in turn has long influenced the structural design of these components and consequently their sound response characteristics. For sound boxes, the frequencies of the eigenmodes of the vibration of a plate that might be part of a sound box require a consideration of geometrical dimensions of a plate, its density and the elastic modulus, and a consideration of support conditions. Use of stronger nanocomposites in boxes could lead to varying thicknesses and hence different sound radiation qualities. Nanocomposites might thus be another tool in trying to either find better composite material replacements for existing natural materials in conventional instruments, or, more interestingly, quite different sound qualities and ranges might be obtainable.

Other Applications

The science and application of acoustics and sound phenomena are highly sophisticated, and there are many other sound-related applications in specialized areas that we have not even touched on in this brief treatment. These applications range from all sorts of sound-based sensory or detection systems through the use of sound as an energy source for many medical and industrial applications. Sound-related phenomena also relate to other environments addressed in this book, such as lighting or thermal environments. Sonoluminescence phenomena, for example, can be expected to find use. (*Sonoluminescence* is the emission of light by bubbles in a liquid excited by sound.) The field of thermoacoustics—the interaction of sound and heat—offers excitement in a different realm. Nanomaterials can be expected to play roles of one type or another in these different areas.

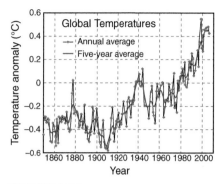

FIGURE 9.49
Global temperatures, 1850–2010.

9.7 ENABLING ENVIRONMENTS AND SYSTEMS: ENERGY

Over the past 100 years, the world's temperature has risen by about 0.6°C due to severe demands in energy levels (see Figure 9.49). Contrary to previous changes in the world's climate due to variations in Earth's rotation angle or its distance from the sun, this time there is a completely different factor involved: manmade "greenhouse gases." At the current rate of increase, the levels of CO_2 and other greenhouse gases will reach 800 ppm by the end of this century. Because CO_2 stays in the atmosphere for up to 200 years, removing those high concentrations will take a long time. If this scenario continues, the world will witness a rise in sea levels, ecosystem changes, and a deleterious effect on public health. Alternative energy technologies are, therefore, no longer an option but a crucial need for sustained economic growth, energy independence, and, ultimately, the protection of our planet and the human race.

As shown in Figure 9.50, the current world's total primary energy supply is still strongly dependent on coal and oil, the two largest producers of greenhouse gases. To address this issue, alternative energy resources such as wind, solar, biomass, nuclear, hydropower, geothermal, and ocean energy must be widely used. In this regard, the development of nanomaterials is crucial for the storage, conversion, and delivery of alternative energy technologies. Among these, the most promising are nanomaterials for solar applications, fuel cells, batteries, and supercapacitors.

Among all energy resources, solar energy is by far the most abundant. In fact, one hour of sunlight on Earth is currently enough to provide energy for the entire world for one year. However, at the moment, capturing and storing solar energy is not a trivial task. Besides the technological challenges, cost is also a problem. The U.S. Department of Energy has established the goal of $0.33/Watt for utility-scale production. Typically, solar energy conversion is

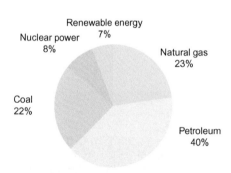

FIGURE 9.50
World total energy supply by source Nole: Sum of components may not equal 100 percent due to independent rounding. Source: EIA, Renewable Energy Consumption and Electricity Preliminary 2007 Statistics, Table 1: U.S. Energy Consumption by Energy Source, 2003–2007 (May 2008).

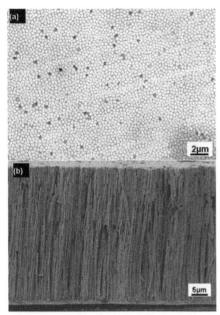

FIGURE 9.51

Scanning electron microscopy (SEM) planar and cross-section view of Bi_2Te_3 nanowires. (Courtesy of Kalapi G. Biswas and Timothy D. Sands, Purdue University.)

FIGURE 9.52

Photovoltaic solar installation. (Courtesy of Argonne National Laboratory.)

done via thermoelectric and photovoltaic materials. Thermoelectric materials convert the heat collected from the sun into electricity. Photovoltaics, on the other hand, convert solar radiation into electric energy. In the case of thermoelectrics, the overall conversion process would operate as follows: First the infrared (wavelength = 800–3000 nm) spectrum of sunlight is collected and turned into heat or stored in a thermal bath. Subsequently, thermoelectric materials convert the heat into electricity. A very important parameter associated with these materials is the conversion efficiency, which is related to a quantity called the *Figure of Merit, ZT,* (see equation 9.1 in Section 9.3)

Most current thermoelectric materials exhibit ZT values of 1. These materials generate conversion efficiencies of around 7% or 8%. To improve the efficiency to around 15%, novel materials with $ZT = 2$–3 are required. In this regard, nanostructured materials seem to be good candidates. One-dimensional nanomaterials, such as carbon nanotubes, molybdenum sulfide (MoS_2), tungsten sulfide (WS_2), titanium sulfide (TiS_2) nanotubes, as well as bismuth telluride (Bi_2Te_3) nanowires (see Figure 9.51), have seen rapid advances. The reason behind this behavior is related to quantum-size effects induced by nanoscale dimensions. In other words, dimensional confinement of electrons may lead to an increase in the electron density of states at the Fermi level, thereby increasing the Seebeck coefficient and consequently the parameter ZT.

In addition to 1-D nanomaterials, nanocomposites filled with Bi_2Te_3 nanotubes are also likely to improve the current ZT values due to enhanced phonon boundary scattering. With regard to solar energy conversion, all thermoelectric nanomaterials would need to work at temperatures close to 700°C. Thermoelectric materials of this type are also discussed in Section 9.3. Besides thermoelectric materials, photovoltaic materials are also an essential technology for solar energy conversion (see Figure 9.52). Currently, the production of photovoltaic materials is primarily associated with crystalline silicon (94% of the market). Devices made of these first-generation materials are basically diodes limited to a conversion efficiency of around 30% and a current cost of $1–2/Watt. Second-generation photovoltaic technologies are mainly those capable of achieving good efficiencies combined with low costs, but these are still in the early phases of commercialization. They include thin films, solar concentrators, and organic photovoltaics. Thin films were initially made of amorphous silicon, representing 5–6% of the total photovoltaic market. These devices are made of multiple layers, some at the nanoscale (see Figure 9.53). Because the techniques used to produce amorphous silicon are the same used for obtaining micro- or nanocrystalline structures, combinations of these amorphous/micro/nanostructures are now being investigated. Despite these developments, the synthesis of these materials is still not fully understood. More recently, solar cells made of polycrystalline Cu(InGa)Se2 (GIGS) and CdTe (see Figure 9.54) have been demonstrated, with costs close to $1.25/Watt.

A large number of companies around the world are now trying to develop processing techniques that would allow the production of high-efficiency, low-cost, and high-yield solar cells made of GIGS and CdTe. However, there are still many material challenges to overcome, such as interactions occurring at the interfaces between the various layers and novel materials with high conductivity and high transparency in the visible spectrum for the transparent conducting oxides (TCO) contacts. Solar concentrators are typically multijunction cells based on gallium-arsenide (GaAs) and related materials. These multilayered structures can achieve theoretical efficiencies above 60%, although to date the best efficiencies have been around 40%. These solar cells are currently quite expensive to be used in wide-area applications. Therefore they are normally part of mirrors or lenses that concentrate the sunlight by up to 1000 times (see Figure 9.55). In this fashion, the overall costs can be minimized.

Another emerging solar technology is the area of *organic photovoltaic cells*. The photoconversion mechanism of organic photovoltaics is based on the generation of excitons. First, an incoming photon induces the formation of an electron-hole pair (exciton). Subsequently, the charge carriers are generated and separated at an interface and then used to do work in an external circuit. This mechanism can operate in devices based on dye-sensitized nanostructured oxide cells, organic-organic composites, and organic-inorganic composites. In the case of dye-sensitized solar cells, an organic dye is used to absorb light and pass electrons to a nanoscale oxide. The hole is picked up by, for example, an organic semiconductor. Organic-organic composites involve a polymer that acts as the transporter of holes and a fullerene derivative, which transports the electrons (see Figure 9.56). Organic-inorganic composites consist of a light-absorbing polymer to transport the holes; nanostructured inorganic materials, such as TiO_2 act as the electron transporter. Besides the fact that these organic small molecules (molecular weights less than a few thousand atomic mass units) and polymers are quite inexpensive, these materials exhibit quite high optical absorption (allowing the use of very thin films), work well with plastic sub-

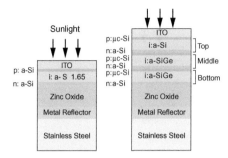

FIGURE 9.53

Single and double layered Si-based solar cells. ITO is the indium oxide layer, p is the p-doped later, n is the n-doped later, i is the intrinsic later, a-Si is amorphous silicon.

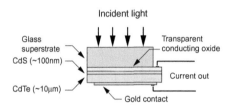

FIGURE 9.54

Architecture of CdTe solar cells.

FIGURE 9.55

Solar hyperconcentrators are devices that focus high-intensity sun radiation on a small area, typically using Fresnel lenses or mirrors. (Courtesy of Stirling Energy Systems.)

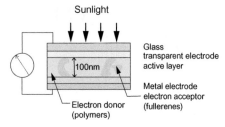

FIGURE 9.56

Organic-organic composites for photovoltaic cells.

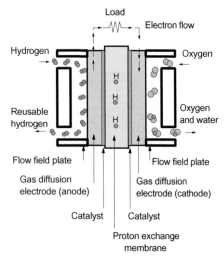

FIGURE 9.57

Schematic of a proton-exchange membrane fuel cell. (Courtesy of the U.S. Department of Defense [DoD] Fuel Cell Test and Evaluation Center.)

strates, and can be manufactured with high throughput rates at low costs. Therefore, despite their moderate efficiencies of around 5%, similar to conventional crystalline silicon, the low cost can make organic photovoltaics an important technology in the near future. Yet several challenges remain for organic photovoltaics to be considered truly matured—in particular the discovery of molecules capable of absorbing a wider solar spectrum, the development of nanoscale interfaces optimized for exciton decomposition, and polymer stability.

Another major alternative energy technology in which nanomaterials will play an important role is the area of fuel cells. Fuel cells are electrochemical devices composed of a conductive ionic membrane, an anode, and a cathode. The latter two consist of catalysts embedded in porous ion-conducting media. At the anode, hydrogen and/or hydrocarbon enters the cell, where it is dissociated by catalysts and ionized into protons and electrons. The protons travel through the conductive ionic membrane (electrolyte) into the cathode. Simultaneously, the electrons flow along an external circuit, producing an electric current that can be used directly or converted to another form of energy. At the cathode, the oxygen from air reacts with electrons and protons. If the fuel is hydrogen, the exhaust byproduct is just water (see Figure 9.57).

In general, fuel cells produce electricity with an efficiency of 40–60%, in contrast to 25% for combustion engines, making fuel cells attractive for automotive applications. Among the types of fuel cells available, polymer membrane fuel cells (PEMFCs) and solid-oxide fuel cells (SOFCs) hold the most potential, particularly because they have the lowest manufacturing costs. Table 9.1 compares these two types of cells.

The estimated cost of PEM fuel cell systems for automotive applications (~$275/kW) is significantly higher than existing internal combustion technologies (~$35/kW for advanced technologies). Another challenge with PEMFCs is in maintaining the catalyst activity of platinum (Pt) nanoparticles during fuel cell operation, due to particle agglomeration, which decreases needed surface area. Third, for the membrane to effectively separate the fuel and the oxidant as well as transport the protons from the anode to cathode, it must meet a set of requirements. In particular, the membrane should be mechanically and chemically stable, exhibit low gas permeability, and provide high proton conductivity. Currently the

Table 9.1 Comparison Between Polymer Membrane Fuel Cells (PEMFCs) and Solid Oxide Fuel Cells (SOFCs) (Adapted from R. Lashway, MRS Bulletin, Vol. 30, No.8, August 2005)

Parameter	PEMFC	SOFC
Electrolyte	Polymer	Ceramic
Conducting ion	H+	O_2^-
Operating temperature	90°C	800°C
Lifetime (h)	>5000	>40,000
Efficiency	40–60%	60%
Power range	p to 150 kW	5–250 kW
Applications	Transportation, PC, laptops	Distributed generation, marine

membranes are quite thick, to prevent fuel crossover. However, for applications in transportation, thinner membranes with higher power density are preferred. In addition, typical membranes for PEMFCs require hydration, which limits the operating temperature of the fuel cell to around 80°C.

To address the problem of cost associated with the use of platinum (Pt) nanoparticles, less expensive platinum alloy nanoparticles have been investigated, in particular binary alloys, such as Pt_3Ni and Pt_3Co. Theoretical calculations as well as experimental results seem to show much higher catalyst activities for Pt-alloy nanoparticles than pure platinum (Pt) nanoparticles (see Figure 9.58). However, the mechanism behind this enhanced activity is still controversial. Several hypotheses describing this behavior have been suggested: (1) a Pt skin develops during heat treatment, (2) a Pt skeleton is produced during an acid-leaching treatment, (3) a Pt-X (X denotes another metal) skin sputters onto the metal M, and (4) Pt-M solid solutions form (see Figure 9.59). With respect to particle agglomeration, one problem seems to be the mobility of Pt nanoparticles along the carbon support, the dissociation of Pt, and the leakage of hydrogen fuel through the membrane.

Finally, to increase the operating temperature of PEMFCs, nanocomposite membranes capable of retaining the water content at

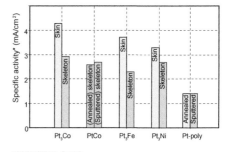

FIGURE 9.58

Platinum (Pt)-alloy nanoparticles show greater catalytic activity than pristine Pt nanoparticles. (Stamenkovic, et al., JACS 128 (2006), 8813-8819.)

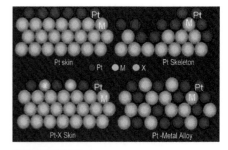

FIGURE 9.59

Mechanisms proposed to explain the enhanced activity in Pt-alloy nanoparticles. (Courtesy of Y. Shao-Horn, MIT.)

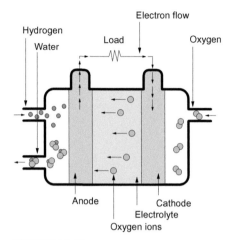

FIGURE 9.60

Operation of a solid-oxide fuel cell. (Courtesy of Earnest Garrison, Illinois Institute of Technology.)

high temperatures are being developed. PEMFCs are very good candidates for transportation, personal electronics, and relatively small stationary and portable power generation. For large-scale stationary applications, SOFCs are preferred (see Figure 9.60). This is because SOFCs operate at high temperatures (800–1000°C) and thus are difficult to use in intermittent applications where on-off cycles are frequent. SOFCs are typically composed of a solid-oxide membrane electrolyte, an anode, and a cathode. The most common materials used for the solid-oxide membrane are ceramic-based materials, such as zirconia doped with yttria and scandia.

The type and concentration of dopant determine the operating temperature and the ionic conductivity of the membrane. In SOFCs, the mobile species are O_2- ions, which travel from the cathode to the anode, to react with H^+ ions to produce water (see Figure 9.60). Conventional SOFCs use a $La_{1-x}Sr_xMnO_3$ metallic perovskite as the cathode and a porous Ni/YSZ cermet as the anode. SOFCs bring some advantages compared to other fuel cells, such as the use of inexpensive metal oxides or composites of metal oxides and metals, as cathode and anode catalysts, instead of Pt, which is expensive and scarce. Higher temperatures also permit, in principle, the direct use of hydrocarbon fuels.

However, conventional SOFCs operating at temperatures around 1000°C suffer from various limitations, namely a restricted selection of materials that can be used as cathode and anode, interfacial reactions between the various layers, and a mismatch in thermal expansion between the electrolyte and the electrodes, with consequences for the mechanical properties. These challenges can be minimized while still maintaining the use of hydrocarbon fuels by operating at temperatures around 500–800°C. However, at these temperatures, novel electrolytes with high oxygen ionic conductivities and efficient catalysts for oxygen reduction and hydrogen oxidations are required. As a result, alternative materials such as $La_{1-x}Sr_xCoO_3$ and $La_{1-x}Sr_xFeO_3$ for the cathode, $Sr_2Mn_{1-x}Mg_xMoO_{6-\delta}$ and rare-earth doped ceria ($Sm_xCe_{1-x}O_{2-\delta}$) for the anode and $Ce_{1-x}M_xO_{2-\delta}$ (M = Gd, Sm and Ca) for the electrolyte have shown, so far, promising results. In summary, fuel cells offer significant advantages as an alternative energy technology, particularly high efficiency, quieter power generation, and clean exhaust products.

Explorations are under way for the storage of energy-rich gases via the use of nanoporous *metal organic frameworks*, or MOFs. These are highly porous organic matrix substances that can store hydrogen and other gases (see Figure 11.3 in Chapter 11).

Batteries, as fuel cells, are also electrochemical power sources. However, in contrast to fuel cells, which produce electricity continuously as long as a fuel and an oxidant are delivered, batteries convert chemical energy into electrical energy by electrochemical reactions occurring between the already existing constituents. A battery is composed of one or more electrochemical cells, which can be connected in parallel or series to deliver power. A typical electrochemical cell consists of three main elements (see Figure 9.61): an anode, which generates electrons to an external circuit and is oxidized during discharge; a cathode, which collects the electrons from the external circuit and is reduced during discharge; and an electrolyte, which allows the transport of ions between the anode and the cathode but not the transport of electrons. The electrolyte is typically a liquid solution in which a salt is dissolved in a solvent.

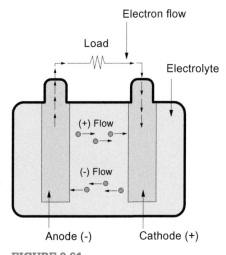

FIGURE 9.61

Basic components of an electrochemical cell. (Encyclopedia Britannica.)

In addition to these three main components, an electrochemical cell also contains current collectors to allow the flow of electrons to and from the electrodes. These materials are normally metals but should not react with the electrodes. There are two types of batteries, primary and secondary, based on their ability to recharge. In the former the chemical reactions are irreversible, whereas in the latter the electrode material exhibits reversible chemical reactions. Since they are not rechargeable, primary batteries are normally used for applications that demand long shelf life and minimum replacement. On the other hand, secondary or rechargeable batteries are usually used in applications that require transient amounts of high energy. Currently, rechargeable batteries are the most common method of storing electrical energy, particularly because they are very efficient (90% energy conversion) and relatively inexpensive. The first rechargeable batteries were lead-acid, eventually evolving to nickel-cadmium and nickel-metal hydride (Ni-MH), and more recently to lithium-ion batteries (see Table 9.2). The Ni-MH batteries were widely used for portable electronic devices, but due to the implementation of Li-ion batteries with high energy density and design flexibility, they have been almost out of circulation for this type of application. Ni-MH batteries are still used in hybrid vehicles, although in the near future Li-ion batteries are also likely to dominate this market.

The development of Li-ion batteries started in the early 1970s, driven by the fact that lithium is both the lightest metal and highly

Table 9.2 Performance of Secondary Battery Systems (Adapted from MRS Bulletin, August 2002)

	Lead-Acid	Ni-Cd	Ni-MH	Li-ion
Anode	Pb	Cd	Metal hydride	C
Cathode	PbO_2	NiOOH	NiOOH	$LiCoO_2$
Electrolyte	H_2SO_4	KOH	KOH	Organic solvent + Li salt
Cell voltage (V)	2	1.35	1.35	3.6
Capacity (mAh/g)	120	180	210	140
Energy density (Wh/Kg)	35	40	90	200

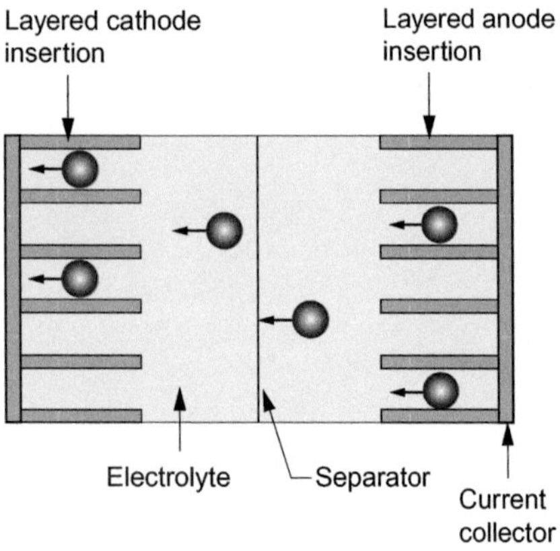

Layered cathode insertion — Layered anode insertion

Electrolyte — Separator — Current collector

FIGURE 9.62

Operation of a Li-ion battery during discharge.

electropositive. In batteries, Li is usually present at the anode as a compound material, typically LiC_6. During discharge, lithium acts as electron donor to the external circuit, whereas the positive Li ions are transported through the electrolyte (Figure 9.62). At the cathode, the lithium ions react with the incoming electrons from the circuit and the cathode material, such as $LiCoO_2$. Since the battery is rechargeable, the reversible reaction must be possible. To facilitate the insertion and extraction of lithium ions from the cathode and anode, respectively, both the $LiCoO_2$ and the LiC_6 compounds have layered structures (Figure 9.63).

Despite the success achieved so far with the development of Li-ion batteries, there are still several factors that limit the performance of these devices, in particular material problems with the anode, cathode, electrolyte, and separator. Carbon-lithium anodes have low energy densities (a factor of 10 lower) compared to pure lithium anodes. However, pure lithium anodes have been shown to cause cell short-circuiting. Further experiments with lithium compounds, such as $Li_{4.4}Sn$ and Li_4Si, resulted in high specific capacities, but the materials suffer from very large volume expansion-contraction (up to 300%) during the charge-discharge process, leading to cracking of the anode. More recently, nano-scale materials have been proposed for the anode—in particular, Si nanowires and an amorphous nanostructured tin-cobalt-carbon material. These materials provide large surface areas and reduced volume expansion-contraction, but many other nanoscale-related properties are still unknown. In addition, these materials are still very expensive.

Aside from the anode, the cathode is a key factor in developing high-performance lithium-ion batteries. Typically, the cathode contains a transition metal, which can be easily reduced and oxidized, must interact with lithium rapidly and without inducing a structural change, should be a good electronic conductor, and should be cheap and nontoxic. Finally, it must react with lithium and produce a high voltage. The most used cathode material is the layered $LiCoO_2$. This compound is easy to process and has a long shelf life. However, it does not exceed a reversible capacity of 140 mAh/g (half the theoretical capacity) and contains toxic and expensive cobalt. The layered $LiNiO_2$ is less expensive and less toxic than $LiCoO_2$, showing also higher capacity, but it is difficult to synthesize, may release oxygen at high temperature, and is likely to exhibit a drastic fade in capacity during cycling.

More recently, $LiFePO_4$ has been suggested as a cathode material for Li-ion batteries. Because Fe is also a transition metal, is more available than Co and Ni, and is less toxic, this compound seems to be a good choice. In addition, it has a good capacity of 170 mAh/g at a cell voltage of 3.4 V, while the phosphate group is very stable. However, this material has a quite low electronic conductivity at room temperature and poor lithium conductivity. As a result, this material is now being studied in great detail, particularly because it is being thought of for use in electric vehicles. In this regard, reducing the size of LiFePO4 particles to the nanoscale and coating them with carbon seem to be very effective in improving the electronic and ionic conductivities.

Another route that has been thought to improve the energy and power of Li-ion batteries is to use layered Li_2MnO_3 and $LiMO_2$ (M = Co, Cr, Ni, Mn). These materials exhibit high capacities of around 250 mAh/g, although they are still in the early stages of development.

Finally, some words about the electrolyte. For lithium-ion batteries, a good electrolyte must exhibit high ionic conductivity, good electrochemical and thermal stability and low reactivity with the other elements of the battery and be cheap and environmentally friendly. Typically, the electrolytes used so far tend to be liquid solutions. However, recently there has been a great interest in replacing liquid electrolytes with gel polymer electrolytes, which are membranes formed by containing a liquid solution in a polymer matrix. Shifting away from the Li-ion system, there

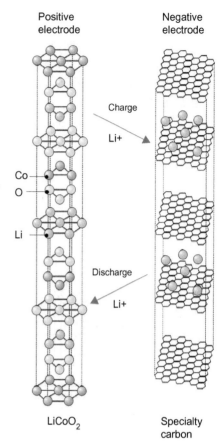

FIGURE 9.63

Layered anode and cathode materials for Li-ion fuel cells. (Courtesy of Electronics-lab.)

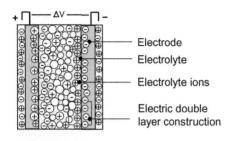

FIGURE 9.64
Electrochemical capacitor.

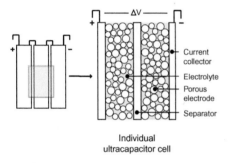

Individual
ultracapacitor cell

FIGURE 9.65
Double-layer electrochemical supercapacitors.
(Courtesy of Maxwell Technologies, Wikipedia.)

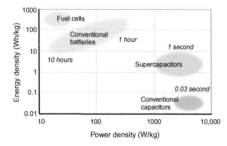

FIGURE 9.66
Comparison of energy density and power density
for fuel cells, batteries, and supercapacitors.
(Courtesy of Maxwell Technologies, Wikipedia.)

has been some work in developing batteries based on sodium. However, the low melting temperature of sodium has generated several practical concerns.

Electrochemical capacitors are also a viable alternative technology for storing energy. However, unlike batteries that store energy by means of a chemical reaction, capacitors store energy as electrical charge on the surface of electrodes. This happens through electrostatic interactions that occur in the electrode and solution interface region by which electric charges of equal magnitude, but opposite sign, accumulate on each electrode (see Figure 9.64). Because the electrodes are not deteriorated by this process, capacitors exhibit good performance over long periods of time. In addition, capacitors are rapidly charged and discharged, making them attractive for applications requiring quick responses. On the other hand, because the energy-storing mechanism is basically a surface process, the energy capacity of typical electrochemical capacitors is very dependent on surface area.

Normally, electrochemical capacitors are subdivided into supercapacitors and pseudocapacitors. Supercapacitors are based on a structure involving an electrical double layer separated by a very thin dielectric wall (see Figure 9.65). Each layer contains a highly porous electrode suspended within an electrolyte. An applied potential on the positive electrode attracts the negative ions in the electrolyte, whereas the potential on the negative electrode attracts the positive ions. A dielectric material between the two electrodes prevents the charges from crossing between the two electrodes. In general, supercapacitors enhance the storage density through the use of activated carbon. This material is composed of nanopores to increase the surface area, allowing a greater number of electrons to be stored in any given volume. Typically, supercapacitors exhibit storage capabilities of less than 10Wh/kg. In addition, because the double layer can withstand only a low voltage, higher voltages are normally achieved by connecting individual supercapacitors in series. Supercapacitors are often used as backups to avoid short power failures and as a supplementary power boost for high-energy demands. Supercapacitors are also ideal for applications that require rapid charging.

In summary, fuel cells exhibit high energy storage but limited power density. Batteries deliver somewhat less energy density, but the power output is improved. On the other hand, supercapacitors, also known as *ultracapacitors*, are high-power devices with low energy storage capability (see Figure 9.66).

9.8 INTERACTIVE, SMART, AND INTELLIGENT SYSTEMS AND ENVIRONMENTS

Responsive Environments

The environments previously discussed in this chapter have all related to fundamental conditions of spatial or surrounding environments (mechanical, thermal, lighting, and so on). There are, however, many other, more behaviorally based ways designers use to characterize environments that are useful to briefly address here. They typically involve the environments already described but go beyond them and focus on interactive, smart, or intelligent behaviors.

In the 1920s the well-known architect Le Corbusier described the house of the future as a "Machine for Living," with its connotations of automation. Subsequently, the 1953 "All Electric House" enabled an occupant in a bedroom to turn on a kitchen coffeemaker at the right time in the morning. These general areas have since evolved into intense and popular areas of study in relation to what are now commonly described as *interactive, smart, or intelligent* environments. Typically these environments seek to improve our experiences in living and work environments in both buildings and vehicles by providing information, aiding in task performance, or more generally improving lifestyle quality. Many "smart environments" in recent years have consisted of spaces with multiple sensors that control lighting and thermal environments and provide ubiquitous communications. Simple commercial systems do many other things, such as informing homeowners whether the pet dog has strayed from the area of the house. Far more sophisticated approaches now seek to have automated systems for detecting and identifying specific human users within a space (e.g., via facial detection) and adjusting the living or work environment characteristics in response to the needs of that specific individual.

Before looking at nano-based applications in this domain, we should lay some background. The three key descriptive phrases—interactive, smart, and intelligent environments—typically mean quite different things to different people. Do the phrases imply equivalent concepts? Can something be interactive without being smart? We already have thermostats for controlling temperature levels and other sensors for light levels. Is a sensor-equipped automated house that automatically shuts lights off or on, depending on whether a room is occupied, truly a smart or intelligent environment? Do common sensor-driven actuation and communication systems *alone* define "smart" or "intelligent" environments? There is a considerable degree of controversy over terminology here, even though common dictionary definitions suggest clear distinctions that largely hinge around the level of cognition involved. At one time in the past, a sensor-controlled lighting system would surely have been thought of as "smart," but the definition bar has now been generally heightened, although involved groups cannot really agree on definitions that really capture the essence of smartness or intelligence. All generally agree on the importance of enabling sensor-based actuation and communication systems, but these may be directly responsive only and not involve any kind of cognitive action that one would somehow think necessary in something called an "intelligent" environment.

These are not easy concepts to untangle, but it is clear that their very fuzziness has engendered a great deal of confusion in what has blossomed into an absolutely enormous area of study that draws from many diverse fields, including materials science, electromechanical engineering, computer science (including expert systems and artificial intelligence), product design, and architecture. Here we seek to only briefly summarize this field. We start by briefly summarizing aspects of interactive, smart, and intelligent environments, as the terms are commonly now used and understood (see Figure 9.67). Impacts of developments in nanomaterial and nanotechnology will then be examined.

Interactivity can assume many forms but is usually considered to involve some type of physical, sound, light, or other response to a directed action on the part of an occupant. Actual applications vary widely. Figure 9.68 shows an interesting and delightful interactive display called Tulipomania. The installation is in one of Torino's shopping galleries. When visitors come close, the animations projected onto the sandblasted glass panels move gently, as though in a breeze. Other kinds of interactive display systems in buildings are becoming more and more common for many purposes: general information and way-finding, entertainment, information about products in the space, and so forth. A common application in a retail setting is for product point-of-sales information to be displayed as a user approaches the product. Proximity or other kinds of sensors are used.

Aspirations can be higher as well. In the American Museum of Natural History in New York City, for example, an interactive display wall helped give visitors a sense of Einstein's theories by responding to the motions of visitors moving relative to the wall. Via displays

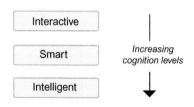

FIGURE 9.67

Terms used to describe systems or environments that are designed to have responsive or cognitive behaviors.

FIGURE 9.68
Tulipomania, an interactive installation in Torino, Italy. (Courtesy of Ivividia.)

that opened up or closed down, visitors were encouraged to feel gravity as a mass-warping time space. Many sensor-based systems of this type involve complex display and sound and audio systems.

So-called *smart systems* have even more of a basis in complex sensor-based electronic and computational systems but frequently involve "smart materials" as well. Again, many different approaches are evolving here. Many of the materials that we have already described have an apparent ability to change one or more of their properties in response to changes in their surrounding thermal, radiant, electrical, mechanical, or chemical environments. Thermochromics, for example, change their apparent color as the temperature of their surrounding environment increases and there is an input of thermal energy into the material. Magnetorheological materials change their viscosity when subjected to varying levels of magnetic fields.

As we will discuss in Section 10.2, these and other materials of this type are often called *smart materials* because they exhibit useful changes in properties in the presence of varying external stimuli. Examples of smart systems based on these kinds of materials abound. For example, there are many self-damping systems used for vibration control, including skis that have self-damping mechanisms for reducing the troublesome "chatter" that can be experienced by advanced skiers in difficult terrains. In the automotive industry, there are "smart tires" or "smart suspension systems" that enhance driving performance (see Section 12.2). In the aerospace industry there has been considerable attention directed toward the development of "smart skins" that incorporate a series of sensors and actuation devices based on smart materials, including shape memory alloys, to enhance wing performance. Nanomaterials and nanotechnologies offer great potential as both tiny sensors and actuators and thus have the potential to be integrated into surface structures quite easily. Section 10.2 addresses smart materials and systems in more detail.

There are many types of *intelligent* systems and environments. Among the most popular are systems that first detect and identify specific human users within a space and then tailor aspects of the living or work environment in response to the needs or tasks of those specific individuals. More advanced systems use facial and gesture recognition systems that seek to enable users to control specific aspects of their living or work environments, without any kind of manual or even voice inputs. These systems invariably involve complex computational infrastructures as well as a host of different kinds of sensory and actuation devices. Characteristics of intelligent systems are examined in more detail in the following section.

Characterizing Interactive, Smart, and Intelligent Environments

Figure 9.69 suggests how intelligent environments can be broadly characterized in terms of their functional descriptions, the way smart behaviors are embedded in the environment or the ways that they are controlled, and the general cognition levels associated with various behaviors. Following this diagram, we broadly use characterizations of smart or intelligent environments that reflect different primary streams of meaning and description. The first is in relation to definitions of systems and environments (use and surrounding). In addition to performance-specific systems that are constituents of larger products, vehicles, or buildings, this area encompasses human occupation and use needs and issues vis-à-vis lifestyle, health, and work. Included are physical comfort and health issues vis-à-vis the surrounding environment. Use issues focus on lifestyle considerations or task performances (e.g., what would an optimal environment for a stockbroker look like?).

The level of component, subsystem, system, or environment to be considered is important here. Many simple components involve only single behaviors, such as a light that goes off or on, whereas other highly complex systems or environments involve multiple

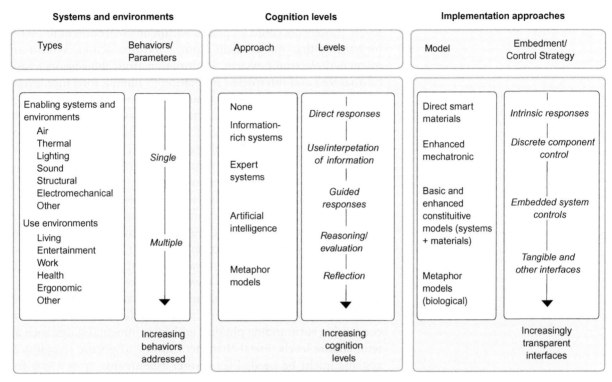

Systems and environments

Types	Behaviors/ Parameters

Enabling systems and environments
  Air
  Thermal
  Lighting
  Sound
  Structural
  Electromechanical
  Other
Use environments
  Living
  Entertainment
  Work
  Health
  Ergonomic
  Other
	Single
	
	↓
	
	Multiple
	
	↓

Increasing behaviors addressed

Cognition levels

Approach	Levels
None Information-rich systems Expert systems Artificial intelligence Metaphor models	*Direct responses* *Use/interpetation of information* *Guided responses* *Reasoning/ evaluation* *Reflection* ↓

Increasing cognition levels

Implementation approaches

Model	Embedment/ Control Strategy
Direct smart materials Enhanced mechatronic Basic and enhanced constituitive models (systems + materials) Metaphor models (biological)	*Intrinsic responses* *Discrete component control* *Embedded system controls* *Tangible and other interfaces* ↓

Increasingly transparent interfaces

FIGURE 9.69

Smart and intelligent environments can be broadly characterized in terms of their functional descriptions and numbers of behaviors, the general cognition levels imparted by the computational environment, and the embedded materials and technologies used for implementation.

and often interactive behaviors that need to be considered. Specific performance or use goals need to be stipulated for each level considered.

A second area is in relation to levels of assisted cognition within the computational world for various system or environment goals. At a low level in this category are the many attempts to provide information-rich environments that support systems or environments, including human physical and intellectual needs. At a higher level, expert systems have been devised to interpret information and provide guided responses. Artificial intelligence approaches seek to provide further cognitive capabilities and even perform reasoning.

The third broad area deals with technological implementation strategies, materials, and devices. It is here that characteristics of sensors and sensor-driven actuators, materials with property-varying characteristics, and various logic control systems, including nanomaterials and nanotechnologies, come into play. Various kinds of technologies could be used to capture more sophisticated information about user states and context awareness via gesture or facial recognition systems.

Consider a first example involving relatively simple cognition levels. Biosensors could be placed within an environment to detect biohazards that might result from a terrorist action. Within an integrated electronic system, information from the sensors could be analyzed and interpreted in relation to human tolerance levels, and various alerts or warnings would be communicated, even direct mitigation systems called into action. In terms of Figure 9.69, this scenario would be illustrative of "single-behavior" sensory detection, coupled with a very simple expert system (a basic "if-then" formulation) and implemented through a direct mechatronic model (discrete component and system control). A more complex example is an "intelligent room" environment designed for individuals with special medical needs or who are recovering from an operation. These rooms might have a whole host of means for monitoring personal vital signs, event recovery measures, general mental and physical condition states (e.g., awareness, ability to function), indicators of whether medicines have been taken regularly, whether the patient has had visits, whether the patient has moved around and how (mobility assessments), and a whole host of information concerning surrounding physical and environmental states, such as temperature levels. Facial characterization and gesture recognition systems might be needed for further assessments or as a way for nonverbal patients to communicate.

The initial sensing goal in the above example is to detect and measure everything needed for an externally located medical group to access a patient's state and progress. If fully automated, expert or artificial intelligence systems would analyze and interpret information and, if necessary, cause specific actions to be evoked (e.g., at a low level this might be a voice communication to a patient to "Take your 11:00 a.m. medicine," or, at a higher level, could cause a medical team to be dispatched). In this scenario, not only are large numbers of sensors involved, but the interactions of multiple behaviors must be analyzed and specific actions evoked. This same room might make use of self-cleaning and antimicrobial surfaces (see Section 10.2).

Though some basic ideas and aspirations are fairly straightforward, not all these kinds of interactions and responses can be achieved at the moment. Limitations exist throughout, including in sensory and computational components that are more the focus of this book than are related cognition algorithms, which remain extremely troublesome. As we move from specific applications, such as the medically oriented room we've described, and more into general intelligent environments, there is less than clarity on what we really want intelligent environments to ultimately do. This latter topic is fascinating but largely beyond the scope of this book. (See Further Reading for a more extensive list of sources on smart or intelligent environments.) Here we focus more on material/technology-related concerns and opportunities.

Nanomaterials and Nanotechnology Applications

Developments in these areas can potentially be greatly enhanced by nano-based approaches. We have already seen several examples of applications, including self-damping systems used for vibration control in cable-supported bridges or buildings, that are based on magnetorheological fluids that can in turn be enhanced via nanoparticles with improved magnetic properties (Section 9.2) and transparency-changing windows (Section 9.5). Behaviors such as these are the basis for many larger approaches to smartness when they are coupled with sensor/control systems that activate them during appropriate conditions. Other approaches that need not include sensor/controls systems include self-healing surfaces or the self-cleaning or antimicrobial surfaces noted previously that are based on nanosized titanium dioxide particles. (These surfaces are discussed in Section 10.2.) Surfaces such as these would be an important component in any kind of smart or intelligent room directed toward health or medical applications or even for everyday cleanliness in kitchens, bathrooms, and other common spaces.

Many interactive, smart, or intelligent environments already use a variety of sensory technologies; nonetheless, it remains surprising how few sensor, control, and actuation devices are actually used in most current applications. Difficulties in precisely locating a few sensors in relation to the measurand remain problematic. More are invariably needed to capture necessary information. Devices that are used are also often quite visibly intrusive due to their size and often appear as "add-ons" to wall, ceiling, or floor surfaces in a room—a fact known to cause occupant behaviors to change (who has not acted at least a bit differently when in front of a Webcam?). An immediate stated desire of designers of these kinds of environments is to have perfectly seamless integration of involved sensors and actuators with the surfaces of products in the space or within surfaces of the surrounding room. Even more speculative approaches that question the very surfaces themselves are discussed in the literature (see Further Reading).

As discussed in Section 9.4, directions for many future nano-based electronic systems are characterized by improved performance, decreasing size, increased multifunctionality, and decreased costs. Many research efforts are explicitly directed toward making improved sensors for a variety of applications, and many are expected to find a ready application area in interactive, smart, and intelligent systems. Expected benefits include not only improved performances but also ever smaller and cheaper sensors. Nanoelectromechanical Systems (NEMS) are on the immediate horizon as well. These trends intersect very well with technological needs in interactive, smart, and intelligent systems.

The idea of being able to have large numbers of inexpensive sensors distributed throughout a spatial environment that are designed to capture specific kinds of information (about users, processes, or conditions within a space) has great appeal with designers involved with conceptualizing not only smart buildings, but smart cities as well. A future can be imagined where distributed sensors (perhaps the "swarms" discussed in Section 9.4, but more likely multitudes of still tiny but more sophisticated inter-connected nanoscale sensors, actuators and other devices) aid in our every living and work activities within buildings and cities. It is here that trends toward smallness, high performance, and multifunctionality in enabling electronic systems intersect with goals of interactivity, smartness, and intelligence. At the moment, this area remains rather speculative, but the potential is clearly there.

9.9 NANOMATERIALS COSTS

The enormous potential for widespread nanomaterial applications to occur inherently depends on the availability of large quantities of nanomaterials at reasonable costs. In the current emerging state of the field, not all nanomaterial forms found in the laboratory are widely available, fewer still as commercialized products. Costs for nanomaterial products are also high. The importance of costs in any application or design context hardly needs explanation. Costs are invariably a driving factor, one often cited as an inhibiting factor in the development of applications involving nanomaterials—and not without good reason. In the early days of nanomaterial exploration, a gram of nanotubes could cost several thousand dollars; even recently, 500 or so dollars per gram was a common price. Even if the amounts used in applications were relatively small, say, 5% by weight in a nanocomposite, final product costs were so high as to prohibit all but the most specialized applications. Costs have since dropped dramatically but nonetheless remain a significant factor.

In the current context, many groups that commercially produce nanomaterials (such as various kinds of nanoparticles or nanotubes) are just now transitioning from their roles as suppliers to the research sector to that of becoming producers of commodity products. This transition, in turn, is being driven by the development of more and more real applications that demand larger quantities, or by real product prospects just on the horizon (see Figure 9.70). The actual number of producers, however, is still relatively few in comparison to the multitude of businesses that produce conventional materials, and outputs remain relatively small. Several claim to be able to scale up to larger volumes quickly. Nonetheless, the industry remains in an emergent state. A sign that further transitions will occur is the increased number of major chemical and traditional material suppliers starting to produce nanomaterials. There is every expectation that increasing demand will in turn lead to increases in production volumes.

FIGURE 9.70

The control room of the new Baytubes® production facility in Laufenburg, showing the top of the fluidized bed reactor. The facility has a capacity of 30 metric tons annually. Market potential estimates for carbon nanotubes (CNTs) in the coming years are several thousand metric tons per year. (Courtesy of Baytubes.)

In the following discussion, the range of currently available nanomaterials is briefly reviewed, as are their general characteristics. Various parameters that affect unit costs—particularly type, treatments, and quantities—are also identified.

Types

Nanoparticles for applications are already commercially available in a wide range of forms that can support a host of applications. Fundamental types include basic elements, compounds, and oxides. Various alloys, carbides, sulfides, and other forms can also be obtained. Materials such as silver (Ag), gold (Au), aluminum (Al), and iron (Fe) are commonly available, as are many fundamental forms. A wide array of compounds (such as tungsten carbide, TiC) and oxides (such as titanium dioxide, TiO_2) are also available. These materials are available in nanoparticle form and some in nanopowder form (a term sometimes used to describe nanoparticles less than 30 nm in diameter). Actual diameters can vary considerably. Common silver nanoparticles, for example, are readily available in diameters ranging from 10 nm all the way up to the micron size (2 um or 2000 nm). Some other materials are currently available only in smaller or larger sizes, depending on their makeup. Titanium dioxide (TiO_2), widely used in its anatase form for photocatalytic applications (self-cleaning, antimicrobial), typically ranges from 5 to 20 nm, whereas in another of its forms (rutile) diameters are often larger (40–50 nm).

Many polymeric nanocomposites are based on dispersed clay nanoparticles (often called *nanofillers*). Montmorillonite is often used.

It is a layered smectite clay mineral that has a platelet structure. Platelets are very thin—for example, 1–2 nm—but have very large surface dimensions, perhaps over 500 nm, that make them highly effective providers of surface areas. These kinds of nanoparticles are relatively inexpensive compared to other types of nanoparticles and can be easily manufactured in bulk quantities. Section 8.8 describes nanocomposites based on smectite materials in greater detail.

Nanoparticles can also come in varying levels of treatments. Various nanoparticles can be obtained that, for example, already have hydrophobic or hydrophilic properties (see Chapter 8) useful in a wide range of applications. Typically, when obtained from a manufacturer, nanoparticle batches may need to be further processed via application of surfactant coatings, ultrasonification, milling, or other processes because of their inherent tendency to agglomerate or attract moisture.

Carbon nanotubes for applications are available in a wide variety of single-walled (SWNT) and multiwalled (MWNT) forms, including different lengths, diameters, and purities. A primary quality and cost issue is purity. Depending on the synthesis process used, purified forms are characterized by descriptions such as "95 wt%" or "60 wt%," which simply refers to the actual percentage by weight of nanotube content within a unit weight of a gram of nanotube material. Other constituents might include ash, amorphous carbon, residual catalyst, or other materials. Some SWNT formulations can also include some percentage of MWNTs. Obviously, higher purity forms have more effective properties. Cost-per-gram values, however, are invariably higher than lower purity forms. Lower purity and lower-cost forms, however, can be suitable for many bulk applications. Most commercialization has been with MWNTs. SWNTs are available but often at higher unit prices. Typical diameters for available SWNTs are on the order of 1–2 nm, and common lengths are on the order of 10–50 um, although there is variation in these numbers as well. Special short nanotube forms are also available that are on the order of 0.3–5 um. Outer diameters of common MWNTs are on the order of 5–15 nm, with inside diameters on the order of 2–8 nm. Several walls are common. Short-form MWNTs often have larger outer diameters (20–50 nm or so) and somewhat larger inner diameters than longer forms.

Incorporating carbon nanotubes into other matrix materials to form nanocomposites has often proven difficult because of the surface chemistry of the carbon. Problems include achieving uniform dispersions and poor adhesion with the primary matrix. Many highly useful polymeric, inorganic, and organic composite forms involving the use of dispersed carbon nanotubes in common liquids, including polymer resins, water, or various solvents, are inherently difficult to make due to the insolubility of carbon nanotubes.

Various kinds of surface treatments have been developed to reduce some of these problems. In particular, many nanotube products are now available in what are called "functionalized" form. Normally these functionalized carbon nanotubes processes involve chemically attaching different molecules or functional groups to the sides of nanotubes that allow various kinds of bonding. Processes vary depending on the need but can include using a variety of oxidation methods or polymeric wrappings. Functionalized nanotubes can develop better bonding in mechanically directed nanocomposites (via covalent rather than Van der Waals bonding). In other cases, noncovalent bonding is preferred.

In the biological or chemical areas, nanotubes can be functionalized to be able to play diverse roles, including as biosensors. Attachment of certain kinds of molecular groups can allow protein adsorption or bindings, which in turn enables drug delivery or other application potentials. Functionalized nanotubes are invariably more expensive than normal types. Not unexpectedly, costs for functional nanotubes used in specialized biological or medical applications can be extremely high.

Figure 9.71 illustrates some general pricing characteristics for carbon nanotubes. Unit costs for high-purity materials are greater than lower-purity materials. Functionalized or treated versions are more costly than basic types. Costs also vary with quantities. In the diagram, the steep curves to the left reflect costs for very small quantities similar to those used in a laboratory or development context. For any real application, quantities acquired would necessarily be quite large, and costs would correspondingly decrease.

Figure 9.72 illustrates how basic unit cost ranges for common nanomaterial products broadly compare to other materials. As with other high-performance materials that are used in composites, however, one should keep in mind that direct comparisons with truly bulk materials, such as steel, should be viewed in the light that quite small quantities of nanomaterials may normally be actually used. Many actual products based on nanocomposites, for example, have only relatively low percentages (for example, 2% to 5%) of nanomaterials by weight. Many nano-based thin films or paints involve only small amounts as well and consequently can be competitively priced. On the other hand, the gross unit costs of

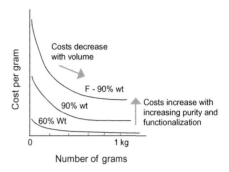

FIGURE 9.71

General cost variables for carbon nanotubes.

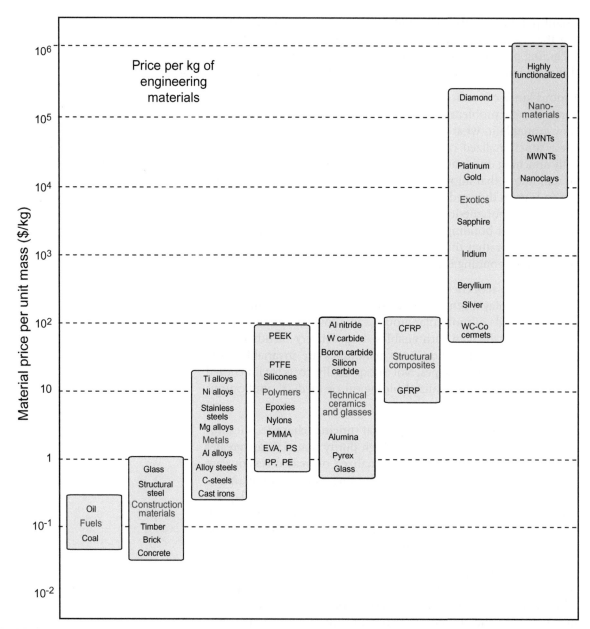

FIGURE 9.72

Material price in US$/kg for common engineering materials and for typical nanomaterials. Costs for nanomaterials vary dramatically with the type of material, the degree of functionalization, and special processing needs for inclusion in products.

many technologically sophisticated and high-value products currently involving the use of nanomaterial or nanotechnologies are normally quite high. Interestingly, the current price of a very high-end hearing aid that has a nanotechnology component is currently just under $1,000,000 per kilogram (mass: 2 grams, cost: $1800), and the price of a cobalt-chrome denture involving nanomaterials is $90,000 per kg, based on a similar calculation (mass: 20 grams)—thus the latter has costs similar to those of the hearing aid. Certainly this is suggestive of where many nanomaterials are most likely to find markets.

FURTHER READING

General

M. Addington and D. Schodek, Smart materials and technologies for the architecture and design professions, The Architectural Press, 2005.

Michael F. Ashby, Materials Selection in mechanical design, 3rd ed., Elsevier Butterworth-Heinemann, 2005.

Michael F. Ashby, Materials: Engineering, science, processing, and design, Elsevier, 2007.

George Elvin, Nanotechnology for green building, Green Technology Forum, 2007

Charles Harper, Handbook of materials for product design, McGraw-Hill, 2001.

Daniel Schodek and Martin Bechthold, Structures, 6th ed., Prentice-Hall, 2007.

Benjamin Stein and John S. Reynolds, Mechanical and electrical equipment for buildings, 9th ed., John Wiley and Sons, 2000.

Ben Walsh, Environmentally beneficial nanotechnologies: Barriers and technologies, Report to Food Department of Environment, Food and Rural Affairs, Oakdene Hollins, UK, 2007.

Environments and Systems

L. Beranek and Istavan Ver, Noise and vibration control engineering, John Wiley & Sons, 1992.

Advanced fenestration technology - Nanogel, Cabot Corporation, 2007.

A.S. Edelstein and R.C. Cammarata (eds.), Nanomaterials: Synthesis, properties and applications, Institute of Physics Publishing, 2002.

A European roadmap for photonics and nanotechnologies, MONA (Merging Optics and Nanotechnologies), 2007.

Paolo Gardini, Sailing with the ITRS into nanotechnology, International Technology Roadmap for Semiconductors, 2006.

Earnest Garrison, Solid oxide fuel cells, Science and Mathematics with Application of Relevant Technology (SMART) program, Illinois Institute of Technology, 2007.

Hideo Hosono, Yoshinao Mishima, Hideo Takezoe, and Kenneth MacKenzie, Nanomaterials: Research towards applications, Elsevier, 2006.

A. Inoue, Acta Mater 2000;48:279.

W. L. Johnson, Material research bulletin 1999;24(10):42.

Michael Moser, Engineering acoustics: An introduction to noise control, Springer-Verlag, 2004.

Microstructured Materials Group, Silica aerogels, Lawrence Berkeley Laboratories, 2004.

A European roadmap for photonics and nanotechnologies, MONA (Merging Optics and Nanotechnologies), 2007.

William Lam and Christopher Ripman (eds.), Perception and lighting as formgivers for architecture, Van Nostrand Reinholt, 1992.

Michel, Lou, Light: The shape of space: Designing with space and light, John Wiley and Sons, 1995

Nanogel—Cabot's Nanogel technology, Cabot Corporation, 2007.

A. Ortega, The engineering of small-scale systems: An NSF perspective, ITherm, San Diego, CA, 2006.

C. H. Smith. In: H. H. Liebermann (ed.), Rapidly solidified alloys, processes, structures, properties, applications, Marcel Dekker, 1993.

Marc G. Stanley, Competing for the future: Unlocking small wonders for tomorrow's nanoelectronic uses, Advanced Technology Program, NIST, 2007.

K. Sobolev, How nanotechnology can change the concrete world, American Ceramic Society Bulletin, Vol. 84, No. 11, November 2005.

Harnessing materials for energy, MRS Bulletin, Vol. 33, No.4, April 2008.

Fuel cells: Advanced rechargeable lithium batteries, MRS Bulletin, Vol. 27, No. 8, August 2002.

Organic-based photovoltaics, MRS Bulletin, Vol. 30, No.1, January, 2005.

Hydrogen storage, MRS Bulletin, Vol. 27, No. 9, September 2002.

CHAPTER 10

Nanomaterial Product Forms and Functions

10.1 CHARACTERIZING FORMS AND FUNCTIONS

This chapter looks at many kinds of nanomaterials or nanotechnologies as they are commonly identified by product type, such as nanopaints, or by their function, such as antimicrobial. Many individuals and firms seek to characterize their research or products for either scientific or marketing purposes in a focused way and naturally make use of everyday descriptions that capture their essence. Indeed, they are often highly useful descriptors that bring alive the exciting intent of a product, such as a self-healing nanopaint, as well as being descriptive of general product form and application arena. These characterizations are thus usually related to highly specific nanoproduct types and applications and much less so to underlying physical or chemical phenomena. As we will see, however, there is considerable overlapping as well as simple confusion in the kinds of terminology used here. Do we really mean *nanopaint*, for example, or something like *polymeric paints containing nanoparticles*? Presumably the latter is a more correct description of what is actually meant—we certainly don't mean "tiny paint"—but the language is surely more cumbersome. In this chapter, we describe research and products by the commonplace, if not always precise, language that is typically used.

We draw a distinction between common types of nanomaterial *properties* and *functional behaviors* that might be the intention or application of a particular type of nanoproduct and the actual *form* of the nanoproduct itself. In the first instance, we have both property characteristics (e.g., mechanical, optical) and functional descriptors (e.g., self-cleaning or antimicrobial). In the second instance of product form, we have descriptors such as *nanocoatings, nanofilms, nanopaints, nanotextiles,* and others. Obviously, different product forms can be described in terms of basic properties as well as possess one or more of the functional behaviors noted. Many nanoproduct forms, such as nanotextiles, possess multiple functional qualities—for example, antistaining and antimicrobial.

There is considerable difficulty and redundancy in describing various nanoproducts and functions in this way because of their overlapping and often ambiguous nature—for example, one person's coating is another's thin film. Figure 10.1 accepts these over-

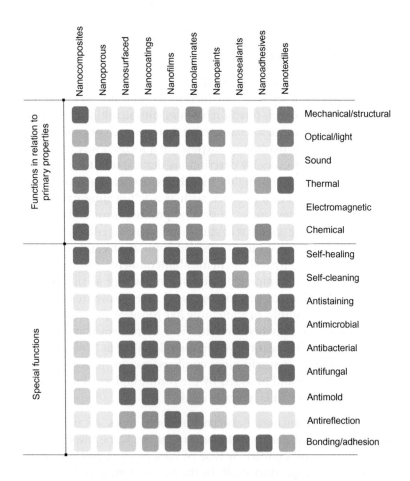

FIGURE 10.1

Typical relationships among common nanoforms and nanoproducts and properties and functionalities that are important to designers and engineers.

The chart is not exhaustive; other product forms or functions could be listed as well, e.g., nanopowders, antistatic behavior, air or water purification.

Application relation

Low   High

laps and ambiguities and is simply a device for organizing the discussion in this chapter.

The figure illustrates common characteristics and functionalities in relation to typical product forms. The vertical axis shows normal attribute or property characteristics of nanomaterials, that is, mechanical (strength, stiffness, etc.), thermal, optical, sound, chemical, and electromagnetic, and various functionalities (antibacterial and other). There has already been considerable discussion of property characteristics of nanomaterials in Chapters 6–9 and their role in product applications. There are also characteristics that are directly related to property attributes that involve words such as *color changing* or *shape changing* or other so-called "smart" actions. This chapter only briefly touches on these characteristics. Rather, we focus more on the kinds of functional behaviors listed in Figure 10.1. In the field there is considerable excitement about how nanomaterials can contribute to imparting different functional qualities of the type shown to various bulk materials. Our list of desired functionalities is abbreviated and could be easily extended to include many others—for example, antistatic, anticorrosive, and seemingly anti-anything. Other applications such as air or water purification could possibly be put here as well. Only selected functionalities, however, are examined in this chapter.

The horizontal axis of Figure 10.1 shows various common product forms that can be obtained in bulk sizes and that can be used in various engineering, architecture, and product design applications. At left on the horizontal axis are primary forms of nanocomposites, in the middle are surface-oriented films and coatings, and on the right are special product forms. Not all product forms are shown. In particular, forms of nanoparticles or nanotubes that are fundamental constituents of other nanocomposite types, such as nanopowders or nanoclays, are not shown. These are indeed product forms, but they are essentially ingredients in more complex bulk forms that have immediate applications and are not treated here.

The chart suggests that different product forms, such as nanopaints, can be described in terms of both their primary properties and intended functionalities. The color intensities on the chart do nothing more than broadly suggest *common relationships* as a guide for thinking about relationships. Thus, we have *self-cleaning nanopaints* or *antimicrobial nanotextiles*. The color variations on the chart are meant to be suggestive of the intensity of relationships only. Thus we see that the exploitation of optical properties and phenomena are naturally done with nanoproducts that are primarily surface oriented, such as nanofilms; or, alternatively phrased, the purpose of many surface-oriented nanoproducts has to do with optical properties—self-cleaning, antimicrobial, and so forth. We also see that primary strength properties are normally a less important consideration in nanocoatings than are other properties. In the following sections, various functionalities and product types—self-cleaning, antimicrobial, and so forth—are explored. Specific forms are then discussed.

10.2 FUNCTIONAL CHARACTERISTICS

The fundamental properties and related applications of nanomaterials and nanocomposites have already been discussed extensively in Chapters 6–9. The first part of this section initially looks at functional characteristics that are primarily, albeit not exclusively, related to *surface behaviors*. In designing or selecting materials, surface properties are of fundamental importance. Surfaces can now be designed to have many specific properties that are fundamentally important in the design and use of buildings and products.

The general need and desire to keep surfaces clean and free of dirt has long been with us. Needs can stem from particular use applications. In medical facilities, the need to keep surfaces from contributing to the spread of bacteria is obvious—hence the need for surfaces that are *easy to clean* or that have *antibacterial* or *antimicrobial* properties. The very act of keeping surfaces clean continues to occupy enormous amounts of human energy; hence the need for *self-cleaning* surfaces. The need for *antistaining* surfaces on everything from walls and automotive seat covers to clothing hardly needs explanation. Problems of fungi or mold growth on surfaces, particularly in buildings, may not initially seem very high on any list of important problems, but many serious and recurrent health problems experienced by building occupants are known to be related to the presence of mold (the "sick building" syndrome) as well as other factors. *Antifungal* and *antimold* material characteristics are needed.

Other functional needs related to surfaces include abrasion and scratch resistance. A successful quest for *self-healing* or *self-repairing* surfaces that recall similar behaviors in the skins of humans and animals would provide an enormous benefit to all. Materials with special behaviors related to light and color also have widespread applicability.

Fundamental Approaches for Cleaning and Antimicrobial Actions

Needs for achieving clean and hygienic surfaces have been with us for a long time, and many conventional material responses to them have achieved varying degrees of success. The use of nanomaterials for self-cleaning, antimicrobial, fungicidal, and related applications for various types of surfaces is more recent but has quickly become widespread. Most nano-based products intended for these applications use coatings, paints, or films that contain nanoparticles and that are applied or bonded to conventional materials. Some applications, however, have surface layers in which nanoparticles have been directly intermixed into the base material. Though many needs, such as cleaning or antifungal, may initially seem quite different, technological responses are often quite similar.

Three primary technologies drive applications in these areas: the ability to make *hydrophobic* (water-repelling) surfaces, *hydrophilic* (water-attracting) surfaces, and *photocatalytic* surfaces. Nano-based approaches that invoke hydrophilic (water-repelling) action to spread water or hydrophobic action to bead water are widely used in achieving clean surfaces (see Figure 10.2). Nanomaterials with special photocatalytic properties that absorb ultraviolet (UV) energy that in turn causes dirt or other foreign particles on the surface to slowly loosen or break down via oxidation can aid in self-cleaning and can provide antimicrobial action. As we will see, actions such as self-cleaning can be accomplished in several ways and can be based on hydrophobic, hydrophilic, or photocatalytic means; these primary approaches can be combined in a variety of ways. Antimicrobial action can also be achieved via photocatalytic action or by making direct use of embedded nanomaterials such as copper that have intrinsic anti-microbial action, as we'll discuss later.

Prior to looking at specific applications in detail, it is useful to look at a common referent in nature to understand how certain types of cleaning actions can naturally occur. The lotus flower has long been a symbol of purity in Asian countries (see Figure 10.3). In the 1970s, two botanical researchers (Barthlott and Neinhuis, from the University of Bonn) noticed that the surfaces of the flower are rarely soiled, even as they arise from muddy lands, and that this effect could be observed in many plants that are covered in tiny wax structures that appeared to make the surfaces unwettable. Interestingly, using a scanning electron microscope (SEM) they found that the surfaces were *not* smooth. Conventional wisdom suggested that the smoother the surface, the less likely dirt was to adhere to it,

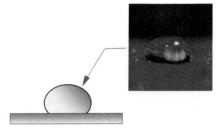

Hydrophobic
(water repelling; drops form beads)

Hydrophilic
(water attracting; drops flatten out)

Photocatalytic
(UV-induced reactions that cause decomposition of dirt molecules)

(a) Basic actions

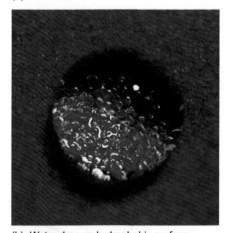

(b) Water drop on hydrophobic surface

FIGURE 10.2
Basic approaches for self-cleaning, easy-cleaning, and antimicrobial functions. Some nanomaterials, e.g., copper, have intrinsic antimicrobial actions as well.

FIGURE 10.3

The lotus flower has long symbolized purity because of its self-cleaning activity, now known to be a result of hydrophobic action. Many plants exhibit this action.

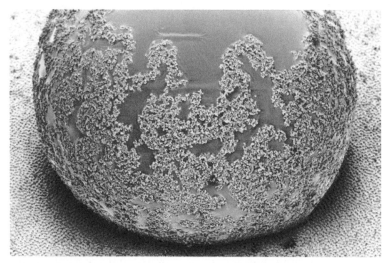

FIGURE 10.4

On superhydrophobic leaf surfaces, water droplets form spherical droplets that roll off easily, removing dirt particles as well. This is happening on the leaves of the Asiatic crop plant Colocasia esculenta, shown here. The papillae that are evident are about 5 to 10 micrometers high and are coated with a nanostructure of wax crystals. The photo was originally taken by Wilhelm's Barthlott's research group. (Courtesy of BASF.)

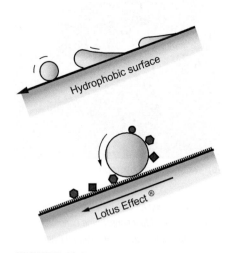

FIGURE 10.5

Self-cleaning action due to the Lotus Effect®. The rolling drops pick up dirt particles.

but, in the case of the lotus leaves, the surfaces had fairly rough micro- and nanostructures that imparted roughness to the surface. Several other plant leaves exhibit similar characteristics (see Figure 10.4). The roughness minimizes contact between a water drop and the leaf itself, and surface tension causes drops to form into beads while reducing the wetting of the surface itself. In the lotus flower leaf, the surface was found to be defined by cells of some 20 to 50 μm in size, with crystals of about 0.5–3 μm (500 to 3000 nm) in size. Barthlott and Neinhuis named this combination of effects the *Lotus Effect.*® The result of this effect is that water droplets bead up into spheres rather than spread. This characteristic is common in hydrophobic (water-repellant) surfaces (see the following discussion) or in oleophilic (oil-repellent) surfaces. These droplets then roll off the surface easily. As they roll, dirt particles cling to the beads and are removed with them as the beads roll off; hence a natural "self-cleaning" action occurs (see Figure 10.5).

The *hydrophobic* (water repellent) action we've mentioned can be contrasted with a *hydrophilic* (water-attracting) action. The amount of wetting that occurs on a surface depends on what action is present, which is in turn dependent on the various interfacial energies between the water droplet, the solid surface, and the surrounding environment. As suggested in Figure 10.6, three primary surface tension factors influence the shape of a water droplet. The relative magnitude of these tensions influence the value of the "contact angle" (θ). If the value of this angle is 90° or more, the surface is generally termed *hydrophobic*. Smaller angles of less than 90° are *hydrophilic*.

Generally, if the surface tension of the solid surface is greater than that of the liquid, drops will spread and surface wetting will occur; if the critical surface tension of the solid is less than that of the liquid, the water droplet will be repelled. Note that the contact area between the water droplet and surface is minimized in hydrophobic surfaces. Relative surface tensions can in turn be influenced by the presence of surface bumps or roughness, which in turn opens the way for using nanomaterials to affect whether the surfaces of a material are hydrophobic or hydrophilic. Surfaces with a contact angle of greater than 150° are called *superhydrophobic* (180° is obviously the upper bound). Making surfaces either superhydrophobic or superhydrophillic are goals of nanomaterial research in this area.

A *first* basic approach to making surfaces self-cleaning is to use nanosurfaces designed with roughness characteristics that serve similar functions to those in self-cleaning plants associated with the Lotus Effect®, which is in turn based on hydrophobic (water-repellant) action. Dirt or other foreign particles are picked up by a water droplet if the force that picks up the particle is greater than the electrostatic force between the particle and the surface. One important variable is the surface roughness. With particular surface roughnesses, the contact area between the particle and the surface is minimized, thus minimizing the force needed to remove the particle. With some roughness patterns, tiny air bubbles can become trapped between the water drops and the surface that can, in turn, enhance the water-repellant properties of the surface and the related cleaning action (see Figure 10.7). This occurs when the energy needed to wet a hydrophobic surface with water exceeds that needed to "wet" the solid surface with air. In this case the liquid drop will try to minimize its contact area with the solid, which in turn can leave air pockets beneath the drop.

Another important variable in a self-cleaning surface is the contact angle (θ) between the water drop and the surface. The lotus flower surface has a localized contact angle of about 160°. Keep in mind

FIGURE 10.6
Factors affecting how much a liquid wets a surface. (Photo courtesy of BASF.)

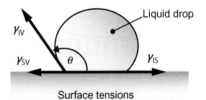

γ_{IV} : Liquid-to-vapor surface tension
γ_{IS} : Liquid-to-solid surface tension
γ_{SV} : Solid-to-vapor surface tension

Surface tensions

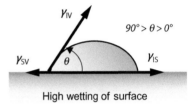

$90° > \theta > 0°$

High wetting of surface

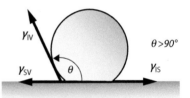

$\theta > 90°$

Reduced wetting of surface

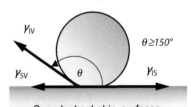

$\theta \geq 150°$

Superhydrophobic surfaces

(a)

(b)

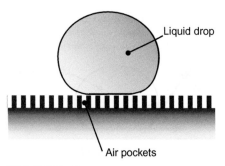

FIGURE 10.7

Hydrophobic (water-repelling) action due to surface roughness. Nanopatterns can reduce the contact area and increase hydrophobic behavior.

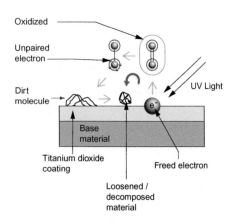

FIGURE 10.8

Photocatalytic action. Photocatalysis can aid in self-cleaning and antibacterial activity and in the reduction of pollutants in the air.

that though our discussion has constantly referred to the lotus flower surface, many materials can be engineered to produce equivalent surface structures. Teflon, for example, intrinsically has a relatively high localized contact angle (115°) for water, which intrinsically gives it some of these properties and can be altered to have even higher contact angle values. Note that though references have been made to water droplets only, the actual contact angles vary depending on what liquid is used as well as what solid surface is present. This dependency is significant when liquids other than water are important. Oleophylic surfaces, for example, repel oil.

A *second* approach involving water-repellant actions involves extremely smooth, surfaces (not rough), which is fundamentally quite different from the Lotus Effect® cleaning action. With highly smooth surfaces, there is a decrease in the surface energy of the surface and consequently a lower force of surface attraction, which again causes water beads to form. Surfaces based specifically on this action are easy to clean and resistant to dirt buildup. They are not intrinsically self-cleaning in the sense we've described, albeit natural actions such as rain can aid in washing away dirt particles from sloped or vertical surfaces. Surfaces of this type are widely used in locations where easy cleaning is important, such as bathroom fixtures. In these approaches, the objective is to use the water-attraction characteristic to develop a thin sheetlike film of water that can be largely transparent and that aids in surface washing and cleaning. Hydrophilic surfaces also form the basis for some "anti-fogging" applications.

The *third* major action of interest here is *photocatalysis*, a naturally occurring process that happens when certain materials are exposed to ultraviolet (UV) UV light. It is one of the most fundamental approaches used to enable many functional behaviors, including self-cleaning and antimicrobial actions. Later we will see that photocatalytic approaches are also used for many other applications, such as air purification.

The photocatalysis process can be initiated by UV light present in normal sunlight or from any other kind of UV light source. The process causes oxidizing reagents to form that in turn cause the decomposition of many organic substances that are deposited on or that form on surfaces (see Figure 10.8). Titanium dioxide (TiO_2) or zinc oxides are often used as photocatalytic materials. They are relatively inexpensive and respond well to UV light. For many self-cleaning applications, titanium oxide (TiO_2) nanoparticles that are in their anatase form (one of the three forms of titanium oxide) are used due to their photocatalytic properties. Exposure to UV radiation produces electron-hole pairs in the material. The free radicals or electrons that result react with foreign substances on the surface and produce chemical reactions that ultimately decompose the foreign substances through oxidation. Organic matter can be quickly decomposed. The loosened and decomposed material can then be washed away by normal rainfall or other means. Since bacteria formation is associated with organic matter, the washing away of this matter also causes the surfaces to have antibacterial qualities. As we discuss more later, the same photocatalytic effect can also reduce pollutants in the air.

Most photocatalytic processes are activated by UV wavelengths that are commonly present in sunlight. Wavelengths in the UV range, however, constitute only a narrow portion of the sunlight spectrum. Increased photocatalytic efficiencies can be envisioned if activated by other wavelengths as well. Considerable research is under way to find nanosubstances that exhibit photocatalytic properties when subjected to a broader range of wavelengths.

The three major actions discussed—hydrophobicity, hydrophilicity, and photocatalysis—form the basis of current easy-cleaning, self-cleaning, and antimicrobial surfaces. The surfaces can form on glasses, tiles, enameled panels, and other hosts (see the following discussion). They also can have overlapping actions. Of particular interest here are photocatalytic surfaces made using titanium dioxide (TiO_2). These same surfaces have high surface energies and hence have been found to exhibit hydrophilic behaviors where water forms into thin films rather than beading. This combined behavior is very useful and has been exploited in many products, including the so-called self-cleaning glasses we'll discuss in a moment.

Specific applications in glasses, tiles, paints, textiles, and other product forms will be discussed shortly. It is important to observe that the use circumstance can dictate which approach is used. Photocatalytic processes now based only on UV light are obviously primarily suitable for exterior, not interior, surfaces where sunlight is present. Products based on the Lotus Effect® are best used where there is frequent general washing from rain or other sources; otherwise streaks can develop. Many surfaces are particularly sensitive to degradation when exposed to the kinds of disinfectants commonly used in cleaning. (This is a particular problem in

FIGURE 10.9

The Beach House, by Malcolm Carver. (Courtesy of Pilkington.)

FIGURE 10.10

A self-cleaning glass that exhibits both photocatalytic and hydrophilic actions is used in the Beach House. (Courtesy of Pilkington.)

medical settings.) Abrasions can also occur that destroy surface characteristics.

Self-Cleaning and Easy-Cleaning Materials

Self-cleaning and easy-cleaning glasses and tiles

In architecture, the primary intended cleaning action is the removal of everyday dirt, dust, or other things that naturally appear on windows or other surfaces in buildings. So-called *self-cleaning glasses* for use in windows in buildings or vehicles as shown in Figures 10.9 and 10.10 are now made by a number of major glass-producing companies (e.g., Pilkington, Saint Gobain, and others). Similar glasses are made for vehicles and other applications. In most products, related objectives include killing bacteria or fungi that develop on surfaces and breaking down pollutants or compounds that can cause discolorations.

In glasses, the self-cleaning action normally comes from coatings with thicknesses at the nanoscale that have particular photocatalytic and hydrophilic (water-attracting) properties. In several applications, the glass is first coated with a photocatalytic material, normally titanium dioxide (TiO_2). The photocatalytic coatings produce chemical reactions (described previously) when subjected to ultraviolet (UV) light that help in oxidizing foreign substances and decomposing them. Coating thicknesses are on the order of 15 nm—and are transparent. Titanium dioxide is normally white, and thicknesses need to be restricted to obtain the needed transparency for glass. When the surface is subjected to rain or simple washing, hydrophilic action then causes the water to form into thin, flat sheets (without large droplets). In hydrophilic action, water droplets hitting the surface of the coated glass attract each other, thus forming thin sheets. Surface drainage is enhanced (as is subsequent surface drying). The water sheet carries off the loosened dirt particles (see Figure 10.11). Since organic matter is decomposed, these same coatings also have positive antistaining and antibacterial properties.

Coatings of the type described have been aggressively developed by a number of major glass product manufacturers and have begun to be widely utilized in real applications. The application to glass shown in Figure 10.9 uses a glass produced by Pilkington. This product is often noted as an application of "nanomaterials." For this product, the company is quite clear about whether or not its product should be described as a nanomaterial or nanotechnology—it should not. This positioning is interesting and worth looking into in light of the definitional and nomenclature prob-lems first noted in Chapter 6. According to a quote provided by Dr. Kevin Sanderson, principal project manager from the On-Line Coatings Technology Group,

> Pilkington Activ™ has been described as an example of nanotechnology. Due to the coatings' thickness, Pilkington Activ™ does fall under certain nanotechnology definitions, e.g., IUPAC. It is critical to understand, however, that the reason we fall under this description of nanotechnology is that the thickness of the coating is <100 nm. However, the unique properties are not derived from the nanotechnology aspect. In fact none of our products contain pre-formed nanoparticles. The films are all based on thin film technology. We understand that some of our products have been used as examples of nanotechnology but this is simply due to them falling under this thickness requirement and not because they contain pre formed nanoparticles, which they do not. We believe that there is a distinct difference between products based on thin film technology (many of which have been out in the market place for 20 years plus) and which now fall under the arbitrary banner of nanotechnology due to their thickness alone and the next generation of nanotechnology involving pre-formed nanoparticulates. Whilst this pre formed nanoparticulate area is a very interesting one, at present we have no products in this area.

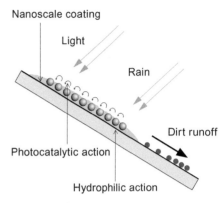

FIGURE 10.11

Thin titanium dioxide coatings exhibit photocatalytic and hydrophilic action. When the coatings are subjected to ultraviolet light, the photocatalysis process oxidizes foreign particles and decomposes them. When the coatings are subjected to washing or rain, the hydrophilic action then causes dirt particles to be carried away.

This quote elegantly captures some of the nomenclature issues and problems discussed earlier. As discussed in Chapter 6, this book has adopted a less narrow approach than described in the quote and indeed considers coatings of "nanoscale" thickness in this and subsequent discussions, a position consistent with common terminology.

Though there are significant advances here, research is still needed into the actual long-term effectiveness of these applications. Certain kinds of matter will not be oxidized by photocatalytic action. Many surrounding environmental variables can also affect efficacies. The amount of solar irradiation received is surely locational and context dependent (from latitude to the shadows produced by surrounding buildings).

Self-cleaning glasses or tiles are by no means a panacea. Some types of thick deposits will not be responsive to photocatalytic action.

Obviously, unless direct washing is done, the cleaning process is quite slow at best and does not work well. Products of this type need UV light and are not suitable for interior applications. They work best for small organic matter and when there is periodic washing via rain or other means. They do not work well for really large deposits of any kind (bird droppings remain problematic). With thick organic deposits, however, normal cleaning is made much easier.

Use of these kinds of self-cleaning surfaces also helps prevent streaking and generally reduces the need for harsh detergents in cleaning. In many senses, this general kind of technology yields surfaces that are perhaps better described as "easy to clean" than "self-cleaning," even though some of the latter action surely occurs. These "easy-to-clean" actions have consequences that go beyond visual appearances. As noted in Chapter 11 on environmental impacts, huge amounts of effort and energy go into cleaning large surfaces in buildings and other forms of construction, and sometimes environmentally doubtful substances are used to aid in cleaning. Any advances in reducing cleaning problems save not only human effort and energy but could potentially help in reducing environmental problems as well.

Self-cleaning paints, textiles, and other materials

Many other material forms use self-cleaning technologies. *Paints* that are relatively thick in comparison to the thin coatings on glass are also touted as having self-cleaning properties and are based on similar technical principles (see Figure 10.12). Titanium dioxide, zinc oxide, and other kinds of nanoparticles are used in paints to provide the photocatalytic action that loosens foreign particles to be carried away by water runoff. Titanium dioxide has long been used as a pigment in paints, but nanosized particles show greater photocatalytic actions than do the normal pigment-sized particles because of their greater relative surface area. The subsequent runoff process can be enhanced using either hydrophilic effect.

These kinds of paints can be applied to many kinds of base materials, including metals. Applications that occur in a factory circumstance, as is common in metal panels used in vehicles, façade elements in buildings, and other products, are invariably better than simple hand painting or spraying. As with self-cleaning glasses, the self-cleaning processes are slow and do not work equally effectively with all kinds of surface deposits, but they do reduce the effort needed to clean the surfaces when necessary. Self-cleaning paints are being widely explored for use in automobile finishes. In all

FIGURE 10.12

The Strucksbarg housing project by Renner Hainke Wirth Associates uses a self-cleaning paint based on the Lotus Effect®. (Courtesy of Christoph Gebler, Photographer, Hamburg.)

these cases, research needs concerning long-term efficacies noted previously are relevant. See Section 10.4 on nanopaints for a further discussion.

Many textiles have been developed that have self-cleaning properties. As previously noted, whether a hydrophobic or hydrophilic state exists between a liquid and a solid surface depends on the relative surface tensions. In everyday textiles such as cotton cloths, the surface tensions of the solid materials are quite high compared to those of water; hence wetting typically occurs. When these same textiles are treated with one or another type of fluorocarbon, the critical surface tensions of the treated textile is less than that of water; hence a water-repellant action occurs. Fluorocarbons achieve this effect by essentially coating individual cloth fibers. This coating can be more successfully achieved with some solid fibers than others. Synthetic fibers are easier to coat than cotton, for example.

Nanomaterial innovations can be helpful here. Coatings with nano-sized whiskerlike elements can create patterns on fiber surfaces with spaces smaller than water drops, thus causing the water droplets to sit on top of the actual fabric. Other methods are based on catalytic self-cleaning approaches similar to those already described. Silver, titanium dioxide, or zinc oxide nanoparticles are typically used. A number of commercial products that are based in this approach are already on the market—for example, photocatalytic acrylic fibers used in yarn forms in various textiles.

Precise techniques for applying nanoparticles onto textiles vary but are typically based on spraying, transfer printing, pressure-related padding, and washing. As the industry looks forward, self-assembly approaches appear to offer promising avenues of development, particularly as a way of achieving more uniform fiber coatings. Self-cleaning textiles are discussed more extensively in Section 10.4 on nanotextiles.

Self-cleaning and antipollutant concrete

A highly interesting application for architects is the exploration of self-cleaning concretes or mortars. Our societies use massive amounts of concrete in the construction of buildings, bridges, roads, and other works. Concrete surfaces are literally everywhere, and one does not need to go far to find dirty, stained, and generally awful-looking ones.

Many current approaches to keeping concrete clean generally rely on either sealants that are preventive in character (see Section 10.4 on nanosealants) or other kinds of coatings that rely on some type of hydrophilic or hydrophobic action for a self-cleaning action. Coatings are often fairly thick since they also typically need to be weather and abrasion resistant. They are also typically constituted to be resistant to degradation induced by the ultraviolet (UV) component of sunlight.

A highly interesting approach mirrors that described in more detail for self-cleaning applications in glass and other materials; it is the use of photocatalytic particles that react to UV light and cause formation of oxidizing reagents that cause the decomposition of many organic substances. In the case of concrete, the photocatalytic nanoparticles are directly mixed into the concrete or mortar mix at their surfaces. The patented process developed by the Italcementi group in Italy, for example, has been used in the exterior concrete façade panels of the Jubilee Church (also known as the Dives in Misericordia) in Rome by American architect Richard Meirs and Partners (see Figure 10.13). There are 256 precast, post-tensioned concrete elements in the building. Similar applications have been made in the new headquarters of Air France at Charles de Gaulle airport and other places, including the Cité de la Musique et des Beaux Arts in Chambéry and the Hotel de Police (see Figure 10.14). In this process, titanium dioxide particles in the anatase form are mixed in and provide the photocatalytic effect that causes dirt and other organic matter on the surface to oxidize and degrade. Hydrophilic

(a)

(b)

FIGURE 10.13

(a) and (b) The Dives in Misericordia in Rome by the architectural firm of Richard Meirs and Partners. The structure uses a self-cleaning concrete based on the use of photocatalytic titanium dioxide. (Courtesy of Italcementi.)

FIGURE 10.14

Self-cleaning concrete is used in the Cité de la Musique et des Beaux-Arts in Chambery, France. (Courtesy of Italcementi.)

action from rains or washings then causes the impinging water to form into thin-sheet forms that carry away dirt particles. Alternative approaches explored by others involve spraying concrete with a photocatalytic coating.

In addition to self-cleaning properties, the concrete reportedly has capabilities for reducing pollutants such as nitric oxides in the air. As discussed in more detail in Section 11.3, air pollution is a major health hazard. Photocatalytic materials exposed to UV light can cause various types of organic and inorganic pollutant particulates to oxidize and be transformed into a more benign form. This possibility has been initially explored by work sponsored by Italcementi via laboratory tests and some on-site experiments. It is worth dwelling for a moment on them, since so little work has been done in this area. In the laboratory tests, nitric oxide (NO_x) particulates in air were blown into an instrumented test chamber with a UV light source and a test specimen (see Figure 10.15). High NO_x abatement rates were recorded within this controlled environment. Subsequent tests on *in situ* conditions were conducted as well. A road-surfacing experiment was conducted in 2002 in Segrate, Milan, on Morandi Street, a heavily trafficked road. A layer of photocatalytic mortar was applied to a 230-meter-long section (about 7000 m^2). Abatement levels were compared to control sections. Levels varied with several conditions, including traffic volume, the amount of light impinging on the surface during the November test period, and wind speed and direction (the latter influenced by surrounding urban conditions as well). Up to 60% abatement levels were recorded.

Another experiment involved road and walkway spaces between buildings (the so-called "urban canyon") where pollutant effects are known to be problematic. A pilot generic site was constructed at a technical center in Guerville, France, as part of an EU project

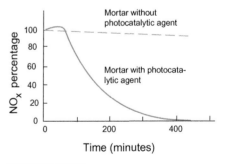

FIGURE 10.15

Reduction in nitrogen oxides (NO_x) as assessed in a controlled test chamber. A preset pollutant concentration is measured for a specified UV light intensity. Abatement rates of up to 91% can be obtained in this controlled environment. (Adapted from The Active Photocatalytic Principle, *from Italcementi, based on work at several European testing centers.)*

FIGURE 10.16
Experimental "street canyon" in Guerville, France, for testing the capabilities of antipollutant concrete. Pavement and building surfaces that use photocatalytic actions can help reduce pollutants in air. Nitrogen dioxide (NO_2) and nitrogen oxide that are dangerous to health are produced by the combustion of fossil fuels typically used in vehicles. Concentrations can be especially high in urban areas. (Courtesy of Italcementi.)

(see Figure 10.16). Pollution conditions from urban traffic were simulated by perforated pipes connected to internal combustion engines that produced emissions calibrated for expected conditions. Sensors at various locations in the canyon measured solar irradiation, humidity, and wind characteristics. NO_x and volatile organic compound (VOC) analyzers were installed as well. Subsequently, 3-D computational fluid dynamics (CFD) models analytically modeled findings. Pollutant abatement levels varied widely, from 20% to 80%, as a function of many factors, especially wind conditions. Cleaning actions on surfaces were measured as well and found highly dependent on weather and solar conditions as well as kinds and intensities of soiling that were present. In other experiments, paving tiles with photocatalytic properties were used in central Antwerp.

In all these cases, relatively little is known about the actual long-term effectiveness of these applications. Considerable experimentation is still needed into amounts and application processes and how external factors of the type previously discussed (e.g., wind, humidity, dirt accumulations) affect them. Common dirt accumulations, for example, can severely inhibit photocatalytic actions that depend on UV light reaching the catalyst. Also, in the long history of concrete, many effective chemical admixtures have been developed that are added to concrete for a variety of reasons, ranging from improving workability during casting to improving its freeze-thaw resistance via "foaming" agents. Studies are particularly needed to understand how these new nanomaterial additives interact with commonly used ones. Further studies are also needed on the long-term performance of these applications, particularly since the expected and needed useful life of concrete products is quite long. Despite the research work that has yet to be done, the prospects are intriguing.

Easy-cleaning materials

A number of products use hydrophobic properties associated with smooth, not rough, surfaces. As noted previously, these are not fundamentally self-cleaning but help prevent buildup of dirt or other organic molecules and are intrinsically relatively easy to keep clean (see Figures 10.17 and 10.18). Reduced surface attractions to these molecules result from the smooth surfaces that exhibit lower surface energies.

Surfaces of this type have extensive application potential anywhere cleanliness and hygiene are important, including kitchens and bathrooms or in medical or health-care facilities or equipment. In the latter, particular care is also frequently taken to be sure that designs do not have places where organic matter can accumulate. This, coupled with the easy-cleaning capability, also helps prevent bacteria buildup and spread. A major material design issue is assuring that common detergents or disinfectants used in cleaning do not cause degradation of these surfaces.

The need for easy-cleaning surfaces goes far beyond the few application domains we've outlined. Additionally, many materials have wide application in fields as diverse as equipment for health-care facilities and cookware.

Antimicrobial materials

Many materials exhibit properties variously described as antimicrobial, antibacterial, antifungal, or antimold. Microbes are minute life forms that include bacteria, fungi, and protozoan parasites. Bacteria are all around us, in water and soils and on the many surfaces we touch every day. Bacteria are microscopic organisms that are typically one celled, have no chlorophyll, and multiply by simple subdivision. Some bacteria aid in fermentation and in other useful processes. Other bacteria are less positive and can cause deterioration, discoloration, or staining of surfaces or can be odor producing (e.g., common body odors). Many microbes can cause disease or aid in its spread, as is the case with pneumonia. Related are various fungal growths that can produce mold. There is a clear need to be able to control and deal with bacterial effects.

Surfaces that possess "antibacterial" properties either kill or inhibit the development of bacteria that have deleterious effects such as discoloration, staining, or odors. Bacterial parasites that cause disease are called *pathogens*, and the term *antipathegenetic* is sometimes used. *Antimicrobial* surfaces target disease-producing microorganisms with the aim of helping prevent the spread of germs injurious to health. The term *antimicrobial*, however, is sometimes loosely extended to include antifungals or even antivirals (although viruses cannot reproduce in the same way as microbes). Clearly these kinds of surfaces are important in the medical field, but they are also important in everyday circumstances for normal healthy living. Positive applications certainly include the multitude of surfaces that proliferate in medical environments (walls, chairs, beds, instruments, and the like) but also in bandages, bedding, and clothing used in care settings as well as in diverse medical products such as filters. In everyday consumer goods, the prospect of antibacterial clothing that can remain "odor free" after wearing is intriguing

(a)

(b)

FIGURE 10.17

The flexible ceramic-coated surfaces shown here provide hydrophobic action that results from very smooth surfaces (unlike Lotus Effect surfaces) that are resistant to dirt buildup and very easy to keep clean by simple washing. (Courtesy of Degussa.)

FIGURE 10.18

Hydrophobic action on a flexible wall covering consisting of a polymer backing and a ceramic coating on the outer surface. (Courtesy of Degussa.)

and is a huge growth industry. Keep in mind, however, that some microflora are needed for human health. Further studies are needed here.

Over the years many kinds of effective treatments to control bacteria on surfaces have been developed. Agents include heat, radiation, or various chemical disinfectants. There is also, of course, a long and positive history of the development of antimicrobial compounds of one type or another—a development that was spurred by the need to treat wounded soldiers in World War II. In looking toward the use of nanomaterials to control bacteria, a key difference is that the antibacterial or antimicrobial properties under development are not simply applied to a surface but are intrinsic to the material itself, either as a coating or embedded in surface layers, so that the treatment cannot be washed off. Several existing materials are already known to possess similar attributes to a greater or lesser extent. Copper has long been known (arguably since early Egyptian periods) to have certain antiparasitic properties and was long the material of choice to be used on ships' bottoms for inhibiting the growth of algae. Copper sulfates have been used on grapevines for many years to prevent fungi from developing. An early study by P. J. Kuhn in assessing the growth rates of streptococci and staphylococci on stainless steel and brass in hospital environments observed heavy growth on stainless steel but only slow growth on brass, which contains copper. (The suggestion was subsequently made that older hospitals with fine old brassware should not replace the pieces with stainless steel.) More recently, various cloths have been coated with different kinds of copper compounds to inhibit microbial growth. The action occurs through copper ions attacking the cell walls of the bacteria. Silver has similar properties.

There are several approaches to making improved antibacterial or antimicrobial surfaces via the use of nanomaterials. One approach is to use copper or silver nanoparticles embedded in some other matrix as a coating on many kinds of base materials or incorporated directly in the surfaces of the base materials. Clearly this approach makes use of the long-known antimicrobial properties of copper, but it is greatly enhanced because of the much larger surface area of the nanoparticles. Similar approaches can use other metal nanoparticles, such as silver (Ag), that are known to have similar properties. Nano-silver-based materials have been used in the design of furniture with antibacterial properties (see Figures 10.19 and 10.20). Many other types of compounds have antimicrobial properties and could be similarly used.

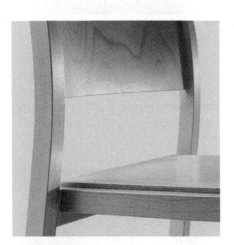

FIGURE 10.19

Surfaces with special hygienic capabilities for use in health care and other environments need to be well suited for easy cleaning, resistant to disinfectants, and have antibacterial action. Surfaces need to be smooth and not have places for dirt to lodge. Coatings with water-repellant action can be used. The finish lacquer used on these wood products consists of a closed-pored antibacterial nanocoating that seals the wooden surface. Several layers are used. (Courtesy of Kusch.)

Relatively little, however, is known about the kinds of preferred types, concentrations, shapes, and size distributions of the nanoparticles for the most effective antimicrobial performance. Not all disease-transmitting microbes, it should be noted, are affected in the same way by similar compositions. Further studies are also clearly needed into exactly what kinds of microbial activities are inhibited and which are less affected for various compositions.

Despite the need for further research, there is great potential and excitement here. Needs in the medical world alone are sufficient to drive research in the area of disease transmission. The vast market for antibacterial clothing or shoes is sufficient to drive research in relation to mitigating other bacterial effects such as odors. The notion of "odor-free" clothing is attractive to all. Antibacterial or antifungal clothing or footwear can be made in various ways, typically either using synthetic polymeric fibers made using antimicrobial additives or by coating fibers. Depending on the application, the intensity of fibers with antimicrobial properties can be increased in certain places (such as in armpit areas for anti-odor applications) and decreased in others as a way of reducing costs (see Figure 10.21). We will return to this topic in a following discussion on nanotextiles in this chapter.

Another process suitable for certain antimicrobial applications is essentially the same one of the approaches described for self-cleaning surfaces—a photocatalysis process. As previously described, photocatalysis is a naturally occurring process that takes place when certain materials are exposed to UV light from either the sun or artificial sources. Titanium dioxide (TiO_2) nanoparticles can be used to impart photocatalytic properties. The photocatalysis process causes oxidizing reagents to form that in turn cause the decomposition of many organic substances that are deposited on or that form on surfaces. Microbial action naturally occurs within organic substances; hence the photocatalysis action breaks down microbial structures as well. Obviously, since the photocatalysis process depends on UV light, this approach is suitable for some applications in which a product or surface is regularly exposed to natural or artificial UV light, but not in others. It is particularly useful in connection with applications in large tents and other types of membrane structures in architecture, where bacterial effects can cause discoloration on the structural fabrics that have large surface areas normally exposed to sunlight (see Figure 10.22).

An interesting avenue of approach for developing antimicrobial or antifungal capabilities is through the microencapsulation of bio-

(a)

(b)

FIGURE 10.20

The metal furniture shown is for use in special hygienic environments. Metal components are coated with a silver-bearing antibacterial powder coating. A ceramic carrier with positive silver ions is used. (Courtesy of Kusch.)

FIGURE 10.21

Antibacterial textiles are widely sought after. The image shows textured nanofiber with microchannels for incrustation of catalysts and antibacterial compounds. See the following section on nanotextiles. (Courtesy of Juan P. Hinestrosa.)

FIGURE 10.22

This fabric membrane structure at the Hyatt Regency in Osaka uses a photocatalytic clear coat based on titanium dioxide, which helps prevent the buildup of organic particles that host stain-causing bacteria. An antistaining action is thus present. (Courtesy of Tayo Kogyo.)

cides or fungicides. The microcapsules are distributed on the surface of a material as a coating and then are designed to have a "timed-release" capability. Though the biocidal effect can be pronounced, it is obviously time limited.

Mold was mentioned earlier. Though some molds are positive (think of certain cheeses), other kinds of mold in buildings can cause extreme health problems to occupants—witness the need to tear down or completely rebuild many water-damaged houses in New Orleans after the Hurricane Katrina disaster, simply because of the mold development problem and its related health hazards to returning occupants. Mold is a type of fungi that breaks down organic materials. Molds grow under damp or wet conditions on many kinds of surfaces. They reproduce by releasing spores into the air, where they fall on other materials and germinate to form large networks. The rate of growth of mold depends on many factors: their type, the organic makeup and moistness of the material on which the spores land, temperature, and other factors. Humans can inhale airborne spores, touch them on surfaces, or even eat them when they land on food.

Certain kinds of molds can very much affect human health by causing allergies, irritations, infections, or various toxicities. Effects depend on the types of mold present, the length of exposure, and the susceptibility of the human in the environment. In buildings with mold problems, there is obviously long-term exposure to occupants. Health effects can be devastating. It is likewise enormously difficult to remove mold in a building once it takes hold. The development of materials for building and other uses that would have effective antimold properties that either killed spores or inhibited their growth would be an enormous contribution to human health.

Many kinds of antimold treatments are available on the market today. Many are simply solution washes of one type or another that kill mold on contact but rarely have any long-term effects. Some coatings or treatments to materials are longer lasting. Many are again based on titanium dioxide. For situations in which UV light sources are present, as from the sun, the same photocatalytic processes (based on titanium dioxide nanoparticles with their large surface-to-volume ratios) that cause organic particles on a surface to break down, as previously described, can be used effectively. Other techniques that use antimicrobial coatings or additives to primary materials offer promising approaches. For some applications, various layering approaches have been suggested. For building applications, an inherent problem in this kind of research is that surprisingly little is actually known about the precise mechanisms of mold growth, not only on but also particularly within the many porous surfaces and material structures that surround us and that hold moisture differently. Without better knowledge of these mechanisms, responses cannot be nuanced but are rather of the "coat it" variety.

Self-Healing Materials

When material scientists and engineers give presentations extolling the future benefits of research associated with advanced materials, the phrase *self-healing* or *self-repairing materials* is seemingly invariably included. Similarly, in the popular press, the same phrase is used over and over again to impart a specific and understandable vision of what nanoengineering might accomplish. Indeed, there is hype here. There is, however, also a very real basis for suggesting that some type of self-healing or repairing is indeed a possibility—and an exciting possibility at that. Cracks in materials or scratches on surfaces are ubiquitous and always problematic. Cracks can either lead to or signify serious performance problems in a material, even foretelling structural collapse. Scratches on surfaces can seemingly be only nuisances, but nuisances that society spends an enormous amount of time and money addressing and trying to prevent.

A number of major automotive companies, for example, have invested huge amounts of research in trying to achieve a really effective self-healing paint to be used on automobile bodies, with every belief that it will pay off in terms of consumer response. Other kinds of cracks or scratches might not be serious in themselves but could lead to other problems, as might be the case with major scratches in painted metal panels that in turn leads to corrosion. The need and value of self-healing materials are clearly present.

Several basic approaches to accomplishing some type of self-healing have been explored for some time. Observation of models in nature, where self-healing is an intrinsic part of life, have naturally led many investigators to look for ways that similar models might be developed for the materials we use. In other cases, astute observers have noted that many materials just seem to naturally heal themselves after cracks develop. In architecture, it has been observed since the Medieval era that many cracks that initially developed in the lime and sand mortars used in historic stone construction seemed to simply disappear over time (see Figure 10.23). The reasons for that phenomenon are interesting. Lime

FIGURE 10.23

Beachamwell Church, Norfolk. Lime mortar was commonly used in churches of this era. Lime mortars were found to have a remarkable "self-repairing" capability that would cause cracks to heal over time.

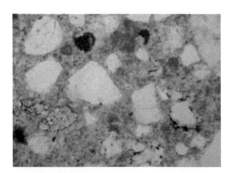

FIGURE 10.24

Lime mortar with blue-dyed resin. There is high porosity present. Moisture can enter and deliver the carbon dioxide needed for carbonation during curing and to help in the crystalline bridging phenomenon that can close larger cracks. (Courtesy of William A Revie, Construction Materials Consultants Ltd., Stirling, Scotland.)

and sand mortars slowly harden over time due to carbonation of the lime. The lime used in historic stone or masonry construction was normally made from burning limestone containing impurities (included either advertently or inadvertently) of silica and clay. A "hydraulic" lime resulted that could even set under water. The resulting mortars were quite porous and allowed moisture to enter. When normal cracking occurred, carbon dioxide dissolved in water could still reach free (as yet uncarbonated) lime to form calcium carbonate crystals that formed and grew in the cracks. A form of crystalline bridging occurred that filled and sealed even quite large cracks.

This behavior is remarkable and a good example of a self-healing material (see Figure 10.24). Lime mortars are still sometimes used for this reason today. The concrete and cement mortars now in common everyday use, however, do not possess this same property. A modern-day related approach for common concrete that is curious but fascinating has been advocated by the biologist Dr. Henk Jonkers, who works with bacteria that naturally produce calcium carbonate. When a crack forms, water and oxygen cause the bacteria to begin working to produce the needed calcium carbonate for crack repair. Finding the right bacterial environment for good growth remains problematic.

Current approaches to self-healing vary considerably. One popular approach is that of embedding into the material small encapsulations (in particle, tube, or layer form) of materials that are released when the encapsulations are ruptured by a crack or scratch. The released materials fill cracks and harden (again involving a chemical reaction) to "self-repair" the crack, or at least inhibit further growth. This approach of using stored components that are released upon rupture has been explored for some time, including the work of Carolyn Dry at the University of Illinois, which was reported as early as 1990 and which used uncured resins to fill cracks and gaps within bulk concrete. Subsequently, similar techniques have been explored for use in polymeric matrix composites. Many of these approaches are at the micron and not nano level but are moving in the latter direction.

A general approach for self-healing polymeric materials is diagrammed in Figure 10.25. In this figure, the components to be released are encapsulated within tiny micron-sized spheres embedded within a matrix containing dispersed catalysts. When a microcrack encounters a sphere and ruptures it, the stored material is released. It interacts with the dispersed particulate catalyst within the polymer matrix and subsequently hardens. Experiments have found that some portion of the mechanical properties lost due to microcracking of the composite can be restored. Use has been made of a vinyl ester resin, which is cured through radical polymerization initiated by peroxides. One approach in the research development phase uses the relatively inexpensive monomer dicyclopentadiene (DCPD) within the microcapsules. DCPD has a relatively low viscosity and works at room temperature. Upon rupture of the microcapsules by a microcrack, the material is drawn along the crack by capillary forces until it comes into contact with the ruthenium based "Grubbs" catalyst, which in turn causes polymerization that fills cracks. Microcapsules are often about 10–100 μm in diameter (obviously not nano sized). A variant of this technique stores the curative material to be released within hollow glass fibers embedded in the primary material.

The approach described is particularly good for resin-only systems. Strength repair is also good. Microcapsules can be placed within bulk polymers. The approach can also be used to form laminates or other kinds of surface coverings on bulk materials. For the system to work, however, the microcapsules must be ruptured. This in turn demands a high-quality and even dispersion of both the microcapsules and the catalyst—not only to assure that microcracks are intercepted but also because the microcapsules and catalyst must always be in close proximity to one another. The catalyst crystals can tend to aggregate rather than disperse evenly. In addition, only limited amounts of material can be stored in the tiny microcapsules, and released volumes may not be sufficient for needs. Voids are also obviously created when microcapsules are ruptured. The use of hollow tubes provides a higher reservoir of healing agents and aids in crack interception, but dispersion and void issues remain (see Figure 10.26). In both cases, it is clear that the various materials used must also be compatible. In some cases the catalyst can be attacked and significant amounts can be lost.Large amounts of catalysts can be used so that polymerization can still occur. Catalysts can also be embedded in micronsized microcapsules to protect them, thus reducing the amount

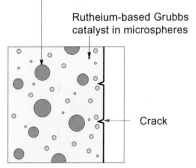

DCPD microcapsules

Rutheium-based Grubbs catalyst in microspheres

Crack

Beginning of cracks in the surface of the matrix material

(a)

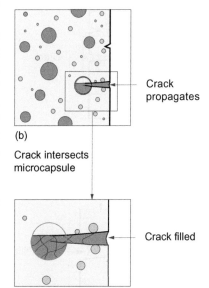

Crack propagates

(b)

Crack intersects microcapsule

Crack filled

The catalyst causes the DCPD monomer to polymerize into a tough cross-linked polymer

(c)

FIGURE 10.25

An approach to self-healing or self-repairing materials based on the use of embedded microcapsules that, when punctured, release healing agents that harden. (a) Beginning of surface crack. (b) Crack intersects microcapsule. (c) The catalyst causes the DCPD monomer to polymerize into a tough cross-linked polymer. (Adapted from R. Trask and I. Bond, Enabling Self-Healing Capabilities—A Small Step to Bio-Mimetic Materials.*)*

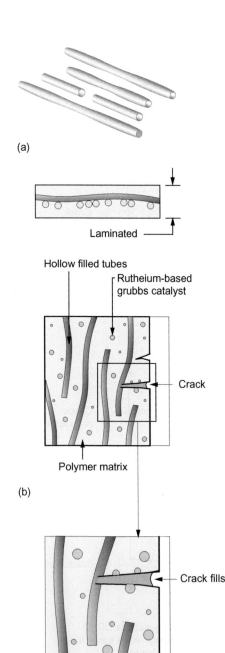

(a)

Laminated

Hollow filled tubes

Rutheium-based grubbs catalyst

Crack

Polymer matrix

(b)

Crack fills

(and cost) of the needed catalysts. The use of wax encapsulation for both agents allows a greater variety of materials to be used in the self-healing process.

Variants of the hollow-tube approach that have been explored include using multiple tubes in composite laminates. Tubes could contain a resin as before with a hardening catalyst in the laminate material, or even a two-part resin and hardener system within alternating tubes. This would allow more variation in the types of self-healing materials that are possible and could be accommodated in many different primary material matrices. Studies have also been made with the objective of determining whether these same tubes can serve as structural reinforcement and even be accordingly aligned.

Other approaches are being developed that require no catalysts or microcapsules. One avenue of investigation for polymers involves mixing thermoplastic healing agents that change viscosity with heat in harder thermoset plastics. Heat is obviously needed. Another self-healing process occurs directly in a material via a reversible polymerization process. One process explored by a number of researchers uses monomers that are mixed, heated, and cooled slowly (a reaction of a furan monomer with a maleimide monomer). Thermally reversible Diels-Alder reactions proceed nearly to completion to form a cross-linked material with significant strength and stiffness (similar to structural epoxies). Under certain conditions, however, these reactions are reversible. As a crack develops in the material, necessary chemical groups are regenerated. If the material is clamped and heated, a "retro-Diels-Alder" reaction occurs again and mends the crack. The material can thus undergo repeated crack-repair cycles. For this to occur, however, the material must be heated—in some instances up to around 120°C—a significant problem for many needed applications. Other self-healing approaches based on silicon carbide (SiC) form protective silica coatings on their surfaces after being scratched, but again, this self-healing occurs only at high temperatures and is thus useful prima-

FIGURE 10.26

A self-healing material based on the used of embedded hollow tubes containing a healing agent. (a) Laminate form; hollow tubes filled with healing agent. (b) Beginning of surface crack. (c) Crack intersects and ruptures tube(s). The catalyst causes the healing agent located inside the hollow tubes to polymerize into a tough cross-linked polymer, filling the crack. (Adapted from R. Trask and I. Bond, Enabling Self-Healing Capabilities—A Small Step to Bio-Mimetic Materials.*)*

rily for components for which high temperatures are commonplace, such as exhaust systems.

Antifogging, Antireflection, and Other Characteristics

A whole host of other kinds of desired material behaviors are possible. Many antifogging applications depend on the hydrophilic action of surfaces; the resulting spreading of water droplets becomes largely invisible. Many films and glasses appear to possess different reflective properties that are highly unique. In the optical area, one of the most common is that of *antireflection*. Nanoproducts with antireflection characteristics are widely used in architecture as well as in product design to reduce troublesome reflections. They are also used in optical devices to increase light transmission. (Decreased reflection leads to increased light transmission.) Many of these optically related products also can block specific types of infrared, visible, or ultraviolet light wavelengths and thus be useful in a wide array of applications. Section 9.5 describes these optical effects in greater detail.

10.3 "SMART" BEHAVIORS

Earlier chapters dealt extensively with the properties of nanomaterials but only marginally addressed many of the dynamic material behaviors—or *smart behaviors*, as they are often called—that designers or engineers need or desire and that are closely related to specific mechanical, thermal, optical, or other property characteristics (also see Section 9.8 on smart environments). The focus in this section is not on properties but on specific desired material behaviors, such as color changing or shape changing. A list of common behaviors is shown in Figure 10.27.

A number of materials have an apparent ability to change one or more of their properties in response to the input of thermal, radiant, electrical, mechanical, or chemical energy from an outside source. Thermochromics, for example, change their apparent color as the temperature of their surrounding environment increases and there is an input of thermal energy into the material, or magnetorheological materials change their viscosity when subjected to varying levels of magnetic fields. The somewhat unfortunate term *smart materials* has been widely used to describe materials that exhibit these kinds of capabilities, as has the term *responsive materials*. In many of these materials, there is an apparent change in certain properties of the material, such as changes in optical prop-

TYPE	INPUT		OUTPUT
Type 1			
Thermomochromics	Temperature		Color change
Photochromics	Radiation (Light)		Color change
Mechanochromics	Deformation		Color change
Chemochromics	Chemical		Color change
Electrochromics	Electric potential		Color change
Liquid crystals	Electric potential		Color change
Suspended particle	Electric potential		Color change
Electrorheological	Electric potential		Stiffness change
Magnetorheological	Electric potential		Stiffness change
Type 2			
Electroluminescents	Electric potential		Light
Photoluminescents	Radiation		Light
Chemoluminescents	Chemical		Light
Thermoluminescents	Temperature		Light
Light emitting diodes	Electric potential		Light
Photovoltaics	Radiation (Light)		Electric p
Type 3			
Piezoelectric	Deformation	↔	Electric
Pyroelectric	Temperature	↔	Electric
Thermoelectric	Temperature	↔	Electric
Electrostrictive	Electric potential	↔	Deformation
Magnetostrictive	Magnetic field	↔	Deformation

FIGURE 10.27

Common smart materials in relation to input and output characteristics.

erties that result in color changes, and hence the somewhat simplistic description as "property-changing" materials is frequently used. What is happening is far more complex and a consequence of a variety of chemical reactions and phase changes that take place internally in a material as a consequence of an energy input from an external source that in turn alters the material's optical, mechanical, or other properties—that is, the property alone does not change, rather the material does, resulting in a property change.

In other cases, a material subjected to the input of one energy type changes it into another energy type as a consequence of induced internal actions—and is thus an "energy-exchanging" behavior. Thus, in a common piezoelectric material, the input of mechanical energy (via a force application) causes an electrical energy output, or vice versa. Many common devices, including common doorbells or speakers and microphones that are based on piezoelectric technologies, rely on these effects. The materials listed are among those often described as having "smart" behaviors. Keep in mind, however, that "smart" is a poorly defined term. Many of the functional behaviors and materials described just previously, such as self-healing or self-cleaning materials, would in the terminology of many be called "smart" materials.

The materials listed in Figure 10.27 are well documented and discussed elsewhere, as is the whole field of "smart" materials. Only a few characteristics are summarized here. It must be emphasized that not all "smart" materials derive their interesting properties from nanomaterials as defined in a narrow sense. Many films, for example, are of nano thickness but might not involve nanoparticles.

Color changing

The apparent ability of a material to change color is a light-related characteristic that is highly sought after in both architecture and product design. Many existing "smart materials" have this characteristic, including *thermochromics, photochromics, chemochromics,* and *mechanochromics* (see Figure 10.28). In all these materials, input of one form or another of external energy causes an apparent color change in the material.

The color of a thermochromic material varies with the temperature of its surrounding environment. In this case, the material is responsive to the input of thermal energy, which is absorbed or released by the material. Many common items, such as the "band thermometers" placed on a person's forehead, use thermochromics. As with other chromics, the process is reversible. Photochromics are widely used and respond to changes in radiant energy in visible or nonvisible spectrums, including sunlight. Common tint-changing sunglasses use photochromic glasses. Color changes can also be caused by the input of mechanical energy (mechanochromics) or via a changing chemical environment (chemochromics). Color-changing materials are widely used not only for a novelty effects but as measurement devices and other applications as well. Specific technologies include electrochromic and other materials. Section 9.5 discusses these materials and characteristics in greater detail.

Shape changing

Few materials attract so much interest as materials that exhibit "memory." If a piece of "shape-memory alloy" is deformed from an initial shape into a new shape, it will remain in the new shape. When heated to a certain critical temperature, however, the material will literally go back to its initial shape without mechanical aid. The material "remembers" its original shape and returns to it. The heat

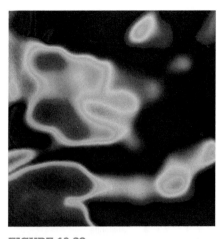

FIGURE 10.28

A thermochromic color-changing film. The blues show hot temperature regions, while the black is the lowest. The film is calibrated for 25–30°C.

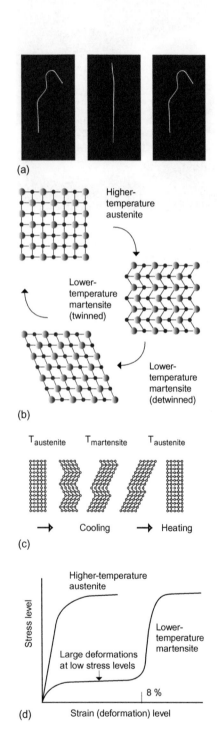

(a)

Higher-temperature austenite

Lower-temperature martensite (twinned)

Lower-temperature martensite (detwinned)

(b)

$T_{austenite}$ $T_{martensite}$ $T_{austenite}$

→ Cooling   → Heating

(c)

Stress level

Higher-temperature austenite

Lower-temperature martensite

Large deformations at low stress levels

8 %

Strain (deformation) level

(d)

can be directly applied or come via an electric current. Materials such as TiNi have been developed for this purpose.

During the bending/heating process, the material undergoes several internal phase changes (martensite to austenite, and vice versa) that are responsible for the memory effect (see Figure 10.29). The process is generally reversible, albeit some shape deterioration can take place if the change and heating process is not carefully controlled. A number of simple "release mechanisms," for example, use shape-memory alloys because they reduce the need for moving parts. The phenomenon of "superelasticity" in metals used in many applications, including eyeglass frames that can be twisted into incredible shapes, stems from this same shape-memory mechanism.

Shape-memory polymers in the form of rigid plastics or relatively porous foams have also been devised to demonstrate similar capabilities. These interesting materials are finding wide application in many kinds of products. Shape-memory polymers have been explored in divergent settings, including unrolling surfaces on NASA spacecraft, knots that tie themselves in the medical arena, and personally customizable furniture for consumers.

Varying electrical, magnetic, and other properties

A host of other desired characteristics sought by designers and engineers include the ability of a material to change its electrical resistance, its magnetic properties, its stiffness, or even its shape via the input of some energy source. A number of existing smart materials possess one or more of these qualities. A magnetorheological fluid changes its rheological properties (including its viscosity and hence its stiffness) as the surrounding magnetic environment varies. An electrorheological material similarly changes in response to an electric current. These kinds of fluids are generally "structured fluids" that contain colloidal dispersions that change phase when subjected to a magnetic or electrical field. Nanoparticles, with their increased surface areas, can serve functions similar to colloidal dispersions.

Either one of these variable rheological materials can go from a very fluid state to a highly viscous state with literally the flip of a switch that changes the electrical or magnetic environment that's present. There are already many practical uses for these kinds of materials.

FIGURE 10.29

Shape-memory alloys. (a) A shape-memory alloy can be bent to a new shape and, on healing, return to its original shape. (b) Shape-memory alloys undergo reversible phase transformations at various temperatures and (c) and (d) change their internal structures.

They are widely used in hydraulically based damping systems for vibration control. In this application, the precise amount of damping that's provided can be fine-tuned for the application by altering viscosity levels (see Section 9.2). These materials have been used in automotive clutches as parts of power transfer devices, replacing mechanical interfaces. The stiffness of tires can also be tuned, thereby proving a way to vary tire stiffness for better cornering or more comfortable straight riding as needed. Surprisingly, little other than exploratory use has been seen in products such as furniture, where these kinds of tunable fluids might be used to vary levels of hardness or softness of seats. Section 9.2 contains a further discussion of these materials.

Smart skins and envelopes

An enormous amount of work has been done in developing various kinds of surface structures (skins or coverings) that exhibit not just one but many smart behaviors. Many of these approaches are essentially equivalent to the "multifunctional" surfaces that are now a buzzword in materials research, which might include functions varying from antibacterial to self-repairing and others. More extended versions also typically include different kinds of sensors that respond to a variety of stimuli and various kinds of response systems, including direct-active actuators. The sensor/actuator systems are in turn linked to a comprehensive control/logic system. Nanomaterials and nanotechnologies offer great potential as both tiny sensors and actuators and thus have the potential to be quite easily integrated into surface structures. Going into detail into these complex systems is beyond the scope of this text, but the subject is widely discussed in the literature (also see Sections 9.4 and 9.8).

10.4   NANOPRODUCT FORMS

Many primary nanocomposite product forms engage the use of nanomaterials. Nanoparticles or nanotubes can be distributed throughout a host matrix made of one or another type of metal, polymer, or ceramic material. The nanoparticles or nanotubes themselves can be made of various primary materials as well. Polymer nanocomposites, for example, can have many kinds of nanofillers (silicas, clays, carbon nanotubes, carbon blacks). The polymer provides the supporting matrix or medium, and the nanofillers largely provide specific improvements in properties (mechanical, electromagnetic, optical, thermal, chemical), or they can provide many different functional behaviors. Nanoclays, for example, are commonly

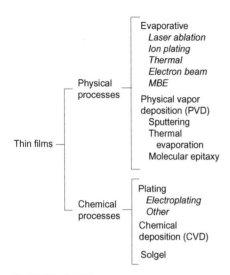

FIGURE 10.30

Thin-film techniques (see Chapter 8).

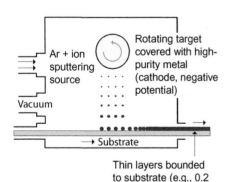

FIGURE 10.31

Sputtering method for making a thin film.

embedded in many bulk polymer materials. Bulk forms were discussed in previous chapters and include many kinds of nanometals, nanopolymers, and nanoceramics (Section 8.8). We have also seen that nanocomposites can take several primary physical forms, such as nanoparticles or nanotubes embedded in a matrix, or films or laminates applied to a bulk material. Figure 10.30 lists several ways of making thin films.

In the following sections, several nanocomposite product forms not already discussed in previous chapters are addressed, including nanocoatings, nanopaints, nanotextiles, and others. There is rarely agreement, however, as to what these various terms actually mean. Some sources lump nanocoatings and nanopaints together, for example; others draw sharp distinctions based on thicknesses and functions. Several of the product forms discussed also reflect industry segments that routinely provide and promote certain materials and products. Nanotextiles or nanopaints provide a good example here. These product forms are highly application driven and lie toward the end of the "raw-material-to-product" sequence. This end-use characterization differs sharply from materials, such as those that are described as nanoporous, that are indeed used extensively but typically as constituents or components, such as filters or membranes, and are thus more often constituent materials in a product rather than the end product itself. Developers of most of these various end product forms normally also seek and claim multifunctional qualities to their products. Thus, many nanopaints are purportedly antistaining and self-cleaning in addition to serving their normal protective or decorative functions.

Products described as thin films, coatings, laminates, and other terms related to imparting new surface qualities to a base material are particularly hard to differentiate. These terminology distinctions are by no means clear and stem from quite different usages (and sometimes passionately held definitions) among materials scientists, product developers, and designers. In the following sections, we use the term *thin film* to refer to a layer of material that is deposited by one or another physical or chemical process onto another material—normally with the intention of imparting some desired surface quality such as light reflectance, electrical conductivity, or other function (see Figures 10.30 and 10.31). Many of these thin films incorporate nanomaterials; others are more simply at the nanothickness scale. The term *nanofilm* is sometimes used to describe these films. Though there is no general agreement about what precise thickness defines a thin film, the general sensibility is that thicknesses lie in the nanometer to the micron range.

An example of a commonly seen thin film is what we observe when a drop of oil spreads across the surface of water. The same term, however, means something quite different in the everyday architecture world. Here many physical films have been developed that have far more appreciable thicknesses than previously described and that are manufactured as products in the form of thin sheets that have their own integrity and can stand alone (e.g., a very thin sheet of plastic). These relatively thick sheets may well have nano-materials incorporated into them or contain multiple layers of nanoscale thin films deposited on them in a laminate form. In this text, we generally refer to these as *thin sheets*.

In this context and in several other common usages, the term *coating* refers to quite thick applications (and could include heavy paints, for example) and is generally more descriptive of more specific materials, application processes, and specific uses.

In many applications, several layers of nano-based materials are built up sequentially. Various layers would normally have different properties and would be intended to serve different functions. These are often described as *laminates* and occasionally *nanolaminates*.

The following sections consider in turn these topics: nanocoatings, nano-based thin films and laminates, nanopaints and nanoseal-ants, nanotextiles, and nanoporous materials. As noted, there is considerable overlap in the surface-based nanoforms, and descriptions should be collectively read.

Nanocoatings

Coatings are thin coverings that are deposited on a base material to enhance its surface characteristics or appearance. This broad definition includes coatings used to improve durability or wearing characteristics, provide corrosion resistance, or otherwise protect the base material. They might also be used for change adhesion qualities, color, reflective qualities, or a host of other reasons. Typical coating forms are shown in Figure 10.32. Coatings are widely used

Nanoalloys   Nanocomposite   Multilayer   Functionally graded   Nanotextured

FIGURE 10.32
Typical nanocoating forms.

because they provide a direct and cost-efficient way of imparting particular surface qualities to a material without having to alter the entire mass of the material. Seemingly everything from metals to textiles can be coated (and probably has been).

As discussed in the previous section, there are also ambiguities in what is meant, exactly, by the term *coating* compared with terms such as *thin films, paints, platings,* and other similar terms that connote surface coverings of one type or another. Nanocoatings and nano-based thin films are quite similar but with the general implication that "coatings" are thicker than films. Many of the applications described in this section apply to both nanocoatings and nano-based thin films. Paints are invariably thicker.

Nanocoatings provide one of the greater overall market potentials for any kind of nano-based product or device. We have already seen that nanomaterials are intrinsically expensive and that production processes for many nanoforms can yield only very small objects due to technological barriers. Many of the production processes for nanocoatings, by contrast, are potentially suitable for covering relatively large areas. Coatings are quite thin, and relatively small quantities of nanoparticles are used in a typical nanocoating, hence reducing basic costs. Using thin nano-based coatings is thus a very effective way to take advantage of many of the remarkable properties of nanomaterials without bearing the production costs of making large components fully out of nanocomposites.

Primary functional uses of coatings include improving the wear, abrasion, or corrosion resistance of a covered material; altering or improving its electrical, magnetic, frictional, or optical characteristics; proving a surface sealing for preventing liquid or gas penetration; and simply being decorative. An important application is in architecture and product design, where various kinds of coatings are used for antireflection or low emissivity glasses. Antimicrobial, self-cleaning, and other actions are also intended uses.

Nanocoatings that provide improvements in mechanical properties were briefly discussed in Chapter 9. There it was noted that several kinds of nanocrystalline powders have been used for making metallic coatings that improve surface hardness. Nanocrystalline nickels have been used in many products, including golf club shafts, as hard protective shells. Polymer-based nanocoatings are in wide use as coatings for corrosion protection of metals. Nanolayered coatings are used for improving the hardness of cutting edges. Coatings are also used to improve the thermal stability of materials. Porous ceramic materials are often used for thermal coatings, particularly in high-temperature applications, and serve a primarily insulating function. Various other alumina/titania coatings also serve thermal functions.

Many nanocoatings have been developed for light or optical reasons. An interesting characteristic of nanoparticles is that their size is smaller than the wavelength of visible light, which ranges from 400 to 700 nm. Nanoparticles in coatings are much smaller, around one tenth or so of the wavelength of visible light. When dispersed in a transparent matrix, these nanoparticles are virtually transparent to the eye. Additionally, nanoparticles can now be made wherein the refractive index of the particles can be matched to those of a cured resin formulation. Translucent or optically transparent coatings can thus be made. Some nanocoatings have been specifically developed to block ultraviolet light or to have particular reflective properties.

A major application is for antireflection or brightness-enhancing applications that are important in many areas of product design, architecture, and industrial products. Most of these applications consist of multiple layers of nanoscale materials built up sequentially. Antireflection films serve to either reduce troublesome reflections (a typical architectural application or product design application) or to enhance the light transmission efficiency of a transparent material. As discussed in Section 9.5, light impinging on a surface is reflected, transmitted, or absorbed. If less light is reflected, more is transmitted. The brightness of a perceived image, for example, can be improved. Contrast is normally improved as well. Hence, many antireflection films are found in everything from computer screens to telescopes.

Not only can high transmission be achieved in complex optical devices, but troublesome ghosting effects can be reduced. Light reflection on a surface is reduced via the use of multiple layers of transparent materials that have different optical qualities, particularly their refractive indices. Depending on various layer thicknesses and the refractive indices present, different destructive and constructive interferences are created that make the optical characteristics of the multilayer coating change with the wavelength of the impinging light as well as its angle. Coatings can be designed to respond to a wide range of infrared, visible, or ultraviolet wavelengths, depending on the needed application. Refractive indices can be adjusted over a wide range by varying concentrations of metal oxide nanoparticles in a polymer matrix. Optical coatings are discussed in more detail in Section 9.5.

Electrical properties can be altered using nanocoatings to improve the conductivity of a surface—for example, by adding carbon nanotubes or other metallic nanoparticles. Even polymeric coatings can be made conductive to a greater or lesser degree. Nanocoatings can be used to alter the magnetic properties of a surface as well. A number of antistatic coatings have been developed as well as coatings that serve other electrical and magnetic functions. The need for a coating to ground a charge or discharge static electricity buildup is surprisingly frequent. Other functions of coatings include acting as sealants or barriers for liquid or gas transmission. In common plastic drink bottles, for example, there is a need for a gas barrier so that pressure can be maintained. Various polyamides with nano-clay particles are often used for this function. (See the following section on nanosealants.)

There is a strong need for surfaces that are self-cleaning, anti-microbial and antibacterial, and scratch resistant or even self-repairing. This topic was discussed earlier in this chapter (see the section on self-healing materials). Many of these actions can be obtained by manipulating surface characteristics via nanocoatings or nanofilms.

For consumer products, nanoscale coatings are used as everything from lubricants on razor blades to watchband protective coatings. Common sunscreens normally incorporate nanoparticles of titanium oxide, known for its light absorption capabilities. Zinc oxides are used as well (see the discussion on nanocosmetics).

Multilayers and Nanofilms

Nanoscale thin films, often called *nanofilms*, can be deposited on the surface of another base material by any one of several physical and chemical processes. Multiple nanoscale layers can be built up to form multilayered or laminate structures. Total thickness often lies in the submicron range. As with similar nanocoatings we've discussed, these layered structures serve an astonishing array of functions. They are relatively easy to deposit on a surface and can impart many qualities to it. Various thin layers can be deposited on one another to build up a variety of functionalities (multicoatings or laminates). Many of these structures provide us with remarkable optical effects and properties and are often used to change light transmission characteristics. They can be conductive or nonconductive. Conductive layers can be deposited on nonconductive base materials and thus serve a variety of purposes (including static discharge). Applications are found in flat-panel displays, touch panels, disk drives, and antireflection glasses. They also serve roles as diverse as catalysts and in decorative applications. There are even thin-film "optical switches" that literally serve a switching function.

Most of the attributes and characteristics of basic films are generally similar to those nanocoatings we have already discussed. In general, many kinds of these structures have been developed to improve the mechanical, thermal, or electrical characteristics of surfaces. Thin carbon nanotube composites, for example, have been layered onto many different base materials to alter one or more of these properties. Here carbon nanotubes are embedded in a thin-film material (e.g., a polymer or ceramic) in one or more layers. This is a common way of improving the electrical conductivity of a normally dialectric (insulating) material. Carbon nanotube composites have also been built up in multiple laminate layers to improve the mechanical properties of surfaces. When this is done, there are normally enhanced properties parallel to the surface covered but sometimes less so with out-of-plane properties. Out-of-plane strength properties of multiple layers of carbon nanotube composite films, for example, are considerably lower than their in-plane properties.

In the optical area, we have already seen many colors and sheen effects. Of particular importance are multilayered structures designed to decrease overall surface reflections or improve surface brightness (see Section 9.5). Layered structures can also be adjusted to control the relative transmissions of light in the infrared, visible, and ultraviolet spectrums. Various kinds of interference filters have also been developed. These ends are accomplished via the use of multiple layers with varying optical properties.

Some of the more exciting applications of these optical approaches in architecture and product design come in the form of either multilayered thin-film coatings on a rigid substrate, such as common glass, or standalone thin sheets that can in turn be used independently or applied to the surface of a substrate. Among the most widely used are dichroic films and filters. Dichroic filters selectively pass certain wavelengths of light while reflecting others; alternatively, they selectively reflect certain colors and not others. In particular, the wavelengths passed or reflected vary with the angle of incidence of the light. Consequently, a variety of colors appear when the material is viewed from different angles, and an iridescent effect can occur.

Many of these same qualities are available in thin-sheet polymer products (see Figure 10.33). These thin sheets have a great many multilayered coatings that can each possess specific optical proper-

FIGURE 10.33

Many multilayered films have unique optical properties. Others serve different functions, including increasing electrical conductivity or providing photovoltaic capabilities. (Courtesy of Ben Schodek, Photographer, Boston.)

ties that collectively, in turn, determine the overall optical characteristics of the material, including the fascinating color variations seen when the material is viewed from different angles and/or the incident angle of the light is changed. Section 9.5 discusses these materials in detail.

Many interesting layered structures are being developed in connection with solar photovoltaic technologies. The use of nanomaterials has enhanced the efficiency of these kinds of solar collectors (see Section 9.7).

To use thin films for any of the applications we've discussed, the actual method of producing the coating is quite important. The type of coating, costs and related production volumes and the size of surface that can be coated are all important variables. Many of these processes are described in more detail in Chapter 8.

Nanopaints and Nanosealants

Nanopaints

Paints that serve protective or decorative purposes are ages old. Their need in our products and buildings remains as important today as in the past. There have been remarkable advances in paints in recent years due to the explosion of work in the field of polymer chemistry. The development of waterborne systems based on polymer dispersions literally revolutionized common interior and exterior paints. Demands for improved performance are still with us. It is no wonder that that there is great interest in nanostructured paints containing nanoparticles or nanotubes in suspension that are applied as liquids and then harden (see Figure 10.34). Today nanopaints come in many forms and have been developed with many different needs in mind. Common objectives include improved scratch and abrasion resistance, hardness, glossiness, and color steadfastness. Self-cleaning and antimicrobial actions are also common objectives.

There are many approaches to making nanopaints, depending on the development objective. Both organic and inorganic nanoparticles find use in relation to pigments. Dispersed inorganic or metallic nanoparticles add hardness and strength. Inorganic nanoparticles aid in elasticity. Well-dispersed inorganic nanoparticles can also impart highly sought scratch-resistant qualities to a paint. Nanosized ceramic particles or nanoclays can be used in various polymeric materials that harden into a cross-linked material when heated in an oven. Various silicates can be used in inorganic-

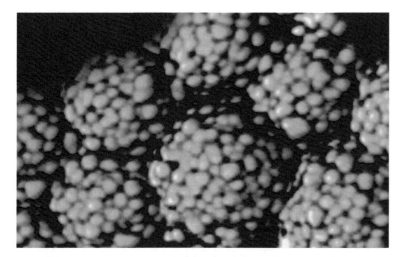

FIGURE 10.34

Nanocomposites have been used for exterior paints. In this example, inorganic nanoparticles have been homogeneously incorporated into organic polymer particles. This takes advantage of the hardness and "breathability" of inorganic binders and the elasticity and water-repellant qualities of inorganic binders. (Courtesy of BASF.)

organic compositions as well as crystalline fillers. Silica materials can help prevent corrosion when applied to metals. The addition of nanoclays can improve the rheological properties of paints (e.g., sagging on vertical surfaces during application or pigment settling). To a limited extent, the addition of nanoclays can help make paints more flame-spread retardant. (The surface spread of flame is a crucial issue in an architectural context.)

In some applications, attention is particularly paid to the interactions among a paint, the way it is applied, and the material to which it is applied. Many paints are electrostatically applied. The paints are electrostatically charged, which in turn can attract them to the body being painted. Hence, by using metallic nanoparticles or carbon nanotube additives in either the paints or the body material, the electrostatic painting process can be greatly enhanced. This approach is being investigated via the thermoplastic bumpers now used in many automobiles. The thermoplastics are not intrinsically conductive, so the addition of nanosized metallic particles helps the painting process.

Many paints are now designed to have self-cleaning actions, wherein dirt and other foreign substances are easily washed away naturally or made loose enough to be easily removed (see Figure 10.35). Many self-cleaning actions are ultimately based on the hydrophobic Lotus Effect® that was described in detail earlier in this chapter. Self-cleaning paints are described there as well. For some applications, a particularly interesting approach for paints uses nanoscale materials that have particular photocatalytic and hydrophilic

FIGURE 10.35

A self-cleaning paint based on the Lotus Effect® is used on the Strucksbarg housing project by Renner Hainke Wirth. (Courtesy of Christoph Gebler, Photographer, Hamburg.)

action. The photocatalysis process breaks down or oxidizes certain materials in the presence of UV light. Common organic substances are decomposed and loosened from the surface. Hydrophilic action causes the water hitting the surface via rain or simple washing to form into thin, flat sheets that in turn carry off the loosened organic particles. This photocatalytic action was described in greater detail earlier in this chapter. Of particular interest is that titanium oxide, a substance long used in paints, possesses marked photocatalytic properties when used as nanoparticles in the anatase phase. When these are used in very thin coatings, the surface can be transparent.

Nanomaterials are also being used in paints to impart antimicrobial or antibacterial properties to either kill or inhibit the development of bacteria that cause common discoloration or staining. Antimicrobial qualities are also needed in paints in many settings to prevent the spread of germs harmful to health. In common usage, antifungal or antimold qualities are also included within the broad term *antimicrobial*. The enormous need for coatings of this type is described earlier in this chapter, where several approaches to imparting these qualities to paint are also described in detail. As discussed there, one of the most common applications for paints again uses the photocatalysis process, which oxidizes common organic substances that form on a surface or are deposited on it. This action, in turn, helps prevent staining, mold, and other problems.

Many paints containing nanoparticles can have multiple properties. Hence many paints are claimed to be self-cleaning, antistaining, antibacterial, and antiscratching. Not all actions can be achieved with equal efficiency simultaneously, but multiple functions can indeed be achieved. Paints with multifunctional characteristics are being developed extensively in the automobile industry with the intent of making finishes glossier, harder, scratch resistant, and resistant to various chemical substances found in everything from rain to bird droppings (see Figure 10.36). For example, Mercedes-Benz has used a nanoparticle clear coat for several years. Small ceramic particles are contained in a polymer matrix that forms a cross-linked matrix when hardened under heat in an oven. The final paintwork on an automobile body commonly consists of five layers, with a total thickness of about 0.10 millimeters (100 micron). The first layer is a zinc phosphate coating to protect the sheet metal from corrosion and provide a basis for the subsequent cataphoretic dip priming layer. The whole automobile body is submerged in a tank. Next comes a filler layer that is electrostatically applied and that has only a minor amount of organic solvent. Electrostatic charges

FIGURE 10.36
Nissan uses a nano-based antiscratch coating on recent automobile bodies. (Courtesy of Nissan.)

attract it to the car body. Its role is to absorb impacts from small stones and the like. A paint layer comes next, with the appropriate color pigment. It is also electrostatically applied. The final layer is formed by the nanoparticle clear coat and is about 40 micrometers thick. It provides the final wearing surface of the paint and, consequently, is exposed to many destructive effects, ranging from abrasion to chemical elements in rain.

Nanosealants

Few products are as humble as sealants. Few products serve simpler but absolutely necessary functions. Few products can rival the number of applications in other products. Sealants are normally applied to the surfaces of other materials to prevent or minimize the excessive absorption, penetration, or passage of liquids or gases. Common sealants are initially fluid or viscous materials that are applied to a surface, bond to it, and then change into a solid via chemical reaction or the use of solvents. Normally they are applied to materials that have numerous small pores. They may be intended to keep liquids or gases either out of something or within something. In some cases sealants might be applied between adjacent materials (in the sense of being a gap filler or serving a joining function) or between surfaces of face-to-face materials. Sealants are normally designed to serve specific needs and are often used on a specified set of host materials.

Common sealants are used to prevent water penetration or absorption in everything from wood floors to fabrics. Many are designed for "breathability" so that water vapor but not water droplets can be let through. In these instances, the surface tension present in surfaces with small pores prevents liquid droplets from passing through. Yet others are designed to prevent gases escaping from containers. Other sealants are designed to prevent staining of materials. Some are intended to prevent the growth of bacteria within the surface pores of the host material. The list is virtually endless.

Each application needs particular properties in a sealant. Generally, the sealant in its solid state must be stable and not break down or dissolve when subjected to the liquid or gas against which it is intended to serve as a barrier. Thus a sealant might need to be resistant to simple water, as is the case in many architectural applications. It might need to be resistant to specific kinds of chemicals, as is the case in many sealants developed to prevent staining of the host material. Corrosion resistance or resistance to UV rays might be needed. The sealant needs to be able to bond securely to the surface of the host material. If the sealant is solvent based, the solvent must not be deleterious to the host material. Many sealants need to have some degree of abrasion resistance, toughness, or scratch resistance so that their sealing properties are not lost during use. Depending on the application, the sealant might need to change from a fluid to a hard solid film (as in common paints) or have an initial state that is viscous enough to be applied by squeezing and that remains highly flexible when hardened. The rheology of the sealant is thus very important. Sealants that serve joining or gap-filling functions often need to be flexible because relative movements often occur in adjacent materials. Many current sealants are made of silicones, polyurethanes, polysulfides, latexes, or vinyl plastisols.

Particle sizes and distributions have long been known to affect the rheology of sealants; hence the use of nanoparticles offers interesting opportunities to improve sealants. Most nanosealants now developed take the form of nanocoatings, nanofilms, or nanopaints, each of which is discussed more extensively elsewhere in this chapter. Many nanosealants take the form of polymer-matrix nanocomposites. Polymers with embedded nanoclay particles are common. Nanosized silicates are often used. Small amounts (around 2% by volume) of silicate nanoparticles to a polyimide resin can increase strengths considerably (around double). Resulting coating thickness can also be reduced in comparison with traditional coatings, since the solids content of nanoadhesives is lower and thinner. Thinner coatings can be produced that have good coating uniformity. As with other nanocoatings (see previous section), nanosealants can also be made optically transparent or translucent due to the small size of the particles relative to the 400–700 nm wavelengths of visible light.

Nanoadhesives

Adhesives are substances that join two surfaces together. The surfaces can be of dissimilar materials. Historically, many adhesives were made from natural materials, including plant resins, gums, and many other substances. Adhesives made of animal glues were widely used in ancient Egypt, for example, in making veneers. Egg whites became widely used in the Medieval art world. Ultimately, glues were made from a wide variety of both organic and inorganic substances (fish, rubber, caseins, and minerals). The introduction of synthetic adhesives based on elastomers, thermosetting and thermoplastic polymers, and other materials has led to many new and improved types. Their use has become amazingly widespread. Adhesives are used in everything from simple consumer products to automobiles. They allow various materials to be adhered or bonded together with comparative ease and speed, thus expediting manufacturing processes for products. Their effects on joined materials are normally benign. Adhesive layers are relatively thin. Adhesives and many sealants are closely related in a technical sense and are often lumped together in discussions and even by promulgating industries. New synthetic adhesives are really quite remarkable. This does not mean, however, that the quest for ever-better adhesives is in any way lessened. The potential of using nanomaterials to enhance existing adhesives or to produce new ones is a major applied research domain, particularly since the potential market for improved products is so huge. There have already been notable successes in this regard. Nanoadhesives that are bio-inspired—notably the "gecko" nanoadhesives—have become meaningful symbols for what nanotechnologies can accomplish.

Before looking at bio-inspired adhesives, we should look at how nanomaterials can yield improvements to existing kinds of adhesives. Many adhesives are targeted for particular surface types and others for more general applications. Mechanisms for adhesion are primarily either mechanical or chemical. Mechanical means involve the adhesive substance providing a type of interlocking mechanism, typically by the substance filling small pores in each of the surfaces. Chemical means are more varied and include direct chemical bonding and the development of intermolecular forces. Corresponding to these several mechanisms are various types of adhesives that have quite different application methods and use properties. Reactive adhesives, including common two-part epoxies, provide direct chemical bonding after being applied on thin layers to a surface and can be quite strong. Light-sensitive adhesives are cured by exposure to light of specific wavelengths (often ultraviolet). Pressure-sensitive adhesives form bonds from molecular interactions, such as Van der Waals' forces. Other types include contact adhesives and simple solvent-based adhesives.

As with sealants and paints, rheological properties (viscosity, prevention of sagging when applied on vertical sheets, and so on) are important. Nanomaterials help maintain good viscosity levels. Since the solids content and viscosity of nanoadhesives are low, thinner and more uniform coatings can be produced. General mechanical properties are improved, including basic strengths of set adhesives. Other properties, such as electrical conductivities, can be imparted or enhanced via the use of nanometallic inclusions. Many of these improvements are the same as for nanosealants and paints. Antimicrobial properties can also be imparted (see previous discussion).

Nanomaterial types used in adhesives vary considerably. Nanosized silicates are often used, since resulting viscosities can be lower than when traditional fumed silica-based materials are used. Adding iron oxide nanomaterials allows electromagnetic heating that can cause adhesive curing to occur very quickly. Carbon nanotubes and carbon nanofibers have been used in adhesives to radically increase their thermal conductivities. Resulting materials are often used in situations in which electrostatic discharges are important, including in electronic or automotive products.

Among the most interesting and well-publicized nanoadhesive approaches are those inspired by the study of nature, particularly geckos and mussels (see Figure 10.37). In Chapter 2 many remarkable behaviors in nature were shown to be dependent on nanosized structures, and the adhesive qualities of mussels and geckos were noted. The ability of a gecko to scamper up walls and even upside down across ceilings has inspired a number of new nano-based approaches to making adhesives. This clinging ability, however, does not come from any kind of natural glue. Glues would normally provide a fixed bond rather than the gripping/releasing cycle needed for the gecko to run about. Rather, it results from the presence of many small hairs that cover their feet. These hairs are much smaller than human hairs and are then divided into even smaller hairs at their ends (called *satae*). At the ends of these multiple tiny split hairs are spatulae that are cuplike and around 200 nm in size. These spatulae radically increase the contact area between the hairs and a surface. There is no real agreement about how many contact points are present, but estimates range from the high millions to the low billions. Adhesion results from intermolecular forces between the spatulae and the surface, including Van der Waals'

(a)

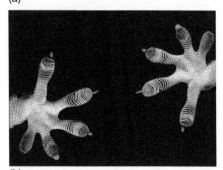

(b)

FIGURE 10.37

Geckos are famed for their ability to scamper up walls and across ceilings. Their clinging abilities are due to the millions of spatulae on the soles of their feet.

(a) (b) (c) (d)

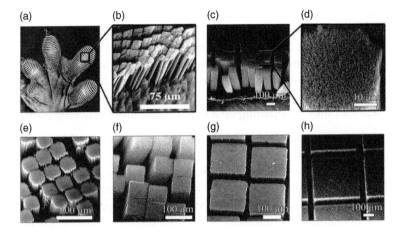

(e) (f) (g) (h)

FIGURE 10.38

Gecko spatulae and a synthetic nanopillared equivalent that can have remarkable adhesive qualities. (a) and (b) are images of the real Gecko and the remainder are synthetic equivalents. (Courtesy of the Dhinojwala Research Group, University of Akron.)

forces and some capillary action (see Chapter 5 that discusses Van der Waals' forces). These are kinds of forces that hold small solids and liquids together. They are relatively weak forces, but the high contact area of the spatula results in an appreciable overall sticking force. Mimicking this adhesive behavior, particularly the reversibility (sticking and unsticking) necessary for the gecko to run up walls, with synthetic materials has long been a goal of material scientists (see Figure 10.38). Several approaches have been explored. Most use some type of nanosurface that is patterned after those of a gecko's foot soles. These include nanopillar arrays made from techniques such as electron beam lithography on thin films or other techniques that use carbon nanotubes. Many of these techniques were quite successful. Several proved to initially mimic the general sticking power of the gecko's foot, but often the sticking power was found to decrease quickly with alternating grip/release cycles. Many were also noted to have decreased sticking power in the presence of high humidity and often did not work on wet surfaces.

During this period, several investigators were working on "wet adhesion" and looking again toward inspiration from nature. Here a primary model was the mussel, which can be seen in water environments sticking to seemingly everything, from rocks to pilings (see Section 2.2). As with the gecko, however, the adhesive property was reversible. Studies of mussels demonstrated that they secrete specialized proteins ("DOPA") that provide these unique adhesive qualities. In a well-known article in *Nature*, investigators reported that by coating gecko-mimicked nanopatterns with a mussel-mimicked polymer containing the needed protein, adhesive qualities could be greatly improved. The particular technique described used electron-beam lithography to create an array of holes in a thin film (PMMA) supported on a substrate. Nanopillars were about 600 nm high and 400 nm in diameter. Experiments in air and water demonstrated increased and repeatable sticking power. Adhesive qualities in wet environments were especially enhanced. Considerable work needs to be done with these bio-inspired adhesives, but it is clear that they are opening up a very positive area of investigation.

All these remarkable new adhesives have great potential in many applications, ranging from the commonplace to the exotic. Product manufacturers are always on the lookout for new adhesives to replace invariably troublesome mechanical connections. In the electronics industry, nanoadhesives have great appeal in systems that do automatic positioning and connecting of components; with adhesives this can be a completely clean manipulation.

Nanoporous Materials

Porous materials are commonplace and are found as constituent elements in a wide variety of products. A porous material has a network of voids (pores) that are distributed through the whole of the primary material. Generally the presence of voids results in materials with relatively low densities. Many porous materials have interconnected voids that allow the material to have a greater or lesser degree of permeability to fluids or gases. Polymers, ceramics, and even metals can come in porous forms. Applications include many kinds of membranes, filters, adsorbents, insulators, ion exchangers, catalyst supports, and others. With polymers alone, many kinds have been developed for various purposes, including thermal insulation material for buildings, common dust filters, and a host of everyday consumer products such as common cups. Porous ceramics also find use in both products and many industrial applications. They can also play a pivotal role in various kinds of environmental applications having to do with air and water environments, including remediation of water and air pollution (see Chapter 11).

Critical characteristics of porous materials include their pore size and shape, pore volume, surface area, and various pore surface properties (e.g., hydrophobic, hydrophilic, other). Porosity is commonly measured in terms of volume percentage. Figure 10.39 illustrates the general relationship between porosity and pore size for several kinds of porous materials and for various typical applications. Note that certain materials, such as activated carbons, have quite small pore sizes, as does xerogel (aerogels have larger pore

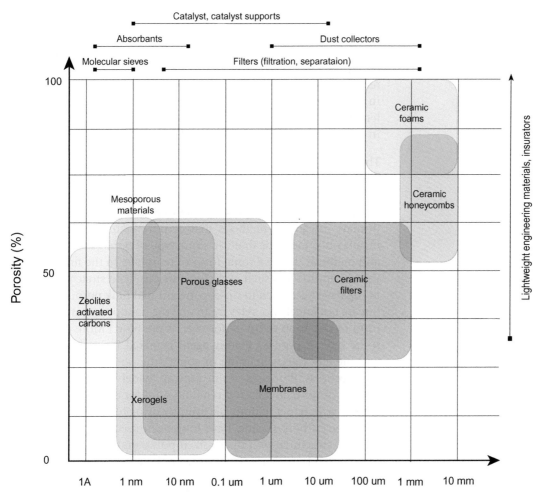

FIGURE 10.39

Relationship between pore size and porosity of typical porous materials. (Adapted from K. Okada and K. MacKenzie.)

sizes). The pore sizes of various porous glasses vary considerably depending on their preparation method. In conventional porous ceramics and polymers, pore sizes are generally 1 um or greater. Relationships among pore size, porosity, and the kind of application are obviously important as we think about the role and uses of nanoporous materials.

With nanoporous materials, as pore sizes (diameters) decrease, the relative pore surface area radically increases. It is this increase in pore surface area that is one of the primary driving forces behind developing nanoporous materials, since many applications are primarily dependent on the surface area. In conventional porous polymers, for example, the smallest pore size commonly used is around 1 um, whereas the intent of developing nanoporous polymers is to have pore sizes 100 nm or less; this in turn means big increases in relative surface pore areas. Actual pore diameters may be important in applications such as filtering. The amounts of solid material and lengths of travel along solid walls surrounding pores are important in thermal applications.

These various geometric and other attributes of porous materials can be literally designed into many materials via choices in the primary matrix materials and in the processing method(s) used to create the pore structure. Processing methods generally fall into three primary categories: methods that employ thermal or mechanical means only, methods that use some type of pore-generating agent, and methods that use some form of template. In the first category, various approaches such as melt extrusions, cold stretching, or sintering have been used to make conventional bulk porous materials but are not largely not relevant to making nanoporous materials. (A form of radiation etching, however, is possible.) More commonly, some type of pore-generating agent is mixed into a host material and literally causes foaming or some other kind of pore development process. Processes of this type are widely used with conventional porous materials and remain useful for making nanoporous materials. Polymer foaming techniques using blowing agents, often based on carbon dioxide, have long been used to produce closed cells for structures with micro-sized pores, and a similar approach can be used to produce nanoporous polymers.

Various solvents can be used to produce a variety of porous morphologies. The last category uses "templates," which can include some type of molecule or phase around which a polymer is formed. These templates are in turn leached out. Templates can also be derived from materials inherently containing nanopores. Terms such as *molecular imprinting* or *micellar imprinting* are used to describe some of these processes. Self-assembly processes are also possible. Several of these approaches are suitable for making nanoporous materials. The choice of which of these several approaches is ultimately used to create a nanoporous material depends on the porous morphology desired, which in turn depends on the application environment.

Porous materials are commonly used for thermal insulation purposes. In insulation materials with small pores, heat transfer occurs in several ways. Conductive heat transfer occurs in the intercon-

FIGURE 10.40
Polymer nanofoam. The nanofoam shown here is produced using prestructured "templates" on the nanoscale rather than using conventional expansion blowing agents. The nanopores reduce the material's ability to conduct heat, making it useful for many insulation purposes in automobiles, buildings, airplanes, and many appliances. (Courtesy of BASF.)

nected cell walls of solid materials but also through the gas molecules that occupy the pores. Radiative transport through the pores also occurs. Convective heat transfer is minimal. In polymer foams, significant contribution to the thermal conductivity results from intermolecular collisions between gas molecules and pore walls. Reducing the pore sizes to nano level increases the collisions between the gas molecules and the pore walls and tends to decrease the total thermal conductivity present. The combination of high porosity to reduce heat conduction through the solids, the long path through the solids, and the small pore size thus all contribute to decreasing the thermal conductivity of the material. Figure 10.40 illustrates a nanofoam. Materials such as aerogels also exhibit these characteristics, as do others (see Section 9.3).

Porous materials are already used in a host of applications in which they serve as *permeable membranes* for filtering or other functions. The purpose of many membranes is to separate permeating gas or liquid materials with various mixes of particle sizes and allow the passage of only certain sizes. This is frequently done by a sieving action. Here the pore sizes must be carefully matched to allow passage or trapping of the target particulates. Another separation effect occurs via an interaction effect with the porous membrane (commonly used for charged particles). Both approaches are often used simultaneously. Specific applications in air and water purification are discussed in Chapter 11.

Many nanoporous materials are being developed for use as *catalysts*, which are materials that increase the rate of a chemical reaction but remain unchanged by the reaction. Many types of basic nanoporous materials can be nanocoated with various kinds of catalytic nanoparticles. The specific morphologies and relatively high surface areas associated with the nanopores improves the catalytic performance of a material. Many membranes of nanoporous materials are being developed for catalytic purposes. They can be used in a wide variety of applications, including water and air purification. They can also be used in a number of *energy-related* applications. Approaches include developing membranes for improved fuel-cell development. Other energy-related applications include nanoporous electrodes that aid the efficiency of batteries and for reversible hydrogen storage in a solid-state medium (see Section 9.7).

Many nanoporous materials are being developed as breathable films for product packaging that have useful chemical and physical properties. Nanopores can impede the passage of bacteria or other microorganisms. Nanoparticles with antimicrobial properties can potentially be integrated into a porous matrix, whereas other material characteristics can provide light shielding and other product packaging needs.

As already discussed, there are many varieties of nanocomposite materials. For some applications, nanocomposites intended to serve some specific function can also be given a porous nanostructure, or a nanoporous polymer can sometimes be infused into other materials.

Nanotextiles

Textiles in design

The use of textiles is pervasive in both product design and architectural applications. The word *textile* is a general term referring to a material made of various kinds of yarns, threads, or other pliable fibers that are interwoven, knitted, or otherwise interlaced or to materials made of fibers that have been bonded together (e.g., felt) by adhesion, cohesion, or friction via some type of mechanical, heat, or chemical process. *Nonwoven textiles* refer specifically to the materials made of bonded fibers that have not been woven or knitted. Usually, nonwoven materials (unless used with backings) are less strong than ones that have been woven. The terms *fabric* or *cloth* are used to describe types of thin, flexible textiles especially suited for apparel. The characteristics of these materials that are important to designers depend both on the type of fibers used and their macro-level organization. Various nanocoatings are already being used to impart many new desirable properties to textiles, and nanocomposite materials are being widely explored as a way of improving the overall properties of various kinds of fibers and hence the overall product itself.

In many product applications, desirable characteristics of textiles range widely according to the intended application. Some level of strength, durability, and tear resistance is important in virtually all applications, albeit specific requirements may vary radically. Needs in apparel, for example, are obviously different from those in a mountaineering tent but are nonetheless still present. In specific applications such as apparel, traditional needs also include flexibility, thinness, softness, breathability, optical properties, stain resistance, heat resistance and flame retardancy, antistatic action, wrinkle resistance, ability to receive different surface treatments (including printing), and many factors related to material workability and ability to incorporate other nontextile inclusions. Recent material advances have made possible new desirable properties. Antibacterial fabrics are a good example here, as are self-cleaning fabrics. Fabrics that slowly release different substances have a wide variety of applications. On the fashion front, interesting experiments have been made with antibacterial fabrics (see Figure 10.41); albeit we should keep in mind that some microflora are needed for human health. Materials that emit light always catch attention (e.g., materials currently used have embedded fiber optics that connect to a light source, electroluminescents, or LEDs). Trends in apparel design now often demand textiles that have many multifunctional characteristics that include combinations of the preceding but can also include sensing or other information acquisition functions, communications, and logic controls. Many of these latter functions are now commonly achieved by embedding electronic elements, such as flexible thin devices, into more basic fabrics. Conductive threads also find wide use. Some of these same basic technologies can find more pragmatic uses as sensory technologies.

In other product applications, strength, toughness, and tear or abrasion resistance are often the dominant design factors. In outdoor sports gear (e.g., backpacks, tents), lightness, breathability, and water penetration resistance can be also be of fundamental importance. Lightness, toughness, and puncture resistance are important in many other situations. In extreme environments, resistance to deleterious effects of the surrounding environment can be extremely important (resistance to degradation). The same general trends as noted for apparels hold true here as well, especially the need for

(a)

(b)

FIGURE 10.41

Nanotextile technologies in fashion. A dress and jacket containing silver (Ag) and palladium (Pd) nanoparticles with antibacterial qualities, designed by Cornell Fashion Design student Olivia Ong with Juan Hinestrosa. (Courtesy of J. Hinestrosa, Textiles Nanotechnology Laboratory, Cornell University.)

multifunctional materials. Many tents, for example, are now embedding various types of flexible photovoltaic materials directly into the basic tent material, to serve an energy supply function. Clearly, the list of needed characteristics is extensive and highly dependent on product type. Interesting applications based on nanomaterials include improved textiles for use as filters and as devices for administering medicines.

In architecture, textiles are widely used in a number of applications. In primary structures, many pneumatic, membrane, and net structures make extensive use of textiles for surface shapes. Pneumatic structures, for example, are widely used in sports applications to cover large arenas. The interior is pressurized and the skin simply inflates to the desired shape. Membrane structures consist of doubly curved surfaces that are stretched via externally applied jacking forces. Net structures use cables to form the shape but have a skin stretched over them. These flexible surface-forming structures carry loads primarily through the development of tension stresses. Surface shapes are invariably pretensioned via either internal inflation pressures in pneumatic structures or external jacking forces in membrane structures, necessary to maintain their shape and stability under fluctuating wind and other forces that act on the surface. Both the pretensioning stresses and the stresses that result from the applied loadings can be quite high. Hence, when textiles are used directly as the skin, they must be able to directly carry significant stresses or have special higher-strength strands embedded in them. Stresses are invariably biaxial and vary considerably throughout a skin.

In architectural structures that use traditional woven textiles, considerable attention is paid to the various relative strengths of the material in the warp and weft direction and in orienting the material so that these strengths best correspond to actual stress levels found in the materials (which can be found by various structural analysis techniques). Additionally, the materials used must be resistant to tearing and abrasion. These same surfaces must typically also be waterproof and not degrade over time due to exposure to UV sun rays. For occupant safety, surfaces must have low flame-spread characteristics and not give off toxic gases during fire situations. Resistance to discoloration is highly desirable. For enclosed structures, some level of thermal insulation capability is always desirable. As with textiles used in products, current design trends pay special attention to multifunctional materials. The idea of using flexible photovoltaic elements in the skin is an obvious one for structures of this type, as a way of generating energy. Skins incorporating phase-change materials are being explored to help provide responsive thermal performances. There is great interest in self-cleaning materials. Increasing attention is also being paid to the material's ability to be recycled.

In addition to textiles being used as primary structures, there are a whole host of less structurally intensive uses in buildings. There are innumerable types of shading devices; most have to do with sun control, but there are others that deal more generally with light control. Other uses include any number of types of privacy devices (for example, recall the wonderful tatami screens in indigenous Japanese architecture). Other textiles find use in ever-present floor, wall, or furniture coverings. Clearly each application demands various kinds of textile characteristics. Sometimes translucency is desired, for example; sometimes not. Levels of required strength and flexibility vary dramatically. All textiles (and other materials as well) that are used in buildings, however, must absolutely meet certain requirements related to the behavior of the material in a fire circumstance. One important measure is the flame-spread rating of the material (how quickly a flame will spread across a material's surface). The type and amount of gases that can be given off during a fire are similarly important. High levels of toxic gases must obviously be prevented. Though not always a requirement, increased attention is also now being paid to the type of normal outgassing that materials exhibit over time in enclosed spaces—a phenomenon that has been demonstrated to contribute to the so-called "sick building syndrome" and deleterious effects on occupants. A list of these and other design parameters is shown in Figure 10.42.

Textiles are also widely used in landscape architecture. The most well known of these applications are *geotextiles*. These materials are permeable fabrics used in roads, airfields, embankments, retaining structures, retention ponds, dams, and other site settings. Though many serve primarily soil reinforcement and stabilization functions, they can also serve as filters or separators. Materials must be strongly resistant to various chemical environments normally found in the ground or associated with heavily polluted sites. Geotextiles are normally characterized by various macro-level geometric shapes that are suitable for different purposes. *Geomembranes* are closely related but are made of nonwoven materials and are typically used for applications such as liners.

Nanomaterials and nanotechnologies in textiles

As we've shown, needed characteristics for textiles vary widely according to application. Actual methods of making textiles utilizing some type of nanotechnology remain of fundamental impor-

Strength
    Warp/weft directions
Stiffness and creep characteristics
Wear and abrasion characteristics
Toughness
Combustibility, flame retardancy,
    toxic gases

Tactility
Surface appearances
Ability to take printing and other
    surface treatments

Breathability
Water-and oil repellancy
Self-cleaning
Antimicrobial

Outgassing (indoor air pollution)
Nonallergenic to skin

Biohazard safety
Biosensors

Wearable electronics
Photovoltaic

Other

FIGURE 10.42

A list of many factors that go into thinking about how to design a textile for a specific purpose.

tance. Textiles are necessarily of large size and, in typical product design or architectural applications, must be of relatively low cost. Many important coatings can be applied via forms of spraying or electrostatically in a fairly cost-efficient way. Self-assembled nanolayer coatings are of great interest but largely remain in the development stage. Nanocomposite fibers can be selectively spun into traditional woven or knitted materials, with densities and directionalities designed in accordance with strength and other mechanical property needs. The use of melt-blown techniques promises to be an excellent way of incorporating nanofibers into certain kinds of limited-size products.

Nano-based approaches can be most effectively used to impart or enhance different needed characteristics. In this discussion, it is useful to distinguish broadly between basic *core* textile materials and the many types of *coatings* that can be applied to them.

Most strength applications are associated with *core* designs, as are many thermal and filtering applications. For core materials in textiles, composite fibers can be used that incorporate various kinds of nanofillers, nanofibers, or carbon nanotubes into some other matrix, typically a polymer, to increase strength, toughness, abrasion resistance, conductivity, and other fiber properties. Nanofillers can be made of clay, various metal oxides, carbon black, and other nanoparticles. Various nanofibers or nanocellular foams can also be used. Clay nanoparticle fillers can be used for improved heat, electrical, water, and chemical resistance (e.g., anticorrosive) and to blocking UV rays. Distortions from heat are decreased. In textile applications, the clay nanoparticles can also create sites for attracting or holding dyes.

Properties in relation to light or color or properties such as antistaining are typically associated with coatings of the several types previously discussed. Many types of coatings can be used to impart needed properties to textiles. Nanoparticles can be used to surround and bond to fibers in the form of nanolayers, thus either providing some special action (e.g., hydrophobic) or simply preventing liquids or gases from penetrating through to the basic core fibers. Coating application techniques, however, must be compatible with approaches used to make the basic fibers themselves, such as electrospinning, which can be greatly complicating.

A whole host of technologies are evolving for improving the finishes of various textiles. The appearance of final finishes is extremely important, as are capabilities for holding dyes or accepting printing. Nanotechniques are proving to provide interesting avenues of approach for improvements here, especially in the area of chemical finishing. One frequent intent is to improve the general quality of surface coverage. Another is to bring various nanoparticles with specific desired properties to different surface locations in a controllable way. Nanoparticles such as ceramics or metal oxides applied via spraying or electrostatic methods are widely used for many applications. Several approaches based on more conventional methods have been developed for emulsification processes to improve surface qualities. These emulsifications can incorporate different kinds of nanomaterials (e.g., nanosized capsules and others) for purposes such as increasing stain resistance or antistatic action. As noted previously, nanocrystalline piezoceramic particles can be applied to textiles so that the mechanical forces associated with the bending or textiles can be converted into electrical signals for various sensing or monitoring roles.

Strength

Strength, durability, tear resistance, and other mechanical properties were seen to be common in all applications, although required levels vary enormously according to application. These properties depend primarily on both the strength of individual fibers and the exact interlacing pattern and/or bonding approach that is used in the core textile material. Mechanical properties of woven or knitted textiles can be improved via the use of nanocomposites by strengthening individual fibers, increasing their density, and creating strong interlacing patterns that respond to the directions of the stress states that are present.

Several types of nanocomposite approaches are suitable for fibers that use nanofillers of one type or another that increase strengths. Given the costs of nanocomposites, these materials are normally selectively used and interspersed with more conventional fibers. Tensile strengths can be significantly increased with even relatively small-volume fractions of nanofillers in the overall matrix. Carbon-black nanoparticles used as fillers are well known for increasing abrasion resistance and toughness of a composite. Carbon nanofibers can increase tensile strengths. Carbon nanotubes can be used in fibers to increase strength, toughness, electrical conductivity, and thermal conductivity as well. Fibers utilizing carbon nanotubes are normally quite small in diameter but can exhibit great strength and stiffnesses. Spun fibers can be much stronger than Kevlar and other well-known materials. Small sizes and high costs normally suggest uses in highly specialized and smaller textile applications rather than in large architectural applications. Strengths of non-

FIGURE 10.43

Nanoparticles less than 100 nm thick on the surface of these textile fibers give them a hydrophobic self-cleaning action. (Courtesy of BASF.)

woven textiles can be enhanced via the use of nanoadhesives of the type previously discussed or by enhancing cohesion induced during bonding processes.

Self-cleaning, antimicrobial, and other surface characteristics

Important properties such as stain resistance or antibacterial characteristics are normally imparted via coatings, sealants, or films of one type or another. Figure 10.43 illustrates a hydrophobic textile. These nanocoatings can normally be applied to the entire textile surface rather to only particular fibers, thus making them quite attractive in many applications. Particularly attractive is the possibility of "self-cleaning" textiles. A common approach for creating self-cleaning, antibacterial, and antistaining coatings, for example, uses nanomaterials such as titanium oxide that have photocatalytic properties that absorb UV energy, which in turn causes organic dirt or other foreign particles on the surface to loosen or break down via oxidation. An associated hydrophilic action causes water striking the surface to develop into thin, flat sheets that in turn aid in washing away the particles. This same kind of action also helps prevent staining of textiles (often caused by organic particles) and imparts antibacterial qualities as well. Several types are promising for antimicrobial or self-sterilizing functions. Methods for achieving self-cleaning, antistaining, antibacterial, and other actions through the use of photocatalytic nanoparticles were discussed more extensively in Section 10.2. Examples of architectural membrane structures using photocatalytic surfaces are shown in Figures 10.44 and 10.45. Nanofibers with silver nanoparticles for antibacterial applications are shown in Figure 10.46.

Fire retardancy

Many textiles used in fabrics for seats and other furniture, carpets, and other large surface areas in buildings, automobiles, and aircraft have long been known to be potentially problematic in fire conditions because of their flame-spread characteristics and toxic gases that might be given off. These characteristics are now highly regulated. It remains fair to say, however, that any advances that can be made in fire retardants would be highly welcomed by this segment of the industry. Many new nanocomposites are being explored, particularly by the automobile industry for vehicle interiors, that have improved fire-retardancy properties. Carbon nanotube composites are also proving interesting from the point of view of improved flame retardance.

Optical qualities

The reflective optical properties of a textile depend on surface characteristics. Hence, the surface properties of the fibers themselves or of any coatings that might be used are extremely important. Their overall reflective properties, however, are also strongly dependent on the macro-level patterning of interlacing or texture of a bonded textile. The specific texture of a surface critically affects whether reflected rays of light are dispersed or not dispersed. Rough textures naturally lead to dispersal and mattelike finishes. Concentrated "highlights" normally result from very smooth surfaces. Thus the material reflectivities and their macro-level patterns both affect reflectance properties. In previous sections, the significant potential of nano-based films and coatings was discussed as a way of manipulating light and color properties. Many of these same approaches can be directly applied to textiles as well (see Section 9.5).

Photovoltaic textiles

Energy-generating textiles normally rely on flexible photovoltaic elements embedded or incorporated in textiles that can in turn be enhanced via the use of nanotechnologies (see Section 9.7). Photovoltaic approaches naturally benefit from extended surfaces to capture solar energy and are best thought of as exterior applications on the exposed sides of textiles. The lack of efficiency of current photovoltaic films currently makes their use of restricted value due to the low amount of power generated by fairly large surface areas, but the potential of nanotechnologies could make a big difference here. Sensing and communication elements can also be more easily

(a)

(b)

FIGURE 10.44

The awning uses a self-cleaning fabric more than 80% covered with a nanoporous layer to provide a hydrophobic self-cleaning action. (Courtesy of BASF.)

FIGURE 10.45

This playground in Japan uses a fabric tensile membrane that is has a photocatalytic coating to aid in self-cleaning and stain resistance (Taiyo Kogo Corporation.)

FIGURE 10.46

SEM image of nanofibers with silver (Ag) nanoparticles for antibacterial gowns and active surfaces such as carpets and upholstery. (Courtesy of Juan Hinestrosa.)

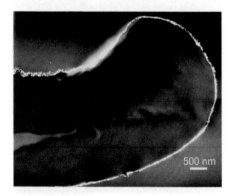

FIGURE 10.47

Widescreen cotton coated with platinum nanoparticles for air purification applications; also see Chapter 11. (Courtesy of Juan Hinestrosa.)

embedded into textiles via nanotechnology-based components and elements. Conductive nanofibers are available for energy distribution and connection purposes. Various kinds of approaches based on nanocrystalline piezoelectric technologies can be used to provide sensor or monitoring functions by converting textile bending into electric signals. The ability afforded by other nanotechnologies to miniaturize various kinds of sensors and communication elements while reducing their power demands makes them quite amenable to inclusion in textiles.

Thermal properties

Thermal properties of various textiles can be enhanced by several different techniques. Nanoporous materials such as aerogels are being designed to serve as internal layers within multilayered fabrics. These materials have remarkable insulating qualities. Various forms of encapsulation of phase-changing materials using nanoparticles have also been used in fabrics for some time, and commercially available product lines (coats, gloves) already incorporate phase-change materials. These products make effective use of the release and absorption of latent heat to modulate heat flow. In architecture, various kinds of coatings that help adjust the emissivity and reflectivity of surfaces can help here as well.

Biofilters and absorbers

As we've mentioned, the properties of porous textiles have always been attractive in filtering operations of one type or another. Interesting new developments include nano-based textiles that can potentially serve as biofilters for a range of bacteria and viruses. One approach explored at Cornell University involves magnetizing nanoparticles and controlling them within a magnetic field during textile fabrication processes. Various kinds of nanolayers can be deposited on fiber surfaces to achieve different functionalities. The chemicals used depend on the biological agent of concern. Similar approaches have been explored for filtering airborne pollution particles or for use in creating biofilters that work on absorption principles as well. The large relative surface areas associated with nanocoated fibers are useful for both filtering and absorption effects. A textile developed for use in air purification is shown in Figure 10.47.

Bioclothing

The approaches we've described are very attractive, not only for filters and absorbers but for a form of bioclothing as well. Here protective clothing is envisioned that can, for example, protect the wearer from hazardous biogases while providing the breathability needed for human comfort. Figure 10.48 shows a nanolayered textile for chemical warfare applications.

The same general approach can also form the basis for improved methods of medicine delivery. This is already done to a certain extent with many existing fabric bandages, but techniques could be made more efficacious. Delivery systems could also be imparted to clothing worn in everyday circumstances.

Sensing

We have already noted the several attempts to use conductive strands and electronic means to make textiles serve as sensor devices. The same capability just noted—of imparting targeted chemicals on nanolayers around fibers that detect specified bioagents—has also been explored to serve as biosensors that yield information or warning alerts.

Nanocosmetics

The use of materials that alter the appearance of or protect human skin has roots that date back into antiquity. Appearance changes served many symbolic or ritual purposes and were certainly a reason for the paints, dyes, tattooing, and scarification (incisions) used by indigenous societies throughout the world. Other societies used body paints or dyes as a means of psychological intimidation in warfare; recall the ancient Britons, who painted their bodies blue before battles. The striking cosmetics used by the ruling classes in ancient Egypt—painted nails, lined eyes and eyebrows, rouged cheeks—undoubtedly served to distinguish nobility from others as well as in relation to many specific rituals (see Figure 10.49). These same ancient Egyptians also used various ointments—oils and creams—to help prevent skin problems caused by exposures to harsh, dry winds. The theme of beautification also runs through the development of cosmetics, including the whitening of faces, starting around the 1400s, to achieve pale appearances—itself a way of signifying class distinctions, that is, emphasizing differences between field workers and others with rough, tanned skins and the upper classes. Achieving pale appearances at the time often required the use of many materials, especially lead oxides, that could eventually cause sickness and even paralysis with prolonged use. Zinc oxides replaced many of these dangerous substances in the 1800s. Today the cosmetics industry is huge and touches nearly everyone. Two of the themes we've mentioned—enhanced appearances or adorn-

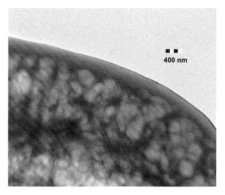

FIGURE 10.48
Polyelectrolyte nanolayers used in protective clothing for controlling the transport of chemical warfare gases and toxic chemicals. (Courtesy of Juan Hinestrosa.)

FIGURE 10.49
Cosmetics were widely used in ancient cultures.

ment and skin care or protection—provide the primary drivers for today's industry.

The term *makeup* is often used to describe products associated with adornment. These products cover or hide wrinkles, blemishes, and other flaws. They add color or smooth skins. They include a vast array of products, from lipsticks to face powders. Makeup can be made from organic or inorganic materials. Those that have a coloring function normally employ a variety of pigments that reflect light in certain ways, to generate many colors and color gradations. Zinc, titanium, and iron oxides embedded in a matrix material are in common use as pigments and come in many particle-size variations, including ultrafine. Various treatments are possible so that final surfaces can be hydrophilic, hydrophobic, or lipophobic (allowing oil to be dispersed). Considerable attention is paid to making applications compatible, stable, and long wearing. Multilayers are common and often used for special color or optical effects, such as iridescence (see Section 9.5).

Skin-care or protection products range from sunscreens to moisturizing agents. They either interact directly with the skin or provide some type of active ingredient that provides a function, such as anti-acne medication. Common objectives include providing safety against harmful effects of UV wavelengths from the sun, hydrating the skin, delivering various antibacterial and anti-fungal agents (e.g., foot powders), and delivering specific medications. Obviously, materials vary widely according to objectives. In the case of sunscreens, objectives are primarily to prevent sunburn, but they also serve to mitigate solar-induced keratoses. Two types exist: chemical blockers that absorb problematic UV radiation and physical blockers that reflect or scatter UV rays. Materials are obviously quite different. Titanium dioxide and zinc oxides are common physical blockers that reflect or scatter a wide range of UV wavelengths. Materials such as benzophenones or cinnamates work by absorbing UV radiation at specific wavelengths.

Both makeup and skin-care products contain active ingredients, such as titanium dioxide, that need to be delivered in a form and intensity to the right skin locations. Usually the active ingredient is contained within a matrix material commonly called the *vehicle* or *carrier*. This material can assume many forms, including emulsions, creams, or aerosols. They may in themselves serve a primary role, such as coloration, or be a carrier for a more specific active ingredient. In either case, getting the right matrix material design is extremely complex. *Emulsions* consist of stable mixtures of small droplets of one liquid dispersed and suspended in a second liquid, with the two liquids not normally being mixable. Common cold creams are often emulsions. Many emulsions are based on lipids or lipophilic compounds (materials that do not dissolve in water but can dissolve in alcohol) and water, or on lipophilic compounds that have an affinity for oil or can combine with it and hydrophilic compounds that combine with water. Either "oil in water" or "water in oil" types are possible. *Aerosols* are dispersions of small solid or liquid particles suspended in a gas. Many perfumes are aerosols, as are some hair products and even colorization mists. (There are also many pharmaceutical aerosols in use, such as local anesthetics or antifugal or anti-inflammatory aerosols). Aerosols can contain quite tiny particles at the nanoscale. Once applied, carriers can undergo many changes due to exposures to other substances, mechanical factors (wear, stress), evaporation of some volatiles, and other factors.

The current efficacy of these products stems from a considerable body of research into both the physical and biological characteristics of skin structures and into processes such as skin aging as well. Skin is known to serve barrier, thermal control, and a host of other functions for maintaining life, such as the creation of vitamins. Skin is composed of several primary layers: the epidermis, the dermis, and a subcutaneous layer. The epidermis is composed of five strata. The outer layer, the stratum corneum, provides a protective barrier of dead cells that are tough and not soluble and that are continuously replaced by new cells growing from below. It does not contain blood vessels. The thicker dermis holds nerves, blood vessels, and sweat glands as well as providing other functions. The subcutaneous layer is essentially fatty. This complex layering has many physical and chemical properties that change and evolve with time (also see Figure 11.2).

Skin necessarily exhibits interesting mechanical properties in its barrier- and shape-maintaining functions that derive from an elastic collagen (long fibrous proteins) network. With time these properties can change, usually leading to relaxations in stiffness and tensions that in turn are manifest as "aging wrinkles" or flaccidness. With respect to light and optical phenomena, many kinds of absorption properties exist at different layers that can again vary with time. Chemical actions are equally remarkable. Needed vitamin D, for example, is produced photochemically in the skin, largely in the stratum basale and stratum spinosum layers of the epidermis.

Many critical issues in both designing and assessing the positive and negative effects of various cosmetic applications hinge around understanding how far some type of applied-on-contact active ingredient penetrates the skin and how it interacts with the penetrated layer. Penetrations can be positive, as in some skin-based drug-delivery systems, or potentially negative if harmful side effects occur. We know from experience, for example, that many particles found in nature can cause allergic reactions in humans, resulting from some type of penetration through the skin (by absorption or other paths). How and whether penetration occurs depend on many factors. Small particles have a greater likelihood for penetrating the skin than do larger ones, but there are many other factors relating to both the body and to the surrounding environment (e.g., temperature) that effect whether penetration occurs.

As we've seen, the field of cosmetics can be exceedingly complex. It is also a huge and secretive industry. Significant amounts of research are ongoing with respect to the many factors we've, including the way various formulations interact with complex skin systems. The large surface-to-volume ratio of nanomaterials makes them particularly attractive for use in as either vehicles or active ingredients, as does the catalysis characteristics of many.

Perhaps one of the most common examples of a widespread use of nanomaterials in cosmetics is that of the ever-increasing use of nanosized titanium dioxide or zinc oxide particles in sunscreens. A nanopowder form of zinc oxide is shown in Figure 10.50. As noted, normal titanium dioxide and zinc oxides have long been used as blocking agents in many sunscreens. Applications have suffered, however, from nonuniformity and a visible whitish discoloration. Titanium dioxide nanoparticles, however, are essentially transparent, since their sizes are near the wavelengths of light. Sunscreens based on these nanoparticles are much clearer, but at the same time they retain their protective power. Concerns about health impacts, however, have been raised (see the following discussion and Chapter 11).

A whole range of nanomaterial types and forms is being explored for use as vehicles or carriers for active ingredients. Use of nanomaterials can allow carrying various types of active ingredients without degradation as well as offering better control of release rates. Small sizes allow them to more easily be incorporated into outer skin layers. Nanocapsules with polymeric membranes are in wide use. Active ingredients are in cores. Release rates can be controlled by

FIGURE 10.50

Nanopowder zinc oxide can offer protection against sunburn by filtering against harmful ultraviolt (UV) radiation. The fine particles act as inorganic UV filters by reflecting the incident UV light (like tiny mirrors). Conventional zinc oxide particles are white and produce a whitening effect on the skin. When we reduce the size of the pigment particles to around 200 nm—near the wavelength of light—they become transparent to the eye. (Courtesy of BASF.)

the membrane design. Nano-emulsions have been developed as well that are sprayed directly onto skins and can have better efficacy than conventional emulsions. Various kinds of nanospheres within host matrix systems are in use.

Various kinds of lipid forms are also important as vehicles. Lipids are organic compounds that are not soluble in water but that dissolve in substances such as alcohol (they are typically "fatty" or "oily"). Solid forms of lipid nanoparticles have been developed to carry active ingredients. More complex forms include nano-sized liposomes. Liposomes are synthetic vesicles with aqueous cores surrounded by one or more phospholipid membranes. They have a hydrophobic cavity and a polar cavity and can encapsulate different active substances. Oil-soluble substances can be carried in their hydrophobic cavities and water soluble ones in the other cavity. These small lipid particles can be absorbed and penetrate skin easily. Simpler nanosomes can carry lipophilic materials. They have single layers with polar heads directed outward and tails inward.

Many issues are of basic importance in using nanomaterials either directly or as carriers. As noted, types of cosmetics vary widely in terms of their functions, modes of application, and target locations for active ingredients. In many situations, functions can be achieved by relatively straightforward surface treatments, as is the case in some makeups and antiwrinkle formulations. In other cases, targets are much deeper and can even have pharmaceutical intents. Others inherently involve interactions with other environmental factors (UV light in relation to sunscreens). Whether time delays are important or not varies.

Always of paramount concern are the types of active ingredients used, their intensity levels, and whether they can inadvertently cause adverse safety or health issues. Kinds of ingredient/body interactions that can occur are a subject of intense study. Products should be tested for many factors. Receptivity to skin irritation varies among groups within the population as a function of many human factors, such as age, that are in turn scientifically understood in terms of keratinization levels and other skin characteristics. Allergic reactions are possible, and conventional cosmetics have long been studied and addressed from this perspective (witness the many products with "hypoallergenic" labels). Various testing means, including the older patch test, are still in common use to test for these reactions. Other formulations that depend on interactions with environmental factors demand different kinds of tests for specific problems, such as reactions induced by interactions with ultraviolet light. The potential for microbial growth in applied cosmetics may seem initially surprising, but many are water based, some quick thick, and applied to warm skin. Depending on the cosmetic, various additives might be needed to prevent the growth of microbes. Others can have inherent antibacterial actions. The potential for other kinds of adverse health effects associated with the deeper penetration of skin layers by tiny nanoparticles is also a concern to both industry and advocacy groups. This general area of concern is by no means unstudied, and some controversies surround it. A relatively few limited-scope studies with varying methodologies have yielded conflicting conclusions for different groups. Caution in these matters is always a watchword. We deal with this issue in more depth in Chapter 11.

In summary, the field of cosmetics is an extremely complex and sophisticated one. The field is not only about adornment that is a basic human need and action that must be respected; it also encompasses formulations that can have powerful protective and rehabilitation functions as well. Opportunities for using nanomaterials as active agents and carriers abound and can potentially lead to improved efficacies. Since these products come into bodily contact, special care must be exercised to assure that no adverse effects occur.

FURTHER READING

Michelle Addington and Daniel Schodek, Smart materials and technologies for the architecture and design professions, Architectural Press, 2005.

P. Brown and K. Stevens (eds), Nanofibers and nanotechnology in textiles, Woohead Publishing, 2007.

Carolyn Dry, Passive smart materials for sensing and actuation, Journal of Intelligent Material Systems and Structures, Vol. 4, No. 3, 420–425, 1993.

A.S. Edelstein and R. C. Cammarata, Nanomaterials: Synthesis, properties and applications, Institute of Physics Publishing, 2002.

Gogotsi, Yury, Editor, Nanomaterials handbook, Taylor and Francis, 2005.

L. Ge, S. Sethi, L. Ci, P. M. Ajayan, and A. Dhinojwala, Carbon nanotube-based synthetic gecko tapes, PNAS 2007, 104, 10792–10795.

Haeshin Lee, Bruce P. Lee, and Phillip B. Messersmith, reversible wet/dry adhesive inspired by mussels and geckos, Nature, Vol. 448, 338–341, July 19, 2007.

Italcementi, TX Millennium, Photocatalytic binders, Technical Report, March, 2005.

Hideki Hosoda, Smart coatings—Multilayered and multifunctional in-Situ ultrahigh-temperature coatings, in H. hosono, et al., Nanomaterials: From research to applications, Elsevier, 2007.

Kiyoshi Okada and Kenneth MacKenzie, Nanoporous materials from mineral and organic templates, in H. Hosono, et al., Nanomaterials: From research to applications, Elsevier, 2007.

Clare Ulrich, Nano-textiles are engineering a safer world, Human Ecology, Vol. 34, No. 2, Cornell University, November 2006.

Nancy Sottos, Scott White, and Ian Bond, Introduction: Self-healing polymers and composites, J. R. Society Interface, 2007 4, pp. 347–348.

Richard Trask and Ian Bond, Enabling self-healing capabilities—A small step to bio-mimetic materials, Final Report, ESTEC, 2005.

Steven Schnitter and Moitreyee Sinha, The materials science of cosmetics, MRS Bulletin, Vol. 32, October 2007, ppf. 760.

Mel Swartz, Encyclopedia of smart materials, John Wiley and Sons, 2008.

CHAPTER 11

Nanomaterials and Nanotechnologies in Health and the Environment

11.1 THE CONTEXT

This chapter initially takes a look a broad look at health and health delivery issues. Individuals and companies in the medical industry are pouring enormous amounts of research effort into understanding how nanomaterials and nanotechnologies might aid in the promotion of human health. The properties of nanomaterials make them attractive for many chemical and biological applications that can assist human health. Nanotechnologies offer an amazing potential in terms of implants with controllable or responsive capabilities. Drug delivery methods appear to be a big opportunity area, since nanomaterial characteristics can be designed for targeted transport and delivery. Traditional areas, such as prosthetics, stand to benefit in a whole series of ways.

Large amounts of evidence confirm the positive value of using nanomaterials and nanotechnologies in health delivery, but there are also reasonable concerns about how these same materials might have detrimental health effects if not used carefully. Adverse health effects may potentially occur from air- or waterborne nanoparticles from natural sources or from conventional materials having nanoscale constituents that have been with us for a long time. Looking carefully at potential impacts of the growing use of even smaller nanomaterials in products all around us is both sensible and necessary. There are indeed health concerns about nanoparticles making their way into humans as a consequence of humans' direct contact with products containing nanoparticles or via surroundings.

Similar observations can be made about nanomaterials in relation to larger environmental issues. There is striking evidence that many nanomaterial and nanotechnology applications can have quite positive environmental benefits, either indirectly, through better use of material and energy sources, or directly via improvements in specific remediation technologies, such as better air or water purification systems or better methods for desalinization. Nonetheless, there are possibilities that nanomaterials could also work their way into our ecosystems by means of accidental discharges or other means. Potential impacts must be understood. These important issues are reviewed in the following sections.

Both the health and environmental fields represent huge areas of research and investigation. It is interesting to note that the scientific community of researchers in the engineering and product fields who deal with nanomaterials and nanotechnologies are not commonly the same as those who deal with biological or medical or drug delivery. Though they share common aims in exploiting the many unique characteristics of nanomaterials, the context of their work is so different that it sometimes appears as though the two fields are emerging in parallel. The following discussion outlines only the major issues in this area.

11.2 MEDICAL AND PHARMACEUTICAL NANOTECHNOLOGY APPLICATIONS

There is little doubt as to the importance of nanomaterials in evolving medical and pharmaceutical applications. Along with the electronics industry and in close cooperation with it, this is one of the biggest and fastest-growing areas of research and development. The potential of improving human health through the use of nanomaterials and nanotechnologies is widely recognized as enormous. It is also a foregone conclusion that the following brief accounting of developments in these areas cannot do justice to the breadth and scope of current activities in these areas, even in summary form. The intent here is only to give a flavor of the overall kinds of directions pursued for the benefit of the specialized part of the design community involved with medical applications such as medical implants or diagnostic devices. This section thus briefly reviews potential medical and pharmaceutical applications of nanomaterials and nanotechnologies in several primary areas, including biosensing, diagnostics, interventions, and drug delivery. We touch on enabling technologies such as nanolabs as well.

Biosensing technologies can use nanomaterials and nanotechnologies to increase our understanding of both the fundamental understanding of cells and a host of disease-specific phenomena. Nanomaterials are being explored for use as both biosensors and biolabels. Biosensors respond to biological molecules. Nanoparticles that are conjugated with various biological ligands can be designed to bind with particular cellular targets. Nanowires, for example, have been coated with substances that only bind to particular proteins that are to be measured. When the proteins bind to the nanowires, their conductivities are changed sufficiently for detection and measurement. As previously discussed, nanoparticles exhibit many extremely interesting properties in relation to light (see Section 8.7), including fluorescence. The use of fluorescent nanoparticles can allow biocolor labelling and subsequent tracking of cells. Other kinds of nanomaterial properties, such as magnetism, are explored for similar pathway tracking. Monitoring techniques such as spectroscopy or magnetic resonance imaging (MRI) can potentially be improved as well. In spectroscopy, certain nanoparticles such as gold can cause chemical signal magnifications. In MRI monitoring, superparamagnetic iron particles are explored for inducting specific signals. Other approaches are used for enzyme and protein analyses.

Diagnostic techniques rely heavily on detection and sensing in relation to various disease markers. Nanoparticles have been explored for use in detecting specific viruses and precancerous cells as well as others. Nanoscale dendrimers are important here. Dendrimers are synthetic, three-dimensional branching molecules that can be of great complexity. Various biomolecules have been explored to functionalize nanosized dendrimers, with the objective of detecting particular kinds of proteins or other substances that characterize certain diseases. They can be useful in diagnostic imaging as well. Dendrimers can also carry fluorescent markers for tracking. Similar approaches use superparamagnetic iron oxides to recognize target molecular markers inside particular kinds of diseased cells, which then can be detected by MRI systems.

The whole realm of diagnostic devices and the role of nanomaterials are clearly beyond the scope of this book, but the exploration of nanosized "laboratories" cannot be passed without notice. The value of smaller and smaller devices with diagnostic capabilities has long been recognized. As noted in Chapter 3, an early targeted exploration of microelectromechanical systems (MEMS) was the lab-on-a-chip (see Figure 11.1). The push toward smallness has continued with explorations of nanoeletromechanical systems (NEMS) devices, not only for convenience but also because of their potential widespread distributed application potential (the tailored "personal diagnostic lab"). The essence of these devices relies on the device sampling liquids (or gases) to which they are exposed

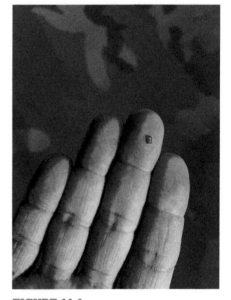

FIGURE 11.1

Among the many applications of nanomaterials and nanotechnologies to medical diagnostic and treatment techniques will be ever-smaller devices, including "labs-on-a-chip." A MEMS-scale medical telesensor chip for measuring and transmitting data is shown. Nanosized NEMS devices are expected to be smaller. (Courtesy of Oak Ridge National Laboratory.)

and then analysing them for specific markers using biosensors or other technologies, such as light-based devices, and communicating results. Doing so requires many nanoscale technologies as well as understanding how gases and fluids behave at tiny sizes.

A characteristic of a nanosized diagnostic device is that tiny channels and reservoirs are needed for the sampled gases and fluids. The latter also must normally be moved around within these tiny channels. Fluid movements at these scales are quite different than are normally understood at the macroscale, are not well understood and there are many problems in inducing movements. The field of *nanofluidics* addresses these many issues. Specific approaches based on electrophoresis (electrically induced movements) are interesting here, as are other techniques. A variety of nanoscale biosensor, nanoelectronic, and communication devices are necessary as well.

Drug delivery involves transporting drugs to specific targeted sites within the body, a fundamental task in medicine and health; this has long been a huge area of development and research. Current delivery systems are well developed, and include oral, nasal, intravenous, pulmonary, and mucosal administration. Specific conditions such as diabetes largely dictate the most effective kinds of delivery. There are many delivery problems with drugs. They can exhibit signs of bioincompatibility, lack of stability, or poor solubility or adsorption characteristics. Some peptides and proteins, for example, degrade in a biological environment, which in turn limits their efficacy. Each year many new drugs are developed that are not water-soluble. This restricts their use because delivery to targeted locations is difficult and their bioavailability is limited. One hope is that when these drugs are in nanoparticle form, they can be delivered more readily without degrading and that their bioavailability will be increased.

Nanomaterials promise to play a significant role in new delivery systems because of the functionalization possibilities associated with their particle types, size and distribution, surface areas and charges, and other characteristics. Nanoparticulate-drug conjugates hold great interest. Very small nanoparticles are thought to be less likely to cause blockages in even the smallest vessels of the circulatory system (such as capillaries, arterioles, and venules) than larger ones and also less likely to exhibit a form of sedimentation. Penetration is easier, as is the endocytosis (cellular uptake) process. Nanomaterials can also be used primarily as carrier vehicles. Polymeric coatings are particularly interesting. They hold the promise of being able to deliver drugs to specified sites and aid in controlled releases upon target triggering mechanisms. Nanocoatings can allow the use of largely insoluble drugs or potentially ones with greater activity or even toxicity. Coatings that have been explored can be quite thin—only several nanometers thick.

For specific conditions such as insulin delivery for diabetics, pulmonary delivery of peptides and proteins are interesting because of their noninvasive character and provision of large surface areas with their potential for increased absorptions. Nasal delivery provides direct access to mucosa surface absorption sites, but its efficacy has proven low and potential mucosal damage problematic. Oral delivery is easy and cost effective, but insulin permeability in the gastrointestinal track is limited and degradation of the drug can occur. Nanoparticles can potentially be designed to be biodegradable and biocompatible and more stable within a biological environment; they may well prove effective as bioactive macromolecule carriers. Polymer-based nanoparticles, such as polyethylene glycol (PEG) coated polylactic acid (PLA) nanoparticles, are a particularly interesting area of study for protein delivery. The PEG coating can help transport the encapsulated protein within and across intestinal or nasal mucosae. Chemical stability is improved. Nanoparticulate-drug conjugates hold promise for improved insulin delivery.

Despite potential benefits, research is only just beginning into the many issues surrounding beneficial ways to utilize nanomaterials for drug delivery. A particularly interesting problem, for example, surrounds the way various nanoparticle types and forms might interact with other cells, including the macrophage cells that play a fundamental role in the body's immune system by attacking foreign substances. Seemingly endless numbers of studies have yet to be done on other issues, including basic efficacy, potentially deleterious side effects, and so forth.

Chapter 2 introduced a few basic ideas about bones in relation to nanomaterials. Bones are indeed fascinating materials that constantly renew themselves at the cellular level. In bones, the basic material (hydroxyapatite) is secreted within nanosized collagen fibers (connective tissue). The body reabsorbs the basic material and secretes it anew in response to changing demands placed on the bone as a function of growth, damage, or other factors. These actions have suggested fascinating interventions for strengthening and repair of damaged or vulnerable bone fabrics through injections of bioactive nanoscale materials that harden under the action of body fluids. Target materials are designed to mimic and be compatible with existing bone fabrics and surrounding body materials.

In many situations, *implants*—for example, strengthening plates or artificial bones—need to be used. Common implant materials are stainless steels or titanium, which do not react to human tissues (biological fluids are quite corrosive). The current state of the art is remarkably high, but issues with implants remain. In many situations, the mechanical wear on these implants is high and periodic expensive replacement can be needed. These materials are also inherently not porous. There should be some porosity so that the surrounding body tissue material joins to the implants. Ideally, the materials used should have some kind of bioactive characteristics.

As discussed previously, nanocomposites can have excellent strength and hardness characteristics that could help address the problem of mechanical wear and reduce the need for periodic replacement. Many metallic nanocomposites could address this problem. Nanoceramics, such as the nanocrystalline zirconium oxide, are also particularly interesting since they can be made extremely hard, wear-resistant, and more porous. They are also biocompatible and do not react to corrosive biological fluids.

The coupling of nanoelectronics systems with implants opens new treatment doors. Opportunities are just starting to be explored, but amazing possibilities, such as a responsive artificial retina, are certainly imaginable. Larger prosthetic devices can also benefit from the sensing and control systems possible via small nanoelectronic devices.

11.3 HEALTH CONCERNS

Our society is already keenly attuned to health and environmental issues, and public health researchers have long struggled with understanding and quantifying the risks associated with known health hazards in relation to the many fine particulates that already exhibit nanoscale dimensions and that are the byproducts of many natural forces or human processes (e.g., volcanic ash or diesel soot emissions). In the eyes of both the public and the scientific research community, it does not take much of a leap to appreciate that engineered nanoparticles that can possibly be ingested or inhaled or that might otherwise find their way into human bloodstreams in some manner might indeed be a potential cause for health concerns. There are already contentious debates arising as to whether particular products containing nanoparticles might or might not have toxicological consequences. This is certainly the current case with cosmetics that use nanoparticles (see the part on nanocosmetics in Section 10.4 and the following discussion) and can surely be expected to soon arise with nanotextiles that might be used for clothing. Use of nanomaterials in agriculture will undoubtedly be highly scrutinized as well.

In general, we can expect that as more nano-based products move into applications that come into contact with ever-increasing numbers of people and exposures escalate, public awareness will further increase and there will be reasonable demands for greater certitude as to whether nano-based products are safe. Needs for this same sense of certitude are similarly felt by the responsible scientific and product development community as well. The question of whether this certitude is there or not is the focus of this section. Those who seek blanket yes-or-no answers, however, are advised to look elsewhere. If there is anything that is clearly suggested by the literature in this area, it is that simple generalizations are not supportable and that many open questions remain. Addressing this subject in depth, however, is simply beyond the scope of this book. Rather, we will seek to provide only a *framework* for thinking about the problem.

Questions concerning health and environmental issues in relation to nanomaterials are not unstudied. There are a number of studies in this area, including review articles and several large symposia or conferences held in the United States and European countries that were specifically directed toward trying to understand the extent to which nanomaterials pose a risk to either human health or the environment (see Further Reading). In summary, several common associated themes run throughout this literature. A first and fundamental theme is that there is indeed great potentially positive technological and societal merit in the use of nanomaterial and nanotechnologies in many application domains that should be valued and respected, but many open questions remain unaddressed concerning health impacts. A second theme is that though it might seem that nanomaterials are entirely new and we have little understanding of their health and environmental implications, there is actually a substantial relevant body of literature to be found in toxicological studies of ultrafine particles with nanoscale characteristics that have long been with us and whose effects have been extensively studied and have been linked to such problems as asbestosis or silicosis, albeit questions of the relevance of many of these studies are often posed.

A third frequently observed theme is that many of the specific claims that unequivocally state that there are, or are not, harmful effects—claims that seemingly often appear to directly contradict one another—are often made on the basis of studies that had relatively limited scopes and small sample sets, and that resulting conclusions may or may not have the general applicability claimed. A fourth theme is that there are research gaps that badly need to be filled in many areas—from types of general exposure risks through the effects of specific compositions and particle sizes and shapes. Recommendations are invariably made that the plethora of scientific and application advancement studies in relation to nanomaterials should be simultaneously balanced by rigorous studies about potential impacts on human health and the environment. A fifth theme that is often present is simply that caution is suggested in any kind of statement suggesting that there are or are not harmful effects.

Related subthemes address the issue that the toxicological effects of nanomaterials are particularly difficult to address for several main reasons. One is that "nanomaterials" are not a particular type or form of material, as is common in many toxicological studies; the term is simply a descriptor for a surprisingly large array of quite different compositions, sizes, embedments, and the like. A second reason is that though prior understandings are available for a limited number of well-studied ultrafine particles, the larger distributions of sizes, relative surface areas, possible types of coatings, and surface reactivities that are associated with many nanomaterials should lead to cautiousness in statements of findings—for example, there are contentions that even well-studied materials of one composition and size can exhibit surprisingly different characteristics at different sizes due to small-size and scale effects (some studies have contended that new chemical behaviors could exist). It has also been generally observed that a primary difference between ultrafine particles (which have been the subject of so many studies) and nanoparticles is that the former are largely "unwanted" results of various activities and processes, whereas nanomaterials are "wanted" and are quite literally engineered for specific purposes; hence there can be greater distributions of sizes, compositions, types of coatings, etc., than commonly encountered with ultrafine particles. This same fact, however, also suggests the possibility of engineering specific nanomaterials for uses in defined circumstances so that any potential harmful effects can be mitigated or eliminated, such as through the use of coatings that change surface reactivities.

Primary Considerations

The scientific community has spent considerable energy on endeavoring to understand the many factors that impact the health risks involved with the effects of the manufacturing and use of new materials as well as those that have been with us for a long time. Ultimately, health consequences can only result from various forms, compositions, and concentrations of materials that have toxicologically adverse effects finding their way into human bodies. As discussed in more detail shortly, most current evidence for adverse effects comes from studies of ultrafine particles that have long been present.

The next question is whether there are indeed adverse consequences associated specifically with newly engineered nanomaterials, and, if there are, of what pathology? The research community is already focusing on the need to treat this question in specific terms. Hence an important set of research considerations hinges around the exact type of nanomaterial that could be associated with an exposure, including material composition, form (e.g., carbon nanotubes, fullerenes, quantum dots), and size and shape distributions, and whether there are or are not adverse consequences associated with each, and, if so, under what precise circumstances (e.g., concentrations, means by which they enter the body). Most studies to date have focused on nanoparticles, with heavy reliance on insights gained from the study of existing ultrafine particles already found in the environment, and only limited attention has been paid to other forms.

Actual risk levels to human populations, however, are also dependent on types and actual exposure levels to any materials identified as having adverse toxicological effects, the pathways by which the particles find their way into bloodstreams, and other factors. Here one set of considerations hinges around effects as they might relate to the means by which humans are exposed to nanomaterials. If there are no exposures, there is no risk; but exposures can potentially occur in the workplace, through the use of products containing nanomaterials, and by other means. Where or how an exposure can occur (including proximity) is a particularly significant factor, since the exposure setting would in turn affect exposure types, intensities, and the forms of nanomaterials likely to be encountered. Many of these exposure types, in turn, depend on how toxicologically adverse materials might find their way into larger life and work environments (more on this in a moment). After exposure, primary paths of nanomaterials into the body include direct ingestion, inhalation, and intake dermally, although other paths exist as well. Ingestion could possibly come from nanomaterial uses in agriculture. Inhalation is a particular danger in workplace environments, especially where nanomaterials are manufactured. Dermal exposures could possibly occur through the use of cosmetics containing nanoparticles. These issues are explored in more depth in the following discussion.

The issue of exposure and possible migration of nanomaterials into human bodies is also tied closely to the way in which nanomaterials are found or used in products, especially in the degree to which they might or might not become unbound. In many nanopolymeric composites, for example, constituent nanoparticles are covalently bound and not easily susceptible to becoming unbound. In a nanometallic composite, nanoparticles embedded in a metal matrix are again less likely to become free than in other situations.

All these factors must be considered in assessing actual risk levels to exposed populations. A block of aluminum with embedded nanoparticles that is part of an automobile chassis poses quite a different risk to a human than does direct ingestion of carbon nanotubes by sipping—or rather inhaling, which is more likely—from a full wineglass of them. The risk never really disappears, however, even when nanomaterials are bound into a matrix. Abrasion or other actions such as the wearing or even washing of nano-based textiles could cause particles to become unbound, as could downstream recycling activities. Full product use conditions and life cycles need to be considered.

Effects

Assuming that exposures occur, the scientific community has long studied the toxicity of ultrafine particles that are commonly found in our workplaces and other environments. Ultrafine particles are usually defined as having diameters in the range 100 nm to 1 micron. Most studied particles, however, are not from engineered nanomaterials but from sources that have long been with us: fly ash, carbon black, and welding fumes. These are normally products of some type of hot processing activity or of a combustion process, and hence they are often associated with workplaces. Carbon black, for example, is used in producing rubber-based products. Other tiny particles, such as fumed silica and other oxides such as titanium, zirconium, and alumina, are used in the manufacturing of various products, such as pigments for paints, that utilize thixotropic agents (see Chapter 10). Other sources of ultrafine particles include diesel emissions. Particles from these emissions are obviously not found in workplaces only but more generally in the air around us. Several epidemiological studies suggest relationships between exposures to airborne ultrafine particles and adverse respiratory and even cardiovascular effects, particularly in susceptible or vulnerable population segments. Combustion-derived particles, for example, show a range of demonstrated pathologies of different consequence in humans or animals. Particles from diesel exhausts, for example, have been related to inflammation in human lungs after short-term high exposures. Dust containing silica that has been inhaled can result in silicosis, a common occupational lung disease. Exposure types, concentrations, and other factors are all important. Actual sizes are important. Particulate matter at the 100 nm scale is an important reference measure (the so-called PM_{10} level).

Completed studies directed specifically toward engineered nanomaterials remain few but are currently the subject of many research efforts. The high surface-to-mass ratio of nanoparticles is often generally noted as particularly important (as it can be in many ultrafine particles as well). In general, surface effects might ultimately prove more important than mass effects.

Additionally, many nanoparticles with diameters less than 10 nm or so have surface properties that are dominated by quantum effects, which potentially can give them quite different properties from their counterparts at the submicron or micron level. Large percentages of atoms in a nanoparticle are at their surfaces, which can then be extremely reactive. For some compositions, oxygen radicals can be generated. The surface chemistry and interparticle forces of nanomaterials depend on whether particles are individually dispersed or form aggregates or agglomerates. For inhaled circumstances, many may be deposited as aggregates, whereas agglomerates might disassociate. One of the positive attributes of many nanomaterials in the medical sphere—that small sizes (say, less than 50 nm) can translocate into cells easily and can interact with enzymes—is a relevant concern if it's happening without control (larger particles do not translocate quite as easily). There are suggestions that nanoparticle sizes might change in interactions with biological settings and change properties.

Many of the existing studies have focused on nanomaterial types already produced in large bulk form. Titanium dioxide, colloidal silica, and several iron oxides have been produced for some time in large production volumes. In some specific instances, for example, lung inflammation was observed in animal studies with prolonged inhalation exposures for titanium dioxide, with more adverse effects present for very small particles compared to larger airborne particulates. Results are often regarded, however, as inconclusive and in need of further study. In the literature, hypotheses have been made that there are more toxicological implications with nanoparticles than with their nonengineered counterparts, but limited data sets again generally preclude making broadly definitive statements about health effects so far. Additionally, it has been noted that the very ability to engineer nanoparticles (including coatings) can potentially radically change their surface chemistries and effects, including toward reducing perceived risks.

A Specific Debate

As discussed in Section 10.4, nanoparticles are finding increasing use in various kinds of cosmetics, including many types of sunscreens, lotions, and moisturizers. In sunscreens, titanium dioxide and zinc nanoparticles are being used to increase the transparency of these creams while being said to offer UV protection. It is interesting to look at this particular application as an example of the complexities of assessing potential health hazards of using nanomaterials and to look at the surrounding debates as perhaps a bellwether of what is in store for other applications in which nanomaterials come into direct contact with humans.

The inclusion of nanomaterials in cosmetics has caused consumer groups and members of the research community to want to know more about the safety of doing so. Questioning groups maintain that nanomaterial applications applied directly to the skin have a direct pathway to the bloodstream via direct skin penetration. They also feel that the research on health effects of the kinds of nanomaterials and exposure types on human health are far from conclusive and that, in the absence of clear findings to the contrary, the large population of cosmetic users should not be subjected to unknown risks. They also contend that these products are poorly regulated by public bodies. Other individuals and groups have challenged these positions, claiming that the positive value of using sunscreens is great and that the weight of evidence points to cosmetic products containing nanomaterials as being safe. There are many nuanced positions on both sides of the debate, and many hinge around the issue of whether nanoparticles do or do not penetrate the skin barrier. If they do, a primary concern among others is that some common nanoparticles (e.g., titanium dioxide) are known to emit photoelectrons when exposed to UV light, which then might cause formation of peroxides, free radicals, and other reactive oxygen species that might interact with lipids and DNA and cause damage. If they never penetrate the skin barrier or light never reaches them, concerns such as this are lessened.

The question of whether or not particles actually do penetrate the skin and enter the bloodstream has been addressed only to a certain extent. Skin is a wonderfully complex and versatile material that serves many functions—a barrier, a device for controlling internal body temperatures, a device for manufacturing vitamins, and other roles. Skin is composed of several primary layers (see Figure 11.2). The major external layer, the epidermis, is quite strong. The stratum corneum, the very outer layer, consists of layers connected by a lipid network. The dermis is beneath and thicker and holds nerves, blood vessels, and sweat glands as well as providing other functions. A layer of fatty subcutaneous layer lies below it.

Controversial early studies have reported titanium in the epidermis following applications of materials containing titanium dioxide to the skin. Other studies have suggested that topically applied titanium dioxide nanoparticles do not cross the outer skin barrier. Another reported that with rubbing or walking, that some penetration of the stratum corneum can occur. The question repeatedly arises of whether studies sufficiently reflect pathway differences

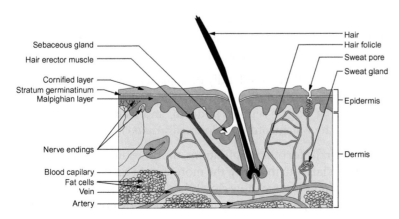

FIGURE 11.2
Cross-section of human skin.

between healthy skin and unhealthy and presumably more vulnerable skin (e.g., with sunburn or psoriasis or abraded), again with findings that many find inconclusive. Actual impacts on health are reported as unclear or suggested adverse effects are controversial (see Further Reading).

As of this writing, most reviews suggest that particles applied to healthy skin not subject to unusual conditions remain largely in outer layers of so-called "dead skin," but comments are invariably cautionary. As noted earlier, *nanomaterials* is a highly general term for a huge array of compositions, types and forms, and size distributions. There are also many skin conditions (healthy, unhealthy) and external circumstances, such as intense pressure, that need to be simultaneously considered. Researchers normally note that not all possible combinatorial situations have been considered and substantially more research is needed. There are also calls for more transparency and openness in scope definitions and limitations surrounding published studies, that is, less generalizing and more specific reporting.

Issues surrounding regulatory control are also controversial. Some representatives of the nanocosmetic industry are resistant to basic labeling, apparently due to their perceptions of the inconclusiveness of health findings. (Keep in mind that others happily promote nanomaterials as an ingredient, some at least presumably doing so as a marketing device). In the United States, consumer groups have claimed that the federal Food and Drug Administration (FDA) has an inadequate position since in the past it allowed the use of nanomaterials without new health impact studies, but studies with the National Institute of Environmental Health Sciences and the National Toxicology Program are under way. In the United Kingdom, a thorough safety assessment has been suggested, along with developing a method, including labeling, that allows consumers to make informed choices. Japan includes risk assessment and health issues in their calls for collaborative research initiatives. Still, specific recommendations largely remain in the future.

The preceding discussion hardly does justice to the many complex health issues surrounding the use of nanomaterials in cosmetics. Ongoing work is serious, in depth, and badly needed. For the moment, the watchword is undoubtedly *caution*.

More generally for other applications, the more direct the pathway of nanomaterials into the body, the more should caution be observed. Nanomaterials that are more closely bound to parent materials (e.g., bicycle frames) and less likely to enter the body pose considerably lesser risks, but some risks might still be present because of life-cycle concerns with products. The workforce in facilties that make nanomaterials should exercise special precautions. Nanomaterials in relation to agricultural products can surely be expected to engen-der controversies. There are, however, potentially great societal benefits to using nanomaterials in many settings—so risk/benefit questions invariably remain.

11.4 ENVIRONMENTAL BENEFITS AND IMPACTS

Our environment has long suffered from harmful effects produced by myriad pollutant-generating sources. Industrialized processes, oil or chemical spills, pesticides, fertilizers, and emissions from vehicles have caused pollution in our air, water, and earth environ-ments. The need to achieve clean air, water, and soil environments is a challenge facing societies throughout the world. It is well known that human health can be directly affected by these issues, as can the health of whole ecological systems. The role of nanomaterials in relation to these environmental issues has been reasonably raised many times—by the scientific research community, by concerned public groups, and by regulatory agencies. A surprisingly large body of literature already exists on the subject, perhaps because it is so controversial.

On one hand, there is every reason to believe that nano-based tech-nologies offer real opportunities for helping to achieve positive envi-ronmental objectives through improvements in technologies that might help reduce our dependency on fossil fuels, such as fuel cells or solar cells—a way of storing energy-rich gases is shown in Figure 11.3, and Section 9.7 more generally discusses this topic—or might directly help purify air and water supplies through the use of nano-porous materials (see below). On the other hand, many point out that the very characteristics that are attractive in nanomaterials may have significant adverse health and environmental consequences. These concerns stem from their very small sizes and correspond-ing transport and penetration abilities as well as their high relative surface areas, which naturally increase surface reactivities. Direct fears include the envisioned prospect of tiny irreducible particles working their way into not only the air environment but into water supplies and soils as well—and hence even into our food supplies.

We have already encountered a similar benefit/risk controversy in the previous section that dealt with nanomaterials in relation

FIGURE 11.3

Developing a way to store energy-rich gases can help our energy systems. Metal organic frameworks (MOFs) are highly porous organic matrix substances that can store hydrogen or natural gas. The cubical nanostructures consist of an "organometallic" framework, whereas the interior of the cube contains numerous nanometer-size pores in an interconnected structure. The nanopores have high surface area. These structures might also be used for energy sources for many devices, such as laptops. Tiny fuel cells could serve as a rechargable storage medium. (Courtesy of BASF.)

to human health—medicine, drug delivery, and the fundamental question of whether nanomaterials entering the human body through means such as ingestion or through the skin via the use of cosmetics can cause deleterious health effects (see Section 11.3). It was observed that human health undoubtedly can and will benefit enormously from the use of nanomaterials in medicine and the general field of treatment, including pharmaceutical applications, but at the same time it was clear that if wrongly used, health risks can develop. Certainly it is well known that there are problematic health effects associated with tiny particulates in the air that are associated with air pollution, and that nanoparticulates can be even smaller. Findings of whether deleterious effects occur because of skin penetration were noted as being less clear. Caution, careful use, and further testing were advised.

The following sections briefly review anticipated environmental benefits and risks of nanomaterials. Considerable excitement surrounds the potential of nanomaterials in the remediation of many environmental problems that affect our society. Many kinds of needs for environmental remediation technologies exist, but two—air and water purification—stand out as having particular importance.

Improvements in these areas would help alleviate many of the pressing health problems facing large population segments throughout the world. Adverse effects associated with pollutants in both water and environments are widely understood and appreciated by both the public and the research community. The subject is huge and well discussed in the literature, so only a broad picture will be painted here.

We begin by looking at how nanomaterials might play a positive role in reducing environmental problems at their sources. Next we briefly look at how nanomaterials can play a role in remediation and treatment measures, particularly for air, water, and soil. Finally we look at risks associated with nanomaterial production and use.

Reducing Environmental Impacts: General Source Reductions

Enormous amounts of energy are consumed in making and using products. In addition to depleting many nonrenewable energy sources, detrimental side effects such as pollution emissions from the burning of fossil fuels are widely understood. Certainly the many contributions nanomaterials are either making or poised to make, with energy sources and systems that were discussed in Section 9.7, are important in this context; they include improved solar cells for renewable solar energy and other energy-related systems that promise higher efficiencies with reduced pollutants, such as fuel cells.

It is also widely appreciated that making products that use less energy to begin with will not only help better utilize limited energy resources, it will have many additional environmental benefits as well. Huge amounts of energy, for example, are used in the operation of lighting in our homes and workplaces. More efficient lighting technologies based on nanotechnologies would have enormous environmental benefits. It has been estimated, for example, that improving the efficiency of lighting in the United States by 10% would ultimately reduce carbon emissions by some 200 million tons per year. Possible improvements via better fuel efficiencies in automobiles are obviously enormous. In earlier parts of this book we also covered an array of other kinds of nano-based products, including thin films and nanoporous insulation materials that can all directly lead to reduced energy consumption. These many product innovations should not be forgotten as we think about the environmental benefits of developments in the nanomaterial sphere.

Energy is expended not only in using products but in producing them as well. Many processes that create products also create enormous amounts of waste in one form or another. The generation of waste is not only energy inefficient—the wastes themselves can be hazardous. Abandoned industrial sites that remain highly polluted because of inappropriate disposal of chemical and other wastes are still with us today, reminding us of the need to pay special attention to the problem of waste. Many industry groups have gone a long way toward mitigating these problems through the adoption of more efficient production means and sophisticated control technologies, but the problems remain present in many parts of the world.

Pollution prevention approaches based on the implementation of practices for improving energy utilization as well as reducing waste creation would go a long way toward improving the health of our environment and hence human health itself. Within this broad arena, nanotechnologies are poised to play a significant and positive role. The improved properties of nanomaterials can allow designers and engineers to create smaller and lighter products that not only use energy more efficiently but that may reduce energy amounts needed for production. Chapters 9 and 10 repeatedly noted that the use of nanomaterials and nanotechnologies allows products to be more efficient. In comparison with conventional approaches, nano-based products for use in connection with thermal environments, for example, can potentially transfer heat more efficiently, insulate more efficiently, or generate heating or cooling more efficiently. With increased efficiency comes reduced product sizes and weights, hence reductions in the energy needed to produce them. Structural nanocomposites are expected to make other products smaller and lighter as well. All these things result in decreased energy expenditures.

The benefits of improved production processes can be very high. A case in point that has attracted considerable interest is the use of nanoparticles in concrete mixes used in massive quantities in construction. As discussed in Section 9.2, manufactured cement is a primary ingredient in concrete. It has been found that the production of cement accounts for between 5% and 10% of the world's total carbon dioxide emissions, a figure that will probably surprise most readers. A research team observed that concrete's strength and durability lie in the way that particles in concrete are organized, thus opening the potential for using materials other than conventional cement to achieve strength and durability but with lower carbon dioxide emission levels. The idea is still in its infancy, but prospects are intriguing.

The potential list of other ways that nanomaterials might help reduce sources from production processes is long. Approaches vary. It is known that catalytic processes can be improved and made more selective via the use of particular kinds of nanomaterials and thus could potentially improve a whole series of production processes related to chemicals. Nanomaterials could start replacing the need for toxic heavy metals in some processes.

Source reductions can also occur in ways that are not commonly thought about. Chapter 10 discussed the emerging potential of self-cleaning materials, ranging from glasses through concrete to textiles. Our society spends an enormous amount of not only effort but direct energy as well in keeping products clean. Washing clothing, for example, is among the most common of society's activities and consumes huge energy resources. Some cleaning processes also use chemicals that can have potentially harmful environmental effects. Decreasing the need for constant cleaning could have a significant environmental benefit; how much, of course, is unclear, since little work has been done in assessing benefits such as these.

Though the opportunities seem high, there is currently little understanding of the overall impacts that nanomaterials might have in improving processing methods as a way of reducing negative environmental impacts associated with society's huge production industries.

General Remediation and Treatment

Many measures seek to mitigate the effects of pollutants already present or recently generated. Treatment measures normally focus on transforming pollutants or other contaminants that are emitted from a source into less harmful products, such as common catalytic converters in automobiles designed to reduce harmful emissions from engines. Remediation measures focus on trying to correct problems that are often longstanding. Many soil remediation efforts, for example, try to address problems with contaminated sites from old industrial locations. Many kinds of treatment and remediation procedures and devices are in use today. Several of the more pertinent in relation to air and water environments are discussed in more detail here.

The particular properties of many nanomaterials offer several broad avenues for improving treatment and remediation techniques. One approach that we'll discuss more later is through catalytic conversion that transforms contaminants from one form to another. Catalytic nanomaterials can be activated in several ways. Many photocatalytic materials that are activated by light are and can be used to transform contaminants, such as titanium dioxide or zinc oxide. They can be relatively inexpensive and thus suitable for widespread use in air, water, and other hosts. Some ongoing research is directed toward finding ways to increase relative surface areas of nanoparticles that may be already large but that could be increased more by various surface manipulations, thus further increasing their surface reactivities. If successful, efforts to increase sensitivity of photocatalytic nanoparticles to a broader spectrum of light (beyond the UV) could vastly improve efficiencies.

Other kinds of nanoparticles are being explored for their use in relation to specific environmental contaminants, including hydrocarbons. Targeted metallic composite nanoparticles for transforming harmful PCBs or pesticides, for example, are being explored— such as various iron-silver for PCBs. Iron-based nanoparticles can help reduce many problematic contaminants, including nitrates. Nanoporous materials of several types have clear applications in air and water cleaning (see the following discussion). Several nanomaterial formulations, including nanotubes, appear to be useful as adsorbents of one type or another and may prove better than commonly used substances, such as activated charcoal. The following sections explore some of these applications in greater detail.

Environmental Monitoring

The use of networked monitoring devices—largely sensors of one type or another—is critical to understanding the kinds, magnitudes, and distributions of environmental pollutants that are present in our air, water, and earth environments. Without these understandings, remediation and treatment efforts may be sporadic or misplaced. Needs are present at the broad scale of the environment and at the very narrow scale of measurements in relation to specific manufacturing processes for control purposes. Many nanomaterials can excel for these needs. As previously discussed in Section 11.3, nano-based biosensors of one type or another can be designed to detect many kinds of biological or chemical substances. Many approaches are possible, depending on the need. Nanomaterials with catalytic properties can not only act as remediators but also as sensors. Chemical sensors based on nanotubes and nanowires have also been demonstrated. Properties used are often unique to nanomaterials. Measurable electrical resistance changes in nanotubes exposed to different chemicals, for example, can be used as the technical basis for detectors of this type. Many other approaches exist.

Nanotechnologies in the form of NEMS devices are under development to serve not only as sensors and monitors but as analyzers and communication devices as well. These are quite similar to "lab-on-a-chip" approaches discussed earlier. Tiny channels can be made to store and transport various gases or fluids through various parts of the device so that specific analyses can be made. The potential to make such devices small can allow their use in many situations not possible before.

Water Cleaning and Purification

Considerable excitement surrounds the potential of nanomaterials in the remediation of many environmental problems that affect our society. Many kinds of needs for environmental remediation technologies exist, but two—air and water purification—stand out as particularly important. Improvements in these areas would help alleviate many of the pressing health problems facing large population segments throughout the world. Adverse effects associated with pollutants in both water and environments are widely understood and appreciated by both the public and the research community. There are, of course, many established technologies that seek to address either air or water pollution. These conventional systems have by no means reached their limits, but the particular characteristics of nanomaterials appear to offer new ways of dealing with the same problems by either offering the potential for greatly improved performance in established technologies or offering entirely new approaches.

Clean or useable water in needed quantities is remarkably hard to find in many parts of the world. Standards vary for water that is considered clean enough for use in a variety of circumstances, with drinking water requiring very high standards. Water might not be considered clean if it contains excess concentrations of contaminants from either natural or artificial sources. Excess salt content is a problem in many situations. Natural sources include algae, silt, or other dissolved natural particles. Artificial sources include pesticide and other chemical runoffs that have found their way into water sources, with heavy metals being particularly problematic. Critical understanding of the need for water purification and cleanliness dates back to the 19th century. Terrible cholera outbreaks decimated urban populations throughout the world. Investigators such Dr. John Snow in the famous Broad Street Pump Outbreak in Soho, at the time an unsanitary place of slaughterhouses, animal droppings, and ancient cesspools beneath houses, found connections between the disease and waterborne microbes transmitted through water obtained from the pump, thus beginning the great movement toward clean water and the rather amazing high level of water purity enjoyed in many countries today. Sadly, however, not in all.

Today there are many conventional means of purifying water, depending on its source and ultimate intended use, such as drinking water versus industrial use. Evaporation processes are employed, particularly to remove salts. Biochemical processes are quite common, as are more direct chemical disinfection and decontamination treatments. Filtration systems of one type or another are also common. These conventional methods do work, but many are highly expensive—for example, chemical treatments, or have limited flow-through capacities or other limitations. Nanomaterials may help improve many of these conventional processes and offer new approaches as well. Recent forward-looking approaches include the use of advanced filtration systems, improved catalytic processes, and electronic processes, all of which are natural candidates for improvements via the use of nanotechnologies.

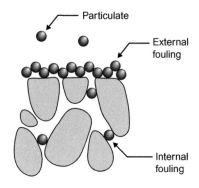

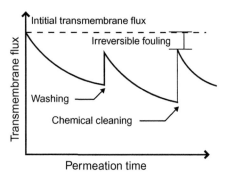

FIGURE 11.4

a) Cross-section of a porous membrane showing fouling and the effect of fouling on membrane performance. (b) Effect of fouling on membrane performance.

Filtering membrane technologies are particularly important. There are many conventional natural and synthetic membranes in use that employ fibrous media to filter undesirable particulates from water. Common filters use various woven and nonwoven types either individually or in layers. Nonwoven types are usually random patterns made from any of several dry- and wet-laid methods. Dry methods include air-laid processes such as spinning and melt blowing. In addition to basic materials, other additives or coatings, such as antimicrobial, may be included during formation processes to provide various kinds of desired behaviors. Most nonwoven filters are commonly used as prefilters in large-scale water treatment activities to reduce demands on subsequent filters. They have high surface areas and can easily trap large quantities of larger particulates. Current nonwoven filters catch particles as low as 10 microns.

Problems are severalfold. Pore size in relation to the filtering objective is obviously important. Pores are often not small enough to catch many contaminating particles. The general problem of fouling—the reduction of flow through a membrane caused by blockages that build up over time—leads to process inefficiencies. Cleaning can be very difficult. The several types of fouling include particulate, colloidal, microbial, organic, and crystalline. Fouling may be either external or internal (see Figure 11.4 for an illustration of simple particulate fouling). Surface fouling can often be cleaned. Internal fouling is harder to remedy since the blocking particles are inside the porous structure of the membrane. Irreversible fouling can occur. These factors can cause a flux decline and also reduce membrane lifetimes. Particular chemicals needed for disinfection can also attack porous materials.

Improvements in membranes or membrane surfaces using nanomaterials can occur in several ways, including changing porosities, changing the hydrophilicity of the surface properties present, changing its electropositivism or electronegativism, and changing surface catalytic properties. Nanofibers promise to provide much smaller porosities and capabilities for catching much smaller contaminants. Internal surface areas are much higher than conventional filter materials. Nanofibrous materials can have interconnected open pore structures and can potentially allow high flow rates. Their use, however, is only just now being explored. Nanoporous materials are discussed in Section 10.4. Of importance here is that pores can be made remarkably small and total pore surface areas increased dramatically. The size gradations possible by incorporating nanoporous materials can help prevent the passage of a range of contaminants and microorganisms.

Polymeric membranes are a particular focus of research. In the early 1960s polymeric membranes designed for reverse osmosis were introduced and have since been used widely for various filtration and separation applications. Reverse osmosis is a process by which water or another liquid is purified of solute impurity particles by being forced under pressure through a semipermeable membrane. The liquids can pass, but not the solutes. Nanoporous materials can be positively used in membranes here (see Figure 11.5).

Porous polysulfone membranes can be prepared by a phase-inversion technique, with porosity sizes and other characteristics achievable by various solvents and coagulant combinations. Polymerics such as polyether sulfone (PES) have been explored because of their many needed characteristics (strength, abrasion resistance, resistance to environmental degradation) and because they can be tailored to specific needs. Many of these membranes, however, can be subject to fouling because of their hydrophobicity. As discussed in Section 10.2, hydrophobic surfaces repel water, whereas hydrophilic surfaces have affinities to water. Hydrophilic surfaces in membranes can promote water transport and hence promote antifouling. The hydrophilicity of polymeric membranes can be improved by various conventional methods, such as use of ozones on surfaces and ultraviolet radiation, but they remain difficult or complicated.

Figures 11.6 and 11.7 illustrate two membrane approaches for water purification. Some newer approaches now center on the use of nanomaterials, such as titanium dioxide (TiO_2) discussed previously, which are known to have high hydrophilicity. They are also known to show photocatalytic effects (see the following discussion) that can cause decomposition of many substances and destroy bacteria. Novel self-assembly techniques have been explored in composite membranes as a way of improving resistance to fouling. Nanoparticles in these composite membranes are on the order of 5 nm to 40 nm in size and made using a type of interfacial polymerization.

Biofouling associated with microorganisms or biofilms remains problematic. Various agents such as chlorine have been used to control biogrowths in feedwater, but these same additives have potentially harmful side effects. Titanium dioxide nanoparticles have been explored as a way of reducing the biofouling of membranes. As previously discussed, titanium dioxide particles are known to show photocatalytic effects that can cause decomposition of many substances and destroy bacteria.

In many parts of the world, *desalinization processes* for converting sea or other salt-bearing water into usable water are of fundamental

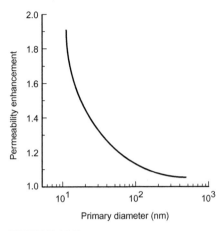

FIGURE 11.5

Permeability enhancement for reverse-selective membranes, as a function of pore size.

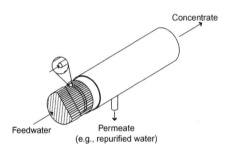

FIGURE 11.6

Hollow-fiber membrane for water purification. (Adapted from G. Tchobanoglous, F. L. Burton, et al.)

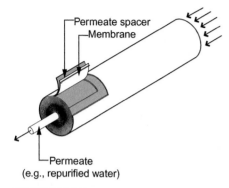

FIGURE 11.7

Spiral-wound membrane for water purification. (Adapted from G. Tchobanoglous, et al.)

importance. Conventional techniques typically rely on evaporation processes and various kinds of activated carbons used in filtration or surface absorption processes. Reverse osmosis processes force seawater under pressure through a membrane. Some fresh water gets through, but salts are stopped. In electrodialysis processes, ions are transferred through membranes because of electric potential differences. All these processes work, but they have either limited flux capabilities or are expensive. Reverse osmosis and electrodialysis processes consume large amounts of energy. Nano-based membrane technologies again suggest promising approaches that are less energy intensive. Selective use of filter gradients can restrict the passage of some particles and allow others to pass through. Carbon nanotubes have also been explored for use with desalinization processes, where reports have suggested reduced energy use of some 100 times less than distillation processes and 10 less than reverse-osmosis processes.

Nano-based *photocatalytic processes* are also emerging as potentially strong means of water purification. As described in more detail in Section 10.2, in photocatalytic oxidation processes a semiconductor material such as titanium dioxide (TiO_2) is exposed to ultraviolet light. On absorption of photons, the material acts as a catalyst, generating reactive hydroxyl radicals that can oxidize organic pollutants and completely change them. Organic molecules are decomposed to form carbon dioxide, water, and mineral acids. Many organic molecules found in pesticides, herbicides, and surfactants can be transformed in this way to less harmful products. Bacteria and viruses can also be killed.

Photocatalytic processes for larger-scale applications currently have limitations. Catalysis particles can either be immobilized on surfaces within the reactor vessel or suspended in the water. For normal flow-through processes, the latter approach has many difficulties with recovering the particles themselves, and surface approaches seem advantageous. Contact areas between pollutants and catalysis particles, however, can then be more limited. The UV light must also reach or penetrate the water and reach contact areas so that the photons can activate the catalysis particles. Penetration depths can be limited by absorption caused by the pollutant particles themselves.

Still, the increased catalytic efficiencies of nanosized catalytic particles because of their increased surface areas makes the approach worth further exploration, particularly since UV light can be generated relatively inexpensively. Design directions for practical water-purification devices based on these principles include developing ways of increasing the contact area between water flowing through and the surfaces holding immobilized catalysis particles as well as ensuring that UV light at the needed intensities bears on these same surfaces. These approaches naturally lead to the use of thin immersed multiple surfaces (fins, tubes) that also tend to increase turbulence, which can also help improve the rate of the pollution-reduction process. At the same time, rate of flow-through problems remain.

There are several types of so-called *electronic water purifiers* based on some type of ion transfer. Flow-through capacitors offer an interesting way of achieving desalinization (see Figure 11.8). One interesting previously tried approach uses a flow-through capacitor. Early versions did not work well, but it appears that nanomaterials can improve efficiencies. These systems work by electrostatic absorption of ions. The capacitor has two conductive plates: a positively charged anode and a negatively charged cathode. Water flows between the plates and ions are attracted to the anode or cathode. The capacitor plates, normally of activated carbon, have large surface areas and very large pore volumes and trap the ions. Conductivity and attractions increase until a point is reached at which the cell is regenerated by discharging it and reversing polarities, thus expelling trapped ions that are carried away by a waste stream. After regeneration, polarities are reversed and the process begins again.

Nanomaterials are known to have unique electric and magnetic properties (see Section 7.4 and 7.5) and be useful in the development of supercapacitors and other devices and are expected to contribute to improvements in these kinds of water purifiers as well.

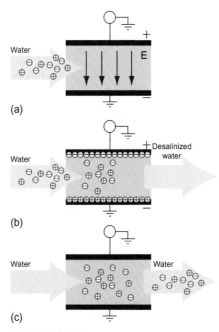

FIGURE 11.8

Desalinization of water—flow through supercapacitors.

Air Cleaning and Purification

Our society is unfortunately all too familiar with air pollution issues, and potential health hazards are widely understood. As with water purification, the use of nanomaterials may help contribute to cleaner air. Smog and high ozone levels develop in response to many substances, particularly nitrogen and sulfur dioxides, as well as many volatile organic compounds (VOCs). Broad approaches to both water and air purification are surprisingly similar. Improving air quality by filtration and adsorbent methods are widely employed, as are catalytic processes of one type or another.

Many types of air cleaning and purification systems are already in use. Specific approaches are highly dependent on the pollutants

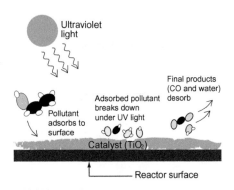

FIGURE 11.9
Photocatalytic air purification and pollutant reduction.

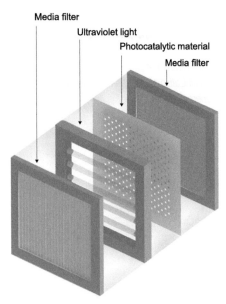

FIGURE 11.10
Typical photocatalytic air purifier.

that need to be removed from the air. Common catalytic converters in vehicles are tailored to remove undesirable byproducts, such as carbon monoxide, nitrogen oxides, and unburned hydrocarbons, from the exhaust of a combustion engine. Adsorption and absorption technologies target combustion byproducts emanating into the atmosphere. Chemical absorption, for example, has been used to sequester carbon dioxide from coal-burning facilities. The air *within* the buildings we occupy is increasingly subject to cleansing through various means. In this architectural domain, purification technology has typically targeted airborne odors, dust, toxic chemical fumes, noxious gases, VOCs, microbes and viruses, and various allergens. There are at least six primary methods of air purification in use in various scales today: catalytic, filtration, adsorption, absorption, electrostatic means, and the use of UV light.

In *catalytic* processes, the form of a contaminant is changed into a benign form through a process that we have seen several times before (e.g., Section 10.2 and the preceding discussion in relation to water purification). Applications are designed with a particular target substance in mind. Traditional catalytic converters use a reduction catalyst of materials such as platinum or rhodium. Photocatalytic UV technology involves the activation of catalytic particles in a medium to attract airborne particles. The most widely cited contribution of nanotechnology currently in use for the purposes of filtering air involves a process known as *photocatalytic oxidation* in which a semiconductor catalyst (usually titanium dioxide or zinc oxide) is exposed to UV light to produce highly reactive airborne hydroxyl radicals that react with VOCs or other pollutants (see Figure 11.9). Various porous oxides coated with catalytic nanoparticles are being explored for use in removing sulfur dioxides and other contaminants. A typical photocatalytic air purifier is shown in Figure 11.10.

In *filtration* systems, air is forced through a filtering membrane, which excludes the passage of certain particles by size. Typical conventional membrane materials include cotton, foam, fiberglass, or synthetic fibers. The application of a filter to an air stream restricts air flow, which becomes an important factor in designing filtration systems. Replacement of the filter is inevitable due to fouling. High-efficiency particulate air (HEPA) filters are those that meet certain governmental standards for removing 99.97% of all particles greater than 0.3 micrometers in size while allowing a specific amount of air flow. Filtration technology is already seeing applications for membranes that can be controlled at the nano level. Nanofibrous materials have been used in many applications for many years, and new types are under development. Their high-surface area and interconnected open pore structure are advantageous for trapping particles but still retaining high air flow or flux levels. Industrial-scale HVAC filtration for large spaces has seen development of filters that incorporate nanofibers to get finer than the HEPA standard. Likewise, facemask filters are being developed to capture particles considerably smaller than the HEPA standard.

In *adsorption* systems, large airborne particles are captured on the surface of a bead, granule, or crystal of an adsorbent material. The adsorbed material is held mechanically on a substrate rather than chemically and can therefore be released (desorbed) rather easily by heat or vacuum. Adsorptive materials are typically used to capture VOCs. The most common adsorbent now in wide use is *activated carbon*, typically used against odors, fumes, and chemicals. The carbon material is not only highly porous but has a high surface-area-to-volume ratio. Preparation processes for this material have an impact on the particles it can best adsorb, which typically include odors, fumes, and chemicals. As mentioned before, nanotubes have been suggested as sorbents, and other types of nanomaterials can serve this function as well. A parameter that characterizes the sorption affinity for carbon nanotubes—the Langmuir constant—has been found to be much higher for carbon nanotubes than for activated charcoal. Costs here for large-scale applications are undoubtedly problematic. Another adsorbent is the aluminosilicate crystal structure known as *zeolite*, which has uniformly sized pores that can be at the nanoscale. All naturally occurring zeolite is hydrophilic and contains aluminum. Removing the aluminum from natural zeolite makes it hydrophobic and gives it an affinity for nonpolar substances, such as many volatile organic compounds. Zeolite has many natural states and can come in nanosized forms having nanodimension pore sizes that have extremely high relative surface areas (see Section 10.4).

In *chemical absorption* systems, a gas is absorbed in a liquid solvent through the formation of chemically bonded compounds. As mentioned, one application involves carbon sequestration, whereby CO_2 is typically recovered from combustion exhaust such as that produced by coal-firing plants. The process, known also as *scrubbing*, uses amine absorbers, most commonly monoethanolamine (MEA), diethanolamine (DEA), or methyldiethanolamine (MDEA) compounds. After removing impurities such as nitric oxide, sulfuric oxide, and hydrocarbons, the gas is channeled into an absorption column, where the amine reacts with the CO_2 and selectively absorbs it. The CO_2-rich amine is then heated, whereby the CO_2 is released from the amine as nearly pure gas. Other kinds of scrubbers are designed for specific application targets. Nanomaterials have been explored as paths to improving these kinds of processes because of their high surface areas and processing ability to obtain a variety of shapes.

In *electrostatic precipitators*, static electricity is used to charge pollutant particles in the air, which are then attracted and bound to a collection plate of the opposite charge. These systems have been typically seen on the scale of centralized HVAC for a single building and at a portable size meant for an individual indoor space. The size of collection plates, air flow rate, debris accumulation, and the charge delivery characteristics influence efficiency and efficacy. Electrostatic systems require maintenance, which consists of wiping accumulated debris from the collection plates, to maintain efficiency. These systems have been used to remove dust particles, fumes, odors, and gases and have also been used in tandem with membrane filtration.

Other approaches are based on *ultraviolet exposure*. Simple UV light is capable of denaturing the DNA of many kinds of viruses, bacteria, and fungi. This technology is common in water filtration as well for the removal of a variety of microorganisms. This technology is often used as part of a hybrid system that employs a mechanical filter and is increasingly being used at the largest scale in building HVAC systems.

Textiles are becoming important in dealing with noxious gases (see Figure 11.11 and Chapter 10).

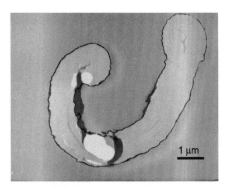

FIGURE 11.11
Cotton textile with nanoparticles for decomposition of noxious gases. (Courtesy of Juan Hinestrosa.)

Special Applications in Air Pollution Reduction

Solving the massive air pollution problems that exist throughout the world is likely to be most directly addressed by source reduction and direct treatment options of the type noted previously. Nonetheless, the problem is so massive that every opportunity must be used to address these issues. Several ideas are evolving that are quite creative and fall outside common approaches. One idea has assumed the sobriquet of "smog-eating concrete." Smog in our cities is commonplace and a consequence of many urban source emissions, including vehicles. Adverse health effects are well known. Nitric oxides, in particular, are problematic for some concerns. These pollutants have even been known to cause degradation in the marbles of historically important buildings and monuments, including the Parthenon in Athens. Our cities and surrounding lands also have literally massive surface areas of concrete in their roadways, sidewalks and parking lots, and buildings and bridges. Innovative investigators have posed the interesting question of whether new concretes could be made with surfaces that somehow contribute to smog reduction. To many this notion might seem odd, but the technological basis for the idea is there.

Chapter 10 discussed the development of "self-cleaning" concretes. The self-cleaning action was described as originating with the incorporation of photocatalytic nanoparticles, typically titanium dioxide, into coatings for concrete or directly embedded in outer surfaces. Self-cleaning action itself works via the photocatalytic conversion of various compounds into other forms. Many pollutants in the air have similar constituents that can be catalytically transformed, in this case by contact with photocatalytically active concrete surfaces. To test this general idea, an innovative Italian company, Italcementi, has tried several experiments involving the use of photocatalytic concrete, including its use on heavily trafficked roadways. (These experiments are described in more detail in Section 10.2 on self-cleaning and antipollutant concrete.) Reductions in pollutant levels were found to occur. Problems in making approaches such as this efficacious are, of course, enormous and include needing to assure long-acting photocatalytic action at a high level. Normal heavy urban dirt cannot easily be eliminated through photocatalytic action, for example, and could easily reduce immediate efficiencies (but it might be washed away more easily). Little is known about long-term behavior. Ideas like this should be explored, however, if our air pollution problems in urban areas are ever to be solved. Even if not applicable in a broad way to existing urban fabrics, perhaps materials like this could be used in the immediate surroundings of our important historical monuments that are threatened by pollutants.

Soil Remediation

Pollutants in our soils and ground waters come from many causes, including all too common accidental discharges of hazardous chemicals from manufacturing or processing centers, or, unfortunately, still deliberate discharges of questionable materials. These discharges can contaminate large areas due to moving ground waters and even find their way into rivers and other water bodies (see Figure 11.12). As noted, many nanoparticles have reactive qualities that interact with various contaminants and transform them into more benign forms. The issue is how to get them into contact with contaminants that are underground.

FIGURE 11.12
Pollutants can find their way into ground water in many ways, including accidental discharges, and be carried to sources of drinking water, rivers, and lakes.

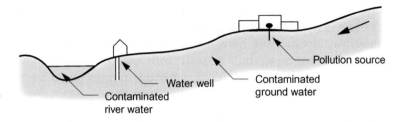

FIGURE 11.13
Experiments have been conducted with dispersing reactive nanoparticles in slurries into contaminated groundwater zones. The nanoparticles react with certain pollutants and render them benign.

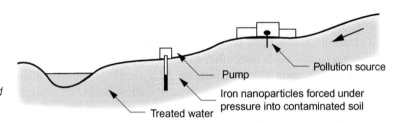

In experiments by a group at Lehigh University with slurries containing reactive nanoparticles, the slurries have been pumped into on-site reactor vessels (typically pipes) and dispersed into contaminated underground zones (see Figure 11.13). Iron nanoparticle slurries containing about 99.9% iron and minor amounts of paladium have been reported to degrade many organic contaminants into benign compounds. Several experiments involved trichloroethene plumes. Large percentages of this material were converted into harmless ethylene and ethane by the underground injections. Instead of being dispersed under gravity or pressure, nanoparticles can also be directly affixed to various other materials. Similar approaches can be envisioned for sedimentations and other deposits, even those containing heavy metals.

General Environmental Risks

We've briefly outlined the many advantages of using nanomaterials, including the positive environmental benefits. The alternative question of how nanomaterials might detrimentally interact with our larger environment, and hence with both humans and ecological communities, however, is a concern expressed by both environmental advocacy groups and the research community. Concern for the potential of adverse impacts comes from the following line of reasoning. Nanoparticles might enter the environment in a number of ways, including the following: accidental escapes during manufacturing, use, disposal of products, or waste containing nanomaterials, such as from nanofluids used in products ranging from cosmetics to agricultural products and general recycling initiatives that involve grinding, abrasion, or heat. In Figure 11.13, for example, the illustrated accidental discharge could potentially be from a large nanomaterial manufacturing facility (at least in Figure 11.13 other nanomaterials are shown treating the discharge). Nanoparticles from any of these sources might then find their way into water, plants, and other materials. Unbound particles might again come into contact with humans and animals and even be ingested or inhaled by them, with potentially adverse health effects (see Section 11.3). If accumulations occur, concentrations might increase with time and ultimately affect animals higher up the food chain. Questions have been raised as to whether the health of whole plant communities themselves could be affected. Though it is broadly acknowledged that current quantities of nanomaterials produced today are relatively small and current impacts not large, the issue arises when nanomaterials become far more widely produced and used in many different products and processes.

Many concerns stem from observations that the very same physical and chemical properties of nanomaterials that often lead to positive use and environmental benefits can also potentially cause health problems in human and animal communities. As we have seen, the tiny sizes of nanoparticles that make them so useful in a host of applications can also be below the sizes of common airborne particulates in our atmosphere that are known to cause adverse health effects. The high surface areas that can also be so positively exploited can potentially also be associated with toxicologically negative effects as surface reactivities can increase. In the discussion on positive uses of nanomaterials in drug delivery, we noted that transport through the body can be enhanced via the use of nanomaterials. This same property can be problematic if the wrong substance is transported that might negatively damage human cells. For negative effects to occur, undesired nanomaterials would have to find their way into bodies via routes such as direct ingestion, inhalation, or dermal means. These routes, in turn, can be affected by the nature of the overall environment that surrounds us.

Section 11.3 outlined the general nature of potential health effects on humans; it will not be repeated here. There it was also noted that a critical component of the health assessment issue was whether or not nanoparticles where strongly bound to some host to the extent that they could not easily become free to potentially enter human bodies. Strongly bound particles were noted as less likely to be problematic than free or weakly bound ones. This issue is a critical one in thinking about more general environmental effects and risks.

Serious research is beginning to emerge in this area. Many questions are simply unanswered or poorly understood. Questions raised are many and include the following: What are sources and means by which nanomaterials might enter the environment (briefly outlined in this chapter but for which more quantitative understandings are needed)? What specific compositions, types, and sizes of nanomaterials are most likely to enter various parts of an ecosystem? If and how are particular nanomaterials subsequently dispersed or accumulated in various parts of an ecological system? What factors influence bioaccumulations? How might nanomaterials in the environment possibly bind to other substances already present, or interact with them? How might they be broadly transported throughout a system? What are specific pathways for nanomaterials in the environment to enter specific types of animal, plant, or human communities, and are there any special concentration or bioinduced accumulations or magnification that could take place? What kinds of chemical or physical transformations occur during these many processes? What specific types of effects, including adverse ones, are there on the many components of an ecological system—for example, how are specific plant or aquatic communities, such as algae, affected?

The research community is beginning to deal with these and other questions. Relatively little is known about them, and individuals seeking definitive and validated "answers" concerning the impact of nanomaterials on the environment in general or on various ecological communities will not find them. Whether nanomaterials of one type or another can make their way into the environment is not the basic question being addressed; rather, it is one of type, extent, and subsequent impact, as quantitatively understood. Many research groups are focusing on one or another of the research needs noted in this chapter, and evaluations of environmental risks are part of many formal research overviews and reports sponsored by various governmental and scientific organizations concerned with the environment (see Further Reading).

Workplace Sources and Exposures

Some settings are expected to be particularly problematic and should receive special mitigation attention. Exposure through inhalation or ingestion of loose nanoparticles was noted earlier as being particularly problematic, even more so if exposures are prolonged. An obvious area in which special precautions should be taken to minimize risks is in the very manufacturing environments where nanomaterials are actually produced. Other workplace exposures and sources are expected to multiply in the future as more and more products are made using nanomaterials or more processes are facilitated by them. Even if discharges are not accidental, methods for treating nanowaste products from manufacturing sources before they are released are still in their infancy.

During production and packaging processes, possibilities exist for exposures to unbound nanoparticles that are airborne or in sprays to be accidentally inhaled or ingested by workers. Many manufacturing methods involve high temperatures; hence containment is a normal approach to reduce exposure risks. Nonetheless, escapes and leaks can potentially occur during various stages of the manufacturing and packaging process. As with any manufacturing process involving hazardous chemicals, special care should be exercised with respect to the production of nanomaterials. Especially important are monitoring systems to detect the presence of airborne particles. Some monitoring approaches exist, but with the variation of nanomaterials types, compositions, and sizes, additional techniques may be needed.

FURTHER READING

Center for Biological and Environmental Nanotechnology, CBEN, Rice University.

Department for Environment, Characterizing the potential risks posed by engineered nanoparticles, Food and Rural Affairs Report, U.K.

Environmental futures research in nanoscale science, engineering and technology, Science to Achieve Results (STAR) Program, National Center for Environmental Research.

S. Kaur, R. Gopal, W. J. Ng, S. Ramakrisha, and T. Matsura, Next-generation fibrous media for water treatment, MRS Bulletin, Vol. 33, January 2008.

T. Masciangioli and Wei-xian Zhang, Environmental technologies at the nanoscale, Environmental Science and Technology, American Chemical Society, 2003.

Friends of the Earth, Nanomaterials, sunscreens and cosmetics: Small ingredients, big risks, report, May 2006.

The International Workshop on Marine Pollution and the Impact of Seawater Desalination, The International Center for Biosaline Agriculture Headquarters, Dubai, UAE, December 1–3, 2003.

National Nanotechnology Initiative: The initiative and implementation plan, NSTC/NSESTS report, March 2001.

The Royal Society and the Royal Academy of Engineering, Nanoscience and nanotechnologies, U.K., 2004.

M. Roco and W. Bainbridge (eds.), Societal implications of nanoscience and nanotechnology, Kluwer Academic Publishers, 2001.

M.C. Roco, R. Williams and P. Alivasatos (eds.), Nanotechnology research directions: IWGN Workshop report, Kluwer Academic Publishers, 1999.

Steven Schnittger and Sinha Moitreeyee, The materials science of cosmetics, MRS Bulletin, Vol. 32, ppf. 760, October 2007.

Mark Shannon and Raphael Semiat, Advancing materials and technologies for water purification, MRS Bulletin, Vol. 33, ppf. 9, January 2008.

K. Shihab, Bino Murad, Marc Andelman, and Ben Craft, Flow-through capacitor technology, In: G. Tchobanoglous, F. L. Burton, H. D. Stensel, Waster water engineering treatment and resuse, 4th ed., McGraw-Hill, 2004.

G. Tchobanoglous, F. L. Burton, H. D. Stensel, Water waste engineering treatment and reuse, McGraw-Hill, 2004.

Wei-xian Zhang and Daniel W. Elliott, Applications of iron nanoparticles for groundwater remediation, Remediation Journal, Vol. 16, Issue 2, pp. 7–21.

CHAPTER 12

The Broader Context

12.1 INDUSTRY PERSPECTIVES

Our economy is made of up many sectors—transportation (automotive, aerospace), sports, textiles, construction, and others—that in one way or another involve design and development activities that are heavily dependent on materials. Though interest in virtually all these industries has been expressed, the actual type, degree, and rate of application development of nanomaterials and nanotechnologies in various industries depend on many factors. This chapter takes a whirlwind tour across several industries to suggest a sense of the role nanomaterials and nanotechnologies are or are not playing in various sectors.

In looking across multiple industries, some of the driving forces behind research and developments in the nanomaterials area are broadly similar and generally include:

- Improved product performances

- Improved use experiences

- Improved reliability and durability

- Achieving safety and health objectives

- Improved production techniques and rates

- Cost reductions

- Improved utilization of material resources

- Energy minimization

- Component miniaturization and multifunctionality

- Improved product choice

Nanomaterials, Nanotechnologies and Design

- Improved appearances (in many industries)

- Development of new products and opening of new markets by taking advantage of emerging technologies

Obviously, not all these drivers are independent. Many are closely intertwined. In many industries they have also long been historical drivers behind the development of any new technology or material innovation, not least because they ultimately translate into the economic health of an industry or the profitability of producing companies. They remain with us today. Other drivers have also recently developed—some because of basic sensibilities concerning societal directions, others because of either demand or imposed regulations. These include:

- Use of more environmentally benign materials and processes, both for production and for material resource recovery, including recycling

- Use of limited energy resources

Older scope definitions of product safety and environmental impact have also now widened considerably and affect many industries because more attention is now paid to the way that adverse effects attributable to a product can be transmitted through interconnected chains and networks to contexts that are seemingly far removed.

Though these drivers are somewhat common at the abstract level, various industries still place differing importance on different factors, as well as having their own specific motivators for potentially using nanomaterials. In the following, we look briefly at a variety of industries in a broad summary fashion. Many of the specific nano-based technologies referred to here were discussed in earlier chapters.

12.2 THE AUTOMOTIVE INDUSTRY

The automotive industry is already one of the world's largest users of nanomaterials, and expectations are that uses will increase. General Motors was an early adopter in 2002 with its use of thermoplastic olefin (TPO) nanoclay composites in running board step-assists. Hundreds of thousands of pounds of the material are now used. Other companies, including Daimler, Nissan, and Ford, are using nanopaints for improved scratch resistance and appearance. Intensive research is in progress on many other fronts, including propulsion and energy systems that use nanotechnologies (e.g., fuel cells and batteries). Some of the primary drivers or motivators for these initiatives include needs for the following:

- Improved overall driving performance and comfort

- Performance improvements in basic systems and components

- Safety improvements

- Improved energy and propulsion systems that are more efficient and environmentally benign (including reduced emissions)

- Use of more environmentally friendly materials (including recyclability considerations)

- Reduced weight

- Improved advanced electronic control and communication systems and accompanying sensor technologies

- Improved appearances and user comfort and amenities

- Improved durability, reliability, and cost efficiency

The factors are not independent. Reduced weights, for example, can lead to improved fuel efficiency, reduced emissions, and other benefits.

The automotive industry is by no means homogeneous, and drivers vary according to product type and market. In some companies with high-value products, the potential of small performance increments can and does justify extensive (and often expensive) design and material innovations. In other companies, cost concerns dominate, and new materials will be less likely to be introduced in a widespread way unless their costs are on a par with or below those of conventional materials for equivalent performance levels.

The role of nanomaterials and nanotechnologies in this area can be rather simply thought of with respect to basic systems and components (including propulsion and energy systems, suspension systems, and others), frame and exterior body components, interiors, and control and communication systems. Surface appearances, including finishing and painting, are invariably important for visible components (see Figure 12.1).

In the *basic systems and components* area, improvements in basic propulsion systems are always sought. As attention turns to hybrid and all-electric vehicles, there is an obvious need for improved energy

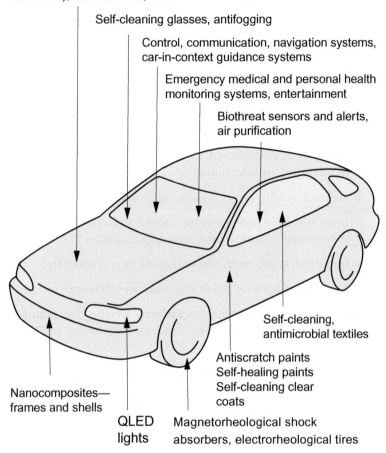

Many electronic and mechanical components: emission controls, fuel cells, batteries, supercapacitors, tribological applications (wear, lubricants), nanocoolants, other

Self-cleaning glasses, antifogging

Control, communication, navigation systems, car-in-context guidance systems

Emergency medical and personal health monitoring systems, entertainment

Biothreat sensors and alerts, air purification

Self-cleaning, antimicrobial textiles

Antiscratch paints
Self-healing paints
Self-cleaning clear coats

Nanocomposites—frames and shells

QLED lights

Magnetorheological shock absorbers, electrorheological tires

FIGURE 12.1

Opportunities for using nanomaterials and nanotechnologies in automobiles.

sources. The "battery barrier" still remains a powerful limitation in the move toward fully electric vehicles or more efficient hybrids. Nonetheless, improvements in lithium batteries are anticipated in both capacity and recharge rates. Attention for automobiles remains primarily focused on the further development of fuel-cell technologies and the role that nanomaterials could play in improving efficiencies. The potential for significant improvements in fuel cells over current technologies is extremely high. A variety of nano-based materials and technologies can contribute here, including new nanomembrane technologies. Various other fuel-cell approaches are also being explored, including solid-state systems. Not all fuel-cell approaches will prove useful for automobiles.

Design demands for all-electric vehicles are high. A very high energy density is needed for the propulsion system, but there are also ever-present pressures to reduce the amount of space occupied by both the fuel cells and supporting fuel tanks for storing the hydrogen. Known sensitivities to humidity variations must be addressed. Durabilily demands are extremely high. Many current fuel cells are relatively sensitive to mishandling, so high levels of ruggedness are required (also see Section 9.7).

The hydrogen storage problem is particularly acute for vehicles. Large quantities of fuel must be able to be stored and replenished safely. Several options are under exploration. Compressed gas hydrogen is possible but quite costly due to the need for expensive tanks. Liquid hydrogen storage has been explored as well, but significant inefficiencies and energy losses occur because of the necessary conversion processes. Combination gas/liquid systems are also being explored. Solid-state materials may ultimately prove to be the best storage medium if needed storage densities and charge and release rates can be achieved. These are difficult problems. Still, materials such as hydrogen-absorbing hydrides or others may prove effective for onboard storage of hydrogen for fuel-cell vehicles.

Despite the interest in electric propulsion, the internal combustion engine is by no means out of the picture. Its power and other advantages will give it a place in vehicle fleets for a long time. The internal combustion engine has also by no means seen the last of improvements and efficiency increases made possible through both design and material advances. Nanocomposites can help meet their well-identified role of providing strength, hardness, and wear-resistance improvements for many components, ranging from cylinder liners and pistons to a host of other parts, including bearings. Efficiencies can also be improved by limiting losses in engine-generated thermal energy—particularly important in diesel engines. Various nanocrystalline ceramics that have both hardness and heat-retention capabilities, such as zirconia, are being explored for use as cylinder liners. The interesting field of nanotribological studies, which deals with friction, has obvious applications in the design of the many mechanisms and linkages present. Even traditional spark-plugs are being rethought to make them more resistant to erosion or corrosion.

The catalytic characteristics of many nanomaterials that have been discussed earlier have the potential to further reduce unwanted emissions. Currently, precious metals such as platinum are used in catalytic converters. This material is costly and subject to shortages.

Catalytic reactivity of platinum can be increased by using platinum nanoparticles because of their much greater surface area. Efficiency increases can lead to less use of platinum. Other kinds of catalytic materials are being explored as well.

Other basic components include suspension systems that help maintain control of an automobile by increasing contacts between tires and roads and minimizing rolling or pitching. These systems also help ensure occupant comfort by mitigating the effects of road bumps. Seemingly humble shock absorbers, constituent elements of a suspension system, are evolving into amazingly sophisticated devices via the use of magnetorheological fluids. The viscosity of these fluids can be altered by changing a surrounding magnetic field (see Section 9.2). Electrorheological technologies can be similarly used. These viscosity changes, in turn, can be used to control vibrations and various road shocks. These absorbers can also be connected to sensor nets and their viscosities electronically controlled to be appropriate for particular road conditions or driving styles. In some automobiles, the driver can also control stiffness for either sport driving, which typically involves intensive cornering, or highway cruising. Another approach uses electromagnetic motors and power amplifiers. Each of these approaches involves utilizing magnetic phenomena that can in turn be enhanced by the remarkable magnetic properties of many nanomaterials (see Chapter 7). Cobalt nanoparticles, for example, exhibit high ferromagnetism. All invariably involved sophisticated electronic sensor-based control systems that include complex computational means of relating needed viscosity levels to specific automobile performance needs (e.g., preventing rolling motions).

Tires, of course, are of fundamental importance in both basic control and to the overall driving experience. They are the subject of intense research, with various nanomaterials found to improve wear and abrasion resistance as well as provide improvements in fatigue resistance. Carbon black, one of the early synthetic nanomaterials, is widely used in rubber tire production to improve mechanical properties and manufacturability. Other nanomaterials, such as silica nanoparticles, are being explored for use as well to improve basic tire properties and overcome many existing problems with carbon black. In other areas, electrorheological approaches are being tried for use to vary tire stiffness and aid in driving control in much the same way as described for active suspension systems. In light of the need to develop more environmentally friendly and recyclable materials, tires are viewed as a target for innovations in this area. Various elastomer and inorganic oxide formulations, for example, are being explored. Much needs to be done here, particularly since the problem of disposing of or recycling used tires is widely viewed as a major environmental issue.

There are far more emerging applications in basic systems and components than can be even briefly touched on here. Polymer nanocomposites with bulk carbon nanoparticles/fillers in a polymer matrix, for example, have interesting electrical conductivity and mechanical properties and are beginning to find use in various components, including electrical connects or for static-dissipative applications. Many of the nano-based coatings described in Chapter 9 that improve surface/edge hardness or toughness are finding wide use as well. Many of these are based on nano carbon-based films. Nano-based films and coatings have been developed for corrosion resistance of parts. Lubricants are being improved through the use of nanoparticles. Heat exchange fluids used in coolants are benefiting from suspended and dispersed nanoparticles, since they can conduct heat better than basic fluids. The list goes on.

There is considerable interest in strengthening basic frame and body components to improve safety and reduce weights. Durability, repairability, and recyclability are also factors, as is receptivity to painting, for body components. Manufacturability considerations are always present. Performance requirements for specific elements vary dramatically. Many frame elements need to have high structural strengths and stiffness as well as energy absorption characteristics for crashworthiness. For some body elements with highly complex geometries, strength and stiffness demands may be minimal, but manufacturing considerations may drive material selections. In other cases, surface finishes may drive considerations. In all cases, however, weight is invariably important and lightweight solutions are always preferred; hence continued research is always ongoing in materials such as aluminum and magnesium. Overall industry drivers also come into play, such as the need to use more environmentally benign materials that can be recycled.

Depending on the component, and hence the mechanical, thermal, or other design requirements placed on it, many types of metal matrix composites, polymer nanocomposites, carbon-fiber products with embedded nanoparticles, and other approaches are being explored, as are more traditional bulk materials with nanocrystalline structures. Many interesting nanocomposites based on carbon nanotubes, for example, are currently considered cost prohibitive for common consumer vehicles, but this could change. For many applications needing light, strong materials, metals reinforced with ceramic fibers (such as silicon carbide) are attractive.

Polymer-based nanocomposites are finding use in many body components. Both thermoplastics and thermosets are in use. Thermoplastic materials can be shaped quite easily. The early use of thermoplastic olefin (TPO) has proven successful for use in certain interior and exterior elements (step-assists, panels). Lightweight nanocomposites of this type use relatively inexpensive nanoclays and nanotalcs. The TPO normally used employs natural smectite clay. It is treated so that exfoliation occurs and molecular sheets are formed. Yet issues of cost, quality, and strength remain. Exfoliation methods normally need high quantities of additives that can be a drawback, although several approaches are alleviating this problem. Several explorations are in progress in relation to thermosets, which can be quite strong, that involve the use of various kinds of nanomaterials in the resins. Carbon nanotubes have been explored for use here.

Paints, coatings, and other surface treatments are obvious targets for the use of nanomaterials in the automotive industry (see Figure 12.2). Many highly important design characteristics of materials relate primarily to their surfaces and less to their bulk or mass characteristics. Nano-based paints were among the first products to make wide use of nanomaterials to improve rheological properties and provide better bonds to substrates. Chapter 10 discusses nano-based paints in more detail. There it is also noted that finishes are smoother and more durable than in the past. Color consistency and clearness are generally improved.

FIGURE 12.2

These pigments are used for automotive coatings. The aluminum flakes are coated with a layer of iron oxide only a few nanometers thick. They are smooth, reflect light well, and help achieve bright colors. Also see Chapter 10. (Courtesy of BASF.)

Nano-based paints and coatings can also provide improved scratch resistance. These kinds of paints are now fairly common, particularly since performance is significantly improved while cost increases are relatively minor due to the kinds of nanoparticles used. In automobile painting, several different layers and application processes are employed. Nano-based materials are often used in final clear coats in a typical five-layer automotive finish (see Chapter 10). Self-cleaning and self-healing actions have also been developed and are also now finding their way into paints. Other kinds of coatings for frame and other components are less about appearance and more about prevention of corrosion. In addition to nano-based paints, the automotive industry has made use of other kinds of coatings that exhibit specific behavioral actions. These include antifogging and antireflection coatings for glasses. Many types of nano-based polymer films are in use, as are nanoclays. Hydrophobic fumed silica nanoparticles, for example, are widely used in many applications.

Interiors also benefit from innovations in nano-based paints and coatings on surfaces and glasses, for similar reasons. The fascinating developments of nano-based textiles discussed in more detail in Section 10.4, including self-cleaning and antimicrobial actions, clearly have relevance here for seating and other soft surfaces. Additionally, the need for fire retardancy is high in all types of hard and soft surfaces. Various kinds of oxide nanoparticles and organic nanoclays are being explored to achieve fire retardancy in various surfaces and for use as fire-retardant fillers in polymers. Many other material properties come into play in the design of interiors. Tactile qualities and whether materials feel cold or warm to the touch are important (see Chapter 5). Seemingly simple objects such as cup holders could benefit from improved insulation materials (see Section 9.3).

The potential impact that nanomaterials can have on the design of electronic systems is simply vast and beyond the scope of this book, and so is the potential impact on control and communication systems used in automobiles. While components grow smaller, processing speeds increase. Real-time control systems and the response times needed are possible now and will only improve in the future. The controllable responses made possible through the active suspension systems we've described are only indicative of future possibilities. Of particular interest is the potential value of the many kinds of sensors and related control systems, with potential for the overall driving experience to be both aided and enhanced. The latter include systems that understand and aid in the behaviors of an automobile within a traffic context and even potentially control them, or, minimally, provide warnings, such as following too closely. Safety can surely be improved by a host of these means. In other areas, display and other communication systems will become better. Driving safety, for example, can be improved via enhanced road views projected onto dye-coated windshields (heads-up or other kinds of displays) that highlight hazards or warnings as part of the viewer experience—for example, crossing pedestrians are highlighted.

Improved way finding and navigation systems are surely expected. Driver alert systems are already being explored and could save many lives. In this unfortunate age of biothreats, biosensors that can provide warnings are certainly feasible. Though only occasionally discussed, the potential of sensors and devices for health monitoring and communication is surely there. Here there are detection and communication of vital signs during crashes; communication of similar information during medical emergencies (heart attacks, stroke, or the like) as well as driver and vehicle status information, and more ongoing personal health maintenance, such as blood pressure levels.

12.3 THE BUILDING AND CONSTRUCTION INDUSTRY

Directions

Large portions of this book have addressed specific needs and potential uses of nanomaterials in the building industry. Industry characteristics were also discussed in Chapter 3. Still, it is useful to view the topic in a summary fashion on a par with other industry discussions in this chapter, simply to put it into perspective. A question always of interest to architects, for example, is whether the essential visible forms of architecture will undergo radical or disruptive changes as a consequence of developments in nanomaterial and nanotechnologies, as was the case with the faddish plethora of "blob" architecture in the last decade, when architects discovered powerful digital modeling tools. As we will see, this is a highly doubtful scenario. Most changes will be largely incremental and many seemingly "invisible" to the eye, although some will indeed have strikingly visible manifestations (the obvious referent here is in relation to light and optical phenomena). Highly positive benefits that have less visible manifestations, however, will accrue through performance increases in the many systems and components of a building. Performance attributes *will* improve. Our buildings *will* be better. They *will* serve society better.

In the list of common drivers to all industries noted in the opening of this section, the ones related to environmental impacts and the use of environmentally benign materials, energy use, health and safety, and costs stand out as very important basic concerns. Achieving good-quality indoor thermal and lighting environments is a primary driver, as is providing buildings that are healthy for occupants. Improved work and task environments are always important. The quality of use experiences that stem from either interesting features or spatial compositions or simply good functional designs remain fundamental as well. Needs or desires to differentiate buildings from one another, in many sectors, remain highly important (witness the success of designs from the office of Frank Gehry, an architect well known for his unusual buildings). Specific material or technological innovations that relate to color, movement, and interactivity are hungrily seized on.

At the broad level, many nano-based materials and technologies are poised to improve basic living and work environments. From an environmental point of view, advances are certainly expected in the quality of our lighting environments. Improvements are expected in both artificial and the provision of natural lighting due to improved light-emitting sources (types, control of wavelengths, illumination characteristics); improved control of light reflection, absorption, and transmission (surfaces, windows), and control systems (see the following and Section 9.5). Improvements are expected in air and water quality through improved cleaning and purification systems based on nanotechnologies. Air environments have the potential to become healthier due not only to improved air-cleaning techniques but the potential of using nanocoatings and other approaches to help reduce outgassing and the generation of airborne particulates from many building elements that cause the well-known "sick building" syndrome. Use of surfaces or materials with antimicrobial behaviors can contribute to healthier buildings in general, certainly in medical facilities. Exterior surfaces and paving made with nanocoatings or nanomaterials can potentially help reduce air pollution. Surfaces may be made more fire retardant and less likely to give off toxic gases in fire circumstances through the use of various nanocoatings and films. Thermal environments can be more energy efficient, and highly improved means of energy production are at hand. Actual impacts on comfort levels, however, may be less dramatic than expected for lighting. Few really dramatic improvements are expected in common passive acoustic environments, although there will be improved vibration and impact noise control and great improvements in sound production systems.

Significant acoustic improvements are expected in spaces dominated by electroacoustic devices.

Advances are also expected in approaches that aid in task performance in workplaces or homes. Many buildings are by now seeming extensions of sophisticated "information and display systems" that respond to the needs of various commerce and manufacturing activities. This trend will surely continue. In home environments, analogous information and display systems are expected for a range of life-enhancing activities, including entertainment. There is great interest in incorporating sophisticated human health monitoring and aid systems into housing and other buildings to meet a number of objectives, including general recovery from medical operations, aiding individuals in rehabilitation programs, and aiding the elderly to healthily "age in place," as many so desire. Advances in sensory technologies, including biosensors for identifying and communicating warnings of chemical or other bioterrorism threats, are also expected to benefit the building industry.

As we've seen, significant improvements are expected in enabling electronic and electromechanical control systems and related sensory technologies. These same technologies are also enabling systems in a larger thrust toward "intelligent buildings" (Section 9.8). Though this concept has been around for quite some time and particularly flowered during recent interests in smart material applications, it is experiencing renewed interest via nanotechnologies, with their possibilities of both easy embedment and extensive distribution. Intelligent building technologies are not only about improved control systems, advanced sensory technologies, or interfaces, however; they also involve a much more complex interaction among technologies, responsive materials, and human use needs that is guided by cognition goals.

Systems

Environments of the type described are ultimately made using many products made for use in the construction industry or by actual on-site construction. The industry that produces buildings, bridges, and other works is enormous and highly fragmented. Buildings utilize a surprisingly diverse array of products. A host of independent producers and suppliers provide these many products that find their way into buildings—far more so than is common with other industries. There are two simple parallel ways of conceptualizing product developments within this confusing context;

one is by building systems and another is to view them in terms of levels within a technology sophistication chain.

Many building products play roles in relation to *monitoring, control, and communication systems*. Many of these systems sense indoor air environment characteristics and in turn provide the logic and controls that allow various building environmental systems to maintain design levels. Other systems similarly deal with lighting or communications. Yet others deal with fire detection, suppression, or other needs. The list is surprisingly long. Many high-value products used in buildings find their uses in these systems, typically in the form of sensors, actuators, or logic controllers. These products are technologically complex and are in the top range of any kind of technology sophistication value chain. Most used within the building industry might have been tailored specifically for building use but usually have underlying technologies that have been developed primarily by other industries and in response to driving forces in those industries. This is often the case with mechanical or electronic products (such as sensor-based control systems used in everything from heating, ventilating, and air-conditioning systems to escalators). Surely the building industry will benefit from the many nano-based technologies associated with the expected rapid advances in electronic and related communication technologies. Certainly there will be enormous benefits from new optoelectronic display systems.

Many buildings are literally all about "information" and can be expected to use new display technologies (Section 9.5). In these general types of monitoring, control, and communication products, nanomaterial-led innovations will benefit the building industry, but the industry will probably only rarely provide the primary technology development leadership for these products. It will, however, provide a big market for tailored products.

Enclosure systems form an essential part of any building. External enclosures provide fundamental thermal, weather, and sound barriers and mediators. Façade elements, such as windows or glass curtain walls, provide not only visual access to the outside world but also are light mediators. Interior walls zone building spaces and provide their own privacy, sound, and light control functions. As observed in earlier chapters, the specific thermal, optical, and sound properties of these many elements are essential in providing comfortable and habitable environments. Both external and internal enclosure elements also provide the visual statement that users and designers associate with the "architecture" of the building; hence appearances, textures, finishes, and the like are highly important as well. Products produced for these systems vary widely in technological sophistication. Glasses and many kinds of panels can be extremely sophisticated, whereas other buildings might use walls made of very low-tech products.

Nonetheless, the huge surface areas associated with enclosure systems make them a natural target for sophisticated products that are surface oriented. If there is one thing that buildings have in comparison with other artifacts, it is surface area. Hence, many applications can be expected in the "surface" or "enclosure" category, such as glasses, paints, and coatings, and have either functional or useful visual properties. As we saw in Chapter 10, many very interesting and useful innovations based on the use of nanomaterials are taking place here, such as various thin films, coatings, and others. Products with self-cleaning, easy-cleaning, antimold, or other behaviors of this type are particularly attractive and can be expected to find more use (see Figure 12.3). Many of these materials are based on photocatalytic, hydrophilic, or hydrophobic properties made possible through the use of nano-based films or other surface treatments. Products for flame retardancy and other mandated safety objectives are also poised to benefit from the use of nanomaterials. Reduced toxicity of emissions during fire circumstances is similarly

FIGURE 12.3
This church in Korea uses a self-cleaning photocatalytic surface. Also see Chapter 10. (Courtesy of Tayo Kogyo).

important. As noted, outgassing from many surface-covering materials, which can contribute to indoor air pollution, can similarly be potentially minimized through nano-based coatings. Despite the importance of these latter topics, there seems to be little current research attention focused upon them.

Many other products are surface-oriented or form part of surface-forming enclosure systems and include thermal insulation materials. Nanoporous materials that can be made into large surface forms, such as aerogels, are expected to continue to make real advances here, as are improved phase-change materials. The long-desired "transparent insulation" material for use in building façades could become a viable reality (or at least a highly translucent version of it); see Section 9.3. Buildings also use extensive amounts of polymeric sheets or coatings, and the many expected advances in polymers through nanofillers or nanoparticles can potentially be beneficial here. The same is true with textiles.

Many of the products made for enclosure systems are quite sophisticated and relatively high up on the technology value chain. Producers making these products respond to building industry needs often lead in developing new technologies for the industry, but they also rely on broader distributions and markets outside the building industry to help support research and development activities. Some of the most exciting uses of nanomaterials in relation to the building industry are found in these products. Products can be designed to respond to many of the currently important industry drivers, such as energy conservation, while uses and markets are big enough to support technologically sophisticated responses.

Lighting and other environmental systems are also natural places for rapid developments in relation to nanotechnologies. Advances enabled by nanotechnologies in the lighting area promise to be highly significant, not only for the quality of lighting but in relation to reduced energy consumption as well. Light-emitting diodes (LEDs) are already in wide use and have become a major product within the building industry. Expectations are that they could evolve into more efficient and frequency-adjustable quantum-based devices (QLEDs; see Section 9.5). Coupled with advances in the optical properties of thin-film or coating technologies based on nanomaterials, highly interesting advances can be expected in the design of lighting environments. Other environmental systems are expected to benefit as well. There is, of course, widespread interest in using renewable or alternative energy sources, as well as making traditional systems more efficient. The idea of a building being independent of common energy sources via the use of solar cells, of course, is so prevalent that it hardly needs mention. Yet widespread energy self-sufficiency is a very long way away. Nanotechnologies will certainly help make solar cells far more efficient than currently (see Section 9.7) and will come in forms that allow more versatile placement and use. Improvements can be expected in the form of other energy sources, such as geothermal. Also important is the potential for improved efficiencies in the plethora of systems that distribute energy throughout buildings, including ubiquitous elements such as heat exchangers or the like (Section 9.3). As noted, improved enclosure systems will reduce heating and cooling demands. Nanoporous materials, or nanomaterials with catalytic properties, are expected to aid in achieving clean air and clean water objectives within buildings.

Structural systems obviously provide a fundamental role in buildings. Included here are structural system elements made from long-used conventional materials, especially steel and reinforced concrete. Aluminum and other metals can be used in certain circumstances as well. These traditional materials have long been the subject of intensive research and development. The basic materials used are already surprisingly sophisticated—advanced metallurgy in steel has certainly had its successes—and nanomaterials can be expected to play a role in their further evolution, such as metal matrix nano-composites. The development of sensor-based structural systems that have either embedded damage detection technologies or active responses to changing loading or force conditions that are enabled by sophisticated control systems are undoubtedly areas that could benefit from nanotechnologies. In these cases, member sizes can be expected to reduce. Many devices are being developed, for example, for controlling damaging effects of seismic ground motions or wind forces. The future is bright here.

In construction, however, it should always be remembered that vast amounts of common building activities use primary materials in massive quantities – often in situations where performance requirements are not particularly high – and in relatively low-value contexts. This is not a picture that fosters widespread technologically based revolutions to improve performances if costs for using innovative new materials escalate as well. This picture differs in comparison with other industries. A driving force in selecting metals for use in the structures for aerospace industry, for example, is to use materials with high strength-to-weight or stiffness-to-weight ratios, so that overall product weights can be reduced. The same driver is present in the automotive industry. Benefits of reduced weights are enormous (improved performance, reduced fuel consumption, etc.) and can warrant expected higher material costs. The same driver is simply not true to anywhere near the same extent in most common buildings. Though interesting and exciting exceptions certainly do exist, this absent primary driver and unfavorable cost/benefit pictures can tend to inhibit innovation. Innovations such as the use metal matrix composites or in response-control systems are thus expected to occur first in specialized or very high-value buildings (see the discussion on this topic in Chapter 3), not your average commercial or residential construction.

Nonetheless, advances are being made and incremental improvements can be expected. With reinforced concrete, for example, improvements based on nanoparticle inclusions not only can improve strengths, they can improve workability and other factors as well (Sections 9.2 and 10.2). Self-cleaning actions or pollution-reducing treatments offer interesting potential. For metal members and various kinds of structural panels, various kinds of nano-based coatings and paints will undoubtedly become more widely used.

Ancillary elements serve elemental but highly necessary functions. A host of products fall into this category, such as common railings, hinges, and myriad others. Most typical products in this area are made exclusively for the building industry and are fairly far down on the technology sophistication chain. Rarely is there appreciable sophistication here in basic bulk material usage, and, consequently, prospects for positive improvements via the use of nanomaterials are currently limited. Exceptions again go toward surface-related interventions such as paints and coatings for these products. Improvements in mechanical properties, such as scratch resistance, are important. Flame retardant and toxicity-reducing coatings are potentially highly important. Nanocoatings can potentially help reduce outgassing from common material such as carpets that can contribute to indoor air pollution. At the very bottom of the technology value chain are products that are widely used in construction but are little removed from raw material product forms. Materials such as gravels fall into this category. Little is expected here.

In summary, nanomaterials and nanotechnologies are expected to contribute to greatly improved living and work environments. More dramatic improvements are expected in specific high-value systems and components. Others will be initially less impacted. The most literally visible of these projected impacts—and keep in mind that creating positive visual constructs is an important and valued province of architects—is expected to be within the realm of light, color, and interactivity. Possibilities are significant here. Improved thermal control means may reduce the more visible manifestations of needs for insulation and the like. Dramatic advances in structural nanocomposites will improve the performance of structures and ultimately lead to component shape changes and size reductions but are not expected to immediately affect the organizational and visual characteristics of passive structural systems to any big degree. Improvements in active nano-based monitoring and control systems and the development of "responsive" structural systems, however, will have greater visible impacts. Surface characteristics of all elements can potentially be positively improved (self-cleaning, scratch-resistant, anti-microbial, and so on). Will any nanotechnology actually be "disruptive" and change the fundamental nature of architecture? Probably not, but improvements *will* indeed occur.

12.4 AEROSPACE, TEXTILES, SPORTS, AND OTHER INDUSTRIES

Developments based on nanomaterials and nanotechnologies that are expected in the electronics industry underpin many other industries (automotive, aerospace, building). Huge research efforts are under way in the *electronics industry*, and many applications are being rapidly developed. It is abundantly clear that enormous performance and other gains can occur via the use of nanomaterials and nanotechnologies in many areas, ranging from chips and communication systems to a host of optoelectrical devices. The potential for performance and reliability improvements through exploitation of the unique magnetic and electric properties of nanomaterials is undoubtedly a primary driver for this industry. Many devices are of high value and command high prices for even incremental performance improvements. Other devices may be of lesser value but are produced in large quantities so that expensive research and development costs directed toward performance improvements can be absorbed. A second primary driver is also the push toward smallness in many products. A huge number of electronic products serve as components in a host of products produced by other industries and serve to provide important enabling or ancillary functions; however, since they are not the final product in themselves, there are always pushes to minimize their size. At the same time, their decreased sizes allow designers to be more aggressive about integrating them into product lines or types where they were not present before, thus increasing demand. (Consider the change in the nature of products produced by the toy industry in recent years as a consequence of the miniaturization of electronics.)

Other industries have different specific drivers. The *aerospace industry*, for example, has long been both a developer and user of advanced technologies in many areas. In addition to obvious drivers related to aircraft operation and safety—the vast potential of nanotechnologies in relation to electronic and electromechanical control systems deserves special remark here—the industry is also always necessarily concerned with both carrying capacity and speed. These factors have obvious positive economic value in several ways. Operating costs, particularly fuel, remain a major industry problem. Various components within the industry seek to achieve different balances among capacity, speed, operating costs, and user pricing, depending on their markets. The demise of the Concorde reminds us that very high-performance aircraft capable of high speeds can fail if capacities are limited and needed price structures to support costs are high.

Fuel is a major cost component in any balance equation in the aerospace industry. Fuel consumption, in turn, is dependent on many factors, including propulsion efficiency—an area in which nanomaterials can contribute in a number of ways that are far beyond the scope of this summary to consider. Fuel consumption is also highly dependent on weight. Hence there is an ever-present press to reduce weights of all system components and infrastructures so that higher payloads can be carried. The emphasis on reducing weight characterizes many of the pushes underlying explorations of many systems. Within supporting electronic and communication systems, the aerospace industry is poised to be a major benefactor of the electronics industry's push toward miniaturization that is producing higher-performance products that are smaller, lighter, and easier to integrate. The same push toward lightness is also undoubtedly a major driver in the development of many nanocomposite materials for mechanical and structural applications. In no other industry do the ideas of strength-to-weight or stiffness-to-weight ratios have equal relevance or importance as design drivers.

As noted in earlier chapters, many nanocomposites can truly play significant roles here. Aggressive exploration and use of nanocomposites, even if still relatively expensive, can potentially pay off due to the high values of final products as well as the fact that surprising numbers of aircraft are actually produced from similar designs. Obviously, all the other unique property characteristics of nanomaterials (thermal, electromagnetic) will be contributors to the improved design of the complex material and propulsion systems and are relevant here as well.

In other areas, environmental comfort levels inside aircraft interiors are far from satisfactory. Interest in improved sound environments is always there, as is interest in improving basic thermal and air environments. There are evolving interests in systems that aid in health monitoring and the communication of health information as well as providing direct aid. Various detection and alert systems for threats from passengers or even crews are being explored. Many of the areas of development already noted for the automobile industry are equally important with respect to aircraft (see Section 12-2).

If we can characterize the electronics and aerospace industries as leading industries in the development of sophisticated technologies, it might be safe to say that not so very long ago the *textiles industry* seemed rather staid. This is interesting in its own right, since historians of technology quite often point to the fabulous early looms of Marie Jacquard and others in 18th century France as fundamental in the development of all subsequent industrial automation but also in the development of the idea of programmable devices (some early looms used perforated cards to control weaving patterns) that contributed to the development of computer logic systems. The incredible rise of synthetic materials, including early nylons and subsequent high-performance textiles in use all around us are inextricably coupled with the textile industry. Current technologies and material use remain sophisticated.

The question facing the industry, however, is where might the next really big level of new development come from? The recent focus of many in the industry on the use of nanomaterials suggests an answer to that fundamental question. As we saw in Section 10.4, applications in textiles are developing rapidly. Many of the drivers common to other industries that were mentioned—needs for improved durability, appearance, and so forth—are the same. In the textile industry, however, the driver of adopting or developing new technologies as a way of opening up new product lines, particularly high-value products, is pushing research and development activities. This is not wishful thinking, either; nanomaterials can help achieve such objectives as textiles that are antimicrobial and can address a whole range of bacterial and fungal agents, or textiles that are self-cleaning and perhaps even self-repairing. Clothing with true antimicrobial actions and some level of self-cleaning would not only promote health objectives—a valued objective in its own right but also an attractive high-value product line—but have useful side effects such as reducing needs for frequent washing and cleaning, which in turn would increase material longevity and decrease environmental impacts. (These types of products are already available, but efficacies could be improved.)

The role of textiles vis-à-vis health promotion and delivery is still in its infancy. The role of textiles in our difficult world of biological threats is important, and expanded roles in relation to personal security threats from bioterrorism are surely anticipated (see Section 10.4 and Figure 12.4). Use of nanocomposites for fibers promises stronger, more durable, and lighter fabrics that are good for all but particularly sought after by the sports industry. The same is true for textiles that are incorporating high-performance thermal and moisture-control materials. Nanostructured aerogels and phase-change materials, for example, have thermal-control properties that are highly attractive. They are already in some products, but their use could be expanded. The recent fascination with "wearable electronics" has its own impetus. Both private and public sectors are promoting research in these many areas.

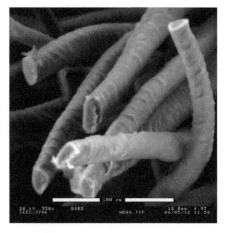

FIGURE 12.4

New textile applications—wool fibers coated with nanolayers for controlling the permeation of warfare and toxic gases. (Courtesy if Juan Hinestrosa.)

For many material scientists, the *sports industry* is one of their most valued connections. This might seem surprising to many, but the high-end segments of the sports industry have long supported innovative developments in materials designed to provide that extra measure of real or perceived performance valued by customers who are willing to pay extra premiums for it. Though many parts of the sports industry cater to mass markets with costs being as critical as in other fields, there has always been a market for products that push performance envelopes that can be paid for by individuals and organizations with largely disposable funds. (In this way this sector is surprisingly similar to the market for luxury goods—an analogy that would no doubt pain many sports lovers.) Examples abound and range from specially designed tennis racquets and golf clubs to high-performance racing bicycles or sailboats. It is open to discussion whether a super-sophisticated golf club actually improves the play and scores of a marginal golfer, but it would in the hands of an expert, and this, in turn, renders the club a desirable artifact for anyone interested in the game (see Figures 12.5 and 12.6).

FIGURE 12.5

This golf ball is engineered with nanoparticles to spin less and reduce slices and hooks. (Courtesy of Nanodynamics.)

Design qualities that are needed in sports equipment vary widely according to the sport and the role of the material object. Driving factors invariably hinge around performance increases, although for some products visual characteristics that suggest high performance are important as well. For many sports products, user safety

FIGURE 12.6

Carbon nanotubes are used to form nanocomposite materials used in hockey sticks, baseball bats, and other sports gear. (Courtesy of Baytubes.)

concerns are a significant driver. Costs are always relevant, but achieving minimum possible costs is rarely a driver. Environmental and other issues are currently by and large less significant than in other industries. (Arguments largely promoted by the industry for not having a high priority placed on these concerns include the basic notion that overall material quantities produced are low compared to products associated with other industries.)

In general, many qualities sought after in materials for the sports industry have to do in one way or another with the mechanical properties of materials. By and large, light weight is desirable—especially so in many products such as high-end racing bicycles. Reduced weight typically means reduced energy expenditure on the part of the sports user, and this available energy can be captured for increased performance by the user. Utilizing high-strength materials is commonly one way to achieve reduced weight, and nanocomposite materials can clearly be appropriate for use here. High stiffness might or might not be a desirable characteristic. It certainly is in some circumstances but is a design variable in others. Depending on the specific golf club type being designed as well as user characteristics, for example, varying degrees of stiffness are usually needed. Thermal properties can be important in many materials—certainly for outdoor trekking and camping gear. Light weights can be also be important in this area. Clothing and coats that have high insulative capabilities while allowing "breathing" in the material fabrics already use high-technology material products (e.g., various materials that are waterproof but allow breathability, or materials that control body temperatures by use of phase-change phenomena). Nanoporous materials such as aerogels are starting to find uses in outdoor coats because of their thermal properties combined with lightness. Depending on the sport, many other qualities can be important.

While the reality of potentially improved performances via the use of sports equipment based on nanotechnologies is positively there, there are also fewer areas that have contributed so much to the hype surrounding nanomaterials. The promise of improved performance via nanomaterials is a common advertising theme for products that might or might not have actually benefited from the use of nano-based materials and technologies. Understanding exactly what is meant can be surprisingly difficult. A material can be advertised as a nanocomposite and highly touted, but finding out the relative percentage by weight of nanomaterials present or what weight or stiffness reductions have been achieved can be virtually impossible and can easily cause informed customers to become suspicious.

12.5 A CLOSING COMMENT

In an opening chapter of this book, a timeline showing various material "ages" was illustrated (Figures 2.1 and 12.7). On the figure, traditionally defined eras such as the Stone Age and the Iron Age were delineated. Near the top of the timeline and beginning in our own time, the term The Nano Age was noted. Anyone picking up a book like this undoubtedly has both an interest and a fascination with nanomaterials, and probably the defining of an age of nanomaterials was not really questioned or given much thought. Perhaps it is now time to revisit that definition.

To name something as an age in this way indicates that the material has assumed the mantle of being the most advanced material technology of the time. The naming also implies widespread use as well as being the material of choice for artifacts that demanded high performance. This by no means suggests that other materials ceased being used, for older materials obviously continued in use. Iron axe heads were a huge improvement over stone or bronze ones because of their ability to retain sharp cutting edges and their overall strength and durability, but humans continued to use wooden hafts for hundreds of years. Only recently has the wooden haft been challenged by newer materials, but even now the wooden haft remains.

Is the naming of our time as the beginning of the age of nanomaterials warranted, or is the mystic of nanomaterials pushing an overly optimistic interpretation of their importance? One of the goals of this book is to demystify nanomaterial and nanotechnologies and to critically examine their real potentials as well as their real limitations. In doing so the unique properties of nanomaterials again and again showed through as having true value. There is little doubt that often-stated expectations about the prospect of ever-accelerating uses of nanomaterials in many industries are on the mark, but certainly more so in some industries than in others. In many instances nanotechnologies were found to be clearly poised to become "disruptive" technologies in the sense that they could potentially literally change the face of a valued process or product. Disruptive technologies normally occur within the context of particularly high-value objects or artifacts for which high performance is at a premium. As it was with the disruptive technology of the iron axe head with its vastly improved cutting capability, so it can be with nanotechnologies in many products and processes in the electronics, medical, and drug delivery industries. Remarkable developments can be expected to occur in other industries as well. Gains

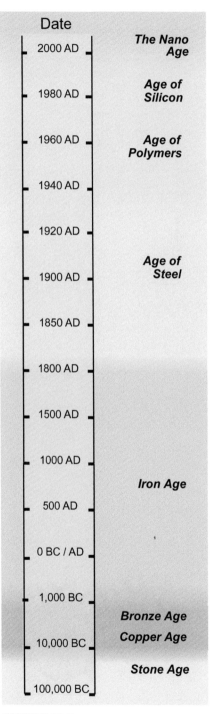

FIGURE 12.7
Materials timeline.

will be selective, however, since not all products will see dramatic changes in their character or basic use of materials. The equivalent of the wooden haft will remain with us.

In many applications across several industries, truly disruptive technologies associated with the use of nanomaterials and nanotechnologies might not occur at all; rather the term *continuous improvement* is perhaps a better way of couching potential impacts. This couching by no means detracts from the importance of nanomaterial and nanotechnology contributions within these contexts. Many valuable and highly functioning technologies have reached mature stages of development where forward progress has slowed and where nanomaterials and nanotechnologies can help achieve new levels of development.

Another aspect of the question of the appropriateness of naming a material age is whether impacts have a long life. Might nanomaterials and nanotechnologies be here today and gone tomorrow? In an earlier discussion on the possible evolutionary path of computing "chips" that are so essential to our technological world as we know it, nanotechnologies were argued as being the next major step forward in increasing speed and computing power while always decreasing in size. Yet in the same discussion it was noted that within a time frame of 20 to 30 years, chips based on ideas of "electron spin" would come to the fore, thus driving the size scale even further downward from the nanoscale to the atomic scale. Does this negate the idea of a nanomaterials age if anticipated near-term technologies might be described in other terms?

This is a difficult question, since predicting the technological future has long known to be a risky enterprise at best. It does appear that some technologies will indeed move further down the size scale into the molecular and atomic scale. Many chemical and biological processes and actions already operate on this scale, to be sure, and there is every expectation that further developments will occur along these lines. The often pictured image shown in Figure 12.8 of the "molecular motor" (a spinning assembly of nanotubes with projecting elements that form elemental gears) provides yet another illustration of this direction and provides an icon of a high-technology future based on very small things indeed—but these might or might not be the technologies needed for the amazingly diverse array of functions, sizes, and operations in the multitude of products and processes that will be used by industries and consumers alike in the future. Many work best at the nanoscale. As discussed in Chapter 3, many holistic products also need to exist at larger

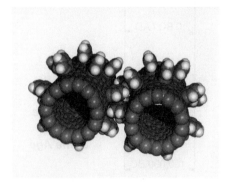

FIGURE 12.8
Fullerene gears. (Courtesy of NASA.)

scales for them to provide their function. For many products, the nanoscale will be the right long term scale. The continued extensive use of nanomaterials to achieve very high levels of performance is anticipated. There is every expectation that nanomaterials or nanotechnologies will become and continue to be a common part of the many things will make our society of the time what it is.

Perhaps one of the most interesting arguments is that the Age of Nanomaterials is indeed here to stay relates to a sea change in the way materials themselves are designed. Until the present, the quest for improved material properties in many of our most commonly used materials has had to be done *indirectly*, such as heat treatments for manipulating the properties of steel to give them nanoscaled structures. Now we have tools for *direct* intervention at the nanoscale. This is a remarkably important point of departure for the future. Until history proves otherwise, suggesting that there is indeed an age of nanomaterials appears a sound bet.

FURTHER READING

Paolo Gardini, Sailing with the ITRS into nanotechnology, International Technology Roadmap for Semiconductors, 2006.

C. Laurvick and B. Singaraju, Nanotechnology in aerospace systems, Aerospace and Electronic Systems Magazine, IEEE, Volume 18, Issue 9, Sept. 2003

NanoRoad: Nanomaterials Roadmap 2015: Roadmap report concerning the use of nanomaterials in the automotive sector, 6th Framework Programme, European Union, 2007.

NanoRoad: Nanomaterials Roadmap 2015: Roadmap report concerning the use of nanomaterials in the aerospace sector, 6th Framework Programme, European Union, 2007.

NanoRoad: SWOT Analysis concerning the use of nanomaterials in the automotive sector, 6th Framework Programme, European Union, 2007.

Juan Pérez, Laszlo Bax, and Carles Escolano, Roadmap report on nanomaterials, 6th Framework Programme, European Union, 2007.

Alan Taub, Automotive materials: Technology trends and challenges in the 21st century, MRS Bulletin, Vol. 31, April 2006, ppf. 336.

Index

527

Printed and bound by CPI Group (UK) Ltd, Croydon, CR0 4YY

03/10/2024

01040325-0010